Studies in Surface Science and Catalysis 169

CATALYSTS FOR UPGRADING HEAVY PETROLEUM FEEDS

Studies in Surface Science and Catalysis

Advisory Editors: B. Delmon and J.T. Yates
Series Editor: G. Centi

Vol. 169

CATALYSTS FOR UPGRADING HEAVY PETROLEUM FEEDS

Edward Furimsky
Ottawa, Ontario, Canada

ELSEVIER

Amsterdam • Boston • Heidelberg • London • New York • Oxford
Paris • San Diego • San Francisco • Singapore • Sydney • Tokyo

Elsevier
Radarweg 29, PO Box 211, 1000 AE Amsterdam, The Netherlands
Linacre House, Jordan Hill, Oxford OX2 8DP, UK

First edition 2007

Library of Congress Cataloging-in-Publication Data
A catalog record for this book is available from the Library of Congress

British Library Cataloguing in Publication Data
A catalogue record for this book is available from the British Library

ISBN: 978-0-444-53084-4
ISSN: 0167-2991

For information on all Elsevier publications
visit our website at books.elsevier.com

Printed and bound in The Netherlands
07 08 09 10 11 10 9 8 7 6 5 4 3 2 1

CONTENTS

PREFACE

The supply of conventional crudes has been dwindling due to decreasing availability in the industrial countries and political instabilities in the regions of the world where sweet crudes may be still available. In the former regions, the abundant reserves of heavy crudes both onshore and offshore, as well as tar sands, have been receiving attention. In North and Latin America, several mega projects on utilization of these resources are now under construction, while others in various phases of preparation. Because of a high viscosity, heavy crudes cannot be pipelined without being upgraded on or near the site of production. Otherwise, they would have to be blended with lighter fractions to decrease viscosity to the level suitable for pipelining. The production of synthetic crude from the Athabasca bitumen derived from tar sands in Northern Alberta in Canada provides an example of the onsite upgrading. In this case, both coking processes (delayed and flexi) and catalytic hydroprocessing have been employed. The primary products from the former processes undergo the secondary upgrading using hydroprocessing method to attain specifications of the synthetic crude which is ready for pipelining to refineries for further processing.

In this book, atmospheric residue (boiling above 350°C) obtained from petroleum is used as a baseline for defining heavy feeds. If not used directly for upgrading, atmospheric residues can be subjected to vacuum distillation to produce heavy feeds such as vacuum gas oil and vacuum residue. Both vacuum residue and atmospheric residue can be deasphalted to produce deasphalted oil. The primary products from coking are distilled to obtain naphtha and heavy gas oil fractions. A large portion of the latter is in the boiling range of the vacuum gas oil.

The configurations of the conventional petroleum refineries have been undergoing continuous changes in order to accommodate the increased volume of distillation residues from crude oils which are being available on the world market. The primary objective is to maximize the production of the most desirable products, e.g., transportation fuels, per unit of the crude oil entering refinery. Thus, vacuum gas oil and heavy gas oil may be converted to middle distillates via commercially available processes. The aim is to increase the overall yield of fuels. The commercial processes for converting vacuum residues to distillates either catalytically or non-catalytically are now available. In both cases, primary products require a hydroprocessing step to meet the product specifications.

Several configurations of systems for catalytic upgrading of atmospheric residues and vacuum residues using hydroprocessing method have been used commercially for decades. The catalysts varying widely in properties have been developed to process residues obtained from conventional crudes. The content of metals such as vanadium and nickel, as well as asphaltenes in similar residues derived from heavy crudes (e.g., Boscan,

Orinoco, Maya and others), may exceed 1000 ppm and 20 wt%, respectively. The development of catalysts and/or catalytic process suitable for such heavy feeds is rather challenging. Although a progress has been made, there are indications that such heavy feeds may require blending with lighter fractions before hydroprocessing, in order to decrease viscosity and/or to attain pumpability required for the introduction of heavy feed to reactors. Otherwise, a carbon-rejecting process may be the method of the choice.

In anticipation of growing volume of atmospheric residues and heavy feeds derived from it (vacuum gas oil, deasphalted oil and vacuum residues) to be processed, the research has been focusing on the improvement of the conventional catalysts and development of novel catalysts. For vacuum gas oil and heavy gas oil, a multifunctional catalysts shall remove contaminants such as sulfur, nitrogen and resins, as well as traces of metals and asphaltenes. It would be ideal if, at the same time, hydrogenation, isomerization, dewaxing and hydrocracking of hydrocarbons occurred to give high yields of desirable products (e.g., diesel oil, lube base oil, and others). However, this cannot be achieved in one stage with the present state-of-the-art processes and catalysts. Depending on the origin of deasphalted oil catalyst formulations differing from those used for vacuum gas oil and heavy gas oil, may be needed. The use of multistage catalytic systems is inevitable for achieving desirable conversions of atmospheric residues and vacuum residues in particular. In this case, selection of a suitable catalyst for every stage requires a special attention.

The cost of catalyst inventory may be an important factor before the final decision between catalytic and non-catalytic upgrading is made. Furthermore, such decision can be influenced by the trends around the world, which indicate that economic performance increases with the increasing size of petroleum refineries. Such situation is favorable for the integration of petroleum refining with the production of other marketable products, e.g., electricity, steam for district heating, ammonia for fertilizers production, olefins to make polymers, alcohols and so on. In this case, the "total extinction" mode of petroleum refining represents a complete conversion of crude entering refinery to usable products. Gasification as an integrated part of the petroleum refinery can convert feeds such as petroleum coke, pitch and refinery sludge, to synthesis gas suitable for the production of a wide range of commercial products. For example, the synthesis gas produced via gasification of petroleum coke may be converted to nearly zero sulfur liquids via Fischer-Tropsch process. A high vanadium and nickel content ash/slag left behind after gasification and/or combustion is suitable for metal reclamation.

Some dilemmas had to be faced while organizing material for this book. In view of the diversity in properties of catalysts suitable for heavy feeds of interest, two options seemed to be available to the author, i.e., one which would include the discussion of different groups of catalyst separately while including all heavy feeds in each group and the other, in which the catalyst development and testing would be organized with focus on to the individual heavy feeds. It was concluded that the latter option provided more concise information. The inclusion of patent literature in the book was based on the belief that practical aspects of the upgrading of heavy feeds are more accurately reflected by such information.

The separate chapter on dewaxing was included because in this case, vacuum gas oil, heavy gas oil and deasphalted oil are the feeds of the primary interest. Dewaxing is usually conducted under the conditions approaching those employed during hydroprocessing. However, the target properties of the products from dewaxing differ from those being attained during the conventional hydroprocessing. Moreover, there might be a marked difference in the catalyst formulations required for dewaxing compared with those of the conventional catalysts.

The processes and catalysts, which are in an early stage of development are discussed in the chapter on non-conventional hydroprocessing. In this case, the focus has been on down-hole upgrading, bio upgrading and the slurry bed upgrading. Some of these processes are approaching a commercial stage, while the potential of others for commercialization is rather remote. It is believed that this chapter puts the emerging processes and catalysts for upgrading heavy feeds in prospective with the conventional processes and catalysts.

LIST OF ACRONYMS

ABC	Asphaltenic bottom conversion
AC	Activated carbon
AGO	Atmospheric gas oil
AMW	Average molecular weight
APD	Average pore diameter
AR	Atmospheric residue
ARDS	Atmospheric residue desulfurization
ASA	Amorphous silica-alumina
ASE	Accelerated solvent extraction
BDN	Bio-denitrogenation
BDS	Bio-desulfurization
CCR	Conradson carbon residue
CSTR	Continuous stir tank reactor
CUS	Coordinatively unsaturated site
DAO	Deasphalted oil
DBT	Dibenzothiophene
DM-AR	Demetallized atmospheric residue
DMDS	Demetallized-desulfurized atmospheric residue
DOC	Dynamic oxygen chemisortpion
EDTA	Ethylenediamine tetraacetic acid
FCC	Fluid catalytic cracking
FT	Fischer - Tropsch
FTIR	Fourier transfer infrared spectroscopy
GC-FPD	Gas chromatography-flame photometric detector
GPC	Gel permeation chromatography
HCR	Hydrocracking
HDAr	Hydrodearomatization
HDAs	Hydroseasphalting
HDM	Hydrodemetallization
HDN	Hydrodenitrogenation
HDNi	Hydrodenickelization
HDO	Hydrodeoxygenation
HDS	Hydrodesulpfurization
HDV	Hydrodevanadization
HGO	Heavy gas oil
HYD	Hydrogenation

ID	Internal diameter
LH	Langmuir–Hinshelwood
LHSV	Liquid hourly space velocity
MAPO	Magnesia-alumino phosphate
MAT	Microactivity test
MCR	Microcarbon residue
MHC	Mild hydrocracking
NMR	Nuclear magnetic resonance
QCR	Quick catalyst replacement reactor
RBC	Ramsbottom carbon
RFCC	Residue fluid catalytic cracking
SAPO	Silico-alumino phosphate
TCLP	Toxicity characteristic leaching procedure
TEM	Transmission electron microscopy
THERMIDOR	Thermal monitoring for iso-performance desulfurization of oil residue
THF	Tetrahydrofuran
THFIS	Tetrahydrofuran insolubles
TIS	Toluene insolubles
TOS	Time on stream
TPO	Temperature programmed oxidation
TPR	Temperature programmed reduction
USY	Ultra stable Y
UV-VIS	Ultraviolet-visible
VGO	Vacuum gas oil
VI	Viscosity index
VR	Vacuum residue
WHSV	Weight hourly space velocity

Chapter 1

INTRODUCTION

Worldwide trends in crude oil supply have been indicating the declining availability of conventional crude. This trend has been offset by the increasing production of heavy crude. For heavy crude, the yield of distillate fractions can be increased by upgrading distillation residues. A number of thermal processes (e.g., visbreaking, delayed-, fluid- and flexi-coking) and asphaltenes and metals separation processes (e.g., deasphalting), the so-called carbon-rejecting processes, have been used on a commercial scale for several decades (1,2). Heavy feeds can also be upgraded by hydroprocessing, the so-called hydrogen addition option (3,4). This requires the presence of hydrogen and an active catalyst. Compared with thermal processes, hydroprocessing operations are more flexible, giving higher yields of liquid fractions. However, the costs of high-pressure equipment, catalysts and H_2 required for hydroprocessing have to be offset by the increased yields and quality of liquid products. The optimum hydroprocessing operation can be achieved by properly matching the type of reactor and catalyst with the properties of heavy feeds. Several types of catalytic reactors, i.e., fixed bed, moving bed and ebullated bed reactors, have been available and used commercially. The efforts to develop entirely new catalytic phases and reactors with the aim to increase the efficiency of hydroprocessing of heavy crudes have been made continuously.

In spite of some similarities, the hydroprocessing of heavy feeds differs markedly from that of light feeds. This results from the presence of high molecular weight asphaltenic molecules and resins, as well as that of the organometallic compounds in heavy feeds. In some heavy feeds, a clay-like mineral matter and water-soluble salts in the form of finely dispersed emulsions can also be present. The difference between hydroprocessing of heavy feeds and light feeds was clearly indicated in one of the first study dealing with this issue published by Beuther and Schmid (5) more than four decades ago. More recently, the topic was reviewed by Oelderik et al. (6) and Beaton and Bertolacini (7) from the point of view of the Shell and Amoco strategies, respectively. It was emphasized that the catalyst design for hydroprocessing applications has to take into consideration the presence of both metals (mostly V and Ni) and heavy species such as resins and asphaltenes. In this regard, the removal of metals represents a rather complex set of reactions and events, as was demonstrated in several studies (8–11) focusing primarily on hydrodemetallization (HDM). A high level of the asphaltenes conversion, i.e., hydrodeasphaltization (HDAs), has to be maintained simultaneously with that of HDM (12–14). Therefore, the selection and design of catalysts for hydroprocessing of heavy feeds is much more challenging than that for light feeds (15,16). Moreover, the methods used for evaluation of the catalysts have to be carefully chosen to obtain meaningful results which can be used for catalyst selection (17). Thus, the methods which have been used traditionally for the evaluation of hydroprocessing catalysts for the light feeds applications cannot be applied directly to heavy feeds without necessary modifications.

In designing the suitable catalysts for hydroprocessing of heavy feeds, special attention has to be paid to the surface properties of catalysts, such as pore size and pore volume distribution. These parameters have to be selected to ensure that the optimal combination of porosity and active surface area is achieved. The optimal parameters are feed dependent suggesting that a combination of the surface properties giving the best performance for a particular feed may change once the feed is changed. The size and shape of the catalyst particles are important parameters for ensuring efficient catalyst utilization, as well as desirable catalyst bed performance. This is much more critical for fixed bed reactors than for moving and ebullated bed reactors.

An active catalyst for hydroprocessing of heavy feeds has to be resistant to deactivation by coke and metal deposits. In the case of the latter, the catalyst has to possess a desirable metal storage capacity. Poisoning of the catalyst by N-bases observed during the hydroprocessing of light feeds occurs during that of heavy feeds also, although to a much lesser extent. However, the poisoning by N-compounds increases in the course of hydroprocessing. For example, when the feed containing asphaltenes and metals is processed in the fixed bed reactor, the poisoning by N-compounds increases from the inlet towards the outlet of the fixed bed. In a multistage system, the poisoning by N-compounds increases towards the last stage of the process.

The duration of a commercial run, before the catalyst replacement is necessary, may be an indication of the difficulties encountered during the hydroprocessing of heavy feeds. For example, when a straight run naphtha is used as the feed, the length of hydroprocessing run may approach 5 years provided the unwanted events have not occurred. As a general rule, the units employing fixed bed reactors processing atmospheric gas oils or vacuum gas oils have typical run lengths of about 2 years. In the case of a high asphaltenes and metals feed, significant modifications of the process employing fixed bed reactors would be required to extend the length of the run beyond 1 year. It should be noted that the structure of catalysts used for hydroprocessing of heavy feeds differs markedly from those used for light feeds. Thus, a total loss of activity would be observed within a few weeks onstream, if the catalyst suitable for hydroprocessing of naphtha is used for hydroprocessing of a heavy feed containing metals and asphaltenes.

Because of the many parameters involved during preparation, the cost of catalysts used for hydroprocessing of the feeds containing metals and asphaltenes is much greater than that for the light feeds derived from conventional crudes. Significant efforts have been made to increase the resistance of catalysts to deactivation, i.e., to prolong their life onstream. These involved modifications of the currently used $Co(Ni)Mo(W)/Al_2O_3$-based catalysts, as well as the development of entirely new types of catalytic phases. The importance of the conditions applied during catalyst preparation and of the different supports on the catalyst performance has been recognized. New information on hydroprocessing of heavy feeds published in the literature will be reviewed including the schemes for regeneration and utilization of spent catalysts. Some problems associated with the hydroprocessing of heavy feeds can be alleviated by using once-through low-cost catalysts. Regarding catalyst design and testing, the focus has been on heavy feeds of petroleum origin. Thus, feeds such as coal-derived liquids, oil shales, bio-crude have not been considered.

Catalytic dewaxing is a process operating under conditions approaching those employed during conventional hydroprocessing. However, the structure of dewaxing catalysts may differ markedly from that of the typical hydroprocessing catalysts. During catalytic dewaxing, heavy feeds such as vacuum gas oil (VGO) and deasphalted oil (DAO) are upgraded to obtain diesel oil and the lube base stock. The latter is used for preparation of lubricants. The primary objective of dewaxing is the enhancement of the cold flow properties of the products. In the case of the lube base stock, improvement in viscosity index may be achieved by converting aromatic structures to naphthenic compounds. The latter possess a higher viscosity index. Also, the long chain paraffins which would affect the pour point and freezing point of lube oils and diesel oil have to be removed. For this purpose, catalytic hydroisomerization of *n*-paraffins to iso-paraffins is the method of choice. The details of catalytic dewaxing of VGO and DAO are given in the Chapter 11 with the aim of identifying the difference between the structure of the dewaxing catalysts and that of the conventional hydroprocessing catalysts.

The methods for unconventional hydroprocessing are in various stages of development. The focus, in this case, has been on the catalysts in either dissolved or micronized form. In these forms, catalysts can be readily slurried with heavy feeds. The catalyst precursors can also be dissolved in an oil and water. These solutions can then be blended with a heavy feed. When added to the heavy feed, catalytically active phase is formed in situ from the dissolved precursor. A nearly molecular size of the catalytically active phase ensures almost complete catalyst utilization in the slurry bed reactor, typically used for such catalysts. These catalysts are dominated by transition metals, particularly those which are part of the conventional catalysts. Unconventional methods such as bio-upgrading are still in the early stages of development, in spite of the decades of research. Although the account of these processes in this book is rather cursory, sufficient information is provided to assess the viability of the emerging hydroprocessing methods in relation to those used commercially. Moreover, references made to numerous articles and reviews published during the last decade may assist in identifying the source of necessary information for those interested in the field.

Among the non-hydroprocessing methods for upgrading heavy feeds, fluid catalytic cracking (FCC) process has been receiving most of the attention. Attempts have been made to use this process for upgrading heavy feeds containing asphaltenes and metals. However, significant modifications of the process configurations, as well as entirely different catalytic phases compared with the traditionally used FCC catalysts, are required. Although some progress has been made, it is unlikely that the upgrading of heavy feeds with metals content approaching 50 ppm of V + Ni, using significantly modified FCC process, i.e., residue FCC (RFCC) process, can be achieved without difficulties. For comparison with the conventional and unconventional hydroprocessing, a chapter on FCC/RFCC (Chapter 13) is added to indicate a significant difference in the ability of these methods to process heavy feeds. Thus, a high flexibility of the hydroprocessing method is indicated by the fact that heavy feeds differing widely in the content of asphaltenes and metals can be processed contrary to that for RFCC. This suggests that it may be inaccurate to refer to the latter as the method for residues upgrading.

Upgrading of heavy feeds using thermal processes (e.g., coking) and separation processes (deasphalting) is discussed in the last chapters. At least a cursory account of these processes was deemed necessary to aid the reader in identifying the non-catalytic options for upgrading heavy petroleum feeds, as well as to determine the limits beyond which catalytic upgrading becomes unattractive compared with the carbon rejection option. Moreover, some heavy feeds under consideration in this book, i.e., heavy gas oil (HGO) and DAO, were obtained using non-catalytic upgrading processes such as coking and deasphalting, respectively.

Chapter 2

PROPERTIES OF HEAVY FEEDS

The simplified schematics of the petroleum refinery (Figure 2.1) indicate the presence of several hydroprocessing units. For the purpose of this book, the atmospheric residue (AR) boiling above 350°C (623 K) was chosen as the primary heavy feed from which other heavy feeds have been derived. Thus, AR can be subjected to several treatments, i.e., deasphalting, hydroprocessing and vacuum distillation. Products of the vacuum distillation are vacuum residue (VR) and VGO. Similarly as AR, VR can be either deasphalted to produce DAO and asphalt or directly hydroprocessed. The VGO is usually used as the feed for production of light and/or middle distillates either via hydroprocessing or via catalytic cracking. Other petroleum products obtained from VGO include lube base stock used for the preparation of lubricating oils. The products identified in Figure 2.1 as distillates are not subject of this book except those obtained by coking, i.e., heavy gas oil (HGO). According to Figure 2.1, the former products were produced during the atmospheric distillation and catalytic cracking.

The FCC and coking of heavy feeds form an integral part of many petroleum refineries. Virgin VGOs, as well as pretreated DAOs and HGOs have been typical feeds for the former process. However, after modifications of the conventional cracking units, RFCC process may also be able to handle the unpretreated DAO and ARs derived from sweet crudes. Hydroprocessing is the process of choice for heavier ARs and VRs, as well as topped heavy crudes. In an extreme case, i.e., for the most problematic heavy feeds, coking processes have been used for upgrading. The simplified flowsheet and brief description of coking processes is given in Chapter 14. These processes operate without H_2 as the reactant, although H_2 is being formed as the by-product. As it was indicated, deasphalting, as the competitive process with the coking process, can produce a low metals and asphaltenes feed by rejecting most of the metals and asphaltenes into the residue using precipitation with the aliphatic hydrocarbon solvents such as heptane, hexane, pentane, butane, etc. Some basic features of this process are given in Chapter 14 as well.

The asphalt and coke, as the final refinery residues, have an impact on the overall performance of petroleum refineries. The yields of such residues are much higher from a carbon-rejecting process compared with a hydroprocessing process. Unless economic outlets for the industrial utilization of coke and asphalt are found, they can be utilized on the site of refinery using combustion and/or gasification routes to produce electricity, steam and H_2 to meet the refinery requirements. The surplus, e.g., of electricity, may be sold to the national grids to meet the public demands. Nevertheless, production of the final residues is believed to be the factor deserving an attention when deciding between the carbon-rejecting option and hydroprocessing option.

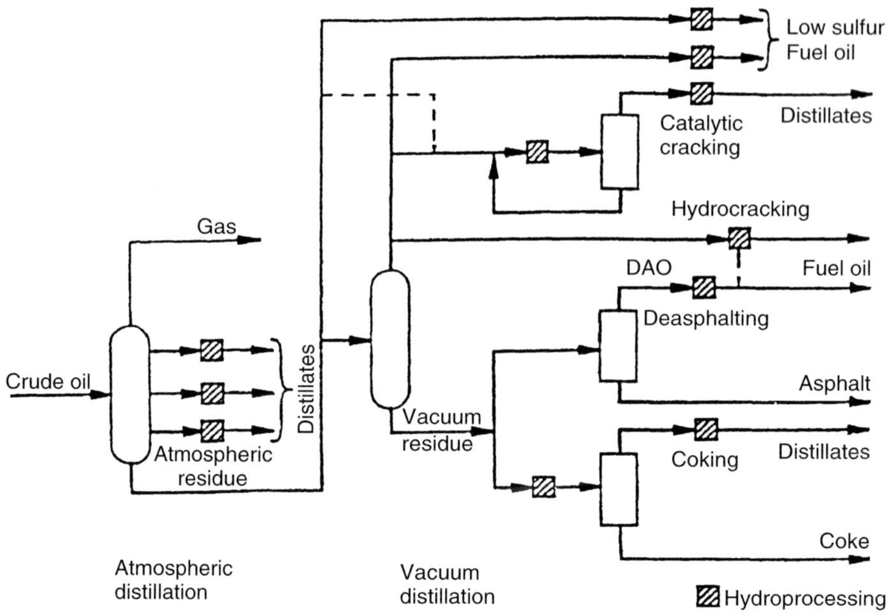

Figure 2.1 Simplified flowsheet of petroleum refinery.

Another type of the heavy feed included in this study includes the primary liquids produced by coking of VR and heavy topped crudes. The liquid products from coking are identified in Figure 2.1 as distillates. Based on the coking temperature, the final boiling point of such liquids may approach 850 K. This suggests that the primary liquids were produced under more severe conditions than VGOs (~623 K). This may indicate on the presence of more refractory compounds and alkenes in the former requiring different conditions for hydroprocessing. For example, in the case of Syncrude plant in Canada, the primary liquids from fluid coking of Athabasca bitumen are transferred to distillation tower to obtain the naphtha fraction and the HGO. Most of the latter is in the boiling range greater than 350 °C. After hydroprocessing, naphtha and HGO are blended and pipelined to refineries providing that specifications of the synthetic crude were attained. Therefore, HGO has been frequently included in the studies on the testing and development of hydroprocessing catalysts. The inclusion of HGO in the present study was based on the fact that fractions boiling above 350 °C account for more than 50% of HGO.

Boiling range is perhaps the simplest parameter which can be used to distinguish among heavy feeds. Thus, the approximate ranges in K (°C) of HGO, VGO, AR and VR are 573 to 800 (300–530), 620 to 850 (350–580), >620 (>~350), >880 (>~600), respectively. The boiling range of DAO depends on its origin, i.e., >620 (350) and >880 (600) K for the DAO obtained from deasphalting of AR and VR, respectively.

Rather extensive information on the properties of heavy feeds can be found in the literature (1,18–22). Therefore, for the purpose of this study, a brief summary will only be given with focus on these properties which are important for hydroprocessing, dewaxing and RFCC. The content of asphaltenes and metals in heavy feeds determine the severity of hydroprocessing, which shall increase in the following order: VGO/HGO

< DAO < AR < VR/heavy crudes. The content of asphaltenes and metals increase in the same order. This may require a different type of catalyst and reactor configurations for every feed. Therefore, several types of hydroprocessing reactors may be present on the site of the same refinery.

Metals are present predominantly in the form of porphyrins. To various degrees, such structures are present in the AR and VR derived from the conventional and heavy crudes, as well as from the bitumen obtained from tar sands. Trace amounts of asphaltenes and metals may still be present in VGO/HGO. Depending on the deasphalting conditions, various amount of asphaltenes and metals can be present in DAO. The contents of metals and asphaltenes are more critical for DAO obtained from VR than that from AR. In some refineries, it may be beneficial to blend VGO with DAO before hydroprocessing to produce either feed for FCC or for dewaxing. Refinery trends indicate on the attempt to utilize DAO directly as the feed for RFCC. In this case, the DAO obtained from the AR derived from sweet crudes may be an acceptable feed.

The schemes in Figure 2.2 provide a summary of the preparation of heavy feeds from the 350 °C + AR, which was used as the baseline for defining heavy feeds. In this case, the AR can be either deasphalted to produce DAO and asphalt (scheme 1) or subjected to vacuum distillation to produce VGO and VR (scheme 2). This may follow by the deasphalting of VR to produce DAO and asphalt. It is expected that properties of the DAO obtained from AR will differ markedly from those of the DAO obtained from VR, particularly when both VR and AR were derived from the same crude. The choice of the scheme depends on the refinery strategy (23–25). The results in Table 2.1 were used to illustrate changes in the content of contaminants in heavy feeds obtained according to the schemes 1 and 2. The significant difference between the metal's content of the DAO obtained by schemes 1 and 2 should be noted. These heavy feeds were derived from the Zuata heavy crude from Venezuela (25). It is believed that similar trends can also be established for the heavy feeds derived from other heavy crudes. The amount of Conradson carbon residue (CCR) and metals in VR and AR in Table 2.1 indicate that hydroprocessing of these residues should be carefully assessed in comparison with the carbon rejecting routes. Thus, depending on the products of interest, more than four fixed

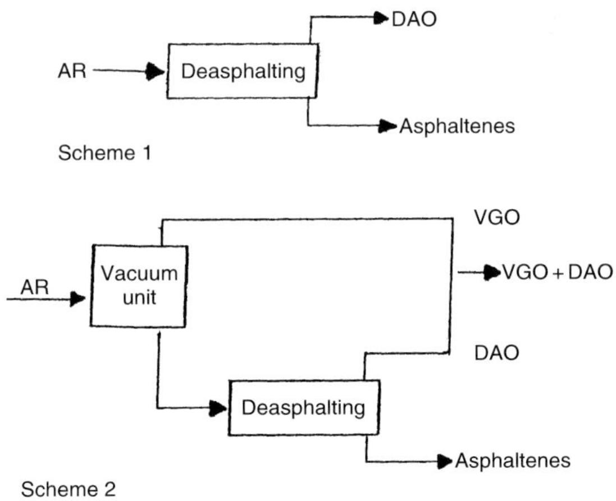

Figure 2.2 Schemes for identifying heavy feeds in petroleum refinery.

Table 2.1 Quality of heavy feeds derived from Zuata heavy crude according to schemes in Figure 2.2 (25)

	Yield, (wt%)	Density, (kg/L)	CCR, (wt%)	V, (ppm)	Ni, (ppm)	S, (wt%)
Heavy crude	100	1.08	13.4	384	81	3.4
Scheme 1						
Atmospheric residue	84	1.02	15.4	448	93	3.8
DAO *n*-pentane	68	0.99	8.3	130	30	3.4
DAO *n*-butane	62	0.98	4.7	13	4	3.0
Scheme 2						
Vacuum residue	51	1.06	28.5	706	162	4.6
DAO *n*-pentane	34	1.02	13.9	175	45	4.3
DAO *n*-butane	21	0.99	7.6	40	14	3.8
VGO	33	0.98	0.1	0.4	0.1	3.4
VGO + DAO[1]	67	0.99	7.2	90	20	3.8

[1] From pentane deasphalting

bed reactors and/or more than three ebullated bed reactors connected in a series may be necessary to achieve desirable conversions. Therefore, the cost of catalyst inventory may be an important factor while making the decision between hydroprocessing and carbon rejecting options.

During deasphalting, most of the metals and asphaltenes from the residues may be rejected in the asphalt. However, when *n*-pentane was used for deasphalting (solvent/feed ratio of 8), the content of contaminants in the resulting DAO (as per cent of those in the feed) was still too high, whereas in the case of *n*-butane, the feed was acceptable for hydroprocessing using commercial processes (e.g., UNIBON process). This was, of course, achieved at the expense of the lower yield of DAO, as it is indicated in Figure 2.3 (25). Therefore, the flexibility of deasphalting process allows the production of DAO, which may be suitable feed either for hydroprocessing or RFCC and dewaxing. Figure 2.3 further shows that the increase in the content of sulfur in DAO with increasing yield of the latter was more pronounced than that of nitrogen. At the same time, a very small amount of metals ended up in DAO before its yield reached about 60%. This confirmed that the predominant part of the metals in heavy feeds was associated with asphaltenes. The results in Figure 2.3 were obtained for an AR. It is anticipated that for the deasphalting of VR, the general trends will be similar, although the curves will be shifted to the lower yield (e.g., less than 50%) of DAO in agreement with the results in Table 2.1. As it was indicated above, it may be beneficial to blend VGO with DAO. As the Table 2.1 shows, such blend could be an acceptable feed for the graded hydroprocessing systems. At the same time, the VGO in Table 2.1 can be used directly as the FCC feed, without removing the trace amount of metals and CCR. However, the hydroprocessing of VGO to remove aromatics from the FCC feed may result in an increased conversion to middle distillates in the latter process. Also hydroprocessing of VGO before dewaxing may be necessary in the case that dewaxing catalysts contain noble metals. It should be noted that for conventional crudes, the content of metals and asphaltenes in AR may be significantly lower than that shown in Table 2.1. For

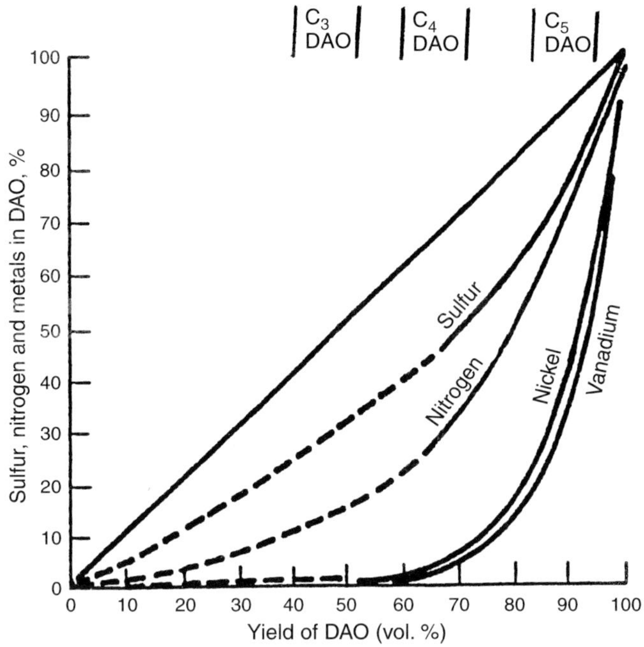

Figure 2.3 Effect of hydrocarbon solvent type on yield and quality of DAO.

example, as Table 2.2 shows, the widely studied AR derived from the Kuwait crude can be hydroprocessed to a desirable level in the ARDS process comprising four trickle bed reactors in a series without requiring deasphalting (22). The AR obtained from the North Sea Ekofisk crude may be used directly as the feed for RFCC. However, the VR derived from the Arab Heavy crude may require severe hydroprocessing conditions such as encountered in the process comprising several ebullated bed reactors connected in a series unless it is subjected to deasphalting step.

The parameters such as CCR have been used to characterize heavy feeds and their products. They are determined by heating the prescribed amount of the sample in an inert atmosphere under the specified conditions, i.e., heating rate, final temperature and hold time. Although the asphaltenes are the main contributor to the carbon residue,

Table 2.2 Properties of atmospheric residue (AR) and vacuum residue (VR) (22)

	Kuwait AR	Arab Heavy VR
Yield (wt%)	30.5	39.4
Density (kg/L)	1.021	1.023
Sulfur (wt%)	5.35	5.10
Asphaltenes* (wt%)	15.4	20.0
Nickel (ppm)	33	40
Vanadium (ppm)	87	142
Viscosity** (cSt)	1105	1540

* pentane insolubles
** at 20 °C

other components (e.g., resins, polyaromatics, etc.) may also be involved. Therefore, the amount of CCR may not be proportional to the content of asphaltenes. Similar parameters estimated under different conditions include the Ramsbottom carbon residue (RBC) and microcarbon residue (MCR). The CCR, RBC and MCR parameters have been widely used to study the effect of catalysts on the hydroprocessing of heavy feeds. In fact, in some kinetic studies, the conversions during hydroprocessing were expressed in terms of these parameters.

2.1 COMPOSITION OF HEAVY FEEDS

From the chemical structure point of view, heavy feeds such as VGO and HGO are characterized by the three main hydrocarbon groups, i.e., paraffinic, naphthenic and aromatic. In addition, HGO contains alkenes because of the more severe conditions applied during coking than during vacuum distillation. The proportions of these groups depend on the origin of heavy crude from which the VGO and HGO were derived. For example, the content of paraffinic and naphthenic groups will be high in the case, that the VGO was obtained from the paraffinic and naphthenic crude, respectively. It is expected that the content of aromatics in HGO is greater than that in VGO, because of the more severe conditions applied during the production of the former. For the same reason, the aromatics in HGO should be less alkylated and/or their alkyl substituents be shorter. Generally, only the trace quantities of asphaltenes and metals are present. A similar situation may be encountered with DAO, providing that a residue (AR and VR) was extensively deasphalted, e.g., by using propane as the solvent. Because of the mild conditions applied during deasphalting, the hydrocarbon groups composition should more closely resemble that of VGO than that of HGO.

The sulfur, nitrogen and oxygen in VGO and HGO are predominantly in the form of heteroring compounds, although small quantities of etheric, thioetheric, amino and other forms can also be present. Based on the boiling point of pure compounds, the alkylated dibenzothiophenes (DBT) and benzoDBT will account for most of the former. In the case of nitrogen, carbazole and acridine, as well as their alkylated derivatives are the main N-containing compounds, though alkylated indols and quinolines may be present as well. The heavier heteroring compounds may include benzocarbazoles, benzoacridine, as well as their alkylated derivatives. Alkylated benzofurans and those with the additional aromatic rings fused with benzofurans are usually the main O-containing compounds. However, if the contact of a heavy feed with air was not avoided, the compounds containing hydroxyl, carbonyl or even carboxyl groups may be formed via autoxidation reactions. The analytical techniques for detailed characterization of VGO and HGO are now available. Based on such information, the properties of catalyst can be tailor-made to suit a particular feed.

Depending on the conditions applied during deasphalting, the boiling range of a large portion of DAO obtained from AR may approach the boiling range of VGO and HGO. Generally, mild conditions employed during the preparation of DAO indicate on little change in its composition due to the thermal effects compared with VGO, i.e., one atmospheric distillation step for DAO, whereas one atmospheric distillation followed by vacuum distillation for VGO. Then, preparation conditions will have the least effect on the composition of DAO obtained from AR compared with VGO and HGO.

The properties of the DAO derived from VR may differ markedly from those of the DAO obtained from AR. For example, much greater quantities of the resin-like molecules comprising polycondensed aromatics and heterorings should be present in the DAO from VR than those from AR. Moreover, the initial boiling point of the latter should overlap with the final boiling point of the DAO obtained from AR. Also the lighter fractions of the DAO derived from VR should overlap with the heaviest fractions of VGO. Table 2.1 shows that a multistage catalytic system may be required for hydroprocessing of the DAO produced by butane deasphalting (scheme 2) of VR, whereas similar DAO obtained from AR may be a suitable feed for the RFCC process.

The extensive characterization studies using petroleum crudes, undertaken by numerous authors form a basis for understanding the composition of heavy feeds. For example, Boduszynski (26,27) studied the effect of boiling range on the distribution of molecular weight, as well as on the content of hydrocarbon groups for various fractions derived from a crude oil. Figure 2.4 shows that the content of heavy components increased with the increasing middle boiling point of the fractions. The severity required for hydroprocessing of these fractions may increase in the same order. Figure 2.5 indicates that

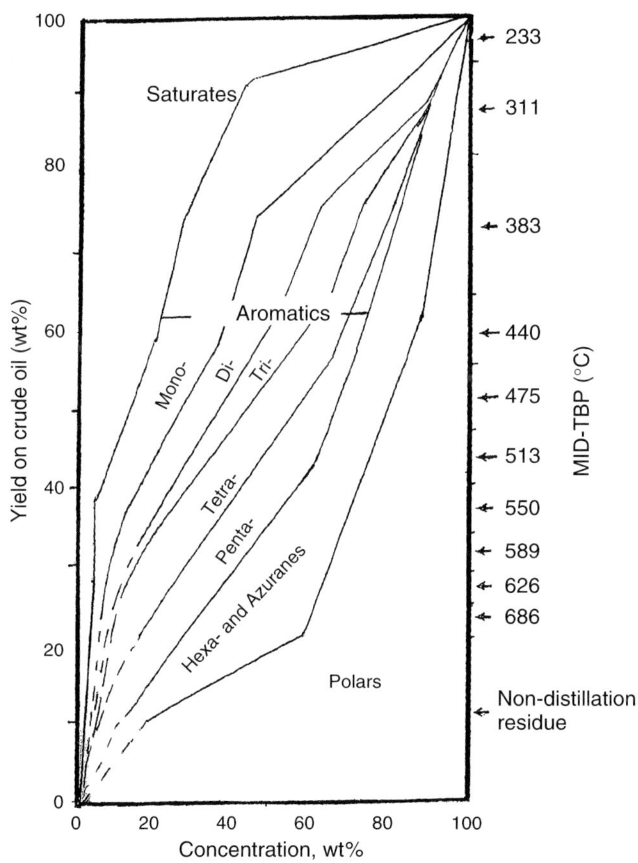

Figure 2.4 Effect of middle boiling point of fractions derived from crude oil on content of hydrocarbon groups (26).

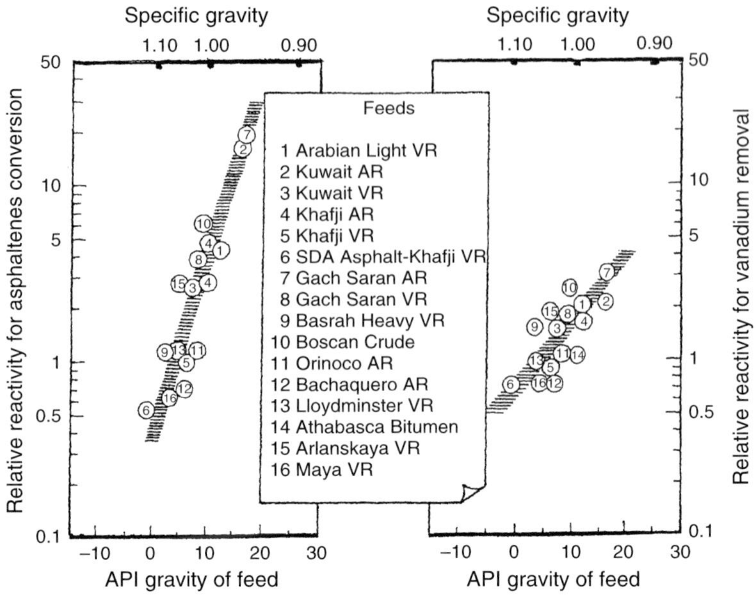

Figure 2.5 Effect of feed gravity on asphaltenes conversion and vanadium removal (28).

the specific gravity of heavy feeds may be another suitable parameter for predicting reactivity for the removal of asphaltenes and vanadium (28). These results confirmed that the reactivity of VRs compared with ARs is rather low. This has to be taken into consideration while selecting the type of catalytic process and a suitable catalyst.

Table 2.3 shows the yields of distillation residues, i.e., AR (345 + C) and VR (565 + C) from well-known crudes, as well as the content of contaminants the removal of which is the primary focus of hydroprocessing (29). It is obvious that the origin of the crude will have an impact on the configuration of petroleum refinery. For example, a conventional refinery such as shown in Figure 2.1 can process the Arab Light crude without requiring

Table 2.3 Yields of atmospheric (345 + °C) and vacuum residues (565 + °C) as volume % of crude (29)

Crude	Arab Light	Arab Heavy	Maya	Boscan	Cold Lake	North Sea Ekotisk	Kuwait Export
Yield							
345 + °C	44.6	53.8	56.4	82.9	83.1	52.6	45.9
565 + °C	14.8	23.2	31.2	63.9	50.0	18.0	21.8
Density (kg/L)	0.86	0.89	0.93	1.04	1.00	0.88	0.89
Sulfur (wt%)	1.8	2.9	3.8	5.2	4.9	0.4	4.1
Nitrogen (wt%)	0.1	0.2	0.3	0.5	0.6	0.2	0.4
Vanadium (ppm)	18	50	273	1220	160	4	55
Nickel (ppm)	4	16	50	120	80	2	20
CCR (wt%)	3	7	15	20	19	4	11

any modifications. On the other extreme, processing the Boscan crude would require a more advanced refinery equipped with the heavy feeds upgrading units. The content of metals suggests that neither catalytic upgrading nor deasphalting of this feed are economically feasible leaving coking as, perhaps, the best alternative. Therefore, a significant difference in the properties of heavy feeds (VGO, DAO, AR and VR) produced in the conventional refinery compared with those produced in an advanced refinery processing the most problematic feeds should be anticipated. The trends around the world show that the upgrading processes operating under severe conditions, as required for such feeds, are becoming part of the advanced refineries.

The structural parameters and properties of heavy feeds which are relevant for hydroprocessing indicate a great variation among heavy feeds. For example, a simple parameter such as the molecular H/C ratio exhibited a decreasing trend from the light crudes toward heavy crudes. A similar decreasing trend has been established from the light fractions toward heavy fractions and residue derived from the same crude (30), i.e., from about 1.9 to 1.7 and 1.4 for the naphtha fraction, gas oil and VR, respectively. The decrease in the H/C ratio is always complemented by the increase in aromaticity. The complexity and/or difficulties, catalyst and hydrogen requirements, as well as the severity of hydroprocessing operation increase in the same order until the non-catalytic upgrading becomes more attractive.

The heavy feeds containing asphaltenes and metals can be described as the colloidal solution comprising three phases in the increasing order of the average molecular weight, i.e., oil, resins and asphaltenes. As the heaviest fraction, asphaltenes represent the group of large molecules which can be separated from the feed by precipitation using paraffinic hydrocarbons such as pentane, hexane and heptane. The essential part of heavy feeds are micelles which are dissolved in the oil fraction. Micelles are the aggregates of asphaltenes and resins held together by the weak physical interactions. Figure 2.6 (31) shows that asphaltenes occupy the core of micelles with resins being adsorbed on the external surface and as such act as the dispersing agent for the asphaltenes in oil. Resins are less polar than asphaltenes but more polar than oil. This ensures the equilibrium between the micelles and the surrounding oil phase. This is required to ensure the stability and homogeneity of the colloidal system. Without resins present, the asphaltenes would coagulate and form sediments during the storage and transportation, as well as during hydroprocessing. Figure 2.6 further indicates that the chemical structure of micelles exhibits a gradual change, e.g., from the nucleus consisting of asphaltenes through the less aromatic and polar resins and even less aromatic layer and a transition zone before getting into contact with the oil phase. A gradual change may also apply within the asphaltenes component of the micelle. Thus, the most centered part of the asphaltenes, termed as carboids, is the least soluble and the most aromatic. Moreover, during upgrading via both hydroprocessing and coking, the carboids are expected to exhibit the lowest reactivity and the highest coke-forming propensity.

The amount of resins required for maintaining colloidal stability varies from crude to crude. This is demonstrated by the results shown in Table 2.4 (32), i.e., the ratio of resins to asphaltenes varied from 1 to almost 4. In this regard, the polarity of resins and asphaltenes may be an important parameter. Table 2.5 (33) shows the difference in polarity of the oil, resins and asphaltenes derived from the Boscan crude, one of the heaviest available crude. Thus, the increase in the heteroatom content from the oil phase toward the asphaltenes was complemented by the decrease in the H/C ratio. This was

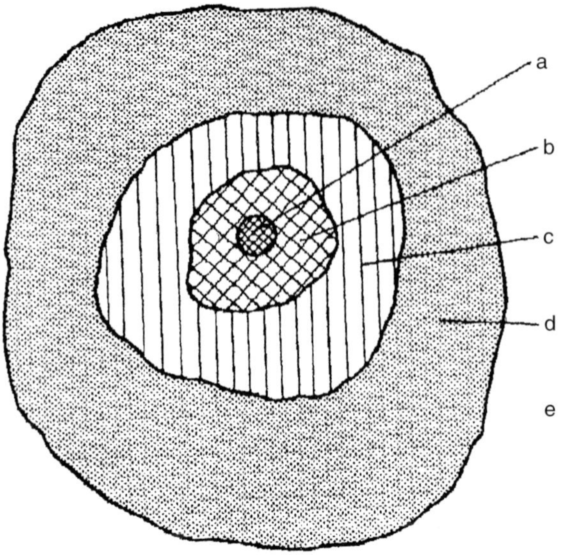

Figure 2.6 Components of micelle in colloid solution (31).

Table 2.4 Oils, resins and asphaltenes in heavy crudes (32)

	Heavy Arab	Hondo	Maya	Gach Saran
Oil (wt%)	59.9	43.9	48.9	64.7
Resins (wt%)	27.5	40.2	25.9	28.5
Asphaltenes (wt%)	12.6	13.9	25.2	6.8
Resins/asphaltenes	2.2	2.9	1.0	4.2
V + Ni (ppm)				
In oils	0	2.4	1	0
In resins	83	238	233	241
In asphaltenes	656	1547	1887	1466
In crude	115	372	496	144

an indication that the aromaticity increased in the same order. As expected, the content of metals exhibited similar trends as did the aromaticity. It is believed that these trends generally apply to the colloidal systems of most of the heavy crudes.

The difference in properties between the oil and asphaltenes phases indicated in Table 2.5 is so evident that the dissolution and/or dispersion of the latter in oil could not be achieved without the aid of resins. Therefore, it is essential that during hydroprocessing the resins are removed from the system at a similar rate as the asphaltenes. Otherwise, a much greater rate of the conversion of resins than that of asphaltenes would result in the incompatibility of the components of the colloidal system. This would lead to the coagulation of asphaltenes resulting in the deposit formation on the catalyst

Table 2.5 Properties of oils, resins and asphaltenes from Boscan crude[1] (33)

Property	Boscan	Oil	Resins	Asphaltenes
Carbon (wt%)	82.50	83.00	82.16	81.34
Hydrogen (wt%)	10.42	11.20	9.85	8.33
Sulfur (wt%)	5.55	5.41	6.75	6.88
Nitrogen (wt%)	0.58	0.26	0.76	1.43
Oxygen (wt%)	0.76	0.20	0.90	1.07
H/C	1.52	1.62	1.44	1.23
V (ppm)	1220	178	1200	4008
Ni (ppm)	117	56	120	400

[1] Pentane solvent (solvent/crude = 8)

surface and in the transfer lines. The deposit formation would be further enhanced under the conditions favoring the hydrogenation (HYD) of the oil phase. This fact has to be taken into consideration while selecting light fractions as the diluents for decreasing the viscosity of heavy crudes either to enable feeding into an upgrading system or to ensure the transportation via pipeline. In this regard, the suitability would improve with the increasing aromaticity of the diluent.

The hypothesis proposed by Gray (19) may describe colloidal behavior of the heavy feeds taking into consideration the conditions encountered during hydroprocessing. In this case, the colloidal behavior of heavy feeds was depicted in terms of the ternary composition phase diagram comprising three pseudo-components such as volatile components, middle distillates and heavy residue. It was determined that the relative concentrations of these pseudo-components as well as their aromaticity depend on the H_2 pressure, temperature and the content of asphaltenes in heavy feed. The regions consisting of the low aromaticity volatile phase and the high aromaticity high density asphaltene phase were identified as having a low stability. It is believed that the form of ternary diagram will vary from feed to feed. This was supported by the different flocculation/precipitation onsets among different heavy feeds (34). In this case, a heavy feed was diluted in the solvent (e.g., toluene) and titrated by the precipitant such as heptane. A number of methods, i.e., IR, microscopic observation, particle size analysis, optical transmission, etc., have been available for identifying the onset of precipitation.

Attempts have been made to develop models depicting the chemical structure of the asphaltenes and resins (1,18–22). This usually involved an extensive fractionation of a heavy feed by chromatographic methods, followed by detailed characterization of the separated fractions. A number of the structural parameters of the fractions could be determined using ^{13}C and ^{1}H NMR spectroscopies, whereas the organic groups present could be identified by the FTIR spectroscopy. This approach was used by Suzuki et al. (21) for the model development of the Athabasca bitumen. The asphaltene dimer proposed by these authors is shown in Figure 2.7a. For comparison, the hypothetical structures of the asphaltenes identified in Maya (35) and Venezuelan crudes (36) were included as well. It is evident that the structures of latter are more complex than that of the Athabasca bitumen requiring more severe conditions for their hydroprocessing conversion to light fractions.

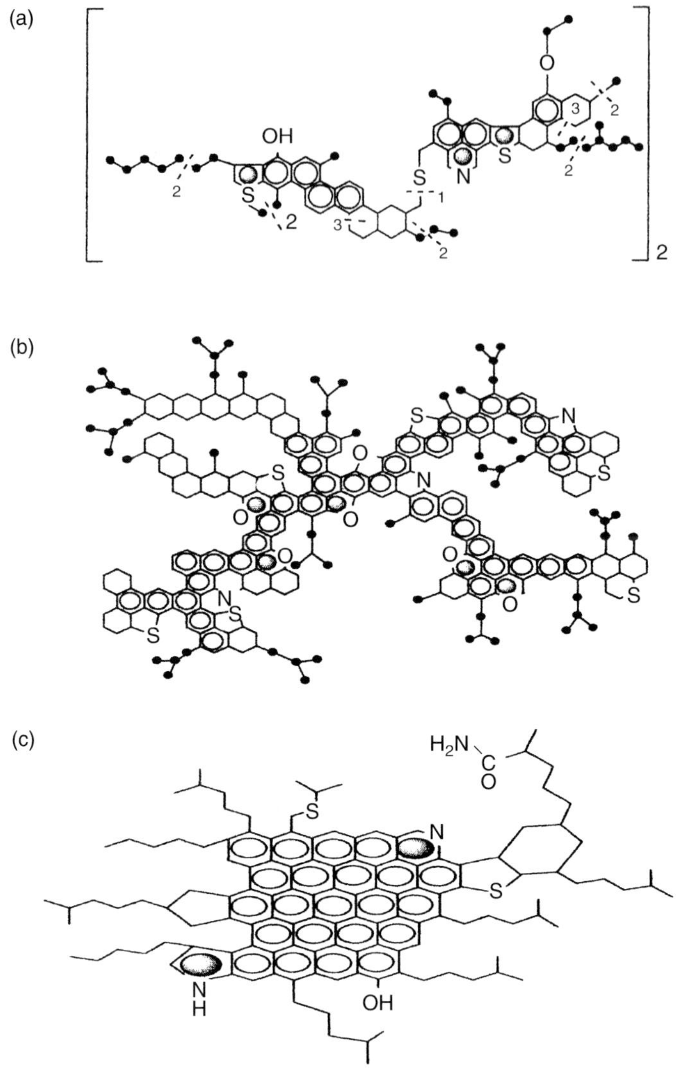

Figure 2.7 Structure of asphaltenes derived from (a) Athabasca bitumen (21); (b) Maya crude (35); (c) Venezuelan crude (36).

In heavy feeds, the micellar clusters may consist of the aromatic sheets of asphaltenic molecules which are stacked and aligned in a planar fashion, as it is shown in Figure 2.8 (5,6). The type of bonding between the sheets is not clear. However, a bonding involving either polar groups or van der Waall's forces may be in the effect. The external surface of the asphaltenes clusters is occluded by resins, the molecular weight of which is smaller by several hundreds and/or almost thousands units than that of asphaltenes. For example, Dolbear et al. (32) reported the molecular weight of asphaltenes and resins to be in the range 5000–10 000 and 500–5000 units, respectively. The clusters may be broken at higher temperatures and by using a vigorous agitation. Similar effects may

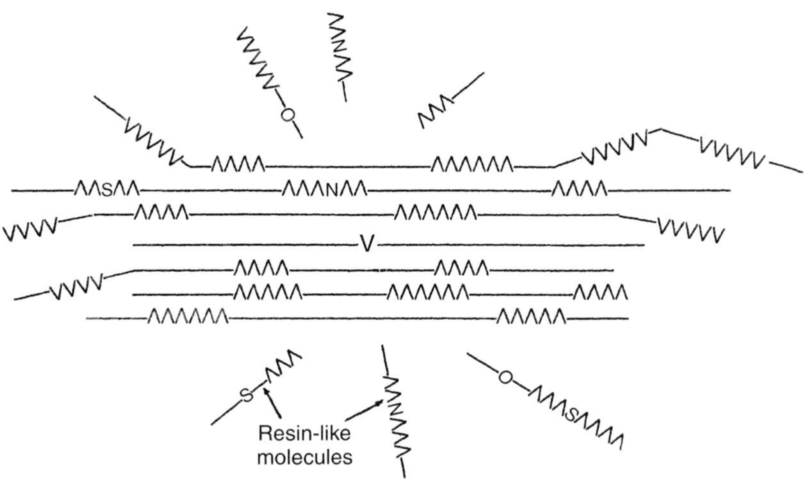

Figure 2.8 Micellar cluster of resins, asphaltenes and V-porphyrin (6).

be achieved by suitable solvents. Therefore, the size of clusters may be significantly diminished under the typical conditions applied during hydroprocessing.

In summary of this section, from the colloidal structure point of view, the heavy feeds such as VGO consist predominantly of the oil phase with resins accounting for a smaller portion. At the same time, resins may account for most of the DAO, although this may depend on the efficiency of the solvent used for deasphalting, i.e., to coagulate asphaltenes. Asphaltenes are the main component of the asphalt by-product separated from heavy feeds, although some resins may be still present. Again, the content of resins in the asphaltenes phase depends on the type of solvent used for precipitation of the latter as well as on the origin of heavy feed.

2.2 METALS IN HEAVY FEEDS

The V and Ni are the predominant metals in the conventional petroleum crudes, heavy crudes and bitumen derived from tar sands. The content of V and Ni in heavy feeds varies between few ppm to more than 1000 ppm. With few exceptions, the content of V is greater than that of Ni. It has been generally accepted that most of the V- and Ni-containing molecules are associated with micellar clusters (Figure 2.8). It is, however, believed that under hydroprocessing conditions these molecules are released because of the clusters' disintegration. To various degrees, Fe, Ti, trace metals and small amounts of other metals can also be present in addition to clay-like mineral matter (37). The "skin" of Ca and Mg on the exterior of the catalyst particles used for hydroprocessing of the heavy feed derived from Canadian tar sands was the confirmation of the presence of fine particles of the mineral matter dispersed in the feed (38). The dewatering and/or desalting of some crudes may be required to remove alkali metals which are usually in the form of chlorides and bromides dissolved in the finely divided water emulsion. The silicon in heavy feeds may originate from the anti-foaming agents. It is not a severe poison during the catalyst utilization cycle; however, its adverse effect may become more pronounced after the catalyst regeneration for subsequent reuse. Arsenic in heavy feeds has been receiving little attention in spite of its detrimental effect on catalyst activity.

Metals in petroleum have been attracting attention because of their detrimental effect on the performance of catalysts during hydroprocessing. If present in petroleum products and by-products (e.g., coke), metals have adverse effects on every utilization option which has been practised commercially (39). From the hydroprocessing point of view, the presence of V and Ni has been of the primary interest (40,41). It has been generally accepted that porphyrins are the predominant form of the V and Ni compounds present in heavy feeds. Several forms of the V-and Ni-containing porphyrins have been identified in heavy feeds (42–44). For example, at least six different porphyrin structures were present in bitumen derived from tar sands (45). The typical structures of porphyrins found in heavy crudes are shown in Figure 2.9. The chlorin structure in Figure 2.9 is not present in heavy feeds but can be formed as the first step during the overall HDM of porphyrins. It is, generally, observed that the relative concentrations of these structures may vary from feed to feed. The frequently investigated tetra-phenyl-substituted porphyrin is less common in heavy feeds but, in its reactivity, it may approach the porphyrins commonly found in heavy feeds (45).

The non-porphyrinic structures were assumed and proposed by some authors (46); however, their presence in heavy feeds was not clearly confirmed. Other studies

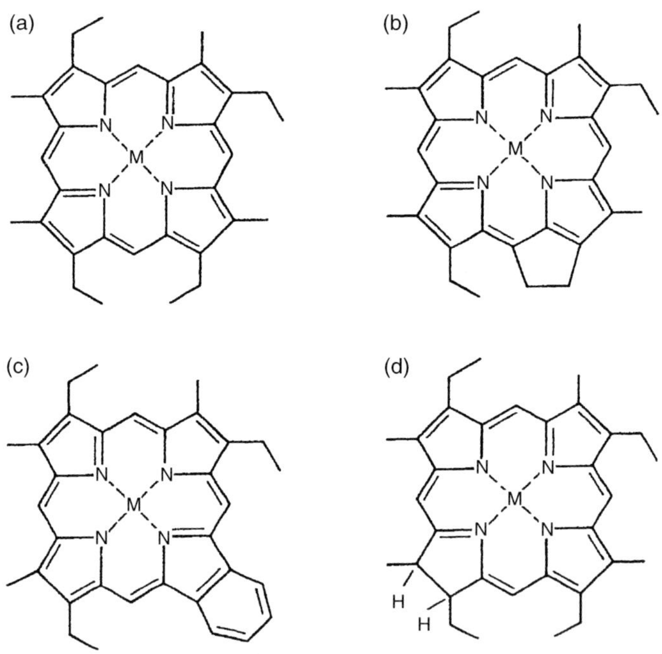

(a) Etioporphyrin
(b) Deoxophylloerithroetioporphyrin
(c) Rhodoporphyrin
(d) Chlorin

Figure 2.9 Typical metal porphyrin structures in petroleum (a–c) and chlorin intermediate (d); M = Ni, V = O.

suggested that porphyrins account for only about half of the V and Ni in petroleum crudes (47,48). Apparently, the unaccounted part of the metals was supposed to comprise less-defined forms which may involve some bonding with heteroatoms in resins and asphaltenes. Some support for this opinion was provided by the study of Sughrue et al. (49) who used the size exclusion chromatography technique for characterizing heavy feeds. The study revealed the presence of large and small V molecules. The latter disappeared after hydroprocessing of the feed. At the same time, the large molecules were more resistant. It is suggested that the large V-containing molecules were still associated with asphaltenes, most likely in the porphyrin-like form because in this case, asphaltenes were not completely removed from the feed, i.e., their presence in the products was clearly confirmed. This was indeed confirmed by Sakanishi et al. (50) who observed that porphyrins may be occluded by the strong non-covalent interaction as part of the asphaltenes aggregates. Once porphyrins were released from the aggregates, i.e., by the interaction with a solvent, the molecular weight typical of the metal-containing species approached that of porphyrins in agreement with the results published by Grigsby and Green (51). It is believed that these observations may not be common to all heavy feeds. Thus, a chemical bonding of the porphyrin structures with asphaltene and resin molecules cannot be ruled out suggesting that at least a partial if not complete conversion of the asphaltene molecules would be required to attain high rates of HDM.

The results in Table 2.4 (32) show that when butane was used as the solvent, most of the V and Ni in the heavy feed ended up in asphaltenes, while less than 20% and a trace amount of the metals could be found in resin and oil fractions, respectively. The same was confirmed by the gel permeation chromatography of Athabasca bitumen (52). In this case, seven fractions varying in molecular weight from 370 to 6680 units were isolated. The content of V and Ni increased linearly and then leveled off when the molecular weight of 5470 was approached. In contrast with Table 2.4 (32), the results in Table 2.5 (33) indicate that when pentane was used for the separation of oil, resins and asphaltenes from the Boscan crude, the content of metal in the oil phase approached and/or exceeded that of the VR derived from conventional crudes.

Hydroprocessing of the V- and Ni-containing porphyrins involves the HYD equilibrium between the porphyrin and the corresponding chlorin as shown in Figure 2.10 (53). Apparently, this step does not require the presence of catalyst if a sufficient H_2 pressure is maintained. The HYD of at least one pyrrole ring induces significant flexibility in the macrocyclic ligand (53). This diminishes the stability of the transition complex. The final step in the HDM of porphyrins involves the cleavage of the metal–N bond. This

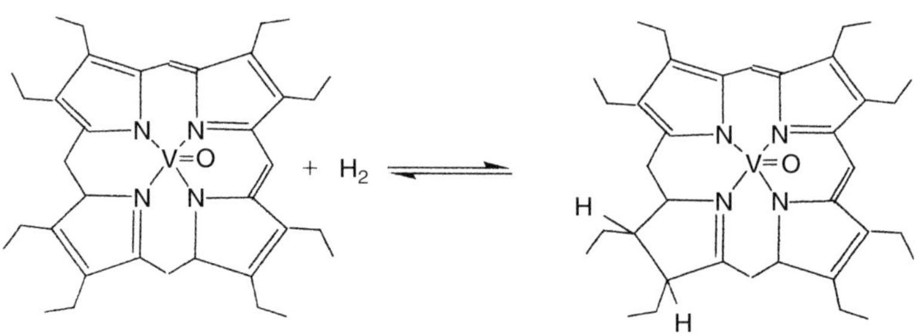

Figure 2.10 Hydrogenation equilibrium between porphyrin and chlorine (53).

step is essential for hydrogenolysis and may require the presence of catalyst. Moreover, Rankel (54) reported that the thermal degradation of porphyrins can be achieved in the presence of H_2 (~4 MPa) above 673 K without catalyst. Similarly, Reynolds et al. (46) observed the non-catalytic demetallization at 673 K, whereas in the presence of catalyst, demetallization began at 473 K. The non-catalytic degradation was enhanced in the H_2 + H_2S mixture. In fact, in this case, rather mild conditions (e.g., near atmospheric pressure of H_2 at about 550 K) were sufficient to achieve a limited porphyrin degradation. The detailed account of these events is given in Chapters 6 and 7 of this book.

During hydroprocessing, the deposition of V on the catalyst surface occurs at much greater rate than that of Ni. This can be attributed to the greater amount of the former in the feed, as well as to the presence of the vanadyl group (V=O) which facilitates strong interaction with catalyst surface. Then, even for the heavy feed containing similar amounts of the V and Ni each, the HDM of V-containing porphyrins shall have a more adverse effect on the HDM of Ni-containing porphyrins than vice versa. Because for most of the heavy feeds the amount of V is much greater than that of Ni, hydroprocessing reactions will be affected to a much greater extent by the V deposition than by the Ni deposition. Moreover, because of the strong interaction, the deposition of V species will occur predominantly on or near the external surface of catalyst particles, i.e., at the first contact with catalyst surface. This will be evident particularly in the case that the porosity of the catalyst was not optimized. This may lead to the restrictive diffusion of large molecules into the pores of catalyst. Once deposited on the surface, the V and Ni are converted to sulfides and/or sulfoxides, which may exhibit some catalytic activity. The overwhelming evidence shows that Ni sulfides were more active than V sulfides. Because of a much more adverse effect on the catalyst performance, it is not surprising that most of the studies on the HDM of heavy petroleum feeds have been focusing on the removal of V compared with that of Ni.

2.3 PHYSICAL PROPERTIES

From the transportation and/or delivery into hydroprocessing reactor point of view, the viscosity appears to be the most important physical property of heavy feeds. Thus, a sufficiently low viscosity has to be maintained to ensure adequate pumpability. The effect of temperature and pressure on viscosity of heavy feeds was studied by Cohen et al. (55). The range of investigated viscosity varied from 10 to 40 000 cP with maximum temperature of 773 K and pressure of 30 MPa. Figure 2.11 shows that the viscosity of a crude depends on the amount of the 350+ residue and/or the yield of 350− distillate (56,57). At an ambient temperature, heavy crudes and distillation residues (VR in particular) are in a semisolid form. In this form, their pumpability is rather low. Increasing temperature leads to the decrease in their viscosity because of the gradual change in the colloidal structure, e.g., solids → semisolids → semiliquids → liquids. This results from the increased mobility of micelles in the oil phase. In refinery practice, the need for reheating heavy feeds is usually avoided by the integration of distillation towers with the upgrading units such as catalytic reactors or cokers. This ensures that distillation residues, while still possessing a sufficiently low viscosity, can be fed directly to the reactor for upgrading.

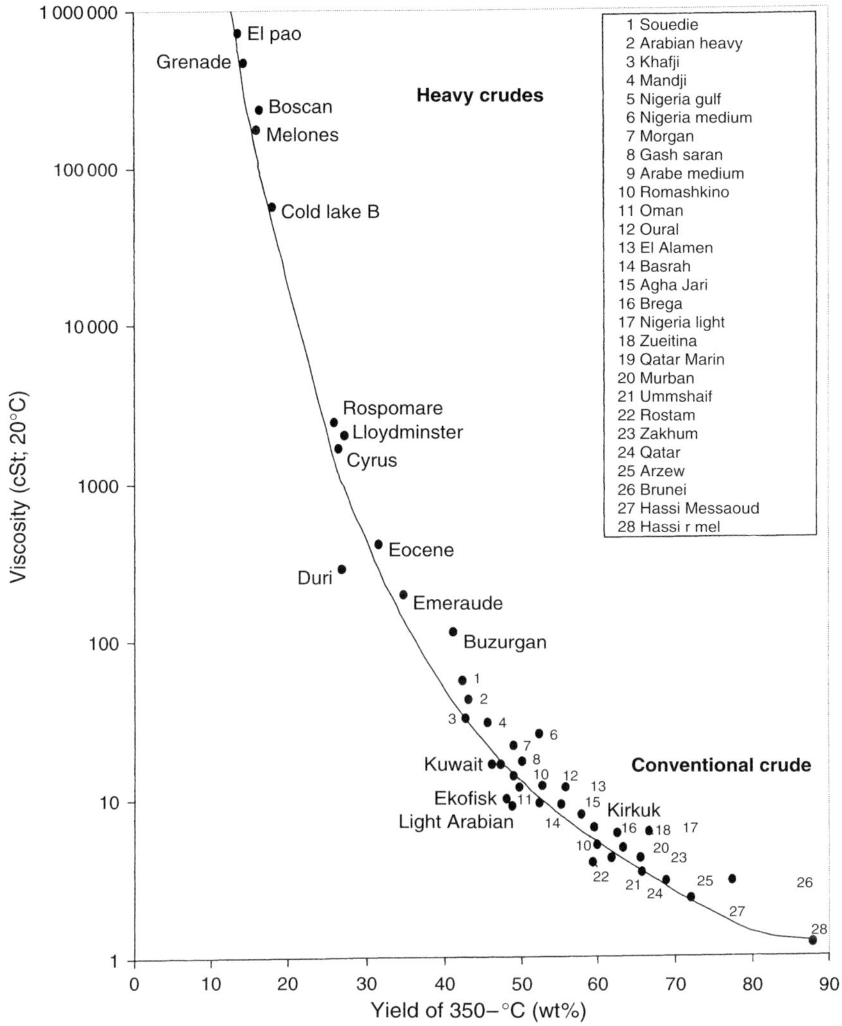

Figure 2.11 Viscosity of crude versus yield of 350 −° C fraction (56).

For heavy crudes, upgrading has to be performed on or near the site of the production well with the aim to achieve sufficient pumpability for pipelining. According to the commonly accepted rule, the viscosity should be lower than 120 cSt at 20 °C, whereas in the Southern Hemisphere a greater viscosity can be tolerated (50). Blending a heavy crude and/or a heavy feed with a lighter fraction may be another option for achieving a desirably low viscosity and pumpability. However, as it was pointed out earlier, in this case, it is essential that the blended materials are compatible. Otherwise the deposit formation in the transport lines cannot be avoided. As a general rule, more aromatic fractions are more suitable diluents for heavy crudes because they ensure a higher dispersion of asphaltenes in the colloidal system.

Figure 2.11 (56) suggests that there might be a correlation between the viscosity, as well as the CCR, and asphaltenes content. However, after closely examining an extensive database, rather poor correlation could be established. Also, there was little correlation between the viscosity and specific gravity. Thus, the feeds varying widely in viscosity may have similar specific gravity. This is not surprising, because different factors are involved in determining the values of viscosity and specific gravity. Therefore, in the case of heavy feeds such as VRs and topped heavy crudes, viscosity is the parameter which has to be carefully monitored.

Chapter 3

PROPERTIES OF CATALYSTS FOR HYDROPROCESSING OF HEAVY FEEDS

The bibliography on the properties of hydroprocessing catalysts is rather extensive and has been the subject of several reviews, which have been periodically updated (58–62). The special attention has been devoted to the hydrogen activation without which hydroprocessing reactions could not proceed (63). Detailed accounts of the methods used for catalyst preparation was given as well (64–67). It was well established that during hydroprocessing the catalyst activity was continuously declining. At certain point, the catalyst has to be replaced and sent either for regeneration or disposal. The extensive reviews on deactivation (40) and regeneration (68) of hydroprocessing catalysts were also published. It has been shown that the safety, utilization and disposal aspects of spent hydroprocessing catalysts require special attention as well (41). It should be emphasized that most of these studies have focused, predominantly on the model compounds and on the light feeds, paying a limited attention to the feeds containing resins, asphaltenes and metals.

It clearly follows from the previous chapter that the design of catalysts for hydroprocessing of heavy feeds has to take into consideration the presence of high molecular weight compounds (e.g., asphaltenes and resins), as well as the heteroatom (S, N and O) containing structures and the metals such as V, Ni, Ti, Fe and others. This suggests that a high activity of the catalyst must be complemented by an adequate tolerance to metals and/or metal storage capacity. A wide range of heavy feeds available suggests that the tailor-made catalysts suitable for hydroprocessing of a particular feed may need to be designed for achieving an efficient operation. In this regard, optimal combination of the chemical composition of catalysts with its physical properties has to be established. Moreover, the size and shape of the catalyst particles must be matched with properties of the feed and the type of catalytic reactor. To certain degree, the problems with catalyst selection for heavy feeds may be alleviated by using the low-cost, throw-away solids disposed from various industrial operations, as well as the naturally occurring minerals and clays containing catalytically active transition metals, e.g., Fe.

3.1 CHEMICAL COMPOSITION

All aspects of the hydroprocessing catalysts have been reviewed in detail elsewhere (40,41,58–68). Therefore, a brief and general account of their chemical composition will only be given with emphasis on the conventional hydroprocessing catalysts. Thus, the improved conventional catalysts and novel catalytic phases, which have been developed for hydroprocessing of heavy feeds will be part of the extensive discussions in the following chapters which are part of this book. This also includes dewaxing catalysts, the structure of which differs markedly from the typical hydroprocessing catalysts, although the conditions employed during dewaxing approach those employed during the conventional hydroprocessing.

The Mo(W)-containing supported catalysts, promoted either by Co or Ni, have been used for hydroprocessing for decades. The γ-Al_2O_3 has been the predominant support; however, other supports, e.g., silica–alumina, zeolites, TiO_2, etc., have also been used with the aim to improve catalyst performance. The enhancement in the rate of hydrocracking (HCR) reactions was the reason for using more acidic supports. The operating (sulfided) form of the catalysts contains the slabs of the $Mo(W)S_2$. The distribution of the slabs on the support, i.e., from a monolayer to clusters, depends on the method used for the loading of active metals, conditions applied during sulfiding and operating conditions. The unsupported $Mo(W)S_2$ catalysts exhibit a hexagonal coordination. It is reasonable to assume that the same coordination will be retained in the supported catalysts. Under hydroprocessing conditions, the corner and edge sulfur ions in $Mo(W)S_2$ can be readily removed. This results in the formation of the coordinatively unsaturated sites (CUS) and/or sulfur ion vacancies which have the Lewis acid character. Double and even multiple vacancies can be formed. Because of the Lewis acid character, CUS can adsorb molecules with the unpaired electrons (e.g., N-bases) present in the feed. They are also the sites for hydrogen activation. In this case, H_2 may be homolytically and heterolytically split to yield the Mo—H and S—H moieties, respectively (63). It is this active hydrogen which is subsequently transferred to the reactant molecules adsorbed on or near the CUS. Part of the active hydrogen can be spilt over on the support and, to a certain extent, can protect slabs of the active phase from deactivation by coke deposits, the size of which on the bare support is progressively increasing (69,70). It will be shown that, in this regard, the protective role of surface hydrogen may be enhanced by optimizing the method of catalyst presulfiding.

The promoters such as Co and Ni decorate $Mo(W)S_2$ crystals at the edges and corner sites of the crystals. In the presence of promoters, CUS are considerably more active than those on the metal sulfide alone. Apparently, this may result from the increased rate of hydrogen activation due to the presence of promoters. The H_2S/H_2 ratio is the critical parameter for maintaining the optimal number of CUS. It has been confirmed that above 673 K, the —SH moieties on the catalyst surface possess the Bronsted acid character (61). The presence of the Bronsted acid sites is critical for achieving a high rate of hydrodenitrogenation (HDN). Otherwise, other hydroprocessing reactions would be inhibited because of the prolonged adsorption of the N-compounds on CUS. Besides preventing other reactants from being adsorbed on active sites, the N-containing species on CUS may slow down hydrogen activation process. These adverse effects are the main reason for catalyst poisoning by N-bases (63). Furthermore, the formation of coke and metal (predominantly V and Ni) deposits on CUS will diminish the availability of active site. In fact, during the later stages onstream, the loss of the catalyst activity during hydroprocessing of heavy feeds will be caused mainly by the deposition of coke and metals in particular. This will result in the restrictive diffusion which will decrease the access of reactants to the active sites in the catalyst pores (40), although during initial stages, the deposited metals can catalyze hydroprocessing reactions. Thus, under typical hydroprocessing conditions, the Ni deposits are expected to have a beneficial effect on HYD reactions, whereas for the V deposits, such effect may be less evident.

During industrial operations, the oxidic form of catalysts is converted to the sulfided form, unless the catalyst sulfidation was conducted prior to the operation. Practical experience supports the catalyst presulfiding prior to contact with heavy feed. The structure of such catalysts is rather complex. In this regard, significant contributions of Topsoe et al. (61) to the understanding of these issues should be noted. In the case of the $CoMo/Al_2O_3$ catalyst, several species could be detected on the γ-Al_2O_3 surface. Thus, presence of the species such as MoS_2, Co_9S_8 and Co/Al_2O_3 was clearly confirmed. Moreover, the Mossbauer emission spectroscopy provided clear evidence for the presence of the phase in which Co was associated with MoS_2, i.e., Co–Mo–S phase. Similar structures were also found in the $NiMo/Al_2O_3$, CoW/Al_2O_3 and NiW/Al_2O_3 catalysts, e.g., Ni–Mo–S, Co–W–S and Ni–W–S, respectively. In this phase, an enhanced concentration of Co and/or Ni promoters at the edge planes of MoS_2 crystals has been confirmed. The occurrence of these promoters in the same plane as that of Mo ruled out the intercalation of the former between the layers of MoS_2. In the Co–Mo–S phase, the Mo−S bond is weaker than in the unpromoted MoS_2. Then, the CUS required for hydroprocessing reactions can be created more readily. Temperature and the H_2S/H_2 ratio are among the important operating parameters for controlling the CUS concentration.

The structure of the Co–Mo–S phase is temperature dependent (61). Thus, the Type I phase formed at lower temperatures was still chemically bound with the support, as it was evidenced by the presence of the Al–O–Mo entities. This phase was favored at low Mo loading on the γ-Al_2O_3. The occurrence of this phase was an indication of the incomplete sulfiding. The sulfiding at higher temperatures facilitated the transformation of the Type I phase into Type II phase. Consequently, the Al–O–Mo entities were not present indicating a diminished interaction of the active phase with the Al_2O_3 support. The existence of the Type II phase was further confirmed in the unsupported Co/MoS_2 system, as well as in the CoMo catalyst supported on carbon suggesting that Type I phase requires the presence of oxygen on the support to facilitate the interaction with the active phase. Because of a lesser interaction with the support, the structure of Type II phase is dominated by the multiple stacks of slubs compared with more less monolayer distribution occurring in Type I phase. Generally, the former phase exhibits a higher catalytic activity. This suggests that the active sites are present at the edges and corners of the Mo(W)S crystallites. The proportion of such sites in the Type II phase is much greater than in the Type I phase.

For hydroprocessing of heavy feeds, an operating time may be gained if the catalyst presulfiding is performed prior to contacting the heavy feed. In this case, a decreased catalyst deactivation was anticipated. Either the H_2S+H_2 mixture or the H_2S-donating compounds can be used as the presulfiding media. Presulfiding can be performed either in situ or off-site. Practical experience suggests that the off-site presulfiding yielded more active catalyst. Figure 3.1 shows that the off-site (actiCAT) process was rather simple compared with the in situ presulfiding (71–73). Absi-Halabi et al. (74) showed that in the case of the heavy feed containing asphaltenes and metals, presulfiding increased the rate of HDM and HDAs, as well as the conversion of heavy fractions to distillates. At the same time, it had little effect on the hydrodesulfurization (HDS) and HDN (74). The in situ presulfiding using a sulfur containing oil was less efficient than the off-site presulfiding method involving the use of the sulfur-donating agents. During the former method, the catalyst activity may have been affected because of the presence of N-bases in the sulfur containing oil. It is essential that presulfiding conditions are optimized to

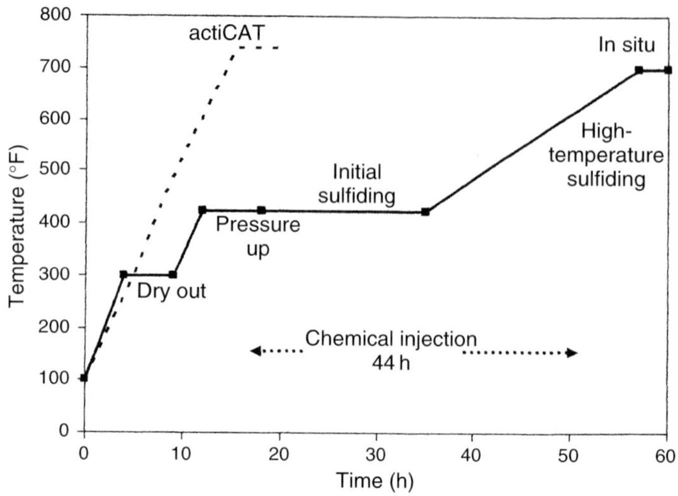

Figure 3.1 Comparison of the ex situ with in situ presulfiding (71).

ensure the predominance of Type II active phase (61). It is believed that this is crucial for both light feeds, as well as for the asphaltenes and metals containing feeds. Thus, this phase can activate hydrogen more readily than Type I phase (63). Then, active hydrogen spilt on the support may react with deposits and as such protect active phase. Inevitably, the loss of catalyst activity caused by fouling will be more evident when the Type I phase is present.

It was indicated that the catalyst performance can be modified by changing the chemical composition of supports. The above discussion may suggest that supports which favor the formation of Type II phase are the supports of a choice. However, the catalyst for hydroprocessing of heavy feeds must possess an adequate HCR activity which can be provided by more acidic supports. The additives such as phosphate, fluoride, boride, etc. can modify properties of the conventional γ-Al_2O_3 support and eventually improve catalyst performance, although in some cases, little effect on the catalyst activity was observed (75). The supports consisting of bi- or even multiple oxides have been evaluated but predominantly in various model compounds and light feeds studies (76). Beneficial effects of such supports resulted from the improvement in the acidic sites distribution. This, in turn enhanced HCR activity, which was desirable for achieving a high conversion of large molecules, produced by the depolymerization of asphaltenes and resins, to distillate fractions. This was indeed confirmed by Corma et al. (77) who compared the conventional NiMo/Al_2O_3 catalyst with a series of the NiMo catalysts supported on the amorphous silica–alumina and on several different zeolites.

More acidic supports may have an adverse effect on the other catalyst functionalities, i.e., resistance to coke formation. Therefore, the activity of the catalysts for hydroprocessing of heavy feeds must be optimized to achieve a desirable level of HCR simultaneously with that of HDAs, HDM, HDS and HDN, while the steady catalyst performance is maintained for as long time onstream as possible. In this regard, Maity et al. (78) showed that the catalyst functionalities can be modified by incorporating about 5 wt% of the TiO_2 into Al_2O_3. In this case, the pore volume and mean pore diameter of catalysts could be controlled by the addition of TiO_2 to Al_2O_3. The catalysts supported on carbon are less sensitive to poisoning by N-bases, however, their HCR activity may be low unless the carbon support was subjected to special pretreatments. It will be shown latter in the review that the potential for improving the catalyst performance during hydroprocessing of heavy feeds can be significantly enhanced by combining several different supports.

The non-conventional catalysts for hydroprocessing heavy feeds have been dominated by the catalytically active metals which are part of the conventional catalysts. In the former case, catalysts are added either in a finely divided form or as the solution either in water or in oil. Bio-catalysts represent another group of catalysts tested for the upgrading of heavy feeds. It should be noted that during bio-upgrading, reported conditions (temperature and H_2 pressure) approach those employed during hydroprocessing, although milder conditions have been anticipated. So far, there is no evidence supporting a commercial utilization of these forms of catalysts for upgrading heavy feeds on the site of a petroleum refinery.

Dewaxing is another process operating under H_2. The feeds used in this process may include VGO and DAO to produce either base oil which is used for the preparation of lubricants or middle distillates for the fuel production. The objective of dewaxing is the removal of the long chain paraffinic compounds to achieve the cold flow properties of products (e.g., pour point) as required by specifications. For this purpose, the catalysts possessing cracking and isomerization activities are necessary. Because of the simultaneous production of alkenes and the content of aromatics exceeding specifications, the lube base oil and middle distillates may require the additional hydroprocessing step, i.e., hydrofinishing and dehazing, to ensure desirable stability of the commercial products. A more detailed account of the dewaxing catalysts is give in the Chapter 11.

A necessity to incorporate RFCC catalysts results from the frequent reference to this process as one being suitable for residues upgrading. In fact, the current state of the art in this process does not allow processing any VR. It is also believed that even the AR derived from most of the sweet crudes may require hydroprocessing step before being suitable feed for RFCC. Therefore, compared with hydroprocessing, the capability of RFCC to upgrade residues is rather limited.

3.2 PHYSICAL PROPERTIES

The physical properties of the catalysts have to be optimized to withstand rather severe conditions applied during hydroprocessing of heavy feeds. In this regard, porosity is of the primary importance for maximizing the catalyst utilization. The chemical composition may not be so important unless the suitable porosity of the catalyst has been established. This is desirable for maintaining a long life of catalyst during the operation. Besides porosity, the optimal size and shape of particles has to be chosen to avoid malfunctioning of the catalyst bed. Furthermore, the catalyst utilization usually increases with the decreasing size of catalyst particles. However, for fixed bed reactors, the size and shape of catalyst particles deserve special attention because of the potential pressure drop developments. The influence of porosity, as well as that of the size and shape of catalyst particles, is evident even for relatively light feeds such as VGO and HGO. Of course, for the asphaltenes and metals containing feeds, the design and selection of the catalysts becomes much more challenging task.

3.2.1 Surface properties

Among the surface properties, pore volume and size distribution, as well as the mean pore diameter, of the catalyst are much more important than surface area when heavy feeds are considered. At the same time, for light feeds, surface area may be a reasonable indication of the catalyst suitability. Figure 3.2 (79,80) shows the difference in pore mouth plugging between small and large pores. The pore size distribution of the fresh and spent catalysts are presented as relative area of the nitrogen adsorption/desorption isotherms for two different mean pore diameter catalysts along the deactivation on the pore mouth. The increases in hysteresis loop were considered as representative of the deactivation by pore mouth coking. In the case of fresh catalysts, the pores are considered cylindrical, whereas for the spent catalyst, the pores are "ink-bottle" types due to the metal and carbon deposition, which resulted in the increase in area of the hysteresis loop. A high surface area and moderate pore volume catalysts are very active for HDS because of the efficient dispersion of active metals in the pores. However, in the case of heavy feeds, these pores become gradually unavailable because they are deactivated by pore mouth plugging. On the other hand, the catalysts with a small surface area and a large

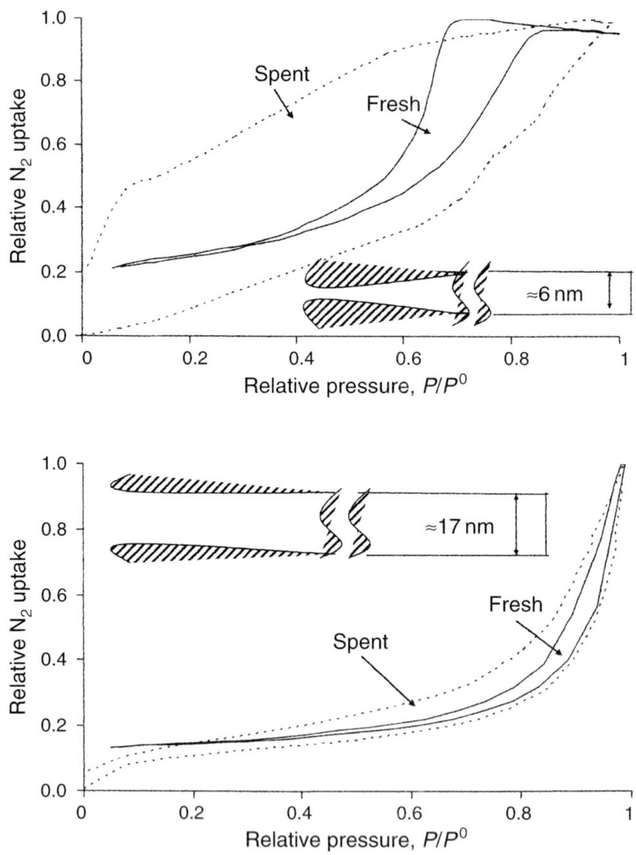

Figure 3.2 Adsorption–desorption isotherms for fresh and spent $CoMo/Al_2O_3$ catalysts (80).

pore volume are less active because of a lower concentration of active sites. However, they are more resistant to deactivation by pore mouth plugging and their metal storage capacity is greater, therefore such catalysts may be suitable for HDM and HDAs. The effect of the surface area and pore diameter and/or pore volume on deactivation of the catalysts are shown in Figure 3.3 (80). These results clearly indicate that the high surface area and low porosity catalysts will deactivate faster than a low surface area and high porosity catalysts. The ranges of the surface area and pore diameter suitable for various feeds are identified in Figure 3.4 (80).

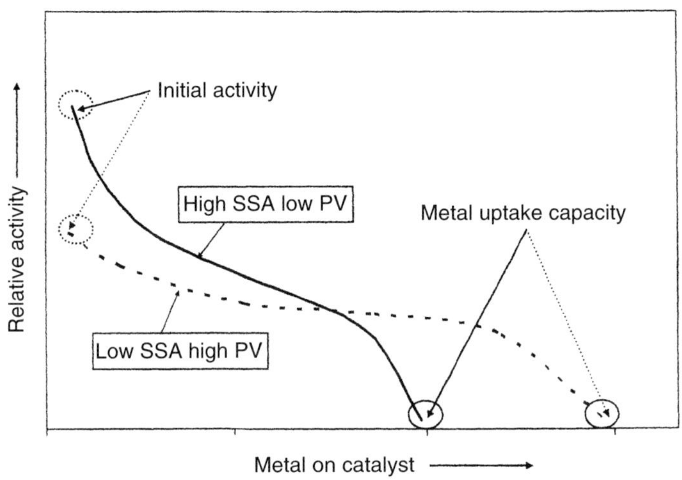

Figure 3.3 Relation between catalyst activity and metal accumulation for high SSA and low PV as well as low SSA and high PV catalysts (80).

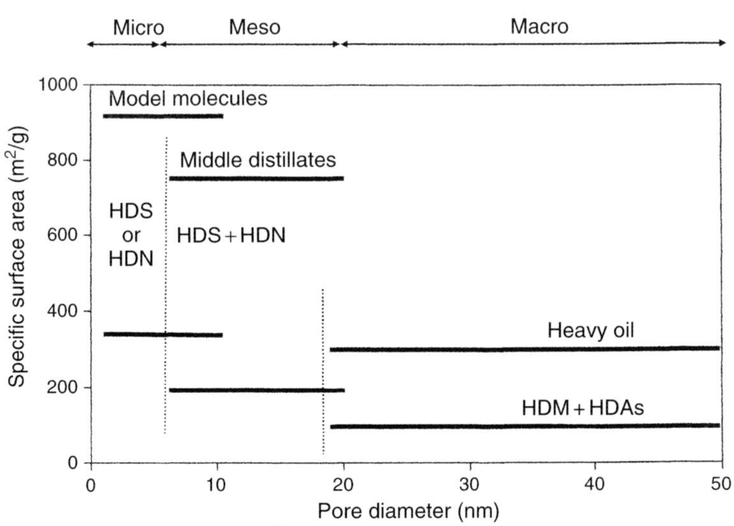

Figure 3.4 Effect of pore diameter and surface area on catalyst functionalities (80).

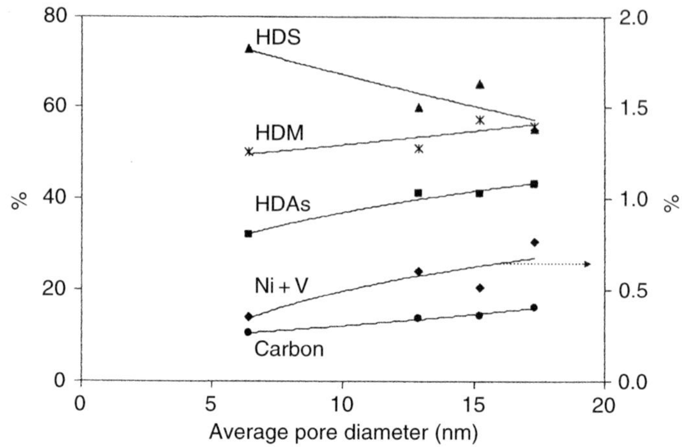

Figure 3.5 Effect of average pore diameter on conversion and deposition of metals and carbon on catalyst after 60 h TOS (81).

The different activity and capacity for carbon and metal depositions are shown in Figure 3.5 (81) as function of the average pore diameter (APD). The opposite trends in the HDS and HDM activities suggest that the HDM catalyst should be macroporous. This is indicated by the low HDM conversion for the low APD catalyst caused by restricted diffusion of the large metal-containing molecules (porphyrins and/or metal-chelating compounds) into the pores. The parallel increase in the HDM and HDAs conversions indicates that these functionalities are dominated by the catalyst pore structure. These trends may also indicate on the necessity of asphaltenes conversion for achieving a high rate of HDM, i.e., the latter depends on the extent of the asphaltenes aggregates disintegration. An efficient dispersion of active phase was essential for achieving desirable HDS activity (5). This was favored by the decreasing pore diameter, i.e., increasing surface area. Therefore, the factors which are important for HDS activity differ markedly from those which dominate HDM and HDAs activities (5,79).

The above discussion suggests that there is an optimal combination of the surface area and pore diameter giving the highest catalyst activity (5). The optimum may be different for different feeds and catalysts. This is evident from the results in Figure 3.6 showing that the optimal pore size for achieving the highest activity during the HDS of the heavy feed differed from those required for the lighter feeds (82). Similarly, the effect of porosity on catalyst performance was confirmed during the hydroprocessing of the AR and HGO over the microporous conventional HDS catalyst of the $CoMo/Al_2O_3$ formulation (6). As Figure 3.7 shows (6), for the HGO, the steady catalyst performance was maintained for an extended period, whereas a continuous catalyst deactivation was observed during hydroprocessing of the AR. For the latter, the catalyst was deactivated both by coke and metal deposits.

It has been generally established that the initial rate of coke formation was high and then gradually attained the steady state. On the other hand, the formation of metal deposits with time onstream was almost linear. This was confirmed by the more gradual increase in the amount of metals deposited on the catalysts shown in Figure 3.8 (6). This suggests

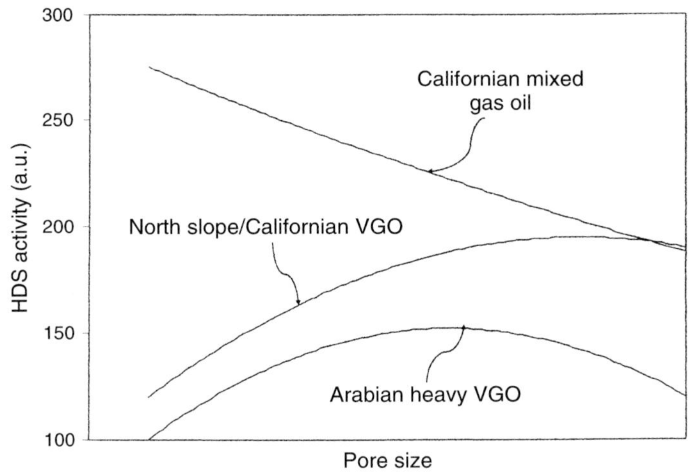

Figure 3.6 Effect of feed origin and pore size on catalyst activity (82).

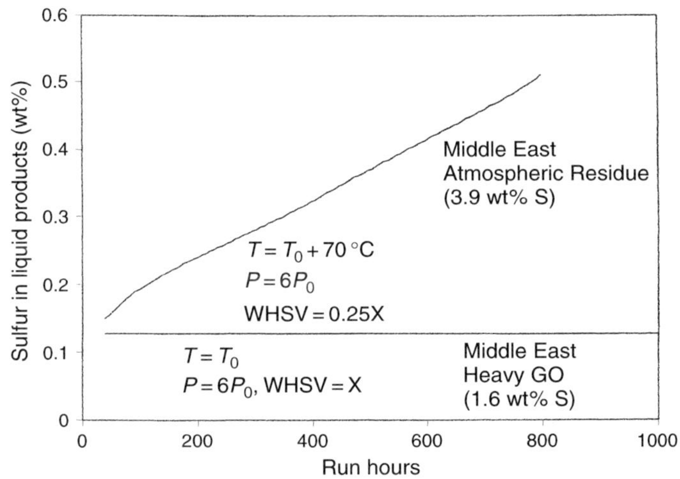

Figure 3.7 Effect of feed origin on HDS activity ($CoMo/Al_2O_3$) (6).

that contribution of the metal deposits to the overall catalyst deactivation will be gaining on importance with time onstream until it will become the main cause of the activity loss.

It is again emphasized that an optimal pore size and volume distribution is critical for hydroprocessing of the high metal content feeds, particularly those derived from heavy crudes. This results from the large molecular diameter of the V- and Ni-containing porphyrin molecules, i.e., for microporous catalysts, the diameter may exceed that of pores. For small pore diameters, most of the metals will deposit on the external surface of the catalyst particles and the diffusion into the catalyst interior becomes the rate-limiting factor. It is, therefore expected that the tolerance of catalyst to metals will increase with the increasing pore diameter as it is shown in Figure 3.9 (83). At the same time, the

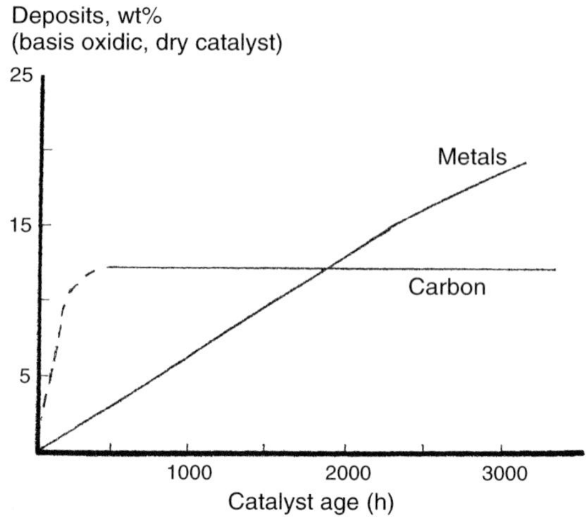

Figure 3.8 Metals and carbon deposition as function of catalyst age (6).

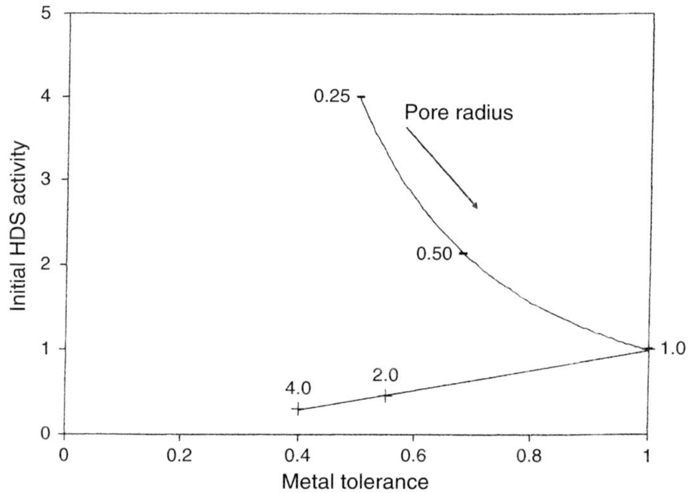

Figure 3.9 Effect of pore radius on metal tolerance and HDS activity (83).

catalyst activity will decrease. At a certain pore radius, the tolerance to metals abruptly decreased, whereas the activity decrease was less pronounced.

3.2.2 Quantification of diffusion phenomena

The effective diffusivity of the reactant molecules into catalyst interior becomes affected when their diameter approaches the diameter of pores. The set of mathematical equations has been developed for quantifying diffusion phenomena. In relation to catalyst deactivation, these equations were reviewed in detail elsewhere (40). However, the primary focus has been either on the model compounds containing feeds or on the light feeds containing no asphaltenes and metals. In this case, theoretically derived equations could closely predict the catalyst performance. For the asphaltenes and metals containing feeds, the situation is much more complex. Moreover, the deposition pattern and diffusion phenomena associated with them differ significantly between the inlet and outlet of the fixed bed of catalyst. They are also influenced by the origin of heavy feed and the type of catalyst. This suggests that the quantification of diffusion phenomena involving a single catalyst particle is not sufficient for describing events in the catalyst bed. Moreover, such evidence cannot be applied generally to a great variety of the feeds and catalysts.

The early attempts to quantify effective diffusivity were affected because of the limited experimental data. Since that time, a significant progress in the development of analytical techniques for analysis of the spent catalyst particles has been made. For example, Tamm et al. (8) used the electron microprobe to determine the radial distribution of metals for the catalyst particles taken from different locations, between inlet and outlet of the fixed bed of catalyst, after the hydroprocessing of several VRs. The validity of mathematical equations derived theoretically could then be verified once such information became available.

For the purpose of this review, it is deemed necessary to give at least a brief summary of the essential equations and factors, which have been used for determining parameters which can relate the catalyst deactivation to restrictive diffusion. The database of various constants and parameters (e.g., bulk and effective diffusivities, Thiele modulus, distribution factor, effectiveness factor, etc.) may be established by using these equations. Various modifications of these equations are discussed and validated in the following chapters using the experimental results from various kinetic studies. Moreover, such information forms a basis for the development of models for predicting the catalyst performance on the reactor scale.

For small reactant molecules, the effective diffusivity (D_e) into the catalyst pores can be expressed as follows:

$$D_e = D_b \epsilon / \tau \times e^{-B\lambda} \qquad [3.1]$$

where D_b is the diffusivity of a reactant in liquid phase (bulk diffusivity), ϵ porosity and τ the tortuosity which accounts for the complexity of pore structure, B the empirical coefficient and λ the ratio of the critical molecular diameter to pore diameter, i.e., d_m/d_p. The diffusivity of reactants in bulk (D_b) can be expressed as:

$$D_b = \mathrm{H}T/6\pi v r \qquad [3.2]$$

where H is Boltzman constant, T absolute temperature, v viscosity and r the radius of diffusing reactant molecule. D_e can also be expressed in the following form:

$$D_e = A(1-\lambda)^z \qquad [3.3]$$

where A is the proportionality constant. To account for the restriction due to the small pores and/or larger molecular size, the following empirical correlation was developed for restrictive factor:

$$\mathrm{F}(\lambda) = D_e t_o^{-4.61} / D_b \epsilon \qquad [3.4]$$

The Thiele modulus (ϕ) has been widely used to describe the diffusion phenomena. It represents the ratio of reaction rate to the diffusion rate. For the Thiele modulus values greater than 1.0, strong diffusion limitations will be present, whereas for the values less than 0.5, diffusion limitations will be negligible. Although there are several forms of ϕ, the one derived by Tamm et al. (8) and used by de Bruijn et al. (83) may describe well the phenomena occurring in the catalyst particles during hydroprocessing of heavy feeds, i.e.,

$$\phi = L_p(k_i C^{n-1}/D_{eff})^{0.5} \quad [3.5]$$

where L_p is the catalyst particle size defined as the ratio of the particle volume (V_p) to geometric surface area (S_p). Other equation of the Thiele modulus has the following form:

$$\phi = R_o(\rho_{cat} A_S k C_{H2} C_o^{n-1}/\rho_{oil} D_{eff})^{0.5} \quad [3.6]$$

where R_o is the effective radius, ρ_{cat} and ρ_{oil} are the densities of catalyst and oil, respectively, A_S is surface area of catalyst, C_{H2} concentration of H_2, D_{eff} the effective diffusivity. The quantity in parentheses is essentially the ratio of the reaction rate to the rate of diffusion of reactant molecules.

The effectiveness factor (E_f) is another parameter used for the model development. It can be estimated experimentally as the ratio of the intrinsic rate constant to apparent rate constant, i.e.,

$$E_f = k_{app}/k_{int} \quad [3.7]$$

As it is shown in Figure 3.10 (84), the porosity of catalyst has a pronounced effect on E_f. Moreover, E_f depends also on temperature. E_f can be related with Thiele modulus (ϕ) using the following equation:

$$E_f = 3/\phi^2\ (\phi\ \coth\phi - 1) \quad [3.8]$$

The value of E_f depends on the catalyst porosity, reactant concentration in the catalyst pores, diffusion of the reactant molecules in porous medium, etc. The following is another form of E_f:

$$E_f = \frac{A \int 4\pi R^2 C^n \mathrm{d}R}{4/3\pi R_o A_S C_o{}^n} \quad [3.9]$$

where A_S is the surface area of catalyst, R the distance from the particle center, R_o the equivalent radius of catalyst particle, C the reactant concentration and n the reaction order. To identify the distribution patterns of the metals in catalyst particles, Tamm et al. (8) used the following distribution parameter:

$$\Theta = \frac{\oint M(r)\mathrm{d}r}{M_{max}\oint r\mathrm{d}r} \quad [3.10]$$

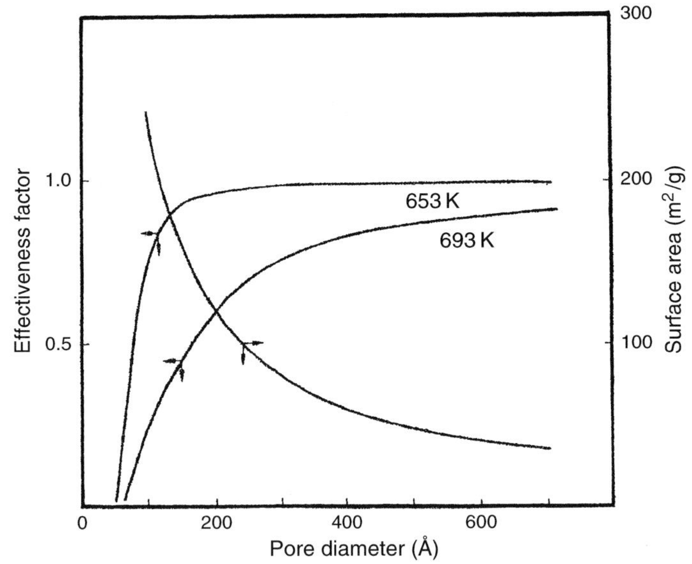

Figure 3.10 Effect of pore diameter on effectiveness factor and surface area (84).

where $M(r)$ is the local concentration of metal deposits in the catalyst particle, r the fractional radius and M_{max} the concentration of metals at the maximum. The Θ values approaching 1.0 indicate an even distribution of metal across particles, whereas the value approaching 0 suggests that metals deposit preferentially on the external surface of the catalyst particles. However, it was demonstrated by Bartholdy and Cooper (85) that in a real situation the value of 1.0 cannot be achieved. Thus, although for large pores the mass transfer is no longer controlled by restricted diffusion, but it is still controlled by the bulk diffusion. Therefore, it is unlikely that the value of distribution parameter could exceed the value of 0.9. The effect of pore diameter on the distribution factor is shown in Table 3.1 (86). It was determined that the life of catalyst whose primary function was HDM increased with the increasing distribution factor. This is supported by the results in Figure 3.11 (8) which indicate that for the lower values of Θ, V deposited predominantly on the catalyst exterior. For fixed bed reactors, the metals deposition patterns depend

Table 3.1 Distribution factors of V and Ni for 454+ residues (86)

Catalyst	MPD[1]	Θ_M for V		Θ_M for Ni	
		Heavy Arab	Maya	Heavy Arab	Maya
A	2.7	0.82	0.68	–	0.77
B	–	0.76	0.64	0.89	0.74
C	2.3	–	0.47	–	0.55
D	1.8	–	0.36	–	0.41
E	1.3	0.41	–	0.56	–
F	1.0	0.30	–	0.34	–
G	0.7	0.13	–	0.17	–

[1] MPD, mean pore diameter (relative scale)

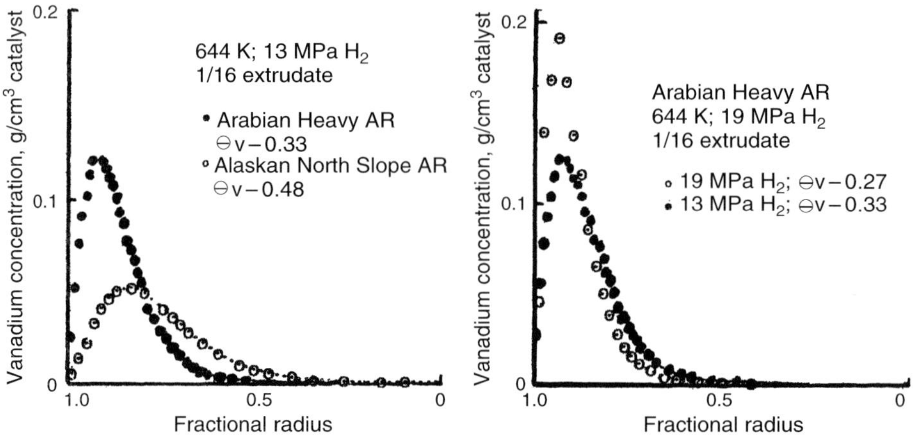

Figure 3.11 Effect of fractional radius on vanadium concentration on catalyst (81).

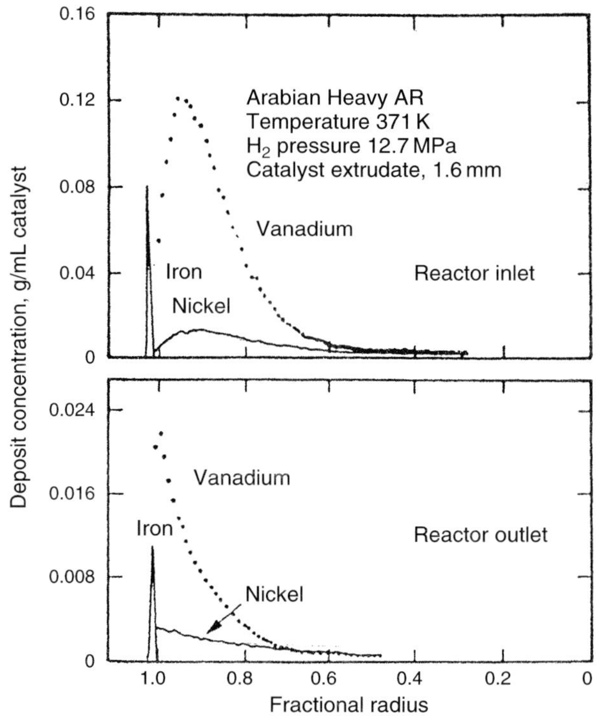

Figure 3.12 Effect of fractional radius on deposition of vanadium, nickel and iron (8).

also on the position of catalyst in the reactor. It was evident from Figure 3.12 (8), that between the inlet and outlet of the fixed bed of catalyst, the deposition patterns changed from M shape to U shape. The Fe distribution pattern shown in Figure 3.12 clearly indicates the non-porphyrinic, most likely a mineral origin of the Fe.

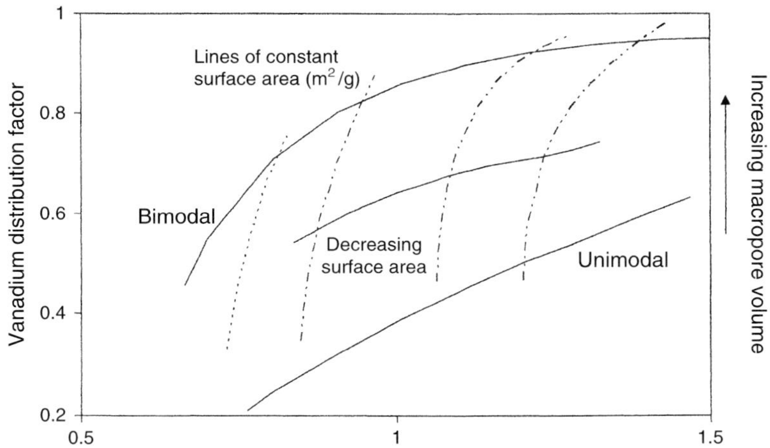

Figure 3.13 Effect of surface properties of catalyst on distribution factor (86).

The correlations between the distribution factor and catalyst properties established using experimental data can be used as the guide for catalyst selection. For example, Figure 3.13 shows that for the same surface area, it was more advantageous to use the bimodal catalysts compared with unimodal catalysts (86). The former catalysts contained different fractions of the micro-, meso- and macropores and could be tailor-made to suit hydroprocessing of various feeds. Because of the higher metal tolerance, bimodal and even polymodal catalysts have the longer life and their metal storage capacity may approach 100% of the original catalyst weight, as it can be achieved for macroporous catalysts. For example, the metal storage capacity of the macroporous NiMo catalyst supported on the "chestnut burr" like Al_2O_3 exceeded 100% of the original catalyst weight (88). As it is shown in Figure 3.14 (87), for the bimodal catalysts with a large fraction of macropores, the radial distribution of metals is more even compared with the microporous unimodal catalyst. An ideal case of the metal distribution represents a polymodal catalyst providing that an optimal combination of the activity and porosity of the catalyst can be attained. The methods used for the preparation of bimodal catalysts have been extensively reviewed several times (64–67). Few examples are used to illustrate preparation of the polymodal support (88) and bimodal γ-Al_2O_3 support (89). For the latter, a simple method involved mixing the γ-Al_2O_3 with combustible solid particles (e.g., carbon black) which were subsequently combusted leaving behind pores, the size of which can be controlled by the size of combustible solids. It was observed that the pore size distribution can also be controlled by chemical composition of the peptizing agents (e.g., by different sugar contents in nitric acid solution) and by the type of forming devices (90).

Equations (1.1–1.10) describe events occurring on the catalyst particle level. The database established on this level is part of the mathematical expressions and models used for simulation of the performance of the catalytic reactors with the primary focus on fixed bed reactors. In most cases, a plug flow mode of operation is assumed. The additional parameters required for this purpose include the volume of reactor and catalyst, liquid holdup, flow patterns, etc. A more detailed account of the model development on the reactor level is given latter in the book.

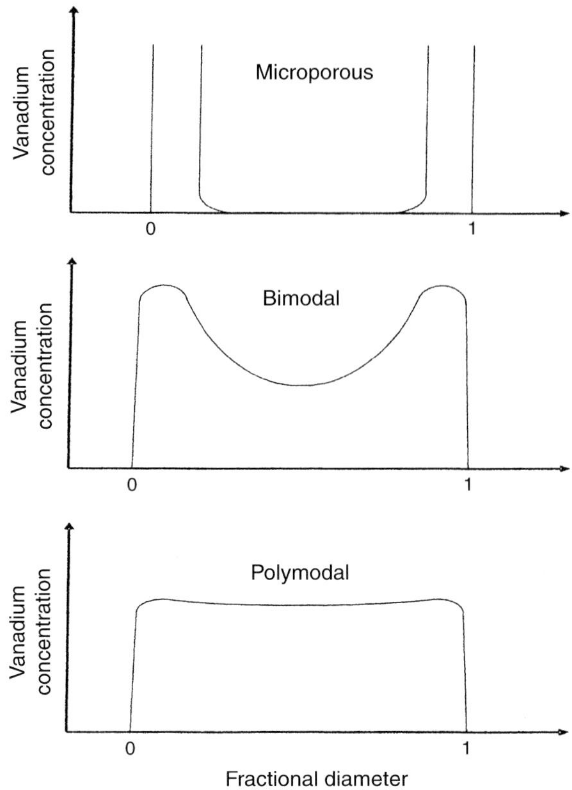

Figure 3.14 Effect of the type of porosity on radial distribution of vanadium (87).

3.3 MECHANICAL PROPERTIES

The mechanical properties of the catalysts must be determined besides the activity, selectivity and physical properties (91). In general, the catalyst activity improves when density of catalyst is decreasing. At the same time, it becomes more difficult to control mechanical properties. In some situations, it becomes an economic tradeoff which must be evaluated to achieve an optimal catalyst performance. Therefore, without adequate mechanical strength, a smooth operation of a catalyst bed cannot be ensured. This is much more critical for heavy feeds than for light feeds and for fixed beds than for ebullated beds. Thus, it is more difficult to maintain the desirable mechanical strength for macroporous catalysts than that for microporous catalysts. This is clearly demonstrated by the decreasing side-crushing strength of the catalyst particles with the increasing pore diameter as shown in Figure 3.15 (92). In the case of fixed bed reactors, cracking of the particles (because of the insufficient mechanical strength) can lead to the unwanted phenomena such as pressure drops along the catalyst bed, creation of the channels causing maldistribution of the feed and even to a collapse of the fixed bed resulting in an unexpected shutdown of the operation. A similar malfunctioning of catalyst bed can be experienced with the catalyst particles possessing an insufficient resistance to attrition. The mechanical properties of catalyst can be controlled during the preparation. In this

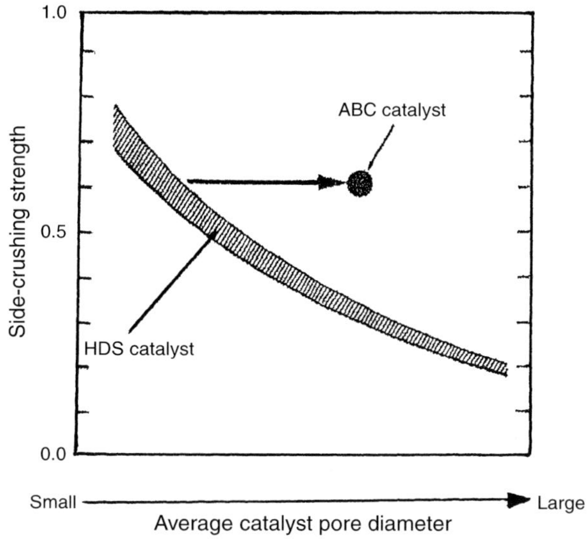

Figure 3.15 Effect of pore diameter on side-crushing strength of catalyst (92).

regard, both the selection of a suitable binder, as well as an optimal temperature, and the duration of catalyst roasting are important. Furthermore, the methods used for the addition of active metals to the catalyst support can play certain role as well (66,67). In the literature, the importance of mechanical properties on the catalyst performance has been generally underestimated even for the most problematic feeds.

3.4 EFFECT OF SHAPE AND SIZE OF CATALYST PARTICLES

In spite of the significant advancements in catalyst development, the diffusion problems encountered during the hydroprocessing of heavy feeds could not be completely eliminated. As a result of this, the catalyst utilization is affected. The size of catalyst particles is another parameter which may at least partially alleviate these problems. For large particles, a near-center part may not be utilized even for a macroporous polymodal catalyst. This problem can be minimized by decreasing the diameter of particles. This will result in a decrease of the diffusion path into the particle interior. But, there may be a limit to the diameter below which the mechanical strength of particles is so low that they can break and disintegrate. Apparently, this limit is being approached when the outside diameter of particles is about 0.8 mm (1/32) and less. However, such small particles may not be suitable for fixed bed reactors because they would cause large pressure drops. Moreover, breaking of such thin particles in fixed bed could not be avoided. This problem may be overcome by selecting the proper shape of particles. Some shapes of particles which may be suitable for hydroprocessing of heavy feeds are shown in Figure 3.16 (80,93). Figure 3.17 (93) demonstrates that, for both the dense and sock loading of catalyst, a selection of an optimal size and shape of particles deserves attention for ensuring the steady performance of the fixed bed. The optimal matching of the particles shape with the different types of the commercial reactors is discussed latter in the review.

The above discussion suggests that the activity of the catalyst having the same chemical composition and structure could be influenced by the size and shape of its particles. To quantify these effects, Cooper et al. (93) defined the size of catalyst particles (L_p) as:

$$L_p = V_p/S_p \quad [3.11]$$

where V_p and S_p are the particle volume and geometric surface area, respectively. Results of these measurements are summarized in Table 3.2. The activity for the HDS of HGO was determined on the equal catalyst weight basis. The trends indicate the improvement in catalyst performance with the decreasing particle size. It was established (83) that the L_p correlated with the Thiele modulus according to the Eqn (3.5) discussed above. The detailed accounts of the effect of these parameters on catalyst performance, particularly its deactivation during the hydroprocessing of heavy feeds, were published elsewhere (40).

Even for the most problematic feeds, the diffusion problems can be almost completely eliminated by using a finely divided form of catalyst. In this case, similar conditions as those applied during the conventional hydroprocessing have been used. In the finely divided form, the catalysts can be readily co-slurried with heavy feeds. Heavy feeds can also be spiked with the water- and/or oil-soluble compounds containing active metals. This ensures that the catalytically active phase formed in situ is of a nearly molecular size. This results in almost complete utilization of catalyst. High conversions

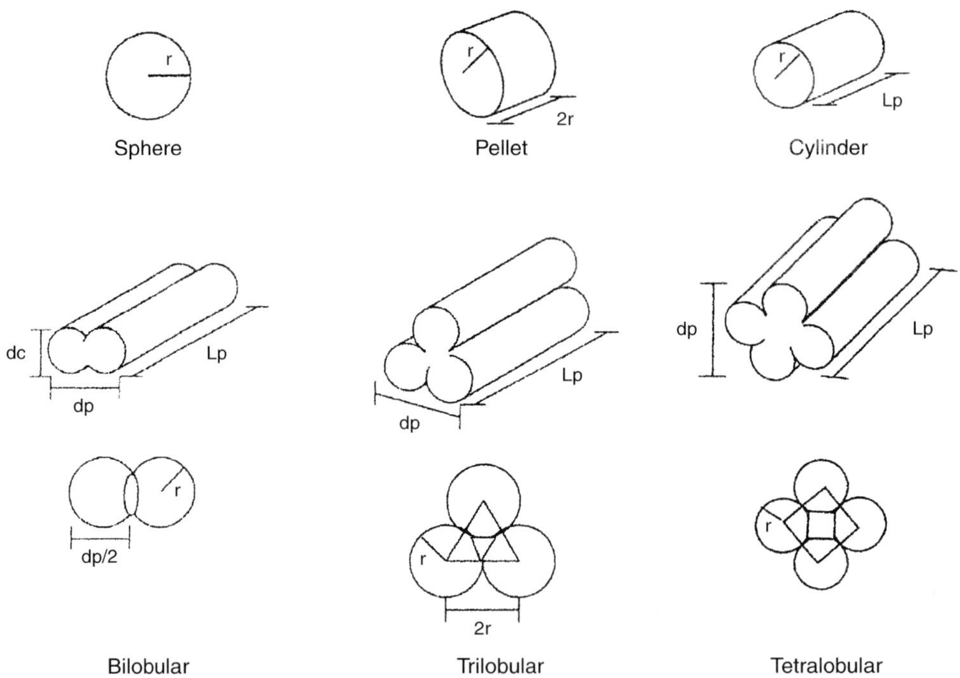

Figure 3.16 Typical shapes of commercial hydroprocessing catalysts (80).

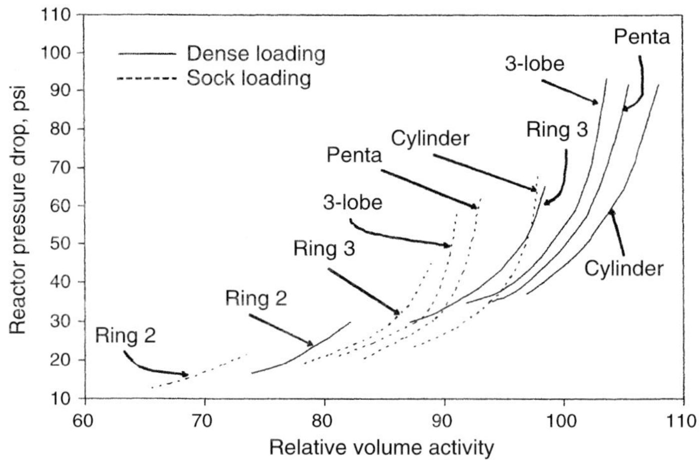

Figure 3.17 Effect of the shape of catalyst particles on reactor pressure drop (93).

Table 3.2 Effect of particle size and shape on HDS activity (93)

Shape	Dimensions (mm)	V_p/S_p (mm)	Activity
Cylinder	0.83 OD × 3.7 length	0.189	9.7
Cylinder	1.2 OD × 5.0 length	0.268	7.9
Cylinder	1.55 OD × 5.0 length	0.345	5.7
Ring	1.62 OD × 0.64 ID × 4.8 length	0.233	8.7
Ellipse	1.9 OD × 1.0 ID × 5.0 length	0.262	8.4
3-lob	1.0 OD × 5.0 length	0.295	8.2
Crushed	0.25 – 0.45	~ 0.04	14.0

OD, outside diameter; ID, inside diameter

of the asphaltenes and metal-containing porphyrins could be achieved by the addition of relatively small quantities of active metals (Mo, Ni, Co, Fe, etc.) in these forms. Design of the slurry bed reactor which is suitable for such application is rather simple compared with the moving and ebullated bed reactors which have to be used for the heavy feeds containing similar amount of asphaltenes and metals. It is also more easy to operate the former reactor. In spite of this, the method has not been used on a commercial scale. This may result from the uncertainty associated with the catalyst recovery. Based on these observations, it is not surprising that a near-commercial stage has reached the slurry bed reactor process employing the finely divided low-cost solids, operating in the once-through mode.

Apparently, the problems associated with catalyst utilization caused by particle size could be completely eliminated by using bio-catalysts. It is, however, believed that for the most problematic feeds, the rate of bio-upgrading reactions is almost certainly affected by diffusion phenomena which decrease the bio-catalyst utilization. Moreover, recovery of the bio-catalysts for reuse is rather difficult. As an alternative to the conventional hydroprocessing and RFCC catalysts, the unconventional catalysts and bio-catalysts are discussed in detail in the latter chapter (Chapter 11) of the book.

Chapter 4

SELECTION OF REACTORS FOR HYDROPROCESSING RESEARCH

It has been generally observed that the primary focus of the studies on hydroprocessing of heavy feeds was on the development of more active catalysts with a high resistance to deactivation. The behavior of heavy feeds under specific hydroprocessing conditions, i.e., type of catalysts and catalytic reactor, the yield and quality of the anticipated products, etc., have been attracting attention as well. A number of the suitable experimental systems have been available for these purposes. These setups are also suitable for establishing database of the parameters which can be used for the selection of catalysts and the optimization of commercial operations. Several types of catalytic reactors have been used for this research. The type of reactor selected for testing influences reliability of the generated data. As it was indicated earlier, the choice of an optimal experimental system for testing the high asphaltenes and metals feed requires an additional attention compared with the system used for testing the VGO and HGO feeds. This issue has been addressed in the series of articles published by Bej et al. (94–97) and Pitault et al. (98).

The reactors for experimental research can be divided into two main groups, i.e., batch reactors and continuous reactors. They are available in various sizes. The cost of the reactors may increase with their size. The differences between the data obtained in the batch reactors and continuous reactors, under otherwise similar conditions, have been noted. In the case of heavy feeds, reliability of the data obtained in small reactors may be somehow limited. It should be noted that there is no ideal catalytic reactor which could fulfill all requirements (98). Therefore, selection of the most suitable reactor is crucial for achieving objectives of the research. Again, this is particularly important for heavy feeds because of the additional factors involved compared with light feeds. It is noted that later in the book, a separate chapter is devoted to the selection of catalysts for catalytic reactors, which have been used commercially.

4.1 BATCH REACTORS

The high-pressure batch reactors (autoclaves) are available in various sizes (volumes), i.e., from few milliliters up to a liter or even a greater volume. In these systems, the contact between catalyst and feed is improved by rolling, agitation or stirring. The high ratio of the reactor diameter/catalyst particles diameter attained in large autoclaves is suitable for studying catalysts having the size and shape used during the industrial operation. Such reactors are more suitable for the investigation of the asphaltenes and metals containing feeds compared with microautoclaves. Microautoclaves can be used for determining the intrinsic activity involving the crushed catalyst particles. The database, which includes the intrinsic activities and the activity determined using the operating size and form of the catalyst particles, provides information on the overall catalyst utilization as well as on the diffusion phenomena which, in the case of heavy feeds, are always present. In this way, the effect of diffusion limitations on the hydroprocessing of heavy feeds can be identified and studied in details. This suggests that the high-pressure autoclaves can be used to generate a valuable information.

The accumulation of H_2S and NH_3 in the system represents a drawback of the batch reactors. The evolution of the former will change the H_2S/H_2 ratio with progress of the experiment. This ratio is an important parameter for maintaining catalyst in an active form. Also, the poisoning of catalyst by NH_3 cannot be ruled out, although among the N-bases, NH_3 is considered a weak poison (76). In an extreme case, H_2S can react with NH_3 and other N-bases to form deposits of ammonium sulfide and/or amine–sulfur compounds (99). Moreover, the catalyst deactivation by N-bases may be more evident than in the reactors from which the products and intermediates are continuously removed. This suggests that in autoclaves, the initial parameters (rate, conversion, deactivation, etc.) will be less affected than those determined in the latter stages onstream. These facts need to be taken into consideration while comparing the results obtained in the batch systems with those obtained in the continuous systems.

4.2 CONTINUOUS REACTORS

In continuous reactors, the primary and secondary products of hydroprocessing reactions are continuously removed from the system unless a long contact time of the feed with the catalyst is maintained. The latter conditions ensure a complete conversion of the feed. Therefore, the modifying effect of the N-bases and H_2S on the catalyst surface is much less evident than in the case of batch reactors. Moreover, continuous supply of the fresh H_2 allows a precise control of its pressure, as well as the control of the H_2S/H_2 ratio. Using an AR as the feed, Gualda and Kasztelan (100) observed rather different catalyst deactivation patterns in the batch reactor than those in the continuous bench scale reactor. The most often used continuous reactors are those employing the fixed bed of catalyst and the continuous stir tank reactors (CSTR). After some modifications, the latter may allow the use of catalyst in the suspended bed.

4.2.1 Fixed bed reactors

The fixed bed reactors operating in a continuous mode are available in various sizes starting with microreactors, bench scale and pilot plant reactors. They differ in the internal diameter (ID), i.e., about 10, 20 and 40 mm, respectively, and the reactor diameter/catalyst particles diameter ratio of about 7, 13 and 27, respectively (94). Fixed bed reactors can be operated either in the down-flow trickle bed mode or up-flow mode. Hydrodynamics and the effects of diluents on the performance of the two reactor modes were discussed extensively by Bej et al. (95,96), whereas the high-pressure trickle bed reactors were reviewed by Al-Dahhan et al. (101). The factors which play a key role during the kinetic studies using trickle bed reactors were described by Korsten and Hoffmann (102). The importance of the liquid holdup on the performance of the trickle bed laboratory reactor was stressed by Garcia and Pazos (103). The increased liquid holdup meant better catalyst wetting which resulted in a more efficient catalyst utilization. In this case, the flow dynamic conditions such as the particle size and liquid space velocity were important parameters.

Apparently, trickle bed reactors have been used more frequently than the up-flow reactors. In the case of the former, a better distribution of the feed through the catalyst bed could be achieved. However, for the up-flow reactors, this problem can be alleviated by using a solid bed diluent, as well as the layer of an inert solid in front and at the end of the catalyst bed. The choice of the size of reactor depends on the objective of research. For example, when the objective is to obtain parameters relevant to a commercial operation, much more reliable and useful data are obtained in the pilot plant than in the bench scale reactor and/or microreactor. Again, the choice of the suitable reactor is particularly important for heavy feeds containing metals and asphaltenes.

The reactors having ID about 10 mm and less are classified as microreactors. It is believed that there are limitations on the use of such reactors in the studies on hydroprocessing of heavy feeds, particularly VR and heavy crudes. Thus, the amount of catalyst which can be used in microreactors is less than 1 g. Moreover, the reactor diameter/catalyst particles diameter ratio allowed the use of catalyst particles of a much smaller size than that of a commercial catalyst (96), i.e., ~1.5 mm. Therefore, for microreactor studies, the catalyst particles used in commercial reactors have to be crushed to less than 0.5 mm size. With such small particle size, the diffusion phenomena present during hydroprocessing of heavy feeds cannot be correctly identified unless the testing in microreactors is complemented by that in larger systems. The rapid formation of deposits on catalyst surface during very early stages is the well-established phenomenon (40). It is believed that in the case of such small particles, plugging and/or developing pressure drops leading to malfunctioning of the catalyst bed would be inevitable. To a certain extent, these problems can be alleviated by diluting the catalyst bed with the particles of inert solids and/or by diluting heavy feed with a lighter fraction. The fixed bed of an inert solid on front of the catalyst bed can improve the feed distribution through the latter. Other factors which limit downscaling of the fixed bed reactors to a microscale were discussed by Sie (104). For some heavy feeds, their colloidal stability may decrease with increasing temperature. This may lead to the excessive sediment formation on the reactor walls causing interruption of the run. Therefore, for some heavy feeds, it may be difficult to perform long runs using micro reactors or even small bench scale units. When microreactors are used, the limitations which may be experienced in the studies using heavy feeds (e.g., distillation residues) are not present for light feeds and model feeds. In the latter case, a valuable information on various aspects of hydroprocessing reactions can be obtained using microreactors.

It was indicated earlier, that the experimental difficulties experienced with heavy feeds in the microreactor systems (both batch and continuous) may be alleviated by diluting the heavy feed with a lighter hydrocarbon fraction. This will decrease the viscosity to enable a continuous feeding of the heavy feed. However, the choice of the diluent must ensure that a colloidal stability of the heavy feed is maintained. Otherwise, the coagulation of asphaltenes leading to the deposit formation could not be avoided. An ultimate result of this would be the unplanned interruption of the experiments, erroneous results and other complications.

Bench scale continuous units (~20 mm ID) have been used successfully in various studies on hydroprocessing of heavy feeds. The volume of catalyst used in these reactors may approach 100 mL and is more than 20 times greater than that in microreactors. In such units, the catalyst particles in their operating form and size can be used without any difficulties. Therefore, much more representative and reliable results can be obtained using such amount of catalyst in its operating form. Some operating parameters determined in bench scale reactors have been successfully correlated with the parameters obtained from the commercial operations. Of particular significance are the studies on accelerating aging, which can generate the database necessary for the development of models used for predicting the catalyst performance in commercial reactors. The amount of catalyst and feed used in bench scale units allow the detailed postrun evaluation of spent catalyst and of liquid products, respectively, compared with a limited evaluation of the experiments conducted in microreactors.

The ID and volume of the pilot plant reactors are at least 2 and 50 times greater than that of the bench scale reactors and microreactors, respectively. Therefore, there are no limitations on the size and shape of the catalyst particles to be used for testing. Moreover, no dilution of heavy feed with a lighter fraction is required to enable a continuous feeding into the reactor. The final decision on the selection of suitable catalyst for a particular feed may require pilot plant testing, although an important information on the preliminary screening of catalysts can be obtained using bench scale units. In pilot plants, the long runs lasting many days can be conducted without any difficulties even for the heavy feeds such as VR and heavy topped crudes. Practical experience confirmed that the extensive database which can be generated in pilot plant reactors can be used for the design of commercial units and the optimization of their operation.

4.2.2 Continuous stir tank reactors

The CSTR systems of various sizes and volumes have also been used to study hydroprocessing of heavy feeds, although to a lesser extent than the continuous fixed bed reactors. Richardson et al. (105) compared the batch reactor with CSTR to test their kinetic model for hydroprocessing of the residue derived from Athabasca bitumen. These results are compared in Figure 4.1. There are some advantages of using CSTR reactors to study the kinetics of hydroprocessing reactions (106). This results from the absence of an internal mass and heat gradients. Moreover, the problems with pressure drops encountered in fixed bed reactors are not present in CSTR units because the catalyst is kept in a motion by continuous stirring. One mode of the CSTR operation includes the continuous recycling of the liquid and H_2. One version of the CSTRs employs a rotating basket of the catalyst held inside the reactor. Another version consists of a static basket of catalyst while the liquid content is agitated by a stirrer. The experiments with the multibasket systems can be conducted as well. The kinetics of the HDM of a heavy residue were investigated in the CSTR reactor, as well as in the same reactor which was operated in a batch mode (107). There was an apparent difference in kinetics, i.e., 1.0-order and 1.5-order kinetics in the CSTR and the batch system, respectively.

It is believed that, after some modifications, CSTRs can be used for the simulation of the expanded/ebullated bed reactors. In this case, catalyst would be in a suspended form rather than being held in a basket. For this purpose, the feeding and stirring rates would have to be matched with the size of catalyst particles and their physical

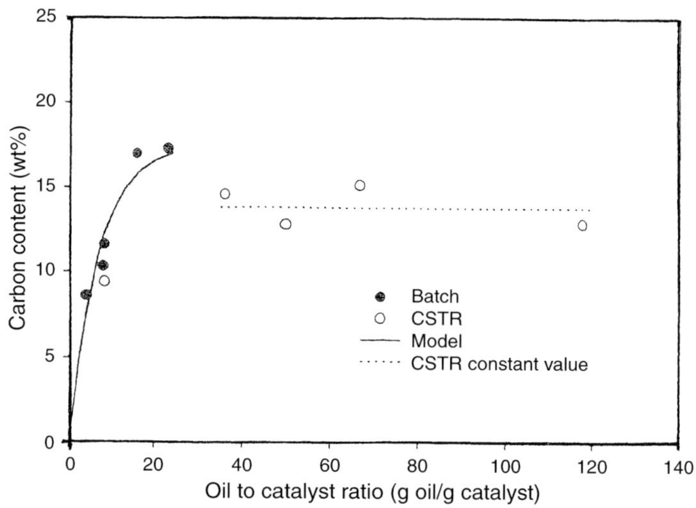

Figure 4.1 Effect of cumulative feed/catalyst ratio on carbon content of catalyst ($NiMo/Al_2O_3$; 703 K; 13.8 MPa) (105).

properties in order to ensure a smooth and continuous operation. Otherwise, the catalyst may be either carried out with the products or settle near the bottom. Apparently, CSTRs could also be modified to enable a periodic addition and withdrawal of catalyst. Because of the continuous operation, the database established using CSTR systems may correlate better with that obtained in the continuous fixed bed reactors than that in batch reactors.

Chapter 5

DEVELOPMENT AND TESTING OF CATALYSTS

Continuous efforts have been made to improve the performance of conventional hydroprocessing catalysts. For this purpose, catalyst performance has been evaluated using both model compounds and real feeds. However, the beneficial effects of the catalyst modification and/or the novel catalysts on activity observed for model compounds were not confirmed for heavy Maya feed (108). The study of Gray et al. (109) can be used as another example of the different effects observed for model compounds and real feeds. In this case, a series of coked $NiMo/Al_2O_3$ catalysts were used to determine their remaining activity. HDS activity decline with increasing amount of deposited coke was more pronounced for the model compound such as DBT than for bitumen used as the feed. Therefore, the use of heave feeds rather than model compounds and light feeds during catalyst development is the essential requirement. In this case, the effect of various additives, different supports, method of catalyst preparation and pretreatment, etc., has been receiving attention. Novel catalytic phases, differing markedly from those present in conventional catalysts, have also been evaluated. The surface properties of catalysts have been receiving much attention because of the presence of resins, asphaltenes and metals in heavy feeds. As was indicated earlier, without suitable porosity a desirable catalyst life could not be achieved. Models have been developed and used as tools for designing and selecting suitable catalysts to match the properties of heavy feeds with the type of catalytic reactor.

5.1 CONVENTIONAL CATALYSTS

It has been emphasized that suitable surface properties such as pore volume, pore size and surface area, as well as the size and shape of catalyst particles, require special attention to achieve a desirable performance of the conventional catalysts and catalytic systems during the hydroprocessing of heavy feeds. This may be illustrated using several examples reported in the literature. Surface properties are the primary focus of most studies on the development and testing of these catalysts. The importance of the shape and size of catalyst particles increases with the increasing content of asphaltenes and metals in heavy feeds. It is believed that for problematic feeds, the chemical composition of catalysts may be of lesser importance unless suitable surface properties are attained, as well as the optimal size and shape of catalyst particles are chosen.

5.1.1 Effect of surface properties

The main objective of this section is to illustrate several examples of how important it is to maintain the optimal combination of porosity and surface area for the catalysts used during hydroprocessing of heavy feeds. As was indicated above, in the series of heavy feeds, the importance of porosity increases in the following order: VGO/HGO < DAO < AR < VR/topped heavy feeds obtained from the high metal and asphaltenes heavy crudes. It has been generally observed that the content of asphaltenes and metals (predominantly V and Ni) increased in the same order. Therefore, the catalysts used for hydroprocessing of the high metal content feeds must be active for HDM and must possess an adequate metal storage capacity. To maintain a high HDM activity, these catalysts must also be active for HDAs which is necessary for achieving a high level of HDM. At the same time, they should exhibit a good activity for other hydroprocessing reactions. In this regard, a significant database has been generated and published in the scientific literature. This information suggests that problems concerning very heavy feeds may be overcome by conducting hydroprocessing in several stages. In this case, every stage may employ a different catalyst.

Reyes et al. (110) studied DAO containing more than 200 ppm of V + Ni and less than 2 wt% of asphaltenes (heptane insolubles). The metal content of this DAO exceeds that of the VR obtained from conventional crudes suggesting that a heavy Venezuelan crude (e.g., Boscan and Orinoco) was the origin of this DAO. Also for this amount of metals, the content of asphaltenes in the VR obtained from a conventional crude should be much greater. Therefore, the DAO used by Reyes et al. (110) appears to be quite a unique feed. The content of contaminants in DAO suggests that the deposition of V + Ni should play a more pronounced role during catalyst deactivation than that of coke produced from asphaltenes. The long run performance of the four catalysts conducted in the trickle bed reactor at 10.5 MPa and between 633 and 693 K is shown in Figure 5.1. The properties of the catalysts are given in Table 5.1 (110). Without any doubt, the largest pore volume of catalyst A was the main reason for its best performance. Also, the mean

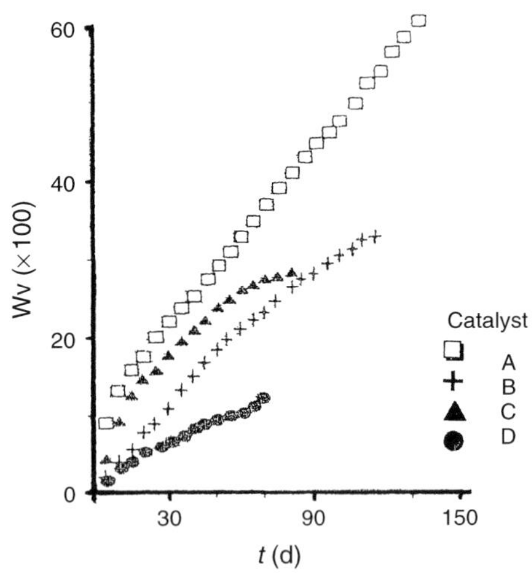

Figure 5.1 Effect of catalyst type on vanadium deposition (110).

Table 5.1 Properties of catalysts (110)

Property	Catalyst/support			
	A/Al_2O_3	B/SiO_2	C/Al_2O_3	D/Al_2O_3
MoO_3 (wt%)	15.0	12.6	16.2	14.5
CoO (wt%)	–	2.5	5.0	5.1
NiO (wt%)	3.5	–	–	–
Part. diam. (mm)	1.5	1.5	1.5	1.5
Shape	Extrud.	Pellet	Trilobe	Extrud.
Pore vol. (mL/g)	1.12	0.75	0.45	0.64
Mean pore diam. (Å)	150	270	90	90
Surf. area (m^2/g)	300	110	202	285

pore diameter of catalyst A was about 150 Å. Figure 5.1 shows that initially, catalyst C was comparable to catalyst A; however, in the middle of the run, the HDM activity of the latter began to decline, most likely due to the continuously increasing restrictive diffusion. This is supported by the lower pore volume and mean pore diameter of catalyst C compared with catalyst A. The trilobe shape of the particles of catalyst C compared with the extrudate shape of catalyst A may be another reason for similar performance during the early months onstream. The former ensured more efficient catalyst utilization than the extrudate form.

In the study of Stanislaus et al. (111) on hydroprocessing of the high metal content heavy feed, the best performance was exhibited by the catalyst with the predominant portion of pores in 100–200 Å range. At the same time, Nunez and Villamizar (112) reported that for DAO containing about 27 ppm of V+Ni and less than 1% of asphaltenes, the ideal value of the mean pore diameter was about 100 Å. This is not surprising when the significant difference between the metals content of the heavy feeds used by Reyes et al. (110) and Nunez and Villamizar (112) is taken into consideration.

The Orinoco heavy crude containing more than 1000 ppm of V+Ni was diluted with a gas oil and used by Galiasso et al. (113) to investigate the progressive change in porosity during hydroprocessing in the trickle bed reactor using either the microporous or macroporous catalyst of similar formulation ($CoMo/Al_2O_3$). Figure 5.2 shows the relative rise in temperature required to achieve the steady rate of HDM and HDS. It is evident that the microporous catalyst (broken line) reaches the critical temperature faster than the macroporous catalyst. This resulted in the earlier discontinuation of the run. During this run, the samples of catalysts were withdrawn from the reactor at the positions indicated as II, III and IV for the determination of asphaltenes content in the products, diffusivity and other analyses. Galiasso et al. (113) concluded that during the first period, coke accumulated in micropores, whereas V in the accessible meso/macropores. During the second period (before III), V deposited at the mouth of mesopores and in the meso/macropore structure. This affected the diffusivity of the reactant molecules. During the last period, V and coke deposited on the external part of the meso/macropores, as well as in the multilayers in the neck of the narrow pores. This blocked the penetration of the large reactant molecules into the interior of the pellet causing rapid catalyst deactivation (between III and IV) in agreement with other studies

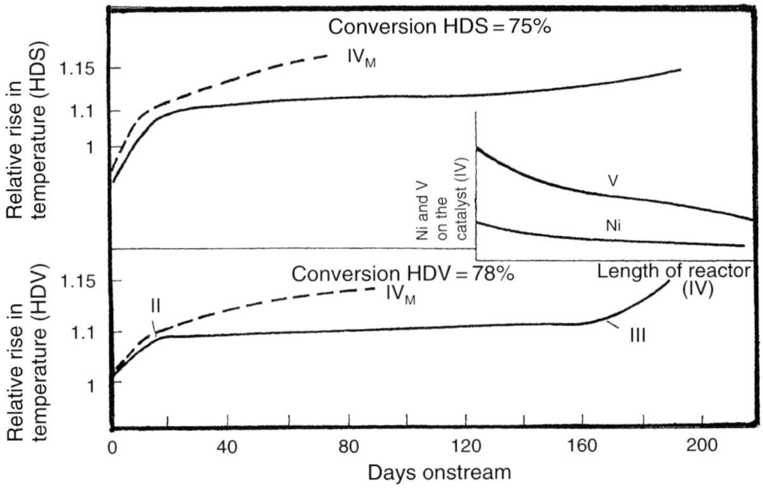

Figure 5.2 Effect of porosity on HDS and HDV conversion (Orinoco crude; 10 MPa): --- macroporous catalyst; — microporous catalyst (113).

(114,115). After stage IV, inlet of the catalyst bed was much more deposited by metals than the outlet of the bed.

Absi-Halabi et al. (116) used the Kuwait VR containing about 120 ppm of V + Ni and about 9 wt% of asphaltenes to evaluate a series of $NiMo/Al_2O_3$ catalysts in the trickle bed reactor at 713 K and 12 MPa. The properties of the catalysts are shown in Table 5.2 (116). An extensive evaluation of porosity should be noted. The HDS activity was compared during the 10-day runs (Figure 5.3). The best performance was exhibited by the PD-M2 catalyst in spite of the lowest pore volume. The meso- and macropore distribution of this catalyst indicated the absence of macropores suggesting

Table 5.2 Properties of catalysts (116)

Property	Catalyst			
	PD-M1	PD-M2	PD-B1	PD-B2
MoO_3 (wt%)	13.2	11.9	11.6	13.2
NiO (wt%)	4.0	2.8	2.5	4.0
Surf. area (m^2/g)	85	228	136	312
Pore vol. (mL/g)	0.60	0.53	0.73	0.76
Mesopore distrib. (nm%)				
3–10	4	35	7	55
10–25	11	60.5	34	8
25–50	27	1.5	19	8
Macropore distrib. (nm%)				
50–100	15	0	6	6
100–300	43	0	16	21
>300	0	0	18	2

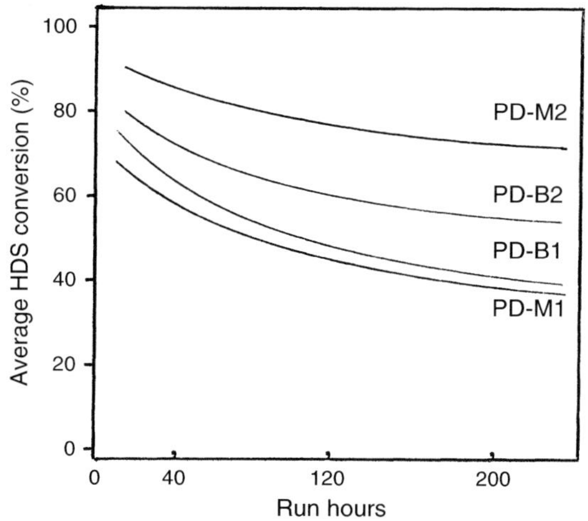

Figure 5.3 Effect of catalyst type (Table 5.2) on average HDS activity (116).

the importance of mesoporosity on the catalyst activity. In qualitative terms, this again supports the conclusion reached by Nunez and Villamizar (112). It is suggested that a 10-day run may not be long enough for evaluating the catalyst life. Then, it is not entirely unlikely that the curves in Figure 5.3 would cross each other at longer time onstream.

For fixed bed reactors, additional operating time can be gained using the grading systems consisting of composite beds of several layers of catalysts possessing different porosities and metal storage capacities (42,114,115). Figure 5.4 shows an average steady-state performance of the three fixed beds during hydroprocessing of the Kuwait AR (~80 ppm of V + Ni). The surface properties of the catalysts and the fixed bed layouts in vol.% of the total bed are given in Table 5.3 (114). The best performance of bed 3 resulted from the optimal combination of the HDM catalyst in the front layer and HDS catalyst in the end layer. The former catalyst protected the HDS catalyst from metal

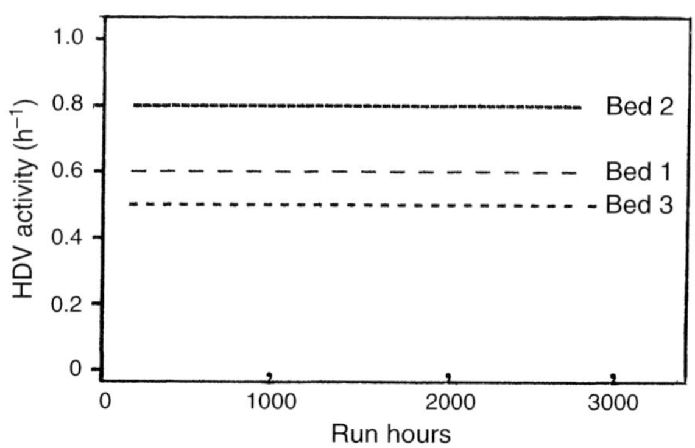

Figure 5.4 Effect of bed composition (Table 5.3) on average HDV activity (114).

Table 5.3 Surface properties of catalysts and bed layouts (114)

Catalyst	Surf. area (m^2/g)	Pore vol. (mL/g)	MPD[1] (Å)	MSC[2]	Layouts of bed (vol.%)		
					1	2	3
HDM	140	0.57	160	High	7	7	38
HDM/HDS	165	0.55	130	Moder.	31	0	0
HDS	200	0.48	100	Low	62	93	62

[1] Mean pore diameter
[2] Metal storage capacity

contamination. Consequently, the catalyst life of the bed 3 composition was extended compared with beds 1 and 2. Yet, for another heavy feed, the best performance of the system was observed for bed 1 (116). Another type of grading system comprises a guard reactor whose main function is HDM followed by several reactors in series. Testing and development of catalysts for such systems will be discussed in detail later in the review.

Not only a suitable initial porosity is crucial for achieving a good and steady catalyst performance, but it is essential that such porosity is maintained for as long a time onstream as possible. In this regard, the study of Zeuthen et al. (117) may be used to illustrate the change in pore volume with time onstream. In this case, the VR derived from the Khafji heavy crude ($\sim$200 ppm of V + Ni) was blended with 10% of a light fraction and used as the feed for experiments conducted in the three-stage expanded bed pilot plant for 120 days. The $CoMo/Al_2O_3$ catalyst contained 4.5% MoO_3 and 0.7% CoO. The samples of catalysts were periodically withdrawn from the first and third reactors for evaluation. Figure 5.5 shows the change in pore volume with time onstream. After 1 day onstream, a noticeable decrease in pore volume of the catalyst from reactor 1 compared with little change for catalyst from reactor 3 was observed. However, after 8 days onstream, the pore volume decrease for the latter catalyst was much more evident. This coincided with the larger amount of coke on the catalyst from reactor 3. The difference in the pore volume loss was even more evident after 120 days onstream. This confirmed that when hydroprocessing conversion was progressing (from reactor 1 towards reactor 3), catalyst deactivation by coke was gaining importance compared with that of metals.

Figure 5.6 illustrates another extreme in catalyst deactivation resulting from different catalyst compositions (28). As expected, the conventional HDS catalyst deactivated rapidly. At the same time, the catalyst of a proprietary nature deactivated slowly. Thus, even at 70% metal storage capacity, this catalyst possessed about 50% of its HDM and HDAs activity. Therefore, in accordance with Figures 5.1 (110) and 5.3 (116), an acceptable performance of a catalyst cannot be achieved without suitable porosity. For the same reason, significant efforts have been made by several research groups to optimize catalyst porosity. Essentially, every serious study on the testing and designing of catalysts for hydroprocessing of heavy feeds, particularly those having high metal and asphaltenes content, has been paying attention to catalyst porosity. Otherwise, the experimental results may only have a limited validity.

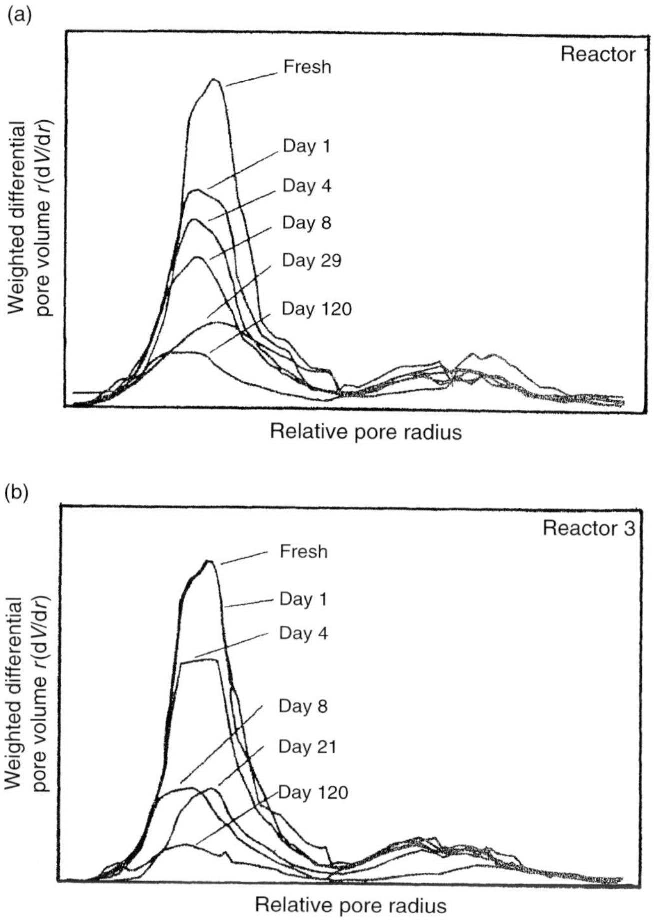

Figure 5.5 Effect of time onstream on pore volume and relative pore radius (117).

5.1.2 Effect of particle size and shape

For heavy feeds, the level of the active metals' utilization can be influenced by the size and shape of the catalyst particles. Thus, the efficiency of catalyst utilization will improve with the decreasing size of catalyst particles because of the shorter diffusion path involved for reactant molecules and a greater external surface per unit weight. Moreover, the relative volume activity and weight activity increase, whereas Thiele modulus decreases, with decreasing size of the catalyst particles (87).

It was indicated earlier that there may be a limit on particle size below which catalyst bed cannot be safely operated. Below this limit, malfunctioning of the fixed bed reactors can occur due to the development of pressure drops, which can enhance the possibility for formation of channels. Detailed account of these events is given by Macias and Ancheyta (88). In moving bed and ebullated bed reactors, very fine particles can be carried out from the reactor with products before being efficiently utilized.

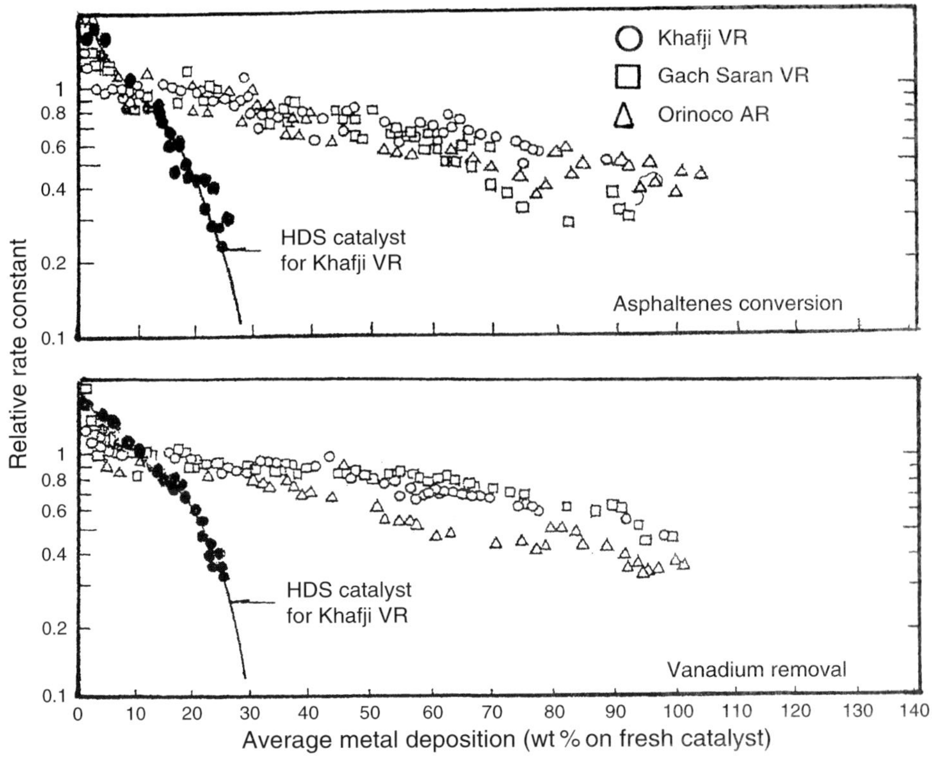

Figure 5.6 Effect of metal deposition on asphaltenes conversion and vanadium removal for HDS and HDM catalysts (28).

A low metal content (V + Ni = 20 ppm) AR was used by Ammus et al. (118) for evaluation of $CoMo/Al_2O_3$ catalysts of similar chemical composition and porosity. The experiments were conducted in the spinning basket reactor at 623 K and 5 MPa. The intrinsic HDS activity was determined under the same conditions using pulverized catalysts. The results are shown in Figure 5.7 (118). As expected, the pulverized form of catalyst was more active during the first half of the run. At the same time, more complex trends were observed for the 2 and 3 mm extrudates. For another catalyst tested by these authors, a higher activity of the pulverized form was more evident than that shown in Figure 5.7. It should be noted that the small difference in activity among the catalysts of different particle size (118) compared with other studies (84,119) can almost certainly be attributed to the relatively low content of metals in the feed used by Ammus et al. (118). Also, the duration of the runs was rather short. It is believed that larger differences should become more evident if the duration of runs was extended beyond that indicated in Figure 5.7.

Contrary to the low metal feed used by Ammus et al. (118), the Khafji AR used by Kobayashi et al. (119) contained more than 110 ppm of V + Ni. Figure 5.8 shows the effect of particle size on V removal for four commercial $CoMo/Al_2O_3$ catalysts. The different particle sizes were prepared by crushing the catalyst pellets and sieving. The same trends were observed for the removal of Ni and asphaltenes. The catalysts had

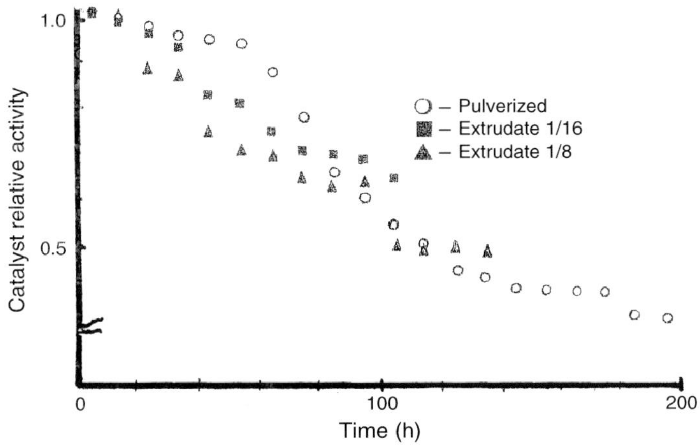

Figure 5.7 Effect of particle size of $CoMo/Al_2O_3$ catalysts on relative activity (118).

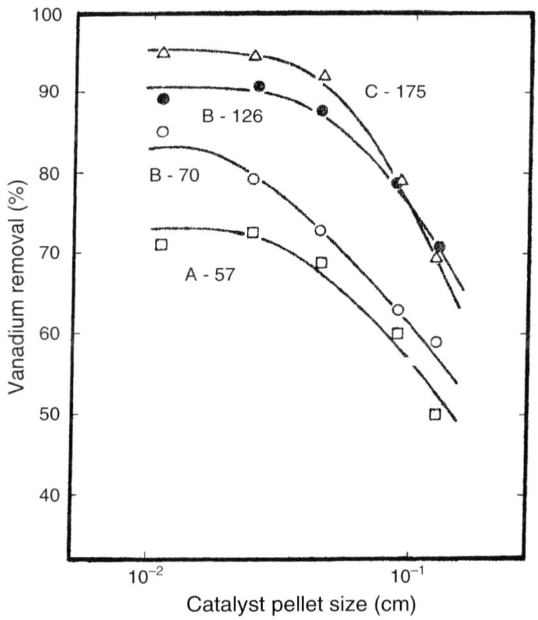

Figure 5.8 Effect of catalyst pellet size on vanadium removal (119).

similar chemical composition but their porosity varied widely, as shown in Figure 5.9 (119). It was evident that less than 0.5 mm size of the catalyst particles was required to attain the region free of diffusion limitations. At the same time, a high level of asphaltenes removal was achieved for particles slightly smaller than 1 mm. In view of the significant difference between the asphaltenes and metals contents in the heavy feeds used, a large effect of the particle size observed by Kobayashi et al. (119) and a small effect observed by Ammus et al. (118) are to be expected.

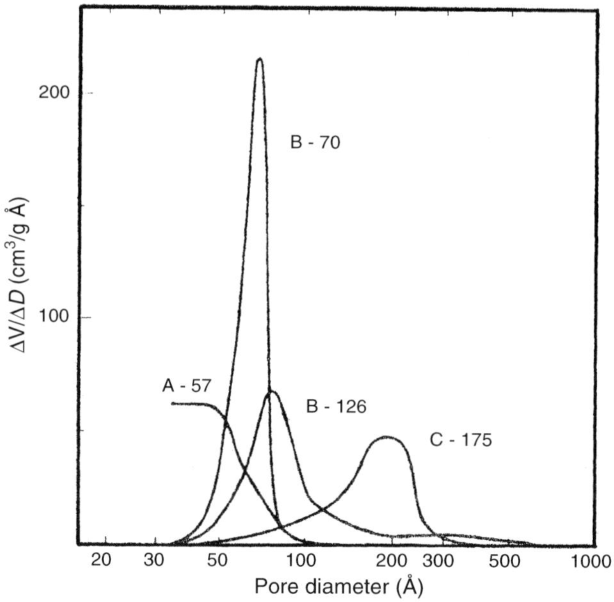

Figure 5.9 Pore volume and size distribution of catalysts in Figure 5.8 (119).

The study published by Koyama et al. (120) gives detailed accounts of the modifications required for accommodating hydroprocessing of a heavy VR in the system used previously for the HDS of an AR. The system comprised fixed bed reactors connected in series. In this case, the problems with the occurrence of hot spots and pressure drops caused by mal-distribution of the feed had to be solved. Among trilobe, spherical and cylindrical catalyst particles employed, the trilobe particles alone were the least efficient in preventing the problems. However, in order to maintain a sufficient HDM activity in the first reactor, a combination of the trilobe with cylindrical particles gave the best performance. Sock loading rather than dense loading method provided sufficient catalyst bed voidage for maintaining an optimal liquid flow distribution.

5.1.3 Design and testing of catalysts

Severe operating conditions in the presence of stable and active catalysts are required to convert heavy feeds into the commercial fuels which meet the stringent regulatory levels of contaminants. Various combinations of the conventional catalysts have been tested in more than one catalytic stage to attain the fuel specifications. Most of the studies emphasized the importance of catalyst porosity even for the relatively light feeds such as VGO and HGO. It was indicated earlier that this becomes more critical for the heavy feeds with high metal and asphaltenes content. Thus, the unpromoted Mo/Al_2O_3 catalyst containing less than 5 wt% of Mo could exhibit adequate HDM activity required for the first stage upgrading because it possessed a high metal storage capacity which was the result of a suitable catalyst porosity (121). At the same time, this catalyst exhibited a low activity for hydroprocessing reactions other than HDAs. The number of catalytic stages required for the upgrading of heavy feeds increases with the increasing content of metals and asphaltenes, i.e., in the order VGO/HGO < VGO/DAO < DAO < AR > VR/topped heavy crude. For some heavy feeds, as many as four catalytic stages are necessary to achieve the targeted conversion and quality of the products. In some cases, these products are subjected to additional processing, i.e., FCC and dewaxing, to produce transportation fuels and lubricants.

In the following, the materials on catalyst development and testing will be organized according to the type of heavy feed. Thus, a significant variability among the catalytic phases and catalyst structures designed for the heavy feeds such as VGO/HGO, DAO, AR and VR has been noted. It is therefore believed that discussion of the material according to the type of heavy feed offers a more concise and clear picture on the catalyst development.

5.1.3.1 VGOs and HGOs

As the lightest feeds among the heavy feeds under consideration, VGO and HGO have been predominantly studied using conventional catalysts in comparison with either modified conventional catalysts or novel catalysts. In this regard, the improved selectivity of catalysts for conversion of the heavy fractions of VGO to diesel oil and/or gasoline fractions with minimum conversion to the gaseous fraction has been one of the objectives (122). This may be achieved by optimizing distribution of the acidic sites of the catalyst support to improve its HCR activity. In cases where hydroprocessed VGO and HGO were used as the feed for catalytic dewaxing and FCC, the HYD, HDN, HDS and HDM functionalities were also receiving attention with the aim to increase the yield of desirable fractions and to prevent catalyst deactivation.

A specific case of hydroprocessing of VGO and HGO is the preparation of the feed for production of the lube base stock. In this case, contaminants such as sulfur and nitrogen have to be removed to protect dewaxing catalysts, usually consisting of an acidic support, noble metals and other additives. During dewaxing, *n*-paraffins are removed from the feed via either hydrocracking or isomerization to attain the cold flow properties specified for diesel oil and lube base oil. Moreover, for the latter, aromatic structures have to be converted to naphthenic compounds to increase the viscosity index to the levels required by the performance standards defined for lubricating oils. Although the operating parameters approach those used during the conventional hydroprocessing, the structure and composition of the dewaxing catalysts may differ markedly from that of the typical hydroprocessing catalysts. Therefore, Chapter 11, Section 11.1 dealing with catalytic dewaxing is added in this book.

The catalytic functionalities of the conventional catalysts for hydroprocessing of the VGOs and HGOs feeds could be influenced by the method of catalyst preparation (123). Thus, catalysts prepared by the impregnation of γ-Al_2O_3 with active metals (Co and Mo) in the presence of EDTA were more active than those prepared without EDTA. Apparently, the former case favored the formation of the more active Type II phase during catalyst sulfidation. In accordance with this, the activity for HDS and particularly for HDN was enhanced. Furthermore, the operating conditions may also influence catalyst performance. For example, the addition of H_2S to H_2 modified catalyst activities during the hydroprocessing of VGO derived from Maya crude, as was reported by Hebert et al. (124). These authors studied the effect of the H_2S/H_2 ratio on the hydroprocessing of VGO to produce the feed for FCC. The experiments were conducted at 11.8 MPa and between 603 and 673 K in the pilot plant fixed bed reactor using commercial

NiMo/Al_2O_3–TiO_2 catalyst. Under these conditions, H_2S had inhibiting effect on HDS, HDN and HDAr (HYD), whereas the rate of HDM reactions was enhanced when H_2S was present. The increased HDM activity of the catalyst in the presence of H_2S may be at least partly attributed to the direct reaction of metal porphyrins with H_2S. It has been noted that the inhibiting effect decreases with increasing H_2S/H_2 ratio until a maximum and then increases again (76). The optimal H_2S/H_2 ratio may depend on the properties of the feed. It is suggested that for the VGO used by Hebert et al. (124), the optimal H_2S/H_2 ratio was not approached.

The importance of the H_2S partial pressure and/or the H_2S/H_2 ratio was also demonstrated by Botchwey et al. (125) who used the HGO derived from Athabasca bitumen. The testing was conducted in two-stage system in the temperature range 613–693 K and from 6.6 to 11.0 MPa of H_2 over NiMo/Al_2O_3 catalyst. The experimental system had a provision for removing H_2S from the stage I products which then entered stage II reactor as the feed. Indeed, the removal of H_2S from the streams increased the rate of HDS and HDN. The most optimal conditions were attained at 653 K and 7.6 MPa. The development of catalysts for the hydroprocessing of VGOs and HGOs will also be discussed later in the review because of the predominance of the modified and novel catalysts in comparison with the conventional catalysts included in the studies. Primary focus will be on the effect of different additives and supports on the catalyst activity and selectivity.

5.1.3.2 Deasphalted oil

As was indicated earlier, solvent deasphalting is applied to heavy feeds (e.g., ARs and VRs) for removing most of the asphaltenes and metals. Figure 2.3 showed that the content of asphaltenes and metals, as well as the yield of DAO, depends on the type of solvent and the solvent/feed ratio. From the colloidal structure point of view, DAO consists of the oil phase and resins with asphaltenes accounting for a small portion. This ensures the homogeneity of the feed. Therefore, the properties of catalysts used for the hydroprocessing of DAO derived from conventional crudes may approach those used for VGO and HGO, rather than those used for AR and VR. However, entirely different catalyst formulations have to be used for the DAO containing more than 200 ppm of V + Ni, which was used by Reyes et al. (110). In some cases, the DAO is hydroprocessed either to obtain the feed for FCC and dewaxing or directly to obtain the products which meet the specifications of the commercial fuels. In the latter case, a multistage system may be necessary. In some situations, it was more beneficial to use the blend of DAO with VGO (Figure 2.2) as the feed for hydroprocessing.

In the study of Nuñez et al. (126), the DAO (27 ppm of V + Ni and less than 1% asphaltenes) obtained from deasphalting of the AR of conventional origin was hydroprocessed in the commercial unit comprising two trickle bed reactors. The initial temperature in the first reactor was ~610 K, whereas in the second reactor 645 K. Both reactors were operated at about 10 MPa pressure. Six NiMo/Al_2O_3 catalysts in baskets were placed in the top of the first fixed bed and the bottom of the second fixed bed. The properties of the tested catalysts are shown in Table 5.4 (126). With this arrangement, asphaltenes and metals in the feed were the predominant cause of catalyst deactivation in the first bed, whereas coke deposition in the second fixed bed. In this regard, the basic N-containing intermediates in the feed (products of the first reactor) in addition to N-bases already present in the feed were the important contributor to deactivation

Table 5.4 Properties of catalysts (126)

Property	Catalyst					
	A	B	C	D	E	F
Mo (wt%)	8.1	10.5	3	8	12	18
Ni (wt%)	2.3	1.5	1	2	2.3	1.8
Surf. area (m^2/g)	320	266	264	146	173	329
Mean pore diam. (Å)	126	99	136	233	83	71
Pellet size (mm)	1.03	3.1	1.0	2.1	2.1	1.3
Surf. area distrib. in differ. pores						
0–60 (Å)	32	16	14	4	9	118
60–100 (Å)	80	94	90	9	134	201
>100 (Å)	208	156	160	133	122	10

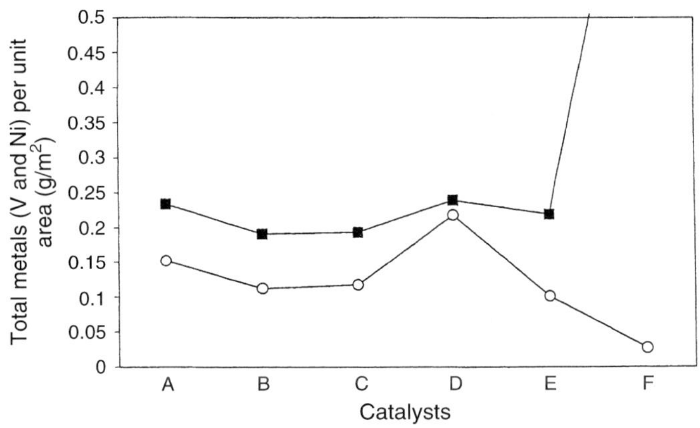

Figure 5.10 Effect of catalyst type (Table 5.4) on total metals retained: ○ per total BET area; ■ per area in pores > 100 Å (126).

(76). The catalysts were withdrawn for evaluation after 7 months onstream. Figure 5.10 (126) shows the ratio of the total metal removed to the unit of the surface area in pores above 100 Å and the total surface area. For the former, the metal/surface area ratio was almost constant, whereas the same was not observed when the total surface area was included. For catalyst F, the deviation can be almost certainly attributed to the negligible surface area in the >100 Å pores. It is believed that some compensating effects, i.e., large particle size for catalyst B and a low NiMo loading for catalyst C, were responsible for the similar metal/surface area ratio. The same study was extended to include a series of catalysts having the same surface area and mean pore diameter with variation in the amount of Mo between 0 and 8.3 wt% (105). It was observed that for the Mo content higher than 4 wt%, the incremental increase in the metal storage capacity was small. Moreover, the addition of Ni to Mo had little effect on the metal storage capacity. The correlation of the catalyst HDM activity versus mean pore diameter exhibited maximum at about 100 Å.

In its composition, the DAO used by Reyes et al. (110) discussed earlier represents another extreme. It contained almost nine times more metals (~230 ppm). Obviously, for this DAO, the metals and asphaltenes had much more adverse effect on catalyst activity than for the DAO used by Nuñez et al. (126). With their properties, the catalysts used for the DAO feed studied by Reyes et al. (110) would approach those used for the hydroprocessing of VR obtained from an intermediate crude. Thus, the content of metals in the VR derived from most of the conventional crudes is expected to be less than that in the DAO used by Reyes et al. (110).

5.1.3.3 Atmospheric residues

The design and testing of catalysts for hydroprocessing of AR has to take into consideration the presence of metals and asphaltenes. Therefore, attention has to be paid to colloidal stability and compatibility of the system, although not to such an extent as in the case of VR. Furthermore, the earlier discussion indicates that depending on the origin of crude oil, a great variability in the properties of AR should be expected.

Because of the presence of metals, the fixed bed hydroprocessing of ARs requires several reactors in series. The primary function of the first reactor, sometimes termed as the guard reactor, is the removal of metals to slow down catalyst deactivation in the subsequent reactors. Takahashi et al. (127) showed that the performance of the guard reactor can be maximized by selecting optimal operating conditions. A low active metal loaded catalyst (less than 5 wt% MoO_3) could efficiently convert the asphaltenes and metal containing porphyrins in the guard reactor. The conversion of asphaltenes was necessary in order to release the V and Ni porphyrins for HDM reaction and subsequent deposition on the catalyst. Otherwise, this catalyst exhibited rather low activity for other hydroprocessing reactions. Because of the protective function of the HDM catalyst in the guard reactor, deep HDS could be achieved in the second graded bed reactor consisting of 20% volume of the same HDM catalyst at the inlet and 80% volume of the active HDS catalyst. Increasing the temperature from 653 to 683 K increased catalyst deactivation in the guard reactor. This affected the long-term performance of catalyst in the second reactor. Therefore, it is important that an optimal temperature in the guard reactor is determined to achieve the deep HDS in the second reactor. In this study, the ARs derived from the Khafji crude, Kuwait crude and Heavy Arabian crude were used. The feeds contained about 100 ppm of V + Ni each.

Sometimes, reference is made to guard reactors which are placed upstream of the first catalytic reactor to remove solids from heavy feeds such as AR and VR. In this case, the guard reactor is filled with low-cost solids (e.g., alumina, clays, kaolinites). The objective is the removal of solids of inorganic origin, which are dispersed in heavy feeds. Therefore, their removal in the guard reactor is physical. This suggests that the catalytic activity of the guard reactor solids may be low. Therefore, to distinguish between the functions, guard reactors filled with a non-catalytic material will be referred to as a guard chamber.

The importance of the average pore diameter for achieving a high HDS conversion during hydroprocessing of the Kuwait AR was stressed by Chen et al. (128). They studied five $CoMo/Al_2O_3$ catalysts with the average pore diameter varying from 70 to 150 Å. At the same time, their surface area varied between 150 and 250 m^2/g. The molecular size distribution of the AR gave the average size of molecules of about 50 Å. The best performance was achieved for the catalyst with average pore diameter of about

100 Å. The decrease in the effective diffusivity values with decreasing pore diameter indicated a large restrictive diffusion effect. It should be noted that the optimal average pore diameter required for the hydroprocessing of DAO used by Reyes et al. (110) was 150 Å. However, the content of metals in this DAO was almost three times greater than that of the AR used Chen et al. (128).

The Kuwait AR (~90 ppm of V + Ni and ~4 wt% of asphaltenes) has been used as the feed in the commercial multistage atmospheric residue desulfurization (ARDS) process. This process consists of four fixed bed reactors operating in a trickle bed mode connected in series. This feed has been extensively evaluated with the aim to optimize the operating parameters and to select the most suitable catalysts for each reactor (129). For this purpose, the typical HDM and HDS catalysts such as Mo/Al_2O_3 (3 wt% Mo, pore volume of 0.7 mL/g and surface area of 200 m^2/g) and $NiMo/Al_2O_3$ (8 wt% of Mo, 2 wt% of Ni, pore volume of 0.5 mL/g and surface area of 230 m^2/g), respectively, were compared in the fixed bed pilot plant (130). The former catalyst was more coked and contained more metals than the HDS catalyst. However, as can be extrapolated from Figure 5.11, for the same amount of coke, the surface area loss on the HDS catalyst was almost twice that on the HDM catalyst. The temperature had a pronounced effect on the catalyst performance (131). For example, for the HDM catalyst, the metal content in products decreased from 90 ppm of V + Ni in the feed to 47 and 10 ppm in the products, when temperature was increased from 633 to 653 and 693 K, respectively. At the same time, the asphaltenes remaining in the products became more aromatic as was indicated by the decrease in their H/C ratio from 1.11 to 0.82. The H/C ratio decrease was much less evident for $NiMo/Al_2O_3$ catalyst than for Mo/Al_2O_3 catalyst because the HYD activity of the former was greater. The HDM catalyst was more active for asphaltenes conversion, whereas $NiMo/Al_2O_3$ catalyst was more active for HDS. The same $NiMo/Al_2O_3$ catalyst was used for hydroprocessing of demetallized AR containing 33 ppm of V + Ni. In this case, the HDM of the AR to obtain demetallized AR was performed over Mo/Al_2O_3

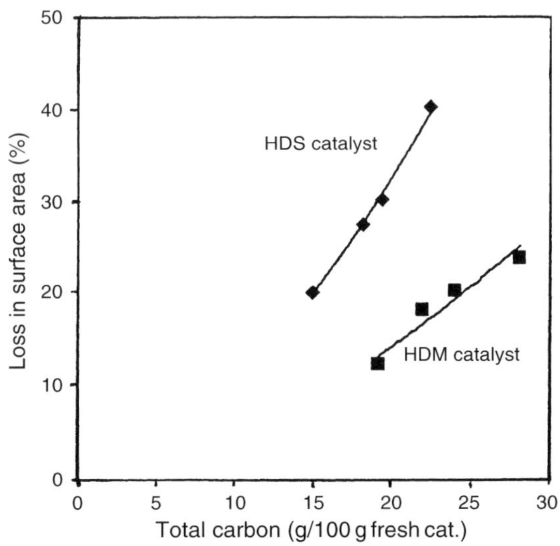

Figure 5.11 Effect of carbon on catalyst on the loss of surface area (130).

Table 5.5 Relative activities of $NiMo/Al_2O_3$ catalyst (132)

Temperature (K)	Functionality					
	HDS	HDV	HDNi	HDAsk	HCR	HDN
633	1.61	1.90	1.00	–	0.94	–
653	2.00	2.03	0.99	4.92	1.25	1.38
673	1.89	–	1.18	4.17	–	1.08
693	–	1.91	–	4.01	1.33	1.33

catalyst at 653 K and 12 MPa. It was observed that the metal content in products could be further decreased by increasing the temperature (130,132). The demetallized feed was a more appropriate feed than the AR when the aim was to simulate the operation of the second reactor of the ARDS process. Table 5.5 shows the ratio of catalyst activities for the demetallized AR to AR (132). The significant enhancement of the catalyst activity for HDAs, HDS and HDV was quite evident, whereas the effect of demetallization on the HDN and HDNi activities was less pronounced. These results confirmed that the selection of an appropriate feed for testing the catalysts to be used in the second, third and fourth reactors of the multistage ARDS process was critical for the overall optimization of the operation.

5.1.3.4 Vacuum residues and heavy crudes

It has been indicated that during hydroprocessing, the compatibility of heavy feeds may be affected because of the loss of colloidal stability caused by decrease in the solubility of asphaltenes as a result of the reduced dissolving ability of resins. Moreover, the asphaltenes' precipitating ability of the oil fraction may be enhanced because their aromaticity may be decreased during hydroprocessing. As a consequence, asphaltenes may coagulate and form sediments/sludges in the transfer lines and/or on the catalyst surface. A similar situation may be encountered under conditions favoring the conversion of resins to oil fraction at a greater rate than the conversion of asphaltenes. Among the heavy feeds, this phenomenon is the most critical for VRs. To a certain extent, these problems may be alleviated by optimizing the operating conditions and designing suitable catalysts. Nevertheless, the experimental system and conditions used for testing the catalysts for hydroprocessing of VRs and topped heavy crudes should ensure colloidal stability. Otherwise, reliability of the results would be affected.

The liquid products from the HDM of an Arabian VR (~200 ppm of V + Ni) and an Arabian AR (~100 ppm of V + Ni) were first demetallized over Mo/Al_2O_3 catalyst to about 55 and 18 ppm of the total metals, respectively (133). Three $NiMo/Al_2O_3$ catalysts of unknown contents of active metals were compared during the next stage of hydroprocessing of the demetallized feeds. A combination of the HDM/HDS catalyst (pore volume 0.67 mL/g; surface area 155 m^2/g) with the HDS catalyst (pore volume 0.57 mL/g and surface area 185 m^2/g) exhibited the best HDS activity, whereas the activity for asphaltenes conversion was similar.

The comprehensive study undertaken by Rana et al. (79,134) showed that significant improvements in the catalyst activity of conventional catalysts could be achieved by optimizing the method of preparation of the γ-Al_2O_3 support. Three methods used by

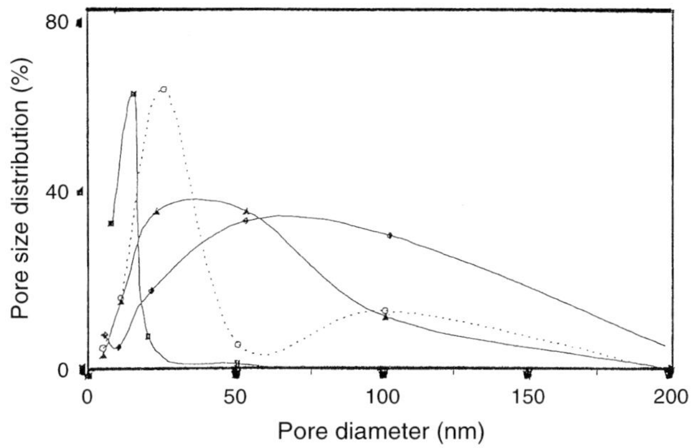

Figure 5.12 Pore size distribution of CoMo/Al_2O_3 catalysts: ♦ cat-urea; ▲ cat-ammonium carbonate; ■ cat-ammonium; ∘ NiMo/Al_2O_3 commercial HDM catalyst (79).

these authors included the hydrolysis of aluminum nitrate with the solutions of urea, ammonium carbonate and ammonium. CoMo/Al_2O_3 and NiMo/Al_2O_3 catalysts were prepared by impregnation methods using these supports. The pore size distribution of these catalysts is shown in Figure 5.12 (79). Catalyst activity for HDS and HDM was tested using the 50/50 blend of Maya crude and diesel oil (151 ppm of V + Ni and 8.4 wt% of asphaltenes), as well as the 50/50 blend of the partially demetallized Maya crude and diesel oil (112 ppm of V + Ni and 6.4 wt% of asphaltenes). Figure 5.13 (79) shows the superior activity of CoMo catalyst supported on γ-Al_2O_3 prepared by the hydrolysis of aluminum nitrate with ammonium carbonate. Thus, both HDS and HDM activities of the catalyst, as well as its stability, were much better than that of the commercial NiMo/Al_2O_3 catalyst. Apparently, the presence of TiO_2 in the commercial NiMo/Al_2O_3 catalyst

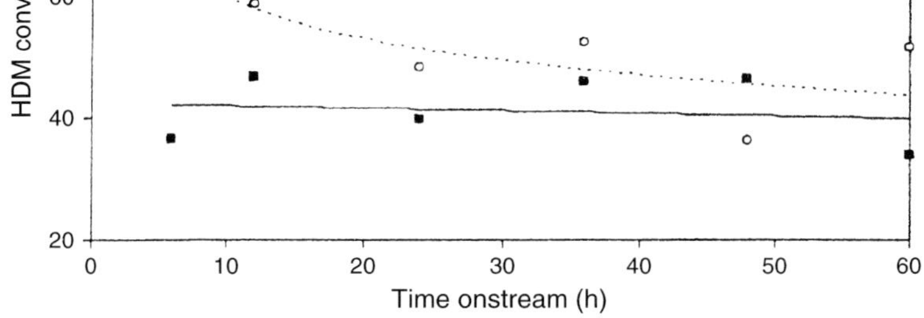

Figure 5.13 HDM conversion versus time onstream: △ CoMo/Al_2O_3 cat-ammonium carbonate; ■ NiMo/Al_2O_3 cat-ammonium; ∘ NiMo/Al_2O_3 commercial HDM catalyst (79).

contributed to faster deactivation because of the more extensive coke deposition caused by the increased acidity of the catalyst. In this case, the blend containing the partially demetallized Maya crude was used. However, for the heavier crude, the advantage of the prepared catalysts compared with the commercial NiMo/Al_2O_3 catalyst was less evident.

Several studies (135–138) focused on the change in the colloidal stability of VR under hydroprocessing conditions (135–138). Mochida et al. (135) observed that aromaticity of the asphaltenes isolated from the products after hydroprocessing of a VR (15 MPa, 653 and 693 K) was significantly greater than that in the feed similarly as was observed for an AR (131). This enhanced the precipitating propensity of asphaltenes. These authors suggested that to a certain extent, this problem may be rectified by the two-stage hydroprocessing. The first stage catalyst, with a high activity for HYD/HDAr, decreased the aromaticity of asphaltenes. The second stage conducted at a higher temperature than the first stage employed the catalyst possessing a high HCR activity.

The precipitation of asphaltenes may be minimized by blending VR with the aromatics-rich fraction. This possibility was explored by Marafi et al. (136–138) using the VR ($\sim$ 140 ppm of V + Ni and 7.5 wt% of asphaltenes) derived from the Kuwait crude. Indeed, significant decrease in the content of sediment in products and coke on the catalyst was observed for the feeds which were blended with a high aromatics content fraction. Thus, even at very high conversion levels, little sediment formation was observed. The best performance was exhibited using NiMo/Al_2O_3 catalyst with more than 50% of pore volume in 100–250 Å diameter range. In this case, both HDS activity and asphaltenes conversion were higher than for the catalyst with more than 80% of pore volume in the pores with diameter greater than 100 Å.

The Mexican Maya crude and/or heavy feeds derived from it have been the focus of extensive studies conducted by Mexican and other researchers. The most detailed evaluations were undertaken by Ancheyta et al. (134,139–141) who established perhaps the one of its kind database on hydroprocessing of heavy feeds. These authors compared three commercial NiMo/Al_2O_3 catalysts in the fixed bed reactor between 653 and 713 K and 6.9 MPa. The Maya feed contained almost 300 ppm of V + Ni and more than 12 wt% of asphaltenes. Table 5.6 shows the porosity as well as the size and shape of the catalyst particles, whereas Figure 5.14 shows the pore size distribution of the catalysts (139). The temperature effects on the catalyst functionalities, shown in Figure 5.15 (139), indicate the complexity in selecting the best catalyst for hydroprocessing of heavy feeds. For example, at 653 K, catalyst M-1 exhibited the highest activity for HDM, HDS and HDAs, presumably because the large portion of its pores was in the 100–250 Å range. However, with temperature increase, the difference became less evident and at 713 K catalyst M-3 was the most active for HDM. The same catalyst was the most active for HDN in the whole temperature range. This may result from the greater external surface available for reaction because of the small size of the catalyst particles. For M-3 catalyst, the content of sulfur and nitrogen in the asphaltenes which were separated from the produced oil was the lowest in the whole temperature range used. At the same time, little difference in the H/C ratio of the asphaltenes in products was observed among the three catalysts.

It should be noted that the smallest particle diameter of the catalyst M-3 could be advantageous compared with the other catalysts in Table 5.6, although the use of the former in the fixed bed reactor may lead to unwanted pressure drops. Figure 5.15 (139) shows that for this catalyst, the HDM rate was consistently greater than that of HDAs,

Table 5.6 Properties of NiMo/Al_2O_3 catalysts used for hydroprocessing of Maya crude (ref. 139)

Property	Catalyst		
	M-1	M-2	M-3
Mo (wt%)	10.7	11.1	8.8
Ni (wt%)	2.9	1.8	2.4
Ti (wt%)	3.7	–	–
Shape	Tetralobe	Sphere	Cylinder
Part. diam. (mm)	2.54	2.54	0.79
Surf. area (m^2/g)	175	120	278
Pore vol. (mL/g)	0.56	0.62	0.60
Pore diam. (Å)	127	180	86
Pore vol. distrib. (vol.%)			
<50 Å	3.1	0.8	11.1
50–100 Å	6.7	2.5	43.4
100–250 Å	69.1	34.2	8.1
250–500 Å	15.0	42.6	3.1
500–2000 Å	6.1	19.9	7.7
>2000 Å	–	–	26.7

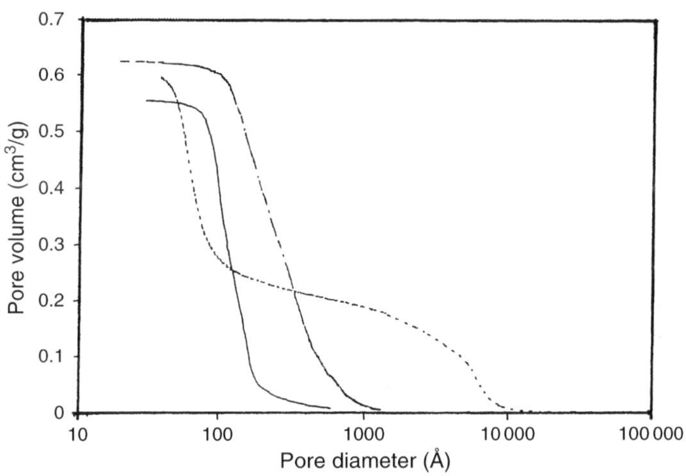

Figure 5.14 Pore size distribution of catalysts in Table 5.6: — cat-1; - - - cat-2; - - - cat-3 (139).

at every temperature. This suggests that for catalyst M-3, the need for disintegration of asphaltenes entities to release porphyrins for HDM was less critical than that for the other catalysts in Table 5.6 (139), unless for this feed, relatively large portion of the porphyrins was associated with resins and oil phase. Otherwise, the larger molecular weight reactants could undergo HDM over catalyst M-3 compared with the other catalysts. It is suggested that this resulted from the small size of the catalyst particles which provided better access to the active surface and ensured a more efficient catalyst utilization. Moreover, between

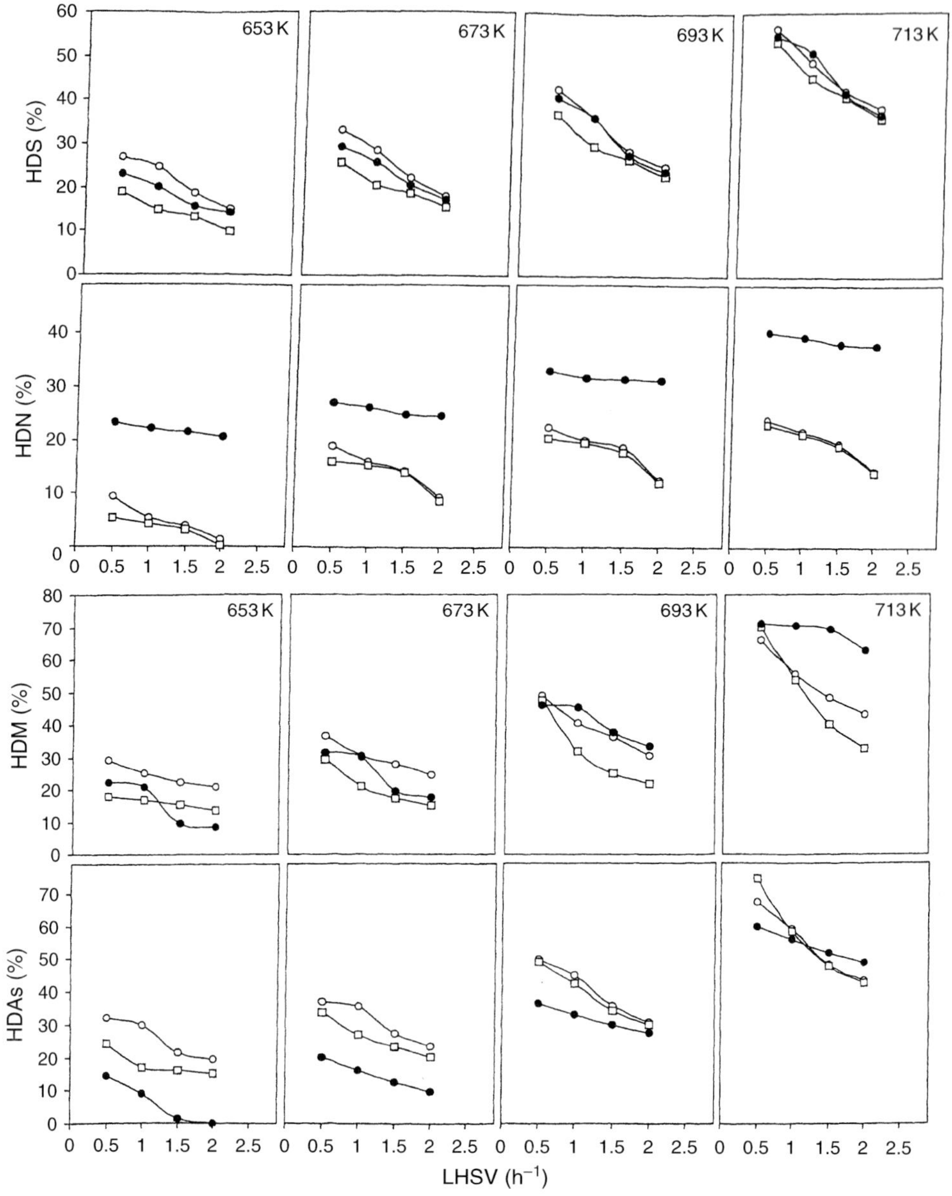

Figure 5.15 Effect of catalyst type, LHSV and temperature on catalyst functionalities: o catalyst M-1; □ catalyst M-2; • catalyst M-3 (139).

653 and 693 K, catalyst M-3 exhibited the lowest HDAs activity suggesting that pores with diameter exceeding 500 Å played little role compared with those in the 100–250 Å range. Thus, in the same temperature range, the catalyst M-1, with the predominant pore volume in the latter range, was the most active for HDAs.

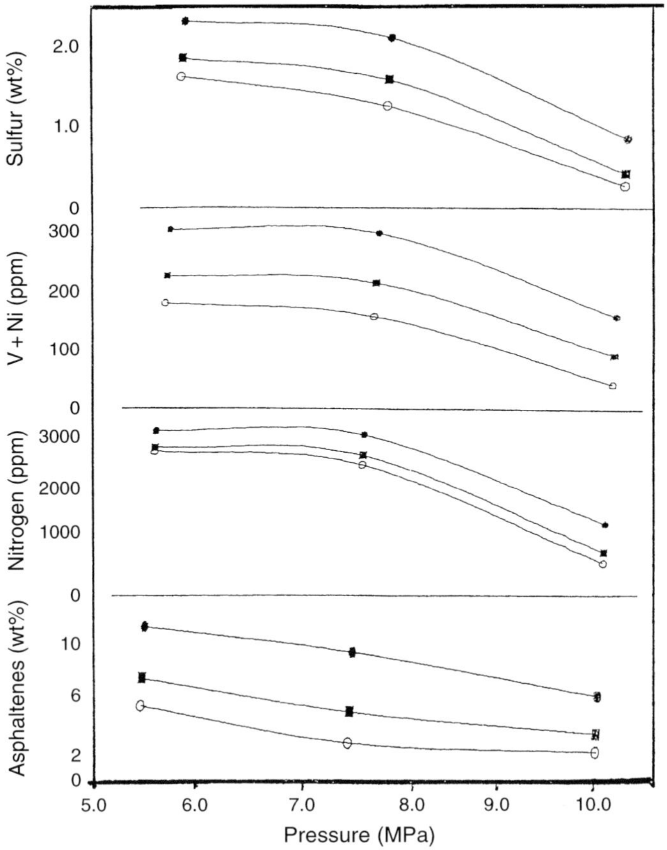

Figure 5.16 Effect of H_2 pressure and temperature on removal of contaminants: • 653 K; ■ 673 K and ○ 693 K (141).

A different Maya crude (~320 ppm of V + Ni) was used to study the effects of H_2 pressure (141). The results in Figure 5.16 show the combined effect of temperature and pressure on the content of contaminants in the produced oils. It is evident that more than 8 MPa pressure was required to achieve the noticeable removal of sulfur, nitrogen, metals and asphaltenes. The properties of liquid products obtained at 10 MPa are shown in Table 5.7 (141). A significant decrease in the contaminant level was achieved by increasing the temperature from 653 to 693 K. In spite of this, the liquid products still contained about 40 ppm of V + Ni. This suggests that such products may need additional upgrading to be suitable as the feed for FCC unless a significantly modified process (e.g., RFCC) is employed. Using another Maya crude as the feed for two-stage hydroprocessing (142), the amount of V + Ni decreased from about 350 to 229 ppm over $NiMo/Al_2O_3$ catalyst in the first stage. The LHSV and temperature had the most pronounced effect on HDS, HDM and HDAs in the second stage over the commercial $CoMo/Al_2O_3$ catalyst. However, for the optimal temperature and LHSV, i.e., 673 K and $0.5\,h^{-1}$, the second stage product still contained more than 100 ppm of

Table 5.7 Properties of Maya crude and products from hydroprocessing (10 MPa; 0.5 h^{-1} LHSV) (141)

Property	Crude	Products at temperatures (K)		
		653	673	693
Sulfur (wt%)	3.52	0.87	0.42	0.27
Nitrogen (ppm)	3006	1320	810	613
RBC (wt%)	11.0	6.6	3.0	0.8
Asphaltenes (wt%)	11.2	6.2	3.5	2.2
V + Ni (ppm)	319	159	90	40

V + Ni. The characterization of asphaltenes in the products from the hydroprocessing of Maya crude (between 653 and 713 K, 7 MPa and $NiMo/Al_2O_3$) revealed a little increase in aromaticity between 653 and 693 K but a marked increase between 693 and 713 K. At the same time, the number of aromatic rings and ring substitution decreased. Most likely, thermal effects were responsible for the marked increase in aromaticity above 693 K. On the basis of the results published by Marafi et al. (129–132), the performance of the first stage catalyst used for the Maya crude (134,141,142) may be further optimized, e.g., by using a low Mo-containing catalyst possessing a suitable porosity. Nevertheless, to obtain liquid products which could be used as the feed for FCC, hydroprocessing of the heavy feeds such as Maya crude may require more catalytic stages than that used commercially.

The studies of Ancheyta et al. (139–143) indicated the complexity involved in selecting the catalysts and suitable reactors for hydroprocessing of the high metal feeds. To overcome some processing difficulties, the 50/50 blend of Maya crude with diesel fraction and 32/68 blend with naphtha fraction were tested over the commercial $NiMo/Al_2O_3$ catalyst (144). For the latter blend, enhanced precipitation of asphaltenes from the feed was observed. This resulted in increased catalyst deactivation. The catalyst activity was significantly enhanced by the partial HDM of the feed. Although the Maya crude/diesel fraction blend could be processed without difficulties, other more aromatic feeds (e.g., gas oil from FCC) may be suitable as well. The 50/50 blend of Maya heavy crude and naphtha fraction was used to compare $CoMo/Al_2O_3$ catalyst with $NiMo/Al_2O_3$ catalyst (145). The $CoMo/Al_2O_3$ catalyst was more active than the latter, presumably because of the more efficient dispersion of the CoMoS active phase, as was indicated by TPR results. The $CoMo/Al_2O_3$ catalyst with the predominant porosity in the mesopore range exhibited the best performance.

The earlier study of Takeuchi et al. (146) deserves attention because of the unique system used to investigate the six feeds including perhaps the heaviest available feeds such as Boscan crude (~1200 ppm of V + Ni), Orinoco AR and (~600 ppm of V + Ni) and Gach Saran VR (~370 ppm of V + Ni). The modified atmospheric bottom conversion (ABC) process (28) used in this study consisted of the deasphalting unit attached to two fixed bed reactors connected in series and had a provision for asphalt recycling. The nature of catalyst was not given due to proprietary reasons; however, its performance is shown in Figure 5.6 (28). In this process, the feed was blended with a light fraction and the asphalt from deasphalting unit before entering the first fixed bed catalytic reactor. The products from the second fixed bed reactor entered the deasphalting unit.

This arrangement resulted in the significant removal of metals and the conversion of asphaltenes to light fractions. For example, for Boscan crude and Orinoco AR, the metal in products decreased to ~130 and 110 ppm of V + Ni from ~1200 and ~600 ppm, respectively.

5.2 MODIFIED CONVENTIONAL CATALYSTS

It has been recognized that further improvements in the performance of conventional catalysts during the hydroprocessing of heavy feeds may be achieved by combining with various additives and different supports, as well as by modifying the conditions used during catalyst preparation. Among the additives, P, F and B were the most frequently tested; however, Ca, Mg and Zn were found to have a modifying effect on the supports as well (147). Available information suggests that further improvements in the catalyst life and selectivity of the conventional catalysts may be achieved. In the case of heavy feeds, the conversion of heavy molecules to distillate fractions is one of the objective. It has been established that this can be achieved by introducing more acidic supports. However, an optimal combination of the surface acidity of catalyst and its activity for hydroprocessing reactions must be established to minimize coke formation.

5.2.1 Effect of alkali metals

The addition of alkali metals to the support decreased the acidity of the latter. This, in turn, should decrease the catalyst deactivation due to coke formation. This was confirmed by Richardson and Gray (148) who added Na and Li to $NiMo/Al_2O_3$ catalyst to be used during hydroprocessing of the heavy feed derived from Athabasca bitumen. In this case, coke formation was decreased from 17.3 to 14.4 wt% by adding 0.26 mmol of the alkali metal per gram of catalyst. As a result of this, HDS activity increased. However, little effect was observed on the HDN activity and on the microcarbon residue (MCR) conversion. This is to be expected, as the latter is favored by the presence of acidic sites which are destroyed on addition of alkali metals. The effect was more pronounced when the metals were added in the form of hydroxide than in the form of nitrate. In the same study (148), addition of K had no beneficial effect on coke deposition. A slight improvement in catalyst activity on the addition of Li to $CoMo/Al_2O_3$ catalyst was also observed by Maity et al. (75) during hydroprocessing of the Maya crude; however, addition of Li had little effect on the long-term performance of the catalyst.

The alkali earth metals such as Ca and Mg were used as the modifiers of the γ-Al_2O_3 support (147). Doping with these metals resulted in significant modifications of the physical properties of the support, e.g., pore size and pore volume, as well as its surface area. It was observed that the type of metal salts added and their concentration in the solution were important factors determining the final properties. The doped $CoMo/Al_2O_3$ catalyst exhibited higher HDS activity compared with the undoped catalyst. However, the activity was only determined using model compounds. Therefore, the potential of the catalyst activity improvement by doping with Ca and Mg in the hydroprocessing applications involving heavy feeds needs to be investigated further.

5.2.2 Effect of phosphorus

Phosphorus has been used most frequently as the additive to modify catalyst activity. Today, some commercial hydroprocessing catalysts contain P. The addition of P to $NiMo/Al_2O_3$ catalyst increased HYD, HDS and HDN activities during hydroprocessing of the HGO of the tar sands origin (149). The experiments were conducted in an autoclave at 683 K and 7 MPa. The optimum activity was attained at about 3 wt% of P in the catalyst. Phosphorus increased the catalyst resistance to deactivation. The results published by Zeuthen and Andersen (150) using a VGO as the feed differed from those published by Jones et al. (149). The former authors observed an improvement in the HDN activity and a decline in the HDS activity on the addition of P. Ferdous et al. (151,152) observed a beneficial effect on HDN but little effect on HDS when 2.7 wt% of P was added to the $NiMo/Al_2O_3$ catalyst. Spectroscopic evaluation of the catalysts at the end of the operation revealed that the addition of P increased stacking of the MoS_2 crystallites (153), i.e., the formation of the catalytically more active Type II phase. Such situation favored the HYD of aromatic rings and N-heterorings which is required before HDN reactions can be completed (76). It is believed that the difference in experimental conditions and origin of the feeds was responsible for the different observations made in these studies (149–153). For example, after closely examining the experimental conditions used in these studies it became evident that the H_2S/H_2 ratio in every study was different, although it was established that with respect to the rate of some hydroprocessing reactions, this ratio is an important parameter to be considered (76).

Chen et al. (154–156) prepared a series of alumina–aluminum phosphate supports of varying atomic Al/P ratios and used them as supports for preparation of CoMo catalysts. The activity of these catalysts was tested during hydroprocessing of the Kuwait AR containing about 70 ppm of V + Ni at 663 K and 7.6 MPa. The metals removal over laboratory-prepared catalysts was much greater than that over the conventional $CoMo/Al_2O_3$ and increased with increasing Al/P ratio of the former. Figure 5.17 (154) shows that HDS rate reached the maximum at Al/P ratio of 8. It was observed that P

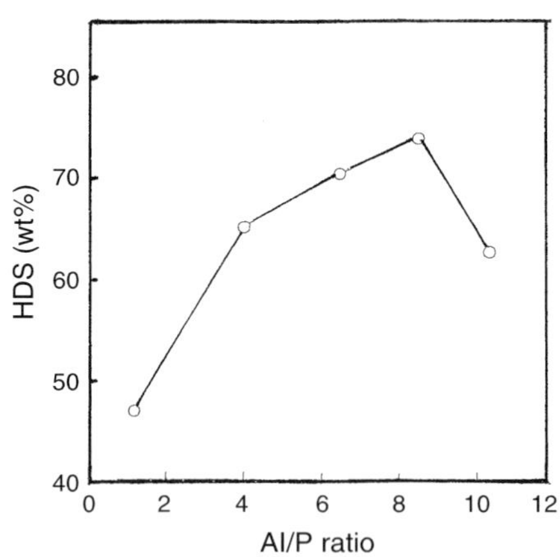

Figure 5.17 Effect of Al/P ratio on HDS activity of $CoMoP/Al_2O_3$ catalyst (154).

could facilitate uniform and large pore diameter without losing surface area. Moreover, the P-doped Al_2O_3 samples were more stable at higher temperatures compared with the untreated Al_2O_3. Also, the presence of P improved reducibility and sulfidability of the active phase, which was then more efficiently distributed.

A blend of the heavy Maya crude and naphtha (177 ppm of V + Ni) was used by Maity et al. (75,78,157,158) to study the effect of the addition of 0.8 wt% of P to $CoMo/Al_2O_3$ catalyst on HDS, HDN, HDM and HDAs in the continuous fixed bed microreactor (653 K and 5.4 MPa). It should be noted that the experiments could not be performed in the microreactor system without diluting Maya crude with the light fraction. The P-doped catalyst had the greatest metal storage capacity and was the most active for HDAs. At the same time, P had little effect on the other catalyst functionalities. Moreover, the deactivation rate of the P-doped catalyst by coke deposition was high. It was assumed that an increased acidity due to the addition of P was responsible for the increased deactivation due to enhanced coke deposition. However, the addition of Li, with the aim to decrease deactivation, had little effect suggesting that factors other than the acidity were involved as well. Under the same conditions, Maity et al. (78) observed that P had a pronounced effect on HDS and HDM when CoMo catalyst was supported on the binary Al_2O_3–TiO_2 support. However, little effect of P was observed for CoMoP catalysts supported on binary oxide such as Al_2O_3 – – SiO_2 (158).

To study the effect of P content (between 0 and 1.2 wt%) on the $CoMoP/Al_2O_3 \cdot TiO_2$ (95/5) catalyst performance, Maity et al. (159) used the 50/50 blend of Maya crude and diesel oil under otherwise similar conditions as used in other studies (78,158). The addition of 0.4 wt% of P slightly increased the HDM rate but decreased the HDS rate. Compared with the P-free catalyst, catalyst deactivation due to coke deposition slowed down when the amount of P approached 1.2 wt%.

Apparently, some differences in the effect of P on catalyst deactivation patterns between the Maya/naphtha blend (78) and Maya/diesel oil blend (159) were observed. In this regard, the contribution of N-bases to the different deactivation patterns should not be ruled out. It has been established that the poisoning effect of the N-bases increased from heavy fractions towards light fractions (63), i.e., from diesel oil to naphtha. Therefore, in the studies where a heavy feed is blended with lighter fractions to decrease viscosity, besides the compatibility aspects, the chemical composition of blending liquids is another factor deserving attention during catalyst evaluation.

Rather unique experiments on the effect of P on hydroprocessing were conducted by Kushiyama et al. (160) who used a series of the heaviest available crudes as the feeds. These authors prepared the Co–Mo–S phase in situ using the oil-soluble Mo-, Co- and P-containing precursors. As Figure 5.18 shows, the origin of the heavy feed had pronounced effect on HDS, i.e., for every feed, the maximum HDS was attained at different amounts of P. At the same time, V removal exhibited different trends than those of HDS, i.e., the V removal increase with the increasing phosphorus content was much less evident before it leveled off. Figure 5.18 may be used to show the dependence of the effect of P on the origin of heavy feed. These results may also clarify the differences in the effect of P on hydroprocessing noted among different studies, i.e., for some heavy feeds, the addition of the same amount of P to the catalyst may decrease HDS activity, while for others it may increase. More detailed account of the method used by Kushiyama et al. (160) will be given later in the review.

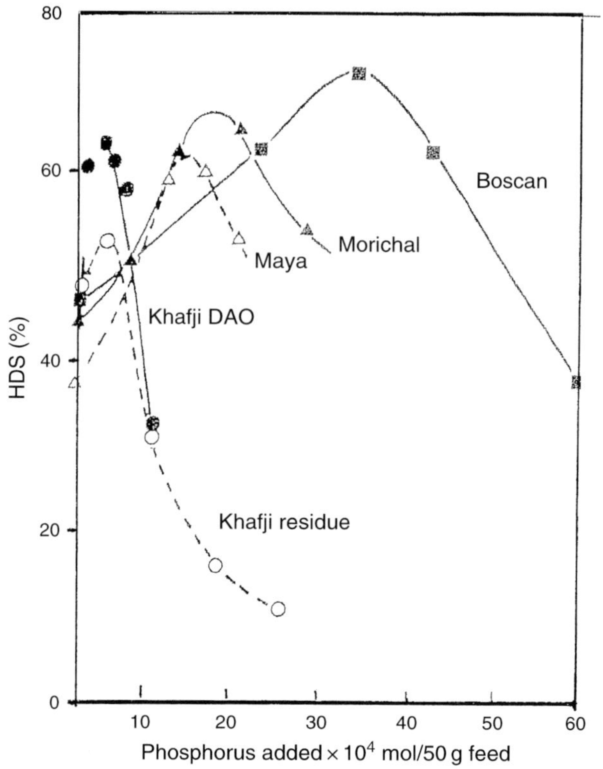

Figure 5.18 Effect of phosphorus on HDS conversion of heavy feeds (160).

5.2.3 Effect of borate

In the study conducted in the continuous fixed bed microreactor (~673 K; 6–10 MPa of H_2), the addition of B to $NiMo/Al_2O_3$ catalysts improved the activity for the HDN of the HGO (FBP ~820 K) derived from Athabasca bitumen by coking (152). At the same time, no beneficial effect on HDS was observed. These observations agree with those made during the addition of P to the same feed (151). Increasing the concentration of B from 0 to 1.7 increased the removal of the total, basic and non-basic nitrogen from 62 to 78%, 79 to 93% and 53 to 70%, respectively. The beneficial effect was attributed to the increased resistance of the catalyst to deactivation. Furthermore, spectroscopic evaluations of the spent catalyst confirmed that B increased the dispersion of the MoS_2 crystallites, thus ensuring a higher availability of the edge and corner S ions (153). Apparently, such ions played an important role during the activation of hydrogen, the presence of which is required for hydroprocessing reactions to occur (63). Then, the beneficial effect of B on catalyst activity differed from that of P. Apparently, the latter enhanced stacking of the MoS_2 crystallites. It is, however, believed that differences in the effects of P and B are dependent on the experimental conditions. In other words, under certain conditions (e.g., origin of feed, type of catalyst and operating parameters), little difference between the effect of P and B may be observed.

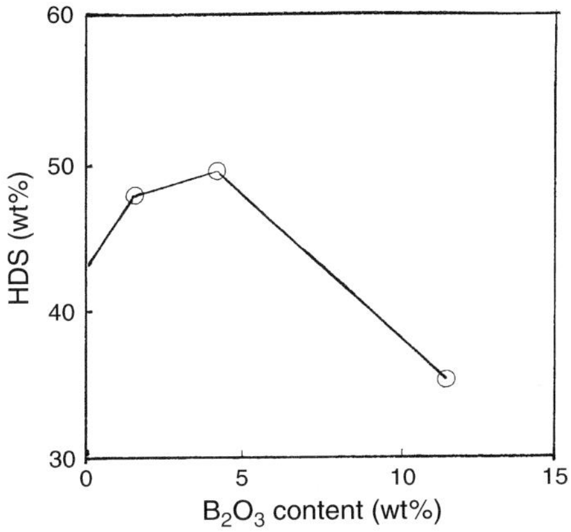

Figure 5.19 Effect of B_2O_3 content on sulfur removal (162).

Extensive studies on the effect of B on hydroprocessing of heavy feeds were carried out by Chen et al. (161–163) in the trickle bed reactor (663 K and 7.6 MPa) using Kuwait AR and gas oil as the feeds. The experiments were conducted under identical conditions as used by these authors to elucidate the effect of P (154–156). They prepared a series of aluminum borates with varying Al/B mole ratios and used them as supports for preparation of $CoMoB/Al_2O_3$ catalysts. The addition of borate resulted in the increase in HDM and HDS activities compared with the borate-free catalyst. The increase in HDM activity was attributed to the optimized pore size distribution in the presence of borate. Similar trends were observed for $NiMoB/Al_2O_3$ catalysts (162). As Figure 5.19 shows, HDS activity versus borate content correlation exhibited volcano curve with the maximum HDS activity at about 4 wt% of B_2O_3. Most likely, the optimal amount of B_2O_3 will vary from feed to feed.

5.2.4 Effect of fluoride

Fluoride has been used extensively in the studies involving model compounds (76). The information on fluoridated conventional catalysts used in the hydroprocessing of heavy feeds is limited. The study of Jones et al. (149) conducted in an autoclave (683 K and 7 MPa) using $NiMo/Al_2O_3$ catalyst showed that fluoride increased the HDS, HDN and HYD conversions of the HGO derived from Athabasca bitumen. The optimum activity was attained at about 1.8 wt% of fluorine. The addition of P to the fluoridated catalyst resulted in further improvement in catalyst activity. Apparently, the bi-additive catalyst was more resistant to poisoning by N-compounds. Also, it improved the retention of fluoride by the catalyst during the experiments.

Figure 5.20 shows the beneficial effect of fluoride on the HDN of the VR containing 1042 ppm of Fe, 361 ppm of V, 133 ppm of Ni and 291 ppm of Ti, as well as about 26 wt% of asphaltenes (164). The experiments were conducted in a fixed bed reactor at 698 K and 10.6 MPa using several $NiMo/Al_2O_3$ catalysts. The porosity of the unimodal catalyst differed markedly from that of the standard (conventional) catalyst, i.e., the former was supported on the macroporous γ-Al_2O_3. Therefore, the presence of macroporosity in the

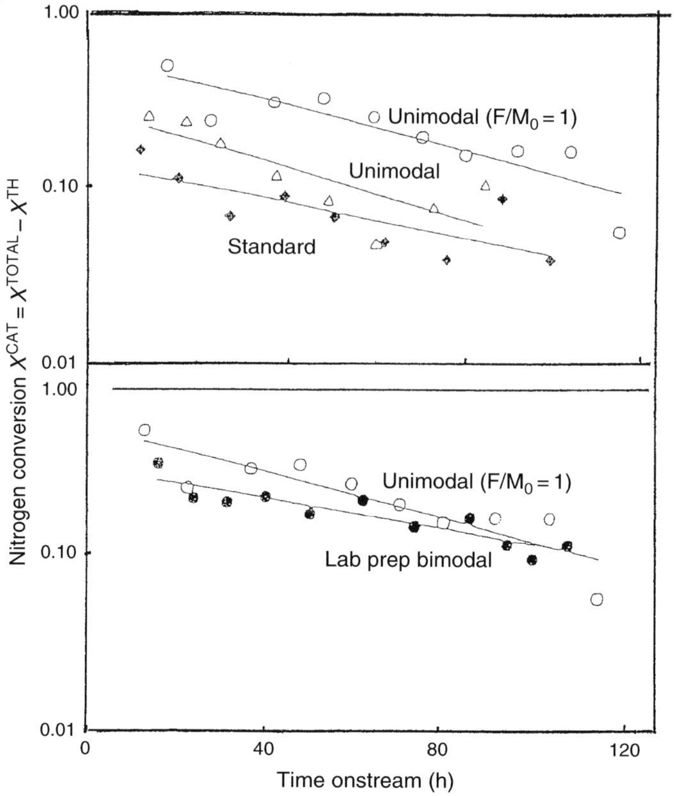

Figure 5.20 Nitrogen conversion versus time: • bimodal, ○ unimodal-containing fluoride (164).

former was the main reason for better performance. HDN conversion was not improved using the bimodal catalyst. The removal of metals and sulfur was not part of this study.

5.2.5 Effect of support

It has been generally known that supports other that γ-Al_2O_3 can have a pronounced effect on the activity and selectivity of hydroprocessing catalysts (165). Attempts have been made to modify catalytic functionalities of the catalysts used for hydroprocessing of heavy feeds by replacing γ-Al_2O_3 with different supports. For example, a suitable acidity of the catalyst for achieving desirable conversions of the large hydrocarbon molecules to light fractions can be maintained with the aid of support. General trends suggest that acidity has been a target parameter in designing the catalysts used for hydroprocessing of VGO, HGO and DAO, whereas porosity has been the target for that of AR and VR. This is not to say that for the former feeds, as well as for AR and VR, porosity and acidity, respectively, can be ignored. Supports, such as carbon, SiO_2–Al_2O_3, zeolites, ZrO_2 and various mixed oxides have been studied using a wide range of heavy feeds. The detailed review of the carbon-supported hydroprocessing catalysts in relation to those supported on conventional supports, i.e., γ-Al_2O_3, was also published (166). The recent information indicates a growing interest in TiO_2 as the support either alone or in combination with Al_2O_3 and SiO_2 (167,168). However, γ-Al_2O_3 modified with a small amount of alkali metals such as Na and Li, as well as alkali earth metals such as Ca and Mg, was also tested as the support for catalysts used during hydroprocessing of heavy feeds (75,154,169). Because of rather extensive information available in the literature on the effect of supports, as well as the diversity of the evaluated catalysts, the following material will again be organized according to the type of feed used for testing catalysts comprising different supports.

5.2.5.1 VGOs and HGOs

Compared with the conventional hydroprocessing catalysts, the improved catalysts for hydroprocessing of the VGO fractions should possess a high selectivity for middle distillates and low selectivity for gas and coke provided the production of diesel oil is the objective. In this case, a catalyst should be active for hydroisomerization and HCR to remove *n*-paraffins. This is necessary for achieving suitable cold flow properties of the diesel oil. When the expected product of the hydroprocessing of VGO is the lube base stock, a catalyst must also be active for HYD to ensure a high rate of removal of aromatics. This is required for achieving desirable viscosity properties of lubricants. Both hydroisomerization and HCR activities may be improved by incorporating supports which are more acidic than the traditionally used γ-Al_2O_3 (170). However, an optimum acidity has to be established to avoid excessive cracking and coke formation. To achieve an optimal level of the catalyst acidity, supports such as amorphous SiO_2–Al_2O_3 (ASA) and zeolites have been frequently tested. In recent years, TiO_2 has been receiving much attention. In all cases, the active metals on these supports comprised those which are part of the conventional hydroprocessing catalysts, i.e., Co(Ni)/Mo(W).

5.2.5.1.1 Acidic supports

Ali et al. (171) prepared a series of NiMo and NiW catalysts supported on various combinations of the ASA with USY and β-zeolite. The catalysts were used for hydroprocessing of VGO between 653 and 683 K and 10 MPa in fixed bed reactor. The best performance, measured as the conversion of the VGO fraction to naphtha and kerosene fractions, was exhibited by NiW catalysts supported either on the ASA/β-zeolite or on the β-zeolite alone. This was attributed to the presence of a suitable proportion of weak and strong acidity. This was confirmed by the results of Ahmed et al. (172) who prepared a series of NiMo catalysts supported either on the ASA or on the mixtures of ASA/zeolite varying widely in surface acidity. Figure 5.21 (172) shows that the VGO conversion increased linearly with the increasing acidity of the catalysts. However, there may be an optimal amount of zeolite in the mixture for which the maximum VGO conversion can be attained. This is supported by the results in Figure 5.22 (173) which showed that the optimum for the formation of naphtha and 350°C fractions was attained at about 32 wt% of zeolite in the mixture with either $SiO_2 \cdot Al_2O_3$ or Al_2O_3.

Corma et al. (77) compared the conventional NiMo/Al_2O_3 catalyst with NiMo catalysts supported on the ASA, USY and dealuminated ITQ-2 zeolites. In this case, a light cycle oil (LCO) was used in addition to VGO. The NiMo/ITQ-2 exhibited the best HCR activity; however, NiMo/Al_2O_3 catalyst was most active for HDS and HDN. The former catalyst showed high selectivity for naphtha and kerosene fractions and low

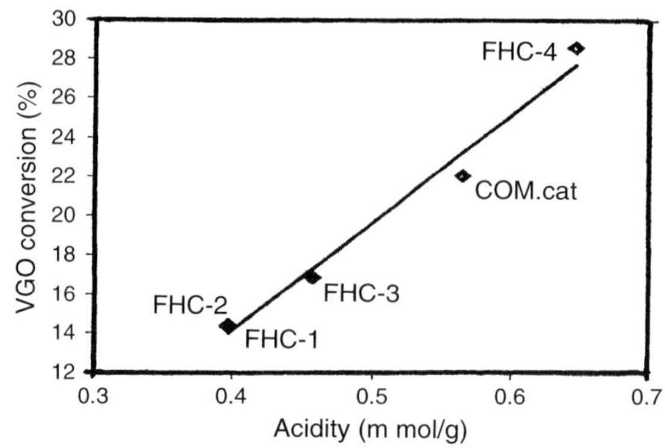

Figure 5.21 Effect of catalyst acidity on VGO conversion (172).

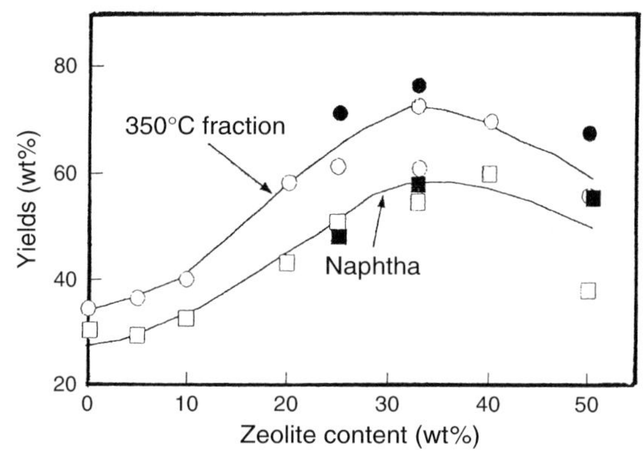

Figure 5.22 Effect of zeolite content of catalyst on yield of distillates (173).

selectivity for gases. Apparently, good performance of the ITQ-2 support could be related to the unique topology combined with a large, well-defined external surface with zeolitic acidity. The observation made by Mann et al. (174) differed from that made by Corma et al. (77). The former authors studied the HGO derived from Athabasca bitumen over NiMo catalyst supported on a combination of the rare earth exchanged Y-zeolite and SiO_2–Al_2O_3 at 7 MPa and between 623 and 698 K. In this case, the novel catalyst was more active than the conventional NiMo/Al_2O_3 catalyst for both HDS and HDN; however, HCR activity was not measured. Under the experimental conditions used by Yang et al. (175), NiMo/USY zeolite catalyst suppressed the formation of naphtha and increased the yield of middle distillates compared with the conventional NiMo/Al_2O_3 catalyst.

The combined effect of temperature and different pretreatments applied to the USY-zeolite was investigated by Abramova et al. (176). This included treatment with dilute solution of NaOH and HCl, as well as ion exchange for rare metal cations. In this case, the zeolite supports were impregnated with similar amount of Ni and Mo. The catalysts were compared with the commercial HYC-642 (Technip) catalyst. The experiments were conducted in the fixed bed reactor at 10 MPa using the typical VGO (FBP ~800 K).

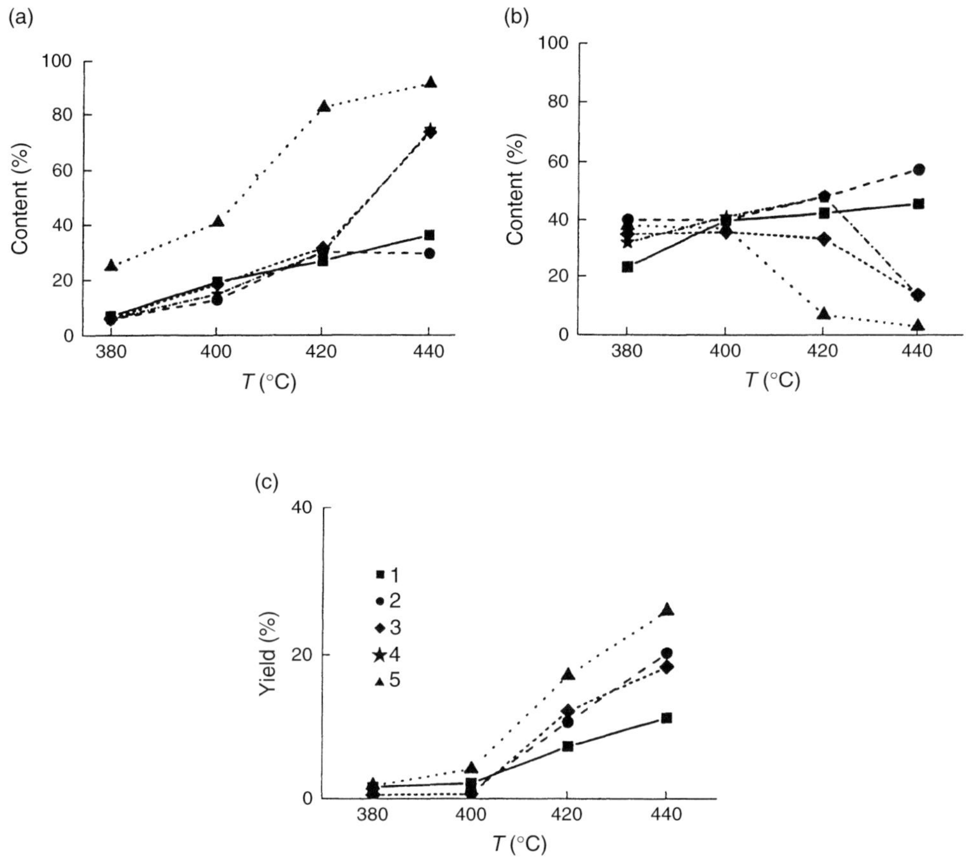

Figure 5.23 Effect of temperature on yield of products from VGO for different NiMo catalysts: (a) gasoline IBP-180°C; (b) diesel fraction (180–350°C) (c) gas (1: USY, 2: USY_{NaOH}, 3: USY_{HCl}, 4: US_{REM}, 5: Technip) (176).

Figure 5.23a–c shows the yields of gasoline (IBP-180°C), diesel oil (180–350°C) and gas, respectively, as the function of temperature and the type of support. Among the five catalysts, Technip HYC-642 had the highest acidity. This translated into the highest gas yields at every temperature. However, below 673 K, this catalyst gave the largest yields of distillates, i.e., almost 80%. Above 400°C, the same catalyst had the highest selectivity for gasoline fraction and the lowest selectivity for diesel oil fraction.

A stacked bed catalytic reactor was designed by Esener and Maxwell (177) to maximize the conversion of the VGO to middle distillates. In this case, the stacked bed reactor consisted of two fixed beds, i.e., the top bed comprising the conventional $NiMo/Al_2O_3$ catalyst, whereas the bottom bed was the NiW/zeolite Y catalyst. With this arrangement, the conversion to middle distillates was significantly improved compared with the single bed of the conventional catalyst. The conversion was further increased by optimizing the procedure for preparation of the zeolite-supported catalyst.

The baskets filled with three NiW catalysts supported on γ-Al_2O_3 and Al_2O_3–SiO_2 containing either 30 or 50 wt% SiO_2 were placed in the fixed bed of the commercial reactor used for hydroprocessing of a VGO at 653 K and 6 MPa (178). Detailed evaluations of the catalysts after 12 months onstream revealed that the agglomeration of WS_2 crystallites causing the loss of active sites was the main reason for catalyst deactivation. The deactivation was less evident in the presence of SiO_2. Apparently, a high level of the WS_2 dispersion was still present on the support containing the 50:50 composition of Al_2O_3–SiO_2 contrary to the other supports.

5.2.5.1.2 TiO_2-containing supports

TiO_2 has been used as the support modifier for the preparation of hydroprocessing catalysts. Both physical properties and the acidity of catalysts could be modified by incorporating TiO_2 in the supports. Two commercial NiMo/Al_2O_3 catalysts were compared with the NiMo/hydrous TiO_2:Si-coated Al_2O_3 and the NiMo/TiO_2-coated Al_2O_3 catalysts by Gardner et al. (179) during the hydroprocessing of VGO in a pilot plant at 655 K and 9.7 MPa. The HDN activity of the coated catalysts was greater by almost 30% than that of the commercial catalysts. At the same time, little effect of coating on the HDS activity was observed. NiMo/Al_2O_3 catalyst prepared by the conventional method was compared with the three NiMo/Al_2O_3·TiO_2 catalysts by Santes et al. (180). The supports for the latter catalysts were prepared by three different methods, i.e., impregnation of Al_2O_3 with Ti-butoxide; co-precipitation of the mixture containing Al-sulfate, Na-aluminate and Ti-sulfate; sol–gel method using alkoxides as precursors. These supports were sprayed with incipient wetness using an appropriate solution of Ni and Mo compounds. During hydroprocessing of the HGO to prepare feed for FCC, the Al_2O_3·TiO_2-supported catalysts prepared by the second and third methods were more active for HDS, HDM and HYD than the conventional catalyst. These methods ensured a close contact between TiO_2 and Al_2O_3, which improved catalyst activity. Extensive spectroscopic evaluations revealed that the addition of TiO_2 increased the support acidity. Also, the method of support preparation influenced the Ti distribution.

Santes et al. (181) prepared a series of NiMo/Al_2O_3·TiO_2 (Ti = 5.6 wt%) catalysts by the incipient impregnation of Al_2O_3·TiO_2 support with the aim to identify the optimal Ni/(Ni + Mo) ratio. For HDM and HDAr activities, the best performance was observed for the ratio of 0.45. However, at this ratio, HDS activity was lower than that of the reference catalyst. The experiments were conducted between 603 and 638 K at a total pressure of 10 MPa, in the trickle bed reactor.

The series of supports prepared by grafting Ti^{4+} into the structure of an alumina-silicate was used for preparation of NiMo catalysts (182). The Si/Ti ratio of the supports varied between 100/0 and 0/100. The catalysts were compared with the conventional NiMoP/Al_2O_3 catalyst during the hydroprocessing of VGO at 623 K and 12 MPa in the batch reactor. At best, the activity of the laboratory-prepared catalysts was comparable to that of the conventional catalyst.

5.2.5.2 Deasphalted oil

It appears that with respect to the effects of support on the catalyst performance, DAO is the least studied feed among the heavy feeds under investigation. However, some observations made in the studies using VGO and HGO may be relevant to hydroprocessing of DAO, particularly when the latter was obtained by propane deasphalting of a residue derived from a conventional crude. For such DAO, the oil phase separated from heavy feed would be in excess of resins. At the same time, only traces of asphaltenes and metals are expected to be present. Therefore, this could be classified as an ideal case of the DAO feed, which in its composition would approach VGO. In contrast, the DAO used by Reyes et al. (110) contained almost 230 ppm of V + Ni and ~2 wt% of asphaltenes. With respect to the properties, this DAO may represent another extreme, requiring entirely different conditions for its conversion, e.g., to the feed for FCC and/or dewaxing.

The DAOs derived from the AR and VR which were obtained from the Heavy Arabian crude were used as the feeds for evaluation of the performance of NiMo catalysts supported on the zeolites such as USY-12 and USY-30 (183). In addition, the TiO_2 (~5 wt%) containing catalyst was prepared by contacting the zeolite USY-30 with aqueous solution of Ti-sulfate. The FT-IR and TEM analyses revealed that part of the Ti was associated with the zeolite lattice while the other part was distributed on the surface as TiO_2. The Ti-containing catalyst exhibited a high selectivity for middle distillates. At the same time, its HDS and HDN activities were comparable to those of the amorphous catalyst. The tests were conducted in the continuous fixed bed reactor at 13 MPa between 643 and 673 K.

5.2.5.3 *Atmospheric residues*

Acidic supports were also used for the preparation of catalysts for hydroprocessing of AR, though to a lesser extent than in the case of the VGO, HGO and DAO feeds. Thus, because of the presence of asphaltenes in AR, it is more difficult to establish and maintain the optimal acidity favoring HCR reactions over the reactions leading to coke formation. These complications may be evident, particularly during the early stages onstream. On the other extreme, the carbon supports possessing little acidity, as well as mixed oxide supports with an intermediate acidity, have also been tested. The unconventional γ-Al_2O_3, not included in the studies involving VGO and DAO, was observed to exhibit some unique behavior as well.

5.2.5.3.1 Carbon supports

A series of carbon-supported CoMo catalysts were compared with the commercial CoMo/Al_2O_3 catalyst during hydroprocessing of the AR containing more than 130 ppm of V + Ni by Rankel (184). The HDM activity of the CoMo/Darco activated carbon was higher than that of the corresponding CoMo/Al_2O_3 catalyst. However, the latter was more active for HDS and HDAs. A similar observation was made during the comparison of NiMo/Al_2O_3 catalysts with NiMo/carbon catalyst (185). In this case, the activity of NiMo/carbon catalyst increased with increasing surface area of the carbon support. This may be attributed to the improved availability of active surface hydrogen on the catalyst made of higher surface area carbon. A higher HDAs activity of the γ-Al_2O_3 supported catalysts is to be expected because carbon can be considered as a neutral support. Thus, γ-Al_2O_3 supplies more acidity necessary for cracking reactions than carbon.

Based on the above evidence, it is believed that unless the acidity is developed by a pretreatment, the potential of carbon-supported catalysts in hydroprocessing of heavy feeds such as AR and VR may be limited to the HDM and HDAs applications. Indeed, CoMo/carbon catalyst with pore diameter in the range of 100–400 Å had a high tolerance to metals, as well as a high metal storage capacity.

Several activated carbons were used as supports for catalysts containing single metal such as Ni, Mo and Fe (186). The activity of the catalysts was tested in the autoclave at

708 K and 7.5 MPa using the Kuwait AR. The addition of metals to activated carbons enhanced the conversion to distillates and decreased coke formation. Activated carbons alone were also active. However, in this case, the yield of gaseous products and coke was greater than those of the metal-containing catalysts.

5.2.5.3.2 Acidic supports

Several NiMo catalysts supported on zeolites were compared with NiMo/Al_2O_3 catalyst during hydroprocessing of the Kuwait AR (187). The results indicated the importance of the mesoporosity of zeolite for achieving a high HCR conversion. However, the zeolite-supported catalyst increased the content of asphaltenes in the products compared with that in the feed. This was attributed to a limited supply of active hydrogen from MoS_2 to zeolite (63). The NiMo catalyst supported on the support which was prepared by the treatment of a parent zeolite with titanium sulfate solution had a high mesoporosity as well. This catalyst exhibited a high HCR activity. Moreover, over this catalyst, the amount of asphaltenes in the products decreased. The catalyst was used by Hashimoto et al. (188) for hydroprocessing of the Kuwait AR in the fixed bed reactor at 13.8 MPa and between 673 and 693 K. The optimization of the preparation procedure resulted in further improvement in the catalyst performance.

The modified zeolite was compared with the USY zeolite as supports for NiMo catalysts on an industrial scale to confirm a high performance observed on the bench scale (189). The lower activity of CoMo catalyst supported on the mesoporous zeolite than that of CoMo/Al_2O_3 observed during hydroprocessing of an AR was caused by the unsuitable pore size distribution of the former (190). Thus, the mean pore diameter of 28 Å was not large enough to achieve an adequate conversion of the asphaltenes in AR. These results suggest that the porosity of catalyst is the critical property regardless of the nature of the support.

The mixed oxide supports such as $Al_2O_3 \cdot SiO_2$ (1–1.5 wt% of SiO_2) and $Al_2O_3 \cdot TiO_2$ (2–6 wt% of TiO_2) were used for the preparation of CoMo and NiMo catalysts by Kang et al. (191). Their activity was tested during hydroprocessing of the Kuwait AR in the semibatch autoclave at 648 K and 10 MPa. The catalysts prepared from the $Al_2O_3 \cdot TiO_2$ supports were more active than those from the $Al_2O_3 \cdot SiO_2$ supports. Also, the former were more resistant to deactivation. For HDS and HDM, the optimum content of the TiO_2 was between 4 and 6 and 1 and 2 wt%, respectively.

5.2.5.3.3 Novel γ-Al_2O_3 supports

The catalysts supported on γ-Al_2O_3 prepared by unconventional methods exhibited some unique behavior. For example, a significant modification of the conventional γ-Al_2O_3 could be achieved by hydrothermal treatment. The Kuwait AR which was discussed earlier (129–132) was used to study the effect of hydrothermal treatment on the performance of NiMo/Al_2O_3 catalyst (192). The treatment resulted in significant enlargement of the narrow pores in NiMo/Al_2O_3. The treatment involved heating the catalyst in an autoclave between 423 and 573 K. For this purpose, 20 g of catalyst was placed in the basket which was suspended in 100 mL of distilled water. The untreated catalyst contained predominantly narrow pores. Thus, 61% of pores had pore diameter less than 100 Å compared with only about 4% for the treated catalyst. For the latter, almost 90% of pores had pore diameter greater than 250 Å. HDM and HADs activities of the catalyst were improved by the treatment, whereas little effect was observed on the HDS and HDN activities.

The AR containing about 70 ppm of V + Ni was used by Ying et al. (193) to study NiMo catalyst supported on the fibrillar alumina in comparison with the commercial NiMo/Al_2O_3 catalyst. The study was conducted in the trickle bed reactor between 643 and 703 K and 10 MPa. The pore volume and mean pore diameter of the former catalyst were 0.87 mL/g and 252 Å, respectively, compared with 0.39 mL/g and 99 Å, respectively, for the conventional catalyst. The HDM activity of the fibrillar alumina-supported catalyst was much greater than that of the commercial catalyst, whereas their HDS activities were similar. At the same time, the commercial catalyst was more active for HDN.

Pillared clays prepared from the commercial bentonite were used as support for NiMo catalysts by Smith et al. (194). The catalysts were used for hydroprocessing of the ARs derived either from the Cold Lake crude or from the Lloydminster crude at 703 K and 15 MPa. The reactors included either the batch stirred-tank system or a continuous CSTR. Among the pillaring agents used, Al had more beneficial effect on the catalyst activity than Fe and/or Zr. However, none of the pillared clay supported catalysts exhibited a higher activity than the commercial NiMo/Al_2O_3 catalyst. This resulted from the significant reduction in pore volume and surface area of the former catalysts, with time onstream.

5.2.5.4 Vacuum residues and heavy crudes

The rapid coke deposition during the initial stages of hydroprocessing of the most problematic feeds such as VR and topped heavy crudes may be the reason that acidic supports (e.g., zeolites) have been receiving little attention. To a certain extent, such coke deposition is of a physical, rather than chemical, nature. At least for the first stage of hydroprocessing of such feeds, a suitable catalyst porosity is one of the essential requirements. Apparently, such supports could be prepared by combining γ-Al_2O_3 with other oxides (TiO_2, ZrO_2, etc.). At the same time, some control of the catalyst acidity could be maintained. Once on the support, the method used for presulfiding of the active metals may determine the catalyst performance during the initial stages onstream. It is believed that if the more active Type II phase is formed, more bare support on which the initial coke deposition takes place would be available.

5.2.5.4.1 Carbon-containing supports

The carbon-containing supports usually receive attention whenever a low catalyst acidity is required. CoMo/carbon catalyst was compared with the conventional CoMo/Al_2O_3 catalyst for hydroprocessing of the VR derived from Athabasca bitumen at 698 K and 13.9 MPa (195). The former catalyst had surface area and mean pore diameter of 116 m^2/g and 286 Å, respectively, compared with 210 m^2/g and 85 Å, respectively, for the conventional catalyst. The catalysts contained the same amount of MoO_3 (15%) and CoO (3%). Initially, CoMo/carbon was more active, particularly for HDM; however, its activity declined with time onstream at a greater rate than that for the conventional catalyst. It was suggested that a stronger interaction of V involving the V=O moiety with the carbon surface, than that with the γ-Al_2O_3 surface, was at least partly responsible for a more extensive deactivation of the former catalyst during later stages onstream.

The activity of NiMo catalyst supported on the activated carbon was improved after the carbon support was treated in CO_2 at 1073 and 1273 K. (196). As expected, the treatment increased the pore volume and pore diameter of the catalyst. For VR containing 180 ppm of V + Ni and 16.5 wt% of asphaltenes, the conversion of asphaltenes to lighter fractions approached 98% after the treatment. The experiments were conducted in an autoclave at 703 K and 20 MPa. Another activated carbon-supported catalyst was tested by Fukuyama et al. (197) using the VR derived from a heavy Mexican crude. In this case, the support was impregnated with Fe, rather than with the metals, which are part of the conventional catalysts.

An alumina–carbon black composite in the form of extrudates was prepared by Lopez-Salinas et al. (198). Before the impregnation with Ni and Mo (3.5 wt % of NiO and 15.0 wt% of MoO_3), the extrudates were calcined in 6% O_2 + N_2 mixture at about 900 K. Depending on the conditions of preparation, pore volume of the catalyst varied between 0.75 and 1.0 mL/g. Also, 11–20% of the pore volume of the catalyst had pore diameter greater than 1000 Å. In spite of the relatively large pore volume, the catalyst had a good side strength (0.67–1.19 kg/mm). The final carbon content of the composite support varied from 8 to 13 wt%. The catalyst was used for hydroprocessing a Mexican VR which was blended with 20% of a VGO. The testing was conducted in the pilot plant at 685 K, 20 MPa of H_2 and LHSV of 1.0 h^{-1}. The novel catalyst exhibited a superior activity for CCR conversion and HCR. Compared with the conventional catalyst, the yields of naphtha and kerosene were 2.6 and 1.3 times, respectively, greater than when the novel catalyst was used. Moreover, the latter catalyst was more resistant to deactivation.

It is believed that for catalysts used in the first stage upgrading, carbon supports may have some advantages over oxidic supports. This results from the lower coke deposition generally observed on the carbon-supported catalysts compared with the latter catalysts. Moreover, carbon alone may exhibit some catalytic activity. This activity combined with the lower coke deposition may be one of the reasons for the higher overall activity of the carbon-supported catalysts compared with the conventional catalysts observed by Lopez-Salinas et al. (198). In addition, the conditions of preparation suggest that the carbon supports varying widely in porosity can be readily prepared. Apparently, there is the potential for preparation of tailor-made carbon supports to obtain catalysts which exhibit high HDAs and HDM rates. At the same time, such catalysts should posses a high metal storage capacity.

5.2.5.4.2 Novel γ-Al_2O_3 supports

The study of Ying et al. (193) on the fibrillar γ-Al_2O_3 as the support for catalyst used for hydroprocessing AR was extended to include VR (199). In the latter case, a series of fibrillar γ-Al_2O_3 was prepared from the fibrillar boehmite. The samples of the different pore size distributions could be prepared by varying the temperature of preparation. Novel supports were used for the preparation of NiMo catalysts. Figure 5.24 shows the removal of V and Ni from the VR containing 510 ppm of V and 66 ppm of Ni over NiMo/Al_2O_3 catalyst and NiMo/fibrillar Al_2O_3 catalyst. For the latter, the metal storage capacity determined at the end of the run was more than twice greater than that of the conventional NiMo/Al_2O_3 catalyst. Better performance of NiMo/fibrillar Al_2O_3 could be attributed to the more suitable pore size distribution, i.e., pore volume and mean pore diameter of 0.58 mL/g and 145 Å, respectively, for the novel catalyst compared with 0.39 mL/g and 98 Å, respectively, for the conventional catalyst. More efficient removal of V using NiMo/fibrillar Al_2O_3 was also confirmed during the aging experiments, i.e., the conventional NiMo/Al_2O_3 reached the end-of-run temperature earlier than the novel catalyst (200). Moreover, the start-up temperature to attain similar V removal during the operation was lower for the novel catalyst.

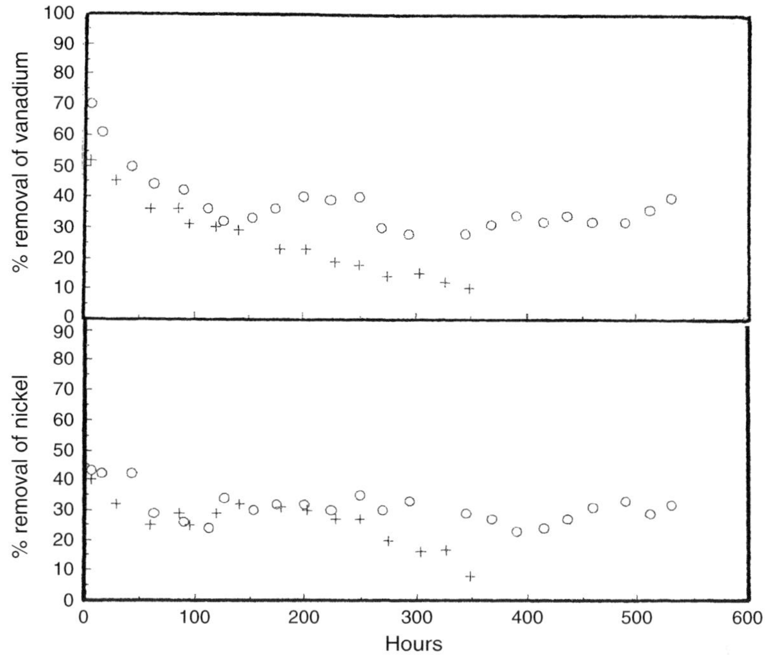

Figure 5.24 Removal of vanadium and nickel over NiMo/Al_2O_3 (+) and NiMo/fibrillar Al_2O_3 (∘) (199).

5.2.5.4.3 Mixed oxides supports

In recent years, TiO_2 has been receiving much attention as the γ-Al_2O_3 modifier for the preparation of catalysts suitable for hydroprocessing of the most problematic feeds such as VR and topped heavy crudes. Depending on the method of preparation and the amount of TiO_2 added, supports varying widely in porosity could be readily prepared. Moreover, the catalyst acidity could also be modified by TiO_2. Several other oxides (e.g., ZrO_2, SiO_2, MgO) were tested either alone or in various combinations as well. Because of the significant effect on catalyst properties, a more detailed account of the procedures used for catalyst preparation is necessary and is given in the following part.

Perhaps, the most detailed study on the effect of the conditions applied during preparation of the TiO_2-containing supports was conducted by Maity et al. (78,157–159). In the first instance, these authors used three different methods to prepare the binary Al_2O_3–TiO_2 (95:5) oxides which were used as supports for preparation of CoMo catalysts. The conditions of preparation had a significant effect on pore size distribution of the catalysts, as shown in Table 5.8 (78). Catalysts A to C were prepared by the sequential incipient wetness technique of the AT-1 to AT-3 (Al_2O_3–TiO_2) supports shown in Table 5.9. They contained similar amounts of CoO (3 wt%) and MoO_3 (10 wt%). Catalyst D was prepared by the co-impregnation method using the AT-3 support and contained 0.8 wt% of P besides the active metals. Catalyst E was the commercial NiMo/Al_2O_3 catalyst containing 10.7 wt% of MoO_3, 2.9 wt% of NiO and 3.7 wt% of TiO_2. This catalyst was used as the reference. The catalysts were used for hydroprocessing of the 50/50 blend

Table 5.8 Properties of catalysts (78)

	CAT-A	CAT-B	CAT-C	CAT-D	CAT-E
Surf. area (m^2/g)	236	216	258	250	156
Pore vol. (mL/g)	0.24	0.25	0.62	0.67	0.55
Aver. pore diam. (Å)	36	44	96	105	142
Pore size distrib.					
>1000 Å	0	0	0	7.4	2.7
1000–500 Å	0	0	0	0.9	4.1
500–200 Å	1.1	0.7	0.5	0.3	23.4
200–100 Å	1.8	1.1	40.1	20.8	50.2
100–50 Å	5.3	7.0	52.9	63.3	15.2
50–17 Å	91.8	91.3	0.6	7.3	4.4

Table 5.9 Properties of binary Al_2O_3–TiO_2 support (78)

Property	Support		
	AT-1	AT-2	AT-3
Surf. area (m^2/g)	292	275	277
Pore vol. (mL/g)	0.30	0.32	0.73
Avr. pore diam. (Å)	37	44	104
Pore size distrib. (vol.%)			
>1000 Å	0	0	0
1000–500 Å	0	0	0
500–200 Å	1.7	0.6	1.1
200–100 Å	0.1	1.0	47.3
100–50 Å	4.7	4.3	44.6
50–17 Å	93.4	94.2	7.0

of the heavy Maya crude and a naphtha fraction in the continuous fixed bed reactor at 653 K and 5.4 MPa. The blend contained 177 ppm of V + Ni. Figure 5.25 shows that compared with the reference $CoMo/Al_2O_3$ catalyst, $CoMo/Al_2O_3$–TiO_2 catalysts were more active. Also, their rate of deactivation was slower. The latter catalysts were more reducible and had the most suitable pore size distribution (158). The improved reducibility resulted in the formation of more active Type II phase. The beneficial effect of P (catalyst D) on HDS and HDM, in particular, should be noted. Under the same conditions, $CoMo/Al_2O_3$–TiO_2 (~3.8 wt%) catalyst was the most active compared with the three $CoMo/Al_2O_3$–SiO_2 catalysts (157). Also, for the latter catalysts, P had little effect on catalyst activity. Apparently, the reducibility and sulfidability of the Al_2O_3–SiO_2 supported catalysts were affected, as was evidenced by the presence of MoO_3 species in the spent catalysts. Under similar conditions, $NiMo/Al_2O_3$–TiO_2 (5 wt%) catalyst exhibited a higher activity for HDS, HDN and HDAs compared with the commercial $NiMo/Al_2O_3$ catalyst, whereas the improvement in HDM activity was less evident. However, catalyst activity was diluent dependent (157). Thus, the higher activity of the TiO_2-containing

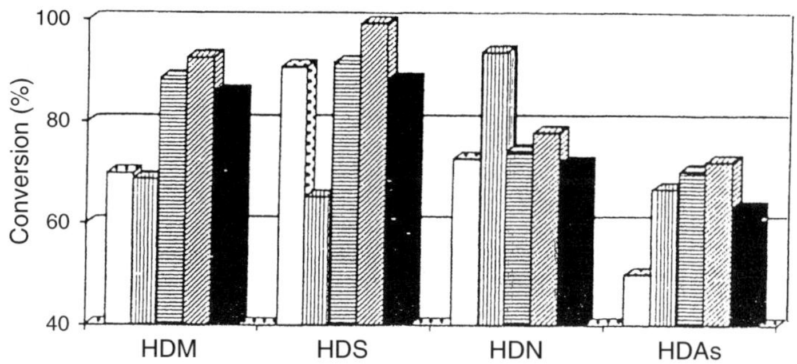

Figure 5.25 Percentage of conversion of catalyst: □ A; ▤ B; ▤ C; ▨ D; ■ E (78).

catalyst was observed for the 50/50 blend of Maya crude with diesel fraction than for the 32/68 blend of the Maya crude with naphtha fraction. Presumably, the solubility of asphaltenes in the former diluent was greater. Also, more inhibitive N-compounds present in the naphtha than those in diesel oil could contribute to the activity difference.

It was indicated above that CoMoP/Al_2O_3–TiO_2 catalyst exhibited a superior performance in comparison with several conventional and modified catalysts (78,157–159). In an attempt to further improve the activity of this catalyst, Maity et al. (201) focused on the preparation of the Al_2O_3–TiO_2 support using five different methods. The summary of these methods is given in Table 5.10. These supports were impregnated with 10% MoO_3, 3% CoO and 1.8% P. After sulfiding using a light oil spiked with DMDS, the catalysts were used for hydroprocessing the 50/50 blend of Maya heavy crude and diesel oil. The experiments were conducted in the continuous up-flow fixed bed reactor at 653 K, pressure of 5.4 MPa and LHSV of 1.0 h^{-1}. The properties of the fresh and spent catalysts prepared using these supports are shown in Table 5.11 (201). The catalysts supported on Al_2O_3–TiO_2 prepared by the methods 4 and 5 exhibited superior HDM and HDS activities compared with the methods 1–3. The former catalysts had a larger pore volume and APD (Table 5.11). Also, the loss of pore volume was less evident and deactivation rate much slower than those for the other catalysts. The post-experiment evaluation was conducted on the catalysts after 7 h onstream. So, promising results as obtained in these studies justify additional investigation (78,157–159). Thus, it may be desirable that the superiority of the catalysts 4 and 5 compared with the other catalysts be confirmed during long runs lasting several months.

Table 5.10 Salts and hydrolysis agents used for Al_2O_3–TiO_2 preparation (201)

Support	Salt	Agent
1	Al-sulfate; Na-aluminate, Ti-isopropoxide	Urea + pH variation
2	Al-sulfate; Na-aluminate, Ti-isopropoxide	Urea + pH variation; slurry aged 7 days in NH_4OH
3	Al-sulfate; Ti-chloride	Urea + NH_4OH
4	Al-sulfate; Ti-chloride	Urea + HCl
5	Al-sulfate; Ti-chloride	HCl

Table 5.11 Properties of Al_2O_3–TiO_2 supports (201)

	Support									
	1		2		3		4		5	
	Fresh	Spent	Fresh	Spent	Fresh	Spent	Fresh	Spent	Fresh	Spent
SA (m^2/g)	128	68	154	119	126	18	177	140	235	117
PV (mL/g)	0.26	0.10	0.23	0.12	0.16	0.05	0.42	0.19	0.41	0.24
APD (Å)	80	58	59	39	52	111	95	55	70	81
<50 Å	37	71	95	79	75	41	27	91	52	30
50–100 Å	57	25	4	17	9	20	48	7	47	43
100–200 Å	3	3	1	4	6	11	17	2	<1	16

The work of Maity et al. (201) showed that there is a possibility for further improvement in the activity of the Al_2O_3–TiO_2 supported catalysts by changing conditions of the support preparation and the content of TiO_2 in the support. This was demonstrated by Rana et al. (202) who prepared a series of CoMo/Al_2O_3–TiO_2 catalysts varying widely in surface area (200–260 m^2/g), pore volume (0.33–0.44 mL/g) and pore diameter (41–74 Å). In this case, the support contained 10 wt% of TiO_2. Among several methods tested by these authors, the precipitation of the $TiCl_4$ + ammonium nitrate solution with either urea or ammonia gave the catalysts which exhibited the highest HDS activity. This was attributed to the beneficial effect of TiO_2 on the morphology of MoS_2 crystallites. However, there was little effect on the HDM and HDAs activities. This is not surprising considering the relatively low average pore diameter of the catalysts. The activity of the catalysts was determined in the up-flow microreactor using the 50/50 blend of Maya crude and diesel oil at 653 K, 5.4 MPa and LHSV of 1.0 h^{-1}.

Two methods of Al_2O_3–TiO_2 preparation, i.e., in which Ti (Ti-isopropoxide) was added either to boehmite or to alumina, yielded supports possessing different properties (203). In the former case, the Ti—O—Al entities were formed after calcination, whereas TiO_2 was the predominant form of Ti when alumina was used instead of boehmite. With such supports, the role of TiO_2 in Al_2O_3–TiO_2 could be clarified. Indeed, there was a significant difference between the properties of NiMo catalysts supported on these supports. The TiO_2-containing catalyst exhibited a high activity for HDS, HDN and HDM. This catalyst had a higher acidity and had more catalytic sites defined as CUS. This study was conducted under identical conditions as used by Rana et al. (202) previously, i.e., a 50/50 blend of the partially demetallized Maya crude and diesel oil, the continuous fixed bed reactor at 653 K, 5.4 MPa and LHSV of 1.0 h^{-1}.

Ramos-Galvan et al. (204) prepared the Al- and Ti-pillared clays from the montmorillonite-type clay by pillaring with the Al- and Ti-containing polyoxo-cations. The pillared clays were used as supports for preparation of NiMo catalysts which were used for hydroprocessing of the heavy feed derived from Maya heavy crude. The feed contained more than 260 ppm of V + Ni. The experiments were performed in the autoclave at 673 K and 8 MPa. The pillared clay supported NiMo catalysts were much more active for HDS and HDAs than the conventional NiMo/Al_2O_3 catalyst. The NiMo/Ti-pillared clay catalyst was more active than the NiMo/Al-pillared clay catalyst. However, the long-term performance of the novel catalysts in a continuous system has not yet been determined.

The introduction of ZrO_2 to γ-Al_2O_3 resulted in the decrease of acidity of the corresponding NiMo catalyst (108). The HDS activity of this catalyst determined using heavy Maya crude was lower than that for the commercial NiMo/Al_2O_3 catalyst. However, for this feed, the addition of ZrO_2 improved HDM activity. At the same time, during the HDS of model compound feeds, the ZrO_2-doped catalyst was more active than the conventional catalyst. This observation confirmed the importance of using real feeds compared with model feeds during the catalyst development.

One communication dealing with the addition of basic oxides, such as MgO, to γ-Al_2O_3 as the support for catalyst used in heavy feeds upgrading was found in literature (169). Similar to ZrO_2 (108), the addition of basic oxides decreased the catalyst acidity. The results in Figure 5.26 confirmed that the HDS activity of the MgO (1 wt%) doped CoMo/Al_2O_3 catalyst (CoMo/AM-0.01) was higher than that of the commercial NiMo/Al_2O_3 catalyst (CC2), whereas the HDM activity improvement was less evident. In this case, the Maya heavy crude was used as the feed at 673 K and 5.4 MPa. Doping with MgO decreased the catalyst deactivation because of the slowing down of coke formation and the decreased poisoning by N-bases which resulted from their diminished adsorption on the catalyst surface.

The decreased coke formation on CoMo catalyst supported on sepiolite ($MgO \cdot SiO_2$) can also be attributed to the presence of MgO in the support (205). In spite of suitable porosity, the HDS, HDAs and HDM activities of this catalyst were low. However, activity improvement was achieved by blending the sepiolite support with the conventional γ-Al_2O_3. The optimal catalyst performance was observed for the blend containing about 25 wt% of the latter.

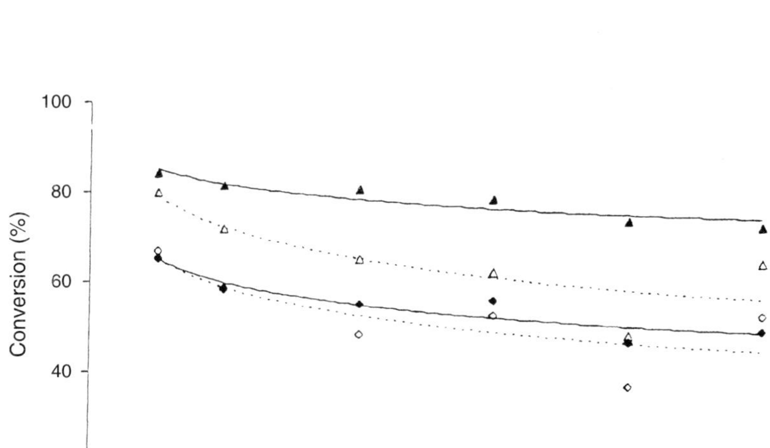

Figure 5.26 HDM and HDS conversions with time onstream. CoMo/AM-0.01: ▲ HDS; ♦ HDM; CC-2: △ HDS; ◇ HDM (169).

5.3 NOVEL CATALYSTS

The novel catalysts discussed in this section consist of the catalytically active phases, which differ from those found in the conventional and modified-conventional catalysts. However, the conventional γ-Al_2O_3 was the usual support used for the preparation of novel catalysts. The development of novel catalysts has been focusing predominantly on the VGO, HGO and DAO feeds with the aim to improve the selectivity to distillate fractions and to decrease the yield of gaseous products and coke. Moreover, the catalyst functionalities such as HCR and hydroisomerization became important whenever the improvement in cold flow properties of middle distillates and lube base oil was the objective. In some cases, hydrofinishing performed over catalysts possessing a high HYD activity was necessary to achieve the desirable quality of the middle distillates and lubricants. Therefore, it is not surprising that other heavy feeds have been receiving much less attention. The catalysts which have been designed specifically for dewaxing of VGO and DAO are discussed in the book in Chapter 11. The catalytic activity of various forms of carbon in the absence of active metals has also been evaluated.

5.3.1 Metal carbides, nitrides and phosphides

Metal carbides and nitrides have been tested for hydroprocessing applications mostly using model compounds. A close examination of the published information suggested that the potential of these catalysts for hydroprocessing is limited because of rather different trends which were exhibited by intrinsic activity (206). For example, it was observed that the intrinsic activity increased with increasing particle size in contradiction to the general trends observed for conventional catalysts. Then, compared with the conventional catalysts, a significantly greater amount of metal carbides and/or nitrides has to be loaded into the reactor to achieve a similar catalyst active surface. For most part, the real feeds used for testing included light fractions. Among heavy feeds, only VGO was investigated (207). In this case, the nitrided Mo/Al_2O_3 and $CoMo/Al_2O_3$ catalysts were compared with the corresponding sulfided catalysts. The nitrided Mo/Al_2O_3 was more active for both HDN and HDS than the sulfided Mo/Al_2O_3. The addition of Co resulted in an enhancement of the activity of the sulfided form compared with a little promoting effect on the nitrided Mo/Al_2O_3.

Using the VGO derived from heavy Maya crude, Melo-Banda et al. (208) showed that the unsupported and supported Mo_2N and W_2N exhibited markedly higher HDN and HYD activities than the reference $CoMo/Al_2O_3$ catalyst. However, their HDS activity was low. The experiments were conducted in fixed bed reactor at 623 K and 8 MPa. It should be noted that for most part, the long-term performance of metal carbides and nitrides as catalysts for hydroprocessing has not been evaluated.

It is believed that because of the high HYD activity under mild conditions (e.g., ~450 K), the potential of metal carbides and nitrides in hydroprocessing applications has not yet been fully explored. For example, suitability of these catalysts for hydrofinishing, i.e., of RFCC products, lube base stock, should be evaluated. Also, the preHYD of the VGO, HGO and DAO feeds over metal carbides and nitrides prior to the use of these feeds in the multistage systems (e.g., dewaxing and RFCC) deserves additional attention. Moreover, the effect of promoters (e.g., Co, Ni and others) and different supports on the activity and selectivity of the Mo/W carbides and nitrides has not been explored in detail.

It appears that with respect to hydroprocessing of real feeds, metal phosphides (209) and phosphates (210,211) are still in the early stages of development. Although they may be promising catalysts, their activity has been evaluated predominantly using model compound feeds. The study on Ni_2P/SiO_2 and $NiMo/Al_2O_3$ catalysts using hydrogenated gas oil conducted by Oyama (209) revealed that the former was more active for HDS and HDN.

Several metal aluminaphosphates (MAPO) were investigated by Ingram and Ghebreyessus (211) during the hydroprocessing of VGO. The materials were synthesized from the reactive aluminophosphate gel containing the metal cations of interest (e.g., Mg, Co and Zn) and tripropylamine as a template. The VGO conversion was tested at 673 K and 5.5 MPa. The following activity order for the conversion of 343–525°C fraction and residue to lighter fractions was established: CoAPO > ZnAPO > and MgAPO.

5.3.2 Transition metals containing catalysts

Noble metal sulfides have been studied for various hydroprocessing reactions using predominantly model compounds mixtures, whereas among the real feeds, only CDL have attracted attention (63,76). Because of the cost, the affordable amount of noble metals in the catalyst must be less than 1%. The problem is even more evident in the case of heavy feeds because of rather complex catalyst regeneration and/or metal reclamation from the spent catalyst for reuse. Apparently, the studies of Escalona et al. (212,213) on hydroprocessing of a gas oil over Re/carbon and Re/Al_2O_3 catalysts are among few found in the literature. The selectivity of the Re/carbon catalyst for HDN relative to HDS was about twice that of the latter catalyst. To a certain extent, this may result from the diminished self-inhibition of HDN because of a neutral nature of the carbon support. The optimum HDS and HDN activities was observed at 2.47 wt% of Re_2O_7. At the optimum, the catalyst had the largest pore volume.

The highly porous saponite (HPS) was used as the support for Pd/Rh–Co/HPS catalysts by Hossain et al. (214) and Al-Saleh et al. (215). The catalyst preparation involved impregnation of the HPS support with noble metal followed by the addition of Co using an ion exchange method. A VGO was used as the feed for determining the catalyst activity for HDS and HCR. Rhodium had a higher promoting effect than Pt; however, the best performance was exhibited by the bi-noble metal catalyst. This resulted from the significantly diminished catalyst deactivation. No conventional catalyst was included for comparison. The Pt-containing catalysts supported on the ASA, USY and ITQ-2 zeolites were prepared and tested by Corma et al. (77). However, a light feed was only tested.

Isoda et al. (216) prepared a series of catalysts by loading the dealuminated Y-zeolite with metals such as Ni, Co and Fe. These catalysts were used for hydroprocessing of VGO between 573 and 653 K and 5 MPa. The best performance was exhibited by the extensively dealuminated Ni-HY-B zeolite (Si/Al = 25). This catalyst gave the highest yield of middle distillates, as well as the lowest yields of gas and coke on the catalyst. This was attributed to the combination of an optimal acidity and a high dispersion of metal sulfides on the support. It should be noted that these catalysts did not contain traditional active metals such as Mo and/or W.

The information on the development of novel catalysts for the feeds which are heavier than VGOs and HGOs is limited. The Ru-promoted and the Ru-free commercial $CoMo/Al_2O_3$ catalysts were investigated during hydroprocessing of AR containing 26 ppm of V + Ni and 13.7 wt% asphaltenes (217). The addition of Ru had little effect on the catalyst performance; however, a similar addition of Ni and W had beneficial effect on the HDS and asphaltenes conversion.

The Y-type zeolite impregnated either with Pd or Ni was used by Nomura et al. (218,219) for the HCR of DAO and heavy asphaltenic materials. For the former feed, the Pd- and Ni-co-loaded Y-type zeolite catalyst showed the highest activity, yielding more than 70% of the gasoline fraction at 623 K and 4.9 MPa. Under more severe conditions, i.e., at 673 K and 6.9 MPa, the conversion of maltenes approached almost 90%. The activity of the Pd- and Ni-co-loaded catalysts could be further increased by modifying the method of catalyst preparation. The improved catalyst was tested for asphaltenes conversion. In this case, the conversion of asphaltenes was about 20% greater compared with the unmodified catalyst. The conversions were dependent on the structure and/or the origin of asphaltenes.

In many instances, catalysts used for the dewaxing of VGO and DAO contain noble metals together with the conventional metals supported on zeolites and other acidic supports. The catalytic functionalities of the dewaxing catalysts differ from those of the conventional catalysts. This results from the objective of catalytic dewaxing which differs from that of hydroprocessing. The detailed account of the former catalysts is given later in the book in Chapter 11.

5.3.3 Carbon catalysts

The catalytic activity of various carbons (e.g., carbon blacks, AC, petroleum coke) in the absence of active metals during hydroprocessing has been known for some time (197). This included carbons containing little mineral matter, as well as those prepared from coal. In the latter case, the content of mineral matter could exceed 10 wt%. Therefore, for such carbons, the involvement of some inorganic oxides and sulfides in catalysis could not be avoided. However, carbon with very low ash content can be prepared from the mineral matter free materials (wood, nut shell, etc.). For such carbons, most of the catalytic activity may be attributed to the carbon sites rather than to some inorganic sites. Beside catalytic activity, finely divided carbon-containing solids when added to heavy feeds act as the seeds of coke formation. In this case, coke precursors which deposited on carbon particles are carried out with process streams, rather than being converted to coke.

Three samples of activated carbon in the form of crushed granules (1.2–1.7 mm) were used in the CSTR system and extrudates (3–6 mm) were used in the fixed bed reactor and in the ebullated bed reactor for the hydroprocessing of several VR derived either from the Mexican or from the Arabian crudes (197). The best performance was observed for the activated carbon with porosity predominantly in the mesopore range. The activities of the crushed granules and extrudate form of the activated carbon determined in the fixed bed reactor were similar. For the VR obtained from Maya crude, the HDS and HDM activities determined in the ebullated bed reactor at indicated temperatures and at pressure of 18.5 MPa are shown in Figure 5.27 (197). Microscopic analysis of the spent activated carbon catalyst particles revealed that the radial distribution of V

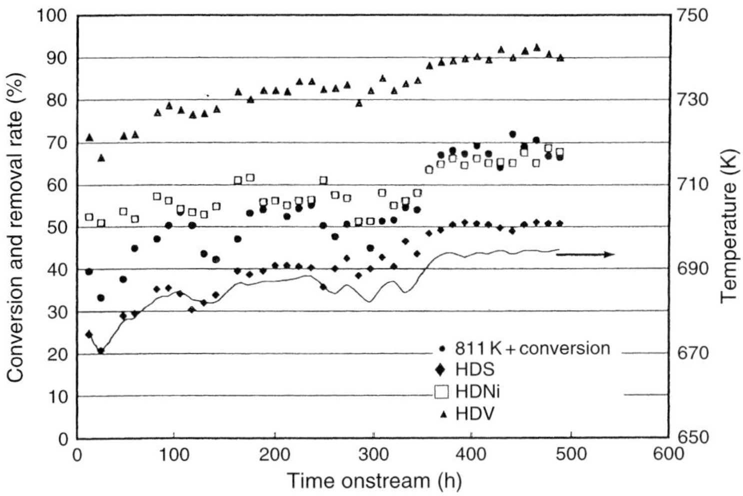

Figure 5.27 Catalyst performance of extrudated AC catalyst in ebullated bed (197).

was much more even than that for the conventional $NiMo/Al_2O_3$ catalyst. Moreover, the latter was more prone to deactivation by coke.

In another study, Terai et al. (220) compared the activated carbon impregnated with Fe nitrate to give about 10 wt% of Fe with the mechanical mixture of activated carbon and pyrite. The Middle East VR used as the feed contained 233 ppm of V + Ni and gave 22.4 wt% of CCR. The experiments were conducted at 7–10 MPa and from 683 to 713 K. Compared with the mechanical mixture, the activated carbon impregnated with Fe catalyst exhibited a superior activity. This was indicated by very high rate of HDM and HDAs. Moreover, it gave much lower yield of gases by preventing cracking of the middle distillate fractions.

A rather unique approach to catalyst preparation from activated carbon was used by Lee et al. (221). In their study, the oil-soluble Mo and Co precursors (naphthenates) were introduced with an AR into the ebullated bed of the granules of activated carbon. The co-dispersed catalyst was very active for HDS, HDM and HDAs. The optimum activity was attained for the Co/(Co + Mo) ratio of 0.3. However, the HCR activity was inhibited by the presence of Co. It is noted that the dispersed/dissolved catalysts used for hydroprocessing of heavy feeds are discussed in Chapter 12 dealing with unconventional/emerging catalysts.

Chapter 6

HYDROPROCESSING REACTIONS

The presence of large molecules indicates a significant complexity of the reactions occurring during hydroprocessing of heavy feeds. Because of the increasing involvement of asphaltenic molecules, the complexity increases from VGO/HGO toward VR and topped heavy crude. In every case, the primary objective is the conversion of large molecules to distillates. This may be accomplished via HCR of resins and asphaltenes simultaneously with the conversion of porphyrin structures. Therefore, for heavy asphaltenic feeds, a high rate of the HDAs is required to achieve a desirable rate of HDM. Thus, the removal of most of the metals cannot be achieved before asphaltenes are depolymerized to smaller entities.

In the case of VGO, HGO and DAO feeds, a high level of HYD of aromatics, i.e., a high rate of HDAr, must be achieved, when the feed preparation for FCC is the objective. Moreover, for such feeds even traces of metals and asphaltenes as well as nitrogen have to be removed to prevent poisoning of FCC catalyst, unless a more advanced process, i.e., RFCC, is used. In the case of catalytic dewaxing of these feeds, catalyst must possess an adequate HCR activity and selectivity to ensure a high yield of middle distillates and lube base oil fractions. For dewaxing catalysts, hydroisomerization of *n*-paraffins to isoparaffins becomes an important catalytic functionality to ensure low freezing point and pour point of the final products. In addition, to be suitable for preparation of lubricants, lube base oil must exhibit good viscosity behavior. For this purpose, aromatic structures must be converted to naphthenic compounds. For most of the VGO, HGO and DAO, desirable properties of the products (e.g., lube base oil and diesel oil) cannot be attained in one stage.

To various degrees, HDS, HDN and HDO reactions occur simultaneously with HYD, HCR, HDM, HDAs and hydroisomerization reactions. Kinetic measurements can be used to quantify the progress of these reactions. The kinetic and mechanistic aspects indicate that the mutual effects of hydroprocessing reactions are rather complex even for less problematic feeds. To certain extent, these effects may be influenced and/or controlled by the properties of catalysts as well as by the experimental conditions.

6.1 KINETICS OF HYDROPROCESSING REACTIONS

Kinetic studies provide valuable information which is part of the database used for the comparison and selection of catalysts. Both, kinetics of the removal of contaminants (e.g., S, N, metals, asphaltenes, resins, etc.) and kinetics of other hydroprocessing reactions (e.g., HCR, HYD, dewaxing, etc.) have been of the interest. Attention has been paid to the kinetics of catalyst deactivation. This topic will be discussed in Chapter 7. The parameters determined during kinetic studies form the basis for the development of mathematical equations and/or models, which can be used for assessing the reactivity of heavy feeds, selection of catalysts and optimizing operation of catalytic systems. Therefore, kinetic parameters are part of the models used for predicting catalyst life. In fact, the determination of such parameters for a particular catalyst may be viewed as the modeling on an activity scale. Because of the complex structure of heavy feeds, kinetics focus mostly on the overall removal of contaminants involving a single power low order or LH equation. Thus, so far detailed characterizations of feeds and products with the aim to identify and quantify components and/or groups of components was only achieved for lighter feeds.

For heavy feeds such as VGO and HGO, the kinetic studies focus on HDS, HDN, HDO, HYD and HCR reactions. In addition to these reactions, for DAO, the kinetics of HDAs and HDM may be of an interest as well. In relation to the first stage conversion in a multistage system, the kinetics of HDAs and HDM are of primary importance for the AR feeds. Similarly, in the case of VR, focus is on the same reactions. However, kinetic parameters expressed in terms of the CCR/MCR conversions have also been determined. In this case, both VR and asphaltenes isolated from VR were used as the feeds for kinetic studies.

Insufficient information on the composition of heavy feeds such as VGO and HGO may be, at least partly alleviated by lumping hydrocarbon groups, as it was proposed by Jaffe et al. (222,223). For example, sulfur can be divided into the non-reactive aromatic, consisting of the thiophenic heterorings and sulfidic structures. The latter is much more reactive than the sulfur heterorings. However, a wide range of reactivity of the thiophenic heterorings suggests that a more detailed characterization of the VGO and HGO is necessary to identify the most resistant structures. Such information is required for understanding the mechanism of the deep and/or ultra-deep HDS. Similarly, aromatics may be divided into reactive and non-reactive. Again, there is a wide range of reactivity among the latter. There is a tendency to divide N-containing compounds into the basic and non-basic structures in addition to the total nitrogen. A distinction of N-compounds on reactivity basis would provide a more reliable information on HDN. Another source of the information includes the kinetic parameters which are usually determined on the basis of the difference in composition of the feeds and corresponding products expressed by the content of hydrocarbon groups such as paraffinic, aromatic and naphthenic, as well as that of contaminants. The difference between true boiling point distillation of the feeds and corresponding products may be another means for obtaining kinetic data.

As it was indicated above, so far the detailed analytical data could only be obtained for light feeds, although among heavy feeds under consideration, a similar level of characterization may already be approached for VGO and HGO and their products. However, the recent review on kinetics of hydroprocessing of heavy feeds published by Ancheyta et al. (224) emphasized that even for the feeds such as VGO and HGO significant complications may be encountered during kinetic measurements because of limited information on the composition of feeds and products.

Kinetic data can also be affected by the conditions in catalyst bed which may not be isothermal, although such assumptions are sometimes being made in kinetic studies. It is important to note that kinetics can be influenced by the type of experimental system. For example, Jaffe et al. (222,223) showed that the kinetic order determined for the system employing the plug-flow reactor (PFR) may differ from that using the back-mixed flow reactor (BMR). The following relationship developed by van Zijll Langhout et al. (226) approximately holds between kinetic orders in PFR and BMR:

$$n_{\mathrm{BMR}} = 1 + 0.5\,(n_{\mathrm{PFR}} - 1)$$

Furthermore, the presence of relatively large inter and intraparticles concentration gradients cannot be completely avoided, particularly for the systems employing fixed bed reactors.

It has been demonstrated that a change in the H_2S concentration in gas phase changed the H_2S/H_2 ratio, which consequently influenced the kinetics of hydroprocessing of VGO (126). Figure 6.1 illustrates a possible change in the H_2S and H_2 distribution between the inlet and outlet of the fixed bed (227). It was noted that the H_2 partial pressure has been included in kinetic equations, although in the case of heavy feeds the H_2 dissolved in liquid phase and adsorbed on catalyst surface plays predominant role during hydroprocessing (228). It is well established that at the same H_2 pressure, the dissolution of H_2 in liquid phase decreased with increasing temperature. Inevitably, this change would influence kinetic data, activation energies in particular, although a decreased H_2 solubility with increasing temperature may be offset by increasing H_2 pressure. This, however, will change the H_2S/H_2 ratio in the system. Consequently, active surface of the catalyst will be affected (76).

The size of catalyst particles used for kinetic measurements deserves a special attention. Table 6.1 shows that the rate constants for the HDS of HGO increased with the decreasing size of catalyst particles (229). At the same time, activation energies increased. There was little effect on kinetic parameters when the size of particles was decreased below 0.25 mm. Apparently, below 0.25 mm, restrictive diffusion was not present. As expected, the effectiveness factor, determined as the ratio of the apparent rate constant to intrinsic rate constant, increased with decreasing particle size. But it decreased with increasing temperature, as it is shown in Figure 3.10 (84). The intrinsic

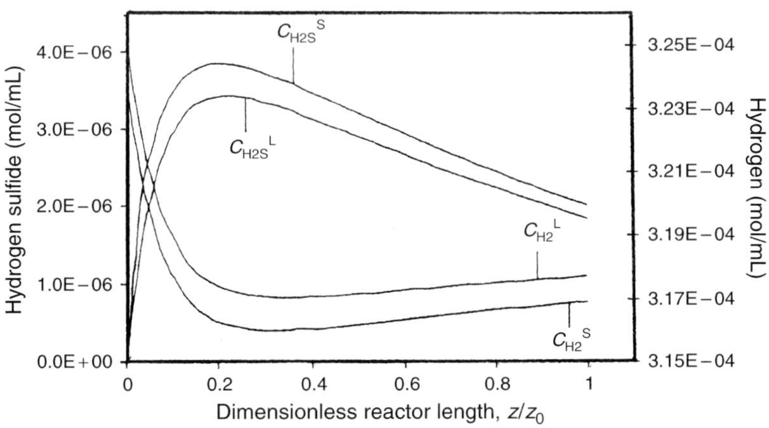

Figure 6.1 Concentration of H_2 and H_2S and along reactor in liquid and solid phases (227).

Table 6.1 Effect of temperature and particle size on rate constants for HDS of gas oil (229)

Temperature (K)	Particle size (mm)			
	2.5	0.83	0.37	0.25
593	0.018	0.041	0.065	0.168
613	0.029	0.065	0.114	0.266
633	0.038	0.085	0.140	0.363
653	0.044	0.102	0.178	0.470

rate constant was determined using the 0.25 mm size catalyst particles. For these experiments, 2.5 mm trilobe extrudate $NiMo/Al_2O_3$ catalyst was crushed to obtain the smaller particle size fractions. It should be noted that these and other effects of particle size on kinetic parameters have to be kept in mind together with the uncertainties mentioned above, when comparing results obtained in different studies. These results show that kinetic data are suitable for estimating parameters such as distribution factor, effective diffusivity and Thiele modulus. From the model development point of view, the estimate of these parameters requires kinetic data determined on the catalyst activity level together with the parameters determined on a single particle level before a model on the reactor level can be developed.

The state of the catalyst surface has a significant effect on kinetic parameters particularly during the initial contact of catalysts with the feed. Thus, because of the rapid catalyst deactivation, the kinetic parameters determined during the early stages on stream will exhibit a great variability with time, until they stabilize in the steady state. Therefore, the reliability of kinetic data is significantly enhanced in the case that they were determined over an equilibrium catalyst. It should be noted that this fact is frequently overlooked and as such may be one of the sources of discrepancies among different studies observed in the literature.

Kinetic parameters also depend on the type of mathematical model used for their estimate. This was demonstrated by the results shown in Table 6.2 (151). The results were obtained over the $NiMo/Al_2O_3$ catalyst in the trickle bed microreactor in the temperature and pressure range of 613–693 K and 6–10 MPa, respectively, using a HGO as the feed. The experimental results were treated using the Langmuir–Hinshelwood (LH) model and the simple power law equation. The difference was attributed to the

Table 6.2 Kinetic parameters for different models (151)

	HDN		HDS	
	E (kJ/mol)	*A* (h^{-1})	*E* (kJ/mol)	*A* (h^{-1})
Power low	94	1.8×10^8	96	140×10^8
LH model	113	750×10^8	137	3900×10^8

E – activation energy; *A* – pre-exponential factor.

competitive adsorption of H_2S on catalytic sites which was considered in the LH model. At the same time, the power law model does not account for this effect.

Because of so many uncertainties involved, a direct comparison of the kinetic data from different studies is affected. For some studies, a comparison is prevented because the information on experimental conditions used during the kinetic measurements is incomplete. In most cases, the reported results reflect very specific conditions, i.e., they were determined during very early stages before an equilibrium surface of the catalyst was attained. Therefore, the significance of such results for the understanding of hydroprocessing reactions may be limited. This chapter has been organized according to the type of feeds, i.e., starting with HGO/VGO and ending with VR and heavy crudes, which were used for kinetic studies. A descriptive summary of the kinetic studies published in the scientific literature can only be afforded, because of so many uncertainties involved in determining kinetic parameters and/or limited information provided.

6.1.1 VGOs and HGOs

The review published by Ancheyta et al. (224) compared the advantages and disadvantages of several kinetic approaches, i.e., lumping techniques, continuous mixture, structure-oriented lumping and single-event models, which have been used to study kinetics of the hydroprocessing of heavy feeds with the primary focus on VGO and HGO. The kinetic models were compared in terms of their accuracy to predict experimental parameters. Because of significant complexity, most of the kinetic studies have been focusing on the overall removal of heteroatoms, aromatics and metals even for the lightest feeds under consideration in this study. Lumping kinetics were used to study the effect of operating parameters on the distribution and composition of the hydrocarbon fractions in products. Few studies indicated difficulties in obtaining kinetic data for individual components of the feed. Thus, so far attempts have only been made to determine the distribution of thiophenic compounds and basic and non-basic nitrogen in the feed and products, for kinetic measurements.

6.1.1.1 Kinetics of thiophenic heterorings

The study of Ma et al. (230) has provided one of the most comprehensive set of kinetic data on the HDS of VGO. It was based on the detailed characterization of thiophenic heteroring compounds in the VGO and corresponding products using the GC-FPD method. Four representative types of thiophenic rings found in the VGO included alkylbenzothiophenes (BTs), alkyldibenzothiophenes (DBTs), alkylphenanthro[4,5-b,c,d]thiophenes (PT) and alkylbenzonaphthothiophenes (BNTs). The VGO used in this study was derived from the Middle East crude and was treated in the batch reactor at 633 K and 6.9 MPa over the commercial $NiMo/Al_2O_3$ catalyst. The pseudo-first-order plots of the representative groups are shown in Figure 6.2 (230). It is evident that the alkyl-DBT with alkyl substituents at the 4 and 6 positions are the most refractory sulfur compounds in the VGO. The presence of these compounds is the main reason for difficulties in achieving the deep and/or ultra-deep HDS of fuels. This was clearly demonstrated in another kinetic study conducted by Ma et al. (231). In this case, a gas oil feed of the Middle East origin was fractionated into five fractions, i.e., F1, F2, F3, F4 and F5 with boiling range of <280, 280–300, 300–320, 320–340 and $>340°C$, respectively. Thus, the fraction F5 was

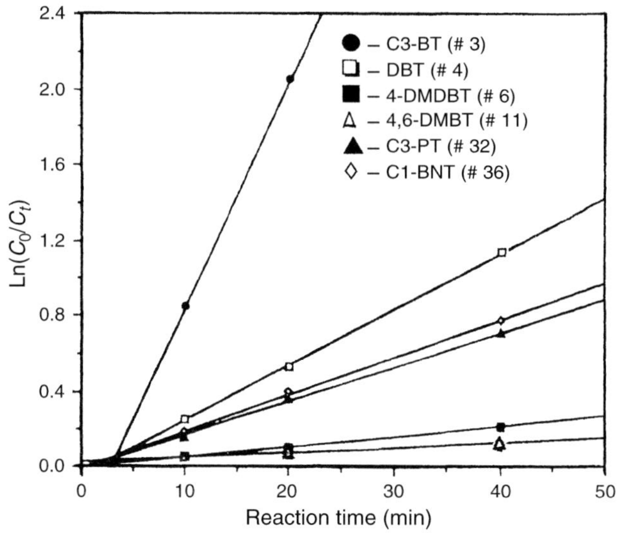

Figure 6.2 Pseudo-first-order plots of sulfur compounds (230).

in the boiling range of VGO. The pseudo-second-order plots for the total sulfur removal from these fractions are shown in Figure 6.3 (231). The lowest reactivity of the fraction F5 was caused by the presence of the alkyl-substituted DBTs. This fraction represented about 25% of the feed, however, its sulfur content accounted for almost 40% of the total content in the gas oil feed. Therefore, it might be representative of the thiophenic heterorings present in VGO. This study confirmed the significant difference between the reactivity of the S-containing compounds in VGO and those in AGO as determined by

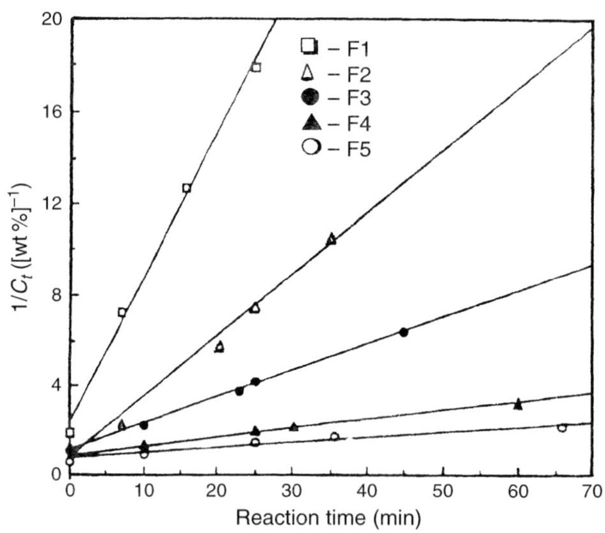

Figure 6.3 Pseudo-second-order plots of fractions (231).

kinetic measurements. Although this difference may be anticipated, the studies of Ma et al. (230,231) provided the convincing experimental evidence confirming the same. These studies also indicate usefulness of the kinetic data determined on the basis of an extensive characterization of the feed and corresponding products.

6.1.1.2 Overall kinetics

Kinetic study of Trytten et al. (232) involved six narrow boiling cuts of 50°C wide each, obtained from the HGO derived from Athabasca bitumen. The HDS and HDN of these cuts were investigated in the CSTR system in the temperature range of 653–698 K and 13.9 MPa of H_2 using the commercial $NiMo/Al_2O_3$ catalyst. The sum of conversions of these fractions was used to determine the overall conversion of the feed, e.g.,

$$X_t = \sum f_i S_i X_i / \sum f_i S_i$$

and pseudo-rate constant was calculated using the following expression:

$$k_t = \frac{\text{LHSV}}{3.6} \times \frac{X_t}{(1 - X_t)}$$

where X_t is the total conversion, k_t the pseudo-rate constant and f_i, S_i and X_i the numbers of the narrow boiling fractions in whole HGO, weight fraction of contaminant in HGO (e.g., sulfur and nitrogen) and conversion of an i fraction, respectively. The pseudo-rate constant and intrinsic rate constant determined for a whole pellet (1.27 mm) as well as the ground and sieved pellet (0.8 mm), respectively, decreased logarithmically with increasing AMW of the fractions. At the same time, the apparent activation energies increased with increasing AMW, while the increase in intrinsic activation energies was less evident. These results are shown in Table 6.3 (232). The effectiveness factor (E_f), determined as the ratio of the apparent to intrinsic rate constants was estimated to verify the extent of the intraparticle diffusion. For both HDS and HDN, E_f increased with increasing AMW and reached the maximum at AMW of about 350, as it is shown in Figure 6.4 (232). The lower effectiveness factors for the heavier fraction indicated the presence of diffusion limitations. However, the decreasing reactivity of the fractions with increasing AMW had more pronounced effect on the decrease in the intrinsic activity than on the decrease in the rate of diffusion. At the same time, effective diffusivity (D_{eff}), mathematically expressed by reaction [1] (Chapter 3.2.2), decreased by an order of magnitude from fraction 1 to fraction 6. The ϵ/τ value of 0.1 was used for this estimate. This database (232) established by kinetic measurements clearly indicates that

Table 6.3 Activation energies (kJ/mol) of fractions (232)

Fraction	HDS		HDN	
	Appar.	Intr.	Appar.	Intr.
Boiling range (K)				
573–623	109	165	78	103
723–773	149	205	109	121

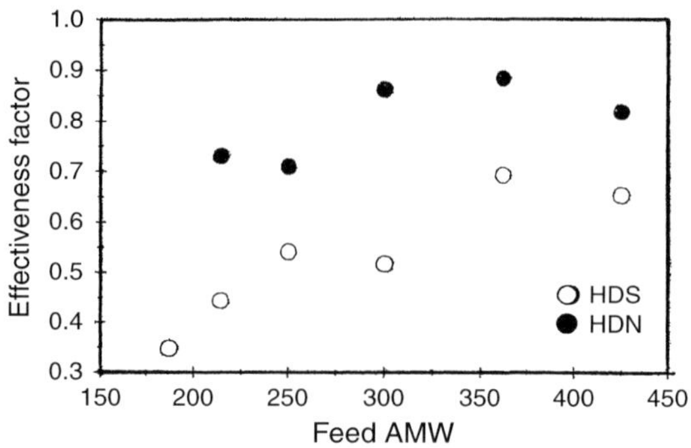

Figure 6.4 Effectiveness factor versus feed AMW at 673 K (232).

the involvement of diffusion phenomena has to be taken into consideration for the relatively light feeds such as VGOs and HGOs and perhaps for even lighter feeds.

The apparent kinetics of HDS and HDN of the HGO derived from Athabasca bitumen were studied by Mann et al. (174) over the NiMo/zeolite-alumina-silica catalyst at 7 MPa, 623–698 K and LHSV varying from 1 to 4. The following integrated equation was used to fit the experimental data:

$$C_o^{1-m} - C^{1-m} = (m-1)k/\text{LHSV}$$

In this equation, C_o and C are concentrations in the feed and products, respectively, k the rate constant and m the reaction order. For HDS, the best fit of the experimental data was obtained for $m = 2$, whereas for HDN, for $m = 1.5$. The activation energies for HDS and HDN were 86.9 and 104.9 kJ/mol, respectively. Bej et al. (95,96) used the similar HGO and experimental conditions as used by Mann et al. (174) to study kinetics of the removal of basic and non-basic nitrogen, as well as total nitrogens over the NiMo/Al_2O_3 catalyst. For the overall HDN, the best fit of data was obtained using the 1.5-order kinetic equation. This gave activation energies for the conversion of the non-basic N-compounds to basic and HDN of the latter, 80 and 74 kJ/mol, respectively, whereas for the total nitrogen removal 80 kJ/mol.

The method of catalyst preparation influenced kinetic parameters (123). Thus, the rate constants for HDS and HDN over the catalysts prepared by impregnation of the γ-Al_2O_3 with active metals (Co and Mo) in the presence of EDTA were consistently greater than those without EDTA. The former method of preparation gave a more active catalyst because of the predominance of the Type II active phase. Apparent activation energies of about 83 and 76 kJ/mol for HDN, as well as 89 and 56 kJ/mol for HDS for catalyst prepared without and with EDTA present, respectively, were obtained. In this study (123), the HGO which was studied by Bej et al. (95,96) was used as the feed.

The studies of Bej et al. (95,96,123), all conducted under the same conditions can be used to illustrate the effect of the catalyst origin on kinetic parameters. Compared with these studies, Mann et al. (174) used a similar HGO but rather different type of catalyst. The role of catalyst type in modifying kinetic parameters was also demonstrated in the study conducted by Ferdous et al. (233) in a trickle bed microreactor (613–693 K; ~6 to 10 MPa; 0.5–2 h^{-1} LHSV) using the same HGO over the NiMo/Al_2O_3 catalyst doped with borate. In this case, the power law model gave the best fit of the experimental data for first-order and 1.5-order kinetics for HDN and HDS, respectively. Moreover, compared with Table 6.2 (151), the addition of borate changed activation energies, i.e., for power model to 75 and 87 kJ/mol for HDN and HDS, respectively, and for the LH model to 110 and 159 kJ/mol, respectively. In the same study, the equations for the overall conversion of nitrogen and sulfur were derived by the regression analysis. These equations were used for identifying the most optimal combinations of the catalyst type, temperature, H_2 pressure and LSHV for achieving high rates of the HDS and HDN of HGO.

The VGO derived from Maya crude was used as the feed for kinetic study conducted by Rodriguez and Ancheyta (227) over the sulfided $NiMo/Al_2O_3$ catalyst. Kinetic parameters for the overall HDS, HDN of non-basic nitrogen, HDN of basic nitrogen and the HDAr were determined using the following set of rate equations:

$$r_{\mathrm{HDS}} = (k_{\mathrm{HDS}}(C_{\mathrm{S}}{}^{\mathrm{S}})C_{\mathrm{H2}}{}^{\mathrm{S}})^{0.45}/(1+K_{\mathrm{H2S}}C_{\mathrm{H2S}}{}^{\mathrm{S}})^2$$

$$r_{\mathrm{HDNnb}} = k_{\mathrm{NB}}(C_{\mathrm{NB}})^{1.5} - k_{\mathrm{B}}(C_{\mathrm{B}})^{1.5}$$

$$r_{\mathrm{HDNB}} = k_{\mathrm{B}}(C_{\mathrm{B}})^{1.5}$$

$$r_{\mathrm{HDAr}} = k\mathrm{p}_{\mathrm{H2}}C_{\mathrm{A}} - k(1-C_{\mathrm{A}})$$

The equations for r_{HDNNB} and r_{HDNB} were adapted from Bej et al. (96), whereas for r_{HDAr} from Yui and Sanford (234). The estimate of the activation energies for r_{HDS}, r_{HDNNB}, r_{HDNB} and r_{HDAr} gave the values of 132, 165, 205 and 121 kJ/mol, respectively. In this case, the reaction temperature varied from 613 to 653 K at the constant H_2 pressure of 5.4 MPa. It was observed that under similar conditions as used by Rodriguez and Ancheyta (227), kinetic parameters could be influenced by the concentration of H_2S in the gaseous phase (125). In this case, the $NiMo/Al_2O_3 \cdot TiO_2$ catalyst was used. The increase in the H_2S concentration above the base case, i.e., from 0 (for base case) to 700 ppv, decreased the rate of HDS, HDN and HDAr and increased the rate of HDM. The same H_2S concentration change increased the apparent activation energy for HDAr and HDN from 31 to 41 and 51 to 76 kJ/mol, respectively, whereas the effect on HDS and HDM was much less pronounced. The decrease in HDS, HDN and HDAr rates suggested that on the addition of H_2S, the H_2S/H_2 ratio was shifted to the inhibition region (76). The beneficial effect of H_2S on HDM may have resulted from its direct reaction with the porphyrin metal entities. The determined kinetic parameters (227) were used for the modeling and/or simulation of the mass concentration dependence on temperature, as well as on the concentrations in liquid and solid phase along the reactor. It was established that the concentrations of sulfur, nitrogen and aromatics were slightly greater in the liquid phase. As one would expect, the profiles of H_2 and H_2S concentrations along the reactor in liquid and solid phases were different in accordance with Figure 6.1 (227), thus confirming that the H_2S/H_2 ratio is an important parameter to be considered in kinetic studies. In fact, it may be one of the main reasons for the difference in kinetic parameters reported by different authors.

6.1.1.3 Lumped kinetics

It has been noted that the HGO derived from Athabasca bitumen has been frequently used as the feed in kinetic studies. The early work on lumped kinetics conducted by Yui and Sanford (235) assumed three lumps, i.e., HGO, LGO and naphtha. The proposed model considered parallel (HGO → naphtha), consecutive (HGO → LGO → naphtha) and combined schemes. The similar HGO was used by Botchwey et al. (236) to study kinetics of the conversion of high boiling fractions to light boiling fractions, as well that of HDS and HDN. In this case, four fractions having boiling range of IBP-573 K (9.6 wt%), 573–673 K (36.1 wt%), 673–773 K (43.1 wt%) and 773–873 K (11.2 wt%) were obtained from the feed. Three ranges of temperature, i.e., a low severity (613–643 K), intermediate severity (643–673 K) and high severity (673–693 K) were investigated using the commercial $NiMo/Al_2O_3$ catalyst at 8.8 MPa in the trickle bed reactor. The kinetic analyses were based on the schemes shown in Figure 6.5 (236). In this case, A is the heaviest and D the lightest fraction. The schemes were developed using the boiling ranges of the products determined by GC. The estimate of the kinetic parameters was based on the approach used by Ancheyta et al. (227,237). A measurable value of k_4 was obtained in the intermediate severity range, whereas k_5 could only be measured in the high severity range. The activation energies estimated from the rate constants in the three regions are shown in Table 6.4 (236). Rather variable effect of

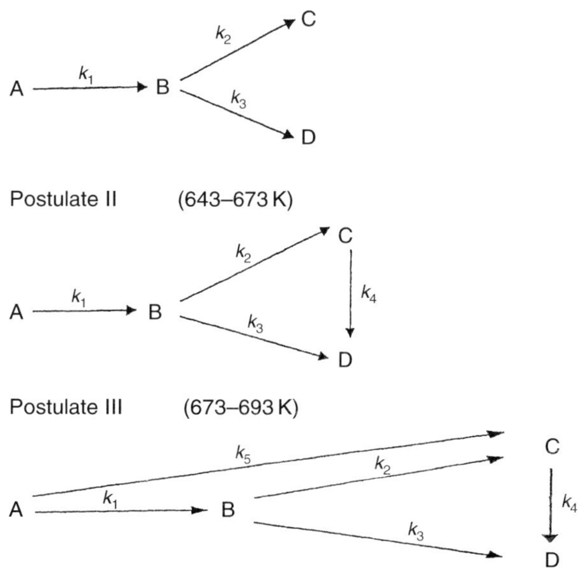

Figure 6.5 Reaction schemes for overall conversion kinetics (236).

Table 6.4 Activation energies (kJ/mol) reaction network in Figure 6.5 (8.8 MPa; LHSV = $1\,h^{-1}$) (236)

	Temperature range (K)		
	613–643	643–673	673–693
*E*1	142	101	107
*E*2	168	88	111
*E*3	37	67	94
*E*4	–	133	122
*E*5	–	–	199

temperature on each of the reactions should be noted. In this case, k_{n1}, k_{n2}, k_{n3} and k_S are the rate constants for removal of the total nitrogen, non-basic nitrogen, basic nitrogen and sulfur, respectively.

Distributions of the products from the hydroprocessing of VGO published by Ali et al. (171) was used by Valavarasu et al. (238) for building the kinetic model which can simulate the performance of commercial units. A discrete lumping approach used in the study assumed four lumps, i.e., gas (0–305 K), naphtha (305–423 K), middle distillates (423–616 K) and VGO feed (616 K+). Middle distillates included kerosene and diesel oil. It was assumed that any lump reacted to give all products boiling below. The kinetic rate constants were determined by minimizing the errors between the predicted and estimated values using the following error function: $\sum(Y_{exp} - Y_{pred})^2$. This function was minimized using the standard optimization procedure. The kinetic model predictions were in line with the operating parameters generally observed in the units used for the HCR of VGOs. The kinetic study using a VGO as the feed conducted over the NiMo/HY-zeolite and $NiMo/SiO_2-Al_2O_3$ catalysts (673–723 K, 0.5–2.0 h^{-1} at 12 MPa) confirmed higher yields of more volatile lumps over the former catalyst suggesting that this catalyst exhibited a higher HCR activity compared with the $NiMo/Al_2O_3$–SiO_2 catalyst (239).

The change in the content of hydrocarbon groups in the feed and corresponding products is a suitable means for studying the kinetics of hydroprocessing. In this regard, the kinetic study conducted by Gray (240) was based on the extensive evaluation of the HGO derived from Athabasca bitumen and its products using ^1H and ^{13}C NMR techniques, as well as the elemental analysis. More than 74% of the feed boiled above 616 K and contained 4.7, 0.32 and 1.35 wt% of S, N and O, respectively. The experiments were conducted in the CSTR system between 653 and 713 K at 13.9 MPa over the commercial $NiMo/Al_2O_3$ catalyst. Of the main interest were the aliphatic and aromatic carbon groups, as well as heteroatoms. The removal of heteroatoms and hydrocarbon groups associated with them followed first-order kinetics. The groups such as C_{AR}-(O,N,S) decreased proportionally with the decrease in the content of heteroatoms confirming that the latter were associated mainly with aromatic structures. The activation energies for the removal of S, (O,N,S), C_{AR}-(O,N,S) and total C_{AR} were 89, 103, 63 and 67 kJ/mol, respectively. Again, similar activation energies confirmed a relation between the removal of C_{AR}-(O,N,S) and C_{AR}. In the recent kinetic study involving a VGO, the products were lumped into paraffinic, naphthenic and aromatic groups (241). Similarly as in the study of Gray (240), the kinetic parameters were determined from the changes in the content of these groups in products compared with the feed.

The discrepancies among kinetic parameters for hydroprocessing of the similar and/or the same HGOs and VGOs reported in different studies should be noted. This can be traced to the difference in experimental conditions such as temperature, H_2 pressure, H_2S/H_2 ratio and the type of catalyst. For example, it is believed that the kinetic data for HGOs and VGOs of different origins will vary even if all these parameters are kept constant. It was noted above, that the time on stream, when kinetic measurements are made is also an important parameter indicating an extent of catalyst deactivation. This parameter is rarely reported in the kinetic studies. Therefore, it is essential that experimental conditions used for kinetic measurements are described in detail. Otherwise, the reasons for different observations made by different authors cannot be identified. As one would expect, the complexity of this issue is further increased for the feeds containing asphaltenes and metals such as ARs, VRs and heavy crude.

6.1.2 Atmospheric residues

Most of the studies on kinetics of hydroprocessing of AR were based on the change in concentration of contaminants (e.g., sulfur, nitrogen metals and asphaltenes) in the feed and corresponding products under variable experimental conditions. Thus, for AR as the feed, only one study involving lumped kinetics could be found in the scientific literature. As one would expect, the concerns and/or complications in determining kinetic data discussed above for VGO and HGO feeds are more evident for AR. This results from the catalyst fouling by coke and metals which for the former feeds was either not present or much less evident.This makes the comparison of kinetic data obtained for VGO and HGO with those for AR even less meaningful.

6.1.2.1 Lumped kinetics

Callejas and Martinez (242) used the kinetic model which was based on the lumping of products obtained during hydroprocessing of an AR. The experimental results were obtained in the CSTR system between 648 and 688 K at 12.5 MPa and the mass space velocities between 1.4 and 7.1 L/h g_{cat}. The products, determined by the simulated distillation, included three fractions, i.e., 343 C+, 343 C− and gases. k_1 and k_2 were the rate constants for simultaneous conversion of the AR to light oils and gases, respectively. The rate constants were related to the concentration changes using the following equations:

$$(C_{10} - C_1/C_{10})(C_2 - C_{20}/C_1) = k_1/\mathrm{WHSV}$$

$$(C_{10} - C_1)C_3/C_{10}C_1 = k_2/\mathrm{WHSV}$$

where C_1, C_2 and C_3 are the concentrations of AR, light oils and gases, respectively, whereas C_{10} and C_{20} are the initial concentrations for AR and light oils, respectively. A reasonable fit of the experimental data with the model could be obtained at 648 and 673 K. However, at 688 K, the fit was affected by the excessive coke deposition and the onset of thermal cracking reactions. Therefore, the reliability of kinetic data obtained above this temperature is limited, unless the contribution of thermal effects to the overall conversion of AR is quantified and subtracted from the overall conversion. This, of course is not an easy task.

6.1.2.2 Overall kinetics

Three commercial catalysts (HDM, HDS and HDS/HDN) were used to study the kinetics of hydroprocessing of the Kuwait AR (~90 ppm of V + Ni and less than 4 wt% asphaltenes) by Marafi et al. (243). The properties of the catalysts are shown in Table 6.5. The work was performed in the up-flow fixed bed reactor by varying LHSV from 0.5 to 4.0 and temperature from 663 to 693 K at 12 MPa of H_2. The apparent reaction order of the HDS, HDAs and HDM reactions as determined using the following rate constants expressions:

$$k = \mathrm{LHSV}\ \ln(C_o/C) \quad \text{for } n = 1$$

$$k = \mathrm{LHSV}\ (n-1)^{-1}\{C^{1-n} - C_o^{1-n}\} \quad \text{for } n > 1$$

Table 6.5 Properties of catalysts (243)

Property	Catalyst		
	A HDM	B HDS	C HDS/HDM
Mo (wt%)	2–3	7–9	9–11
Ni (wt%)		2–3	2–4
P (wt%)			2–4
Surf. area (m^2/g)	150–200	200–250	170–200
Mean pore diam. (Å)	150–200	80–100	80–100

where C_o and C are the concentrations of sulfur, asphaltenes and metals in the feed and corresponding products, respectively, and n is the reaction order. Activation energies were determined using the usual form of the Arrhenius plot expression:

$$k = Ae^{-E/RT}$$

Figures 6.6 and 6.7 show the second-order plot and 1.5-order plot of kinetic data for HDAs and HDM, respectively. Activation energies are shown in Table 6.6 (243).

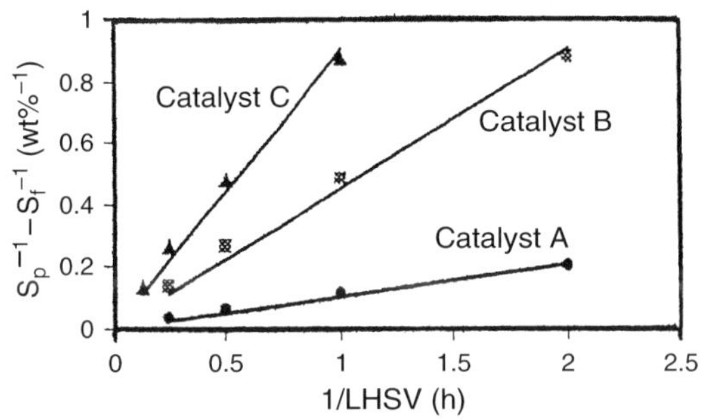

Figure 6.6 Second-order plots for HDAs of Kuwait AR (243).

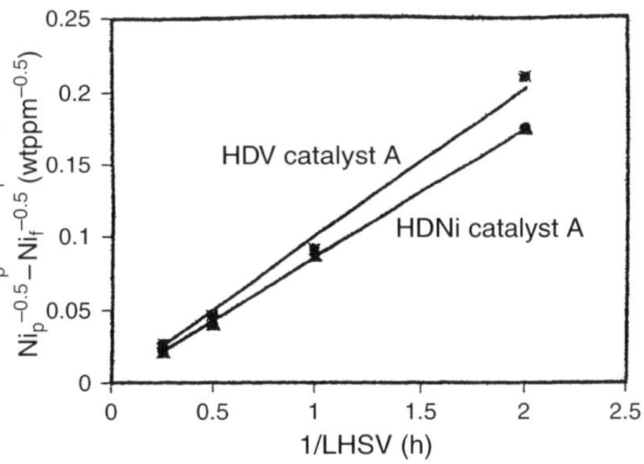

Figure 6.7 The 1.5-order plots for HDV and HDNi of Kuwait AR (243).

Table 6.6 Activation energies (kJ/mol) (243)

	Catalyst		
	A	B	C
HDS	109	107	120
HDN	–	121	126
HDAsp	98	131	63
HDV	116/260[1]	123	98
HDN	73/121[1]	81	106

[1]Activation energy at >673 K

A variability in these values among the catalysts should be noted. Marafi et al. (244) extended their kinetic studies to include two feeds, one containing 33 ppm of V + Ni and 2.5 wt% asphaltenes (DM-AR) and the other 23 ppm of V + Ni and 0.9 wt% asphaltenes (DMDS-AR), both derived from the same Kuwait AR by the two-stage hydroprocessing, i.e., DM-AR was the product from the first stage and the feed for the second stage to obtain DMDS-AR. In this case, catalysts B and C were only studied at 653 K, while the other experimental parameters were the same as in the previous study (243). For catalyst B, except for HDN, all reactions exhibited second-order kinetics. At the same time, HDN followed first-order kinetics when the DM-AR feed was used. The rate constants for the AR and DM-AR feeds for catalysts B and C (Table 6.5) are summarized in Table 6.7. Again, the significant variability among these values, indicating different effects of the catalysts and the feeds, was quite evident. The rate constants for HDNi are consistently greater than those for HDV. This contradicts the results obtained by Bhand and George (245) for the Kuwait AR containing 73 ppm of V + Ni and about 12 wt% of asphaltenes over the HDM catalyst of a proprietary composition. As Figure 6.8 shows, in this case, the rate constants for the removal of V were greater than those for the removal of Ni. This should be expected in view of the greater stability of the Ni porphyrins compared with the V porphyrins. However, the relative rates may change for partially deactivated catalyst. In this regard, the difference in metal contents of the heavy feeds used in the studies of Marafi et al. (243,244) and Bhand and George (245) should be noted as well. Therefore, conditions (e.g., state of catalyst surface) may influence relative values of kinetic parameters. Also the activation energies for HDV were consistently greater than for HDNi suggesting that the temperature increase will have more pronounced effect on

Table 6.7 Rate constants for different feeds and catalysts (243)

Catalyst/feed	HDS	HDV	HDNi	HDAs	HCR	HDN
Catalyst B						
AR	0.711	0.031	0.077	0.155	0.063	0.210
DM-AR	1.420	0.063	0.076	0.763	0.079	0.290
Catalyst C						
AR	1.076	0.027	0.050	0.291	0.095	0.071
DM-AR	2.239	0.048	0.057	1.250	0.075	0.290

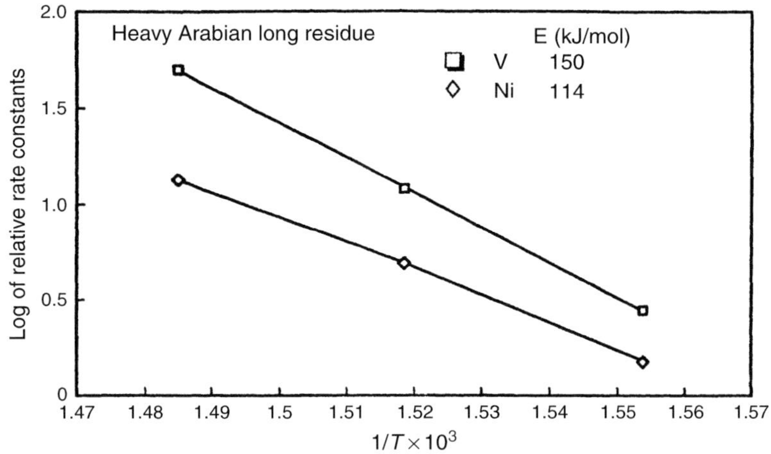

Figure 6.8 Arrhenius plots for HDV and HDNi of long residue (245).

the former. The H_2 pressure at which kinetic data were obtained is important as well, although its significance is sometimes underestimated.

The studies of Marafi et al. (243,244) demonstrated the importance of the properties of feed and catalyst on kinetic parameters. This was confirmed by the values of activation energies and rate constants in Tables 6.6 and 6.7, respectively. Definitely, this factor is much more important for ARs compared with VGOs. This has to be taken into consideration while designing and/or selecting catalysts for graded systems such as the ARDS and HYVAHL processes. Thus, the investigation of the same feed or the same catalyst for testing in different reactors of a graded system may result in erroneous conclusions. Therefore, a catalyst for the first stage must be suitable for hydroprocessing of AR, whereas for the properties of the last stage catalyst may approach those required for hydroprocessing of VGO and HGO.

The Kuwait AR was used in the studies of Chen et al. (161–163) focusing on the effect of phosphate on the activity of conventional catalysts. Kinetic analysis approached that used by Marafi et al. (243,244). For the $CoMo/Al_2O_3$ catalyst modified with P, the best fit of data was obtained for second-order power equation for both HDM and HDS. The activation energies for HDM and HDS were in the range of those shown in Table 6.8 for catalysts in Table 5.6 (139) for an AR derived from Maya crude, i.e., about 121 and 100 kJ/mol, respectively.

Table 6.8 Activation energies (kJ/mol) for catalysts in Table 5.6 (139)

Reaction	Catalyst		
	M-1	M-2	M-3
HDS	73	100	89
HDM	75	98	161
HDAs	77	112	201

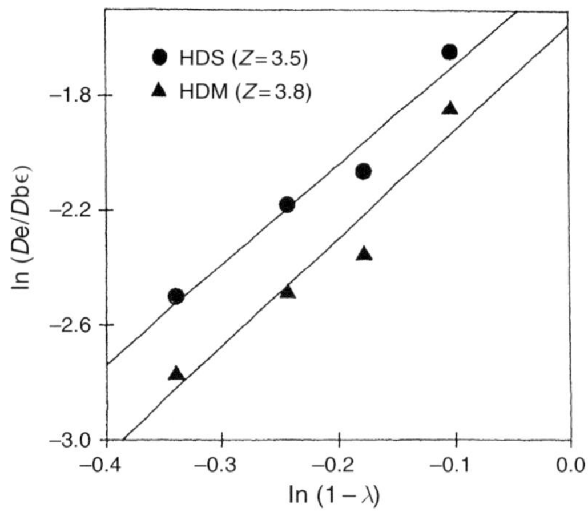

Figure 6.9 Determination of "Z" from experimental data (156).

It was indicated earlier that the $CoMo/Al_2O_3$ catalysts varying widely in porosity could be prepared by doping with P (155). Consequently, the effective diffusivity (D_{eff}) of these catalysts for HDS and HDM reactions varied as well, i.e., from 9.03×10^{-7} to 17.88×10^{-7} and from 6.91×10^{-7} to $14.75 \times 10^{-7} cm^2/s$, respectively. The z parameter in Eqn. (3.3) (Section 3.2.2) was estimated from the experimental results shown in Figure 6.9 (156) as 3.5 and 3.8 for HDS and HDM, respectively.

Some kinetic parameters for HDM and HDAs of the Kafji AR (120 ppm of V + Ni and ~12 wt% of asphaltenes) were estimated by Kobayashi et al. (84,119) over four commercial catalysts. The experiments were conducted in the autoclave at 673 and 10 MPa. The log C_o/C versus time correlation (Figure 6.10) showed the first-order kinetics for HDV, HDNi and HDAsph. It should be, however, noted that only two experimental parameters were used for the estimate. Apparently, for the 100/220 mesh catalyst, the kinetics were predominantly in chemically controlled region. Therefore, much lower values of the rate constants for the 12/14 mesh catalyst confirm the involvement of diffusion phenomena. The estimate of the apparent and intrinsic rate constants allowed the determination of the effective diffusivity for all three reactions, i.e., 7.86×10^{-7}, 7.22×10^{-7} and 18.03×10^{-7} cm^2/s for HDV, HDNi and HDAsph, respectively. The estimate was made for the $CoMo/Al_2O_3$ catalyst having surface area, pore volume and mean pore diameter of 160 m^2/g, 0.7 cm^3/g and 175 Å, respectively at 673 K. The effective diffusivity can be related to molecular diameter (m_d) and pore diameter (p_d) using the modified form of the Eqn. (3.3) in Section 3.2.2 such as:

$$D_{eff} = K(1 - m_d/p_d)^4$$

Using the estimated value of D_{eff}, the molecular diameter of 50 Å and pore diameter of 175 Å, the value of the proportionality constant K was determined to be 40×10^{-7} cm^2/s.

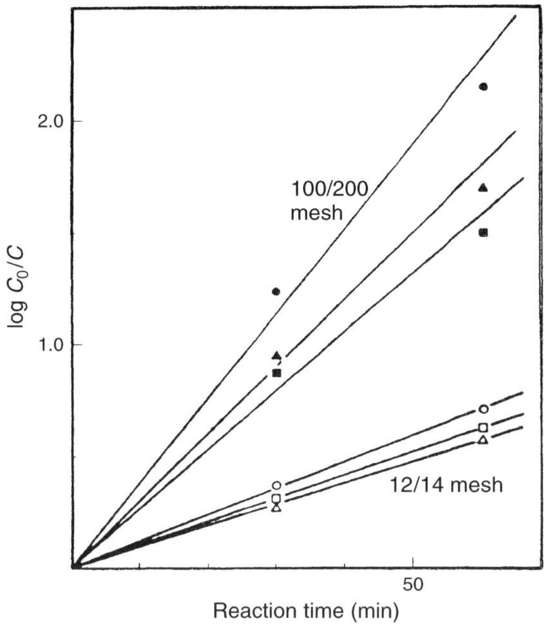

Figure 6.10 First-order plots; (○●) HDV, (△ ▲) HDNi, (□ ■) HDAs (119).

The kinetics of HDV of the Kuwait AR (83 ppm of V + Ni and 3.4 wt% of asphaltenes) studied by Bartholdy and Hannerup (114) assumed the two sequential steps, i.e.,

$$A \xrightarrow{k_1} B \xrightarrow{k_2} C$$

in which the V-containing porphyrins were hydrogenated in the first step and demetallized in the second step. The study was conducted in the fixed bed reactor and it can be used to illustrate the use of kinetic data for the estimation of the effectiveness parameter and Thiele modulus. These authors assumed that for intrinsic rate constants, $k_2 > k_1$. In this case, the ratio of concentration of B to A (C_B/C_A) should approach an equilibrium value p. In equilibrium, $Q_v = \eta = k_{1e}/k_1$, where Q_v is the distribution parameter for V, η the effectiveness parameter and k_1 and k_{1e} the intrinsic and apparent rate constants, respectively. For maximum metal concentration at the particle surface (M profile) the distribution parameter was expressed as:

$$Q_v = p \times C_A/C_B \times k_{1e}/k_1 + (1 - p \times C_A/C_B) \times k_{2e}/k_1$$

The correlations involving the distribution parameter Q_v, as well as the HDV activity and Thiele modulus are shown in Figures 6.11 and 6.12 (114), respectively. Based on the observations made by Long and Gevert (200), the B intermediate in the sequence proposed by Bartholdy and Hannerup (114) may be the tetrahydrogenated porphyrin. The former authors had to incorporate this intermediate rather than dihydrogenated porphyrin in their kinetic scheme in order to obtain a reasonable fit of experimental data. Thus, the hydrogenolysis of the tetrahydrogenated porphyrin was the slowest step in the overall

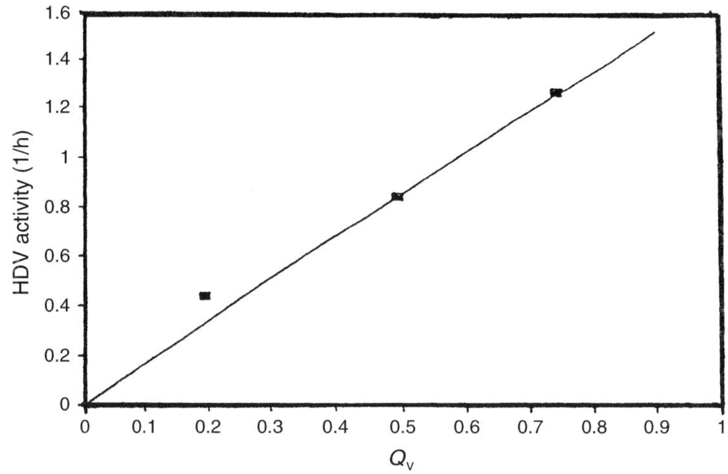

Figure 6.11 Effect of distribution parameter on V removal (114).

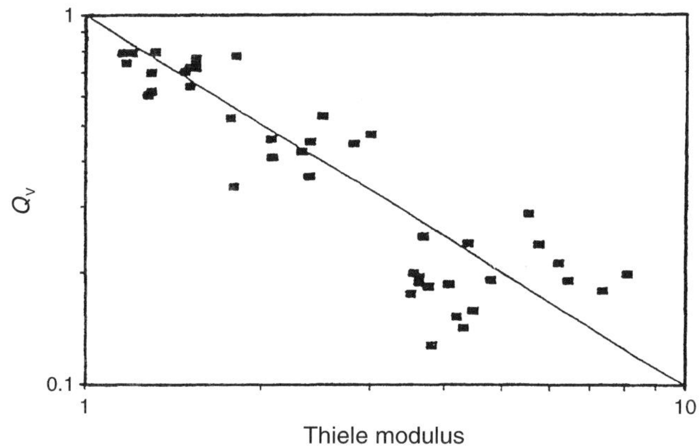

Figure 6.12 Correlation between distribution parameter and Thiele modulus (114).

kinetic scheme. This would suggest that under conditions used by Long and Gevert (246), k_1 in the scheme proposed by Bartholdy and Hannerup (114) may be greater than k_2.

6.1.3 Vacuum residues and heavy crudes

Besides very complex chemical composition, the physical properties such as viscosity and pumpability become important properties when the VRs and heavy crudes are used as hydroprocessing feeds. These problems can be alleviated by blending the heavy feeds with lighter fractions. This, however, raises the issue of compatibility and/or colloidal stability of the system. In other words, it must be ensured that the desirable solubility of asphaltenes in blending fraction is maintained to prevent sediment formation which would interfere with the determination of kinetic parameters. The colloidal stability of the VRs and heavy crudes can also be affected by the products which, as the result of being hydrogenated, may enhance the flocculation of asphaltenes leading to the sediment formation.

It is obvious that a special attention must be paid to the selection of experimental system and conditions for studying kinetics of the hydroprocessing of VRs and topped heavy crudes. Without any doubt, the reliability of kinetic data obtained in the continuous microreactors may be limited, although the situation may be improved by diluting these feeds before the experiments. Similarly, it is believed that microautoclaves may have only a limited application because of the small sample amount of the catalyst and feed allowed by the system. Therefore, CSTR units and bench scale reactors adapted for heavy feeds are recommended systems. Of course, the most reliable kinetic data would be obtained in pilot plant units.

Compared with other heavy feeds, kinetic studies involving VR focus on the HDM, HDAs and CCR conversion. Therefore, such results are relevant to the selection of catalysts for the first-stage reactor of the multistage operation. This is not to say that other hydroprocessing reactions are not important. However, the conversion of asphaltenes and resins to distillates, as well as the removal of metals from VR are the most essential requirements of the upgrading heavy feeds such as VR.

6.1.3.1 Overall kinetics

The kinetics of CCR conversion of the VR derived from Maya crude have been studied in detail by Spanish researchers (247–252). Most of the experiments were conducted in the CSTR system between 648 and 699 K and 10 and 12.5 MPa. The feed contained almost 290 ppm of V + Ni and about 9 wt% of asphaltenes. In these ranges, the CCR conversion varied between 21 and 98%. The following set of kinetic expressions was considered for deriving a pseudo-kinetic rate constant:

$$r = (C_o - C)\,\mathrm{LHSV}$$

$$r = k^* P_{H2}{}^{m} C^{n}$$

$$C_o - C = k^* C^{n} P_{H2}{}^{m}/\mathrm{LHSV}$$

$$k = k^* P_{H2}{}^{m}$$

In this set of equations, r is the rate of CCR conversion, C_o and C the initial and outlet concentrations of CCR, k^* the intrinsic rate constant, k the pseudo-kinetic rate constant, m the order dependence in H_2 pressure, n the kinetic order and P_{H2} the H_2 pressure. The experimental data in Figure 6.13 (250) show that the CCR conversion fit half-order kinetics. At the same time, no H_2 pressure influence on rate constant was observed. This would support the observation that the H_2 pressure had little effect on the total amount of active surface hydrogen, however, the rate of attaining this amount increased with increasing H_2 pressure (63). It is proposed that at the relatively high H_2 pressure (12.5 MPa) used by these authors, this amount of surface hydrogen was approached at every temperature. Although the participation of the active surface hydrogen was of the primary importance, thermal effects may be another important contributor to the CCR conversion, particularly at temperatures approaching 700 K. At about such temperatures, the reliability of kinetic results is usually affected by the contribution of thermal cracking.

In another work, Callejas and Martinez (248,249) prepared a series of the feeds containing different amounts of asphaltenes to study the kinetics of their conversion. For

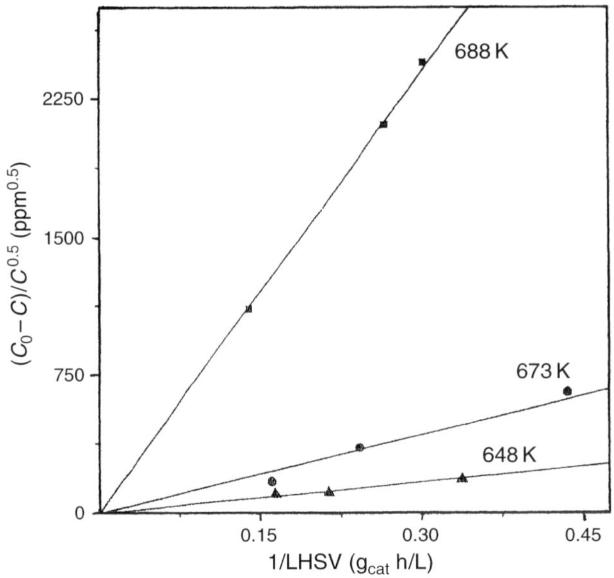

Figure 6.13 Half-order plots for nitrogen removal at 12.5 MPa of H_2 (250).

this purpose, asphaltenes were precipitated from the VR derived from the Maya crude and dissolved in the gas oil derived from the same crude. To study intrinsic kinetics, the commercial NiMo/Al_2O_3 catalyst was crushed to obtain particle size between 50 and 530 μm. The experiments were conducted under similar conditions as those used in the study of Trasobares et al. (247). The data for the removal of asphaltenes fit half-order kinetics giving an activation energy of 173 kJ/mol. The half-order kinetics were also observed for the removal of CCR suggesting that most of the CCR has originated from asphaltenes (247). It is believed that kinetic parameters are dependent on the origin of both asphaltenes and diluent oil. The detailed analyses of asphaltenes in the feeds and products were the basis for determining kinetic parameters for the removal of S, N, V and Ni from asphaltenes (249). In every case, half-order kinetics were observed with the values of activation energies of 149, 164, 189 and 199 kJ/mol, respectively. This suggests that the conversion of asphaltenes to lighter fractions was required before the HDS, HDN and HDM reactions became evident. Apparently, for high asphaltenic feeds, the conversion of asphaltenes to lighter fractions may be the rate-determining step during the overall hydroprocessing. In this regard, the disintegration of asphaltenes aggregates is believed to be an essential requirement to allow hydroprocessing reactions to proceed.

For the overall hydroprocessing of the VR derived from Maya crude, Callejas and Martinez (250) found a second-order, half-order, first-order and half-order kinetics for HDS, HDN, HDNi and HDV, respectively, whereas the activation energies were 288, 180, 347 and 226 kJ/mol, respectively. In contrast to Trasobares et al. (247), these authors found a 0.4 order in H_2 pressure dependence on HDS. At the same time, H_2 pressure had little effect on HDN and HDM. For resins derived from a heavy crude, the kinetic orders with respect to H_2 pressure were 0.5–1.5 for HDM and 0.8 for HDS (251). The VR used in these studies (250,251) was partially hydroprocessed, i.e., metals content decreased from ~250 to ~150 ppm and used to study kinetics under similar conditions by Trasobares et al. (252) and Martinez et al. (253). The CCR conversion followed first-order kinetics compared with half-order for the untreated VR. For the former, the value of activation energy was ~278 kJ/mol (247) compared with ~152 kJ/mol for the treated feed (252). In this case, the Arrhenius expression used by Marafi et al. (243,244) was used for the estimation. For the treated VR, the rate of HDNi and HDV followed second and first-order kinetics and activation energies of ~393 and ~192 kJ/mol, respectively (253). This suggests that under these conditions, the rate of HDNi would gain on importance relative to the rate of HDV with increasing temperature. The order of the H_2 pressure dependence of HDNi and HDV was about 1.4 and 1.8, respectively. The difference between kinetic parameters of the VR and partially upgraded VR indicated the involvement of asphaltenes in catalyst deactivation. Thus, it was shown that kinetic order decreased with the increasing concentration of contaminants in the feed and increasing catalyst deactivation. The latter was continuously changing the surface structure of catalyst (76). Apparently, the results were not obtained for an equilibrium catalyst.

The effect of H_2 pressure on the kinetics of asphaltenes and vanadium removal from the Khafji VR was investigated by Takeuchi et al. (28). Figure 6.14 indicates that the second-order plot and first-order plot gave the best fit of the experimental data for the removal of asphaltenes and vanadium, respectively. The experimental results were obtained in the continuous fixed bed reactor at 678 K. These studies indicate that difference between the kinetic parameters obtained by Callejas et al. (247–250) and those obtained in other studies (28,252,253) can be attributed to different experimental setups (e.g., CSTR system versus continuous bench scale reactor or an autoclave) and conditions, as well as the different properties of the heavy feeds used for investigations.

Maya crude (~350 ppm of V + Ni and 12.4 wt% asphaltenes) was included in the kinetic study of Ancheyta et al. (139–141). The study was conducted in the fixed bed reactor at temperatures from 653 to 713 K and pressure of 6.9 MPa over the catalysts

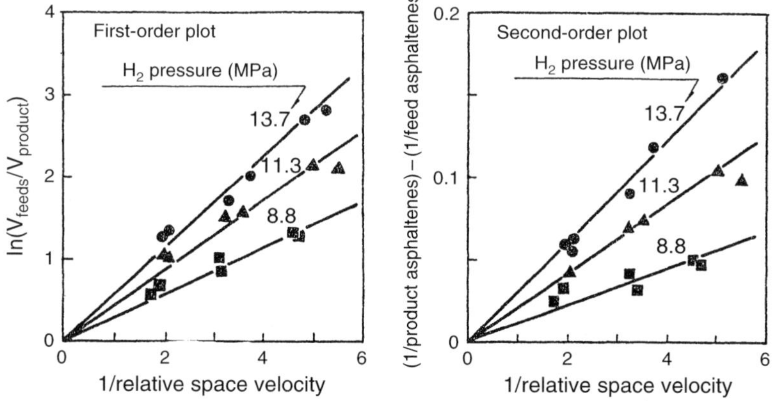

Figure 6.14 Effect of space velocity and H_2 pressure on removal of vanadium and asphaltenes at 678 K (28).

shown in Table 5.6 (139). The best fit of the experimental data was obtained using the following second-order equation:

$$X/(1-X) = kC_o(1/\mathrm{WHSV})$$

where X is the conversion expressed as $X = (C_o - C)/C_o$, C_o and C the concentration of parameters of interest in the feed and products, respectively, and k the apparent rate constant. The data for second-order HDM and HDAs rate equations are shown in Figure 6.15 (139). The activation energies estimated from the usual Arrhenius plots are summarized in Table 6.8. The difference between activation energies for catalyst M-1 and M-3 should be noted. For the former, all values were lower than for catalyst M-3. This was attributed to the influence of intraparticle mass transfer caused by the different particle size of the catalysts. The relatively low H_2 pressure (6.9 MPa) used in these studies should be noted (139–141). Therefore, the effect of the rate of hydrogen activation on the kinetic data may have been quite evident. Moreover, it is believed that the contribution of thermal effects to the overall conversion of asphaltenes was quite significant because of the relatively high temperature (713 K) employed.

Athabasca bitumen used by Nagaishi et al. (254) contained more than 90% of 350+ and more than 50% of 524+ fractions, as well as almost 14 wt% of the microcarbon residue (MCR). This feed was investigated in the system comprising four continuous flow-stirred reactors connected in a series. The experimental conditions included the commercial NiMo/Al_2O_3 catalyst at 703 K and 13.7 MPa of H_2. Each reactor used the same catalyst which was placed in an annular screen basket. With this arrangement, the product from the first reactor was the feed for the second reactor and so on. The experimental results for HDS, HDN, HDM and MCR conversions were fitted using the first-order kinetics expression. The apparent rate constants for HDS and HDN decreased between the first and third reactor and then increased after the fourth reactor. At the

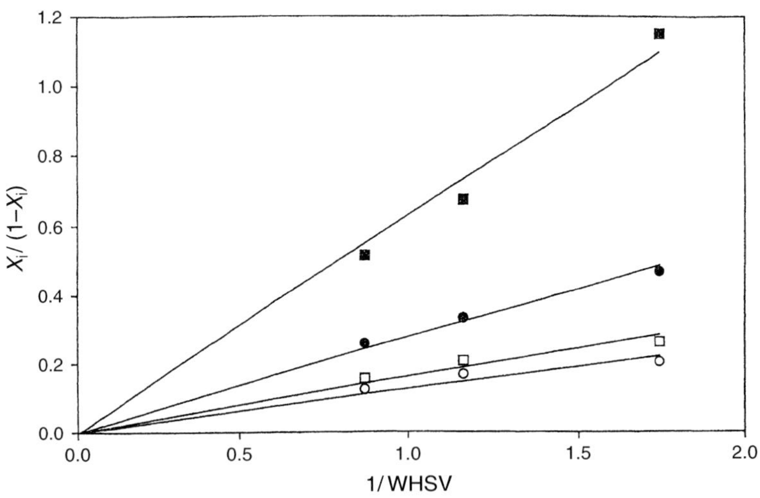

Figure 6.15 Second-order plots for HDAs of Maya crude at 8.9 MPa of H_2; ○) 653 K; □) 673 K; ●) 693 K; ■) 713 K (139).

same time, the apparent rate constant for HDM exhibited a steady increase from the first reactor towards the last reactor, whereas the apparent rate constant for MCR conversion increased significantly after the second reactor. The product analysis indicated that before the fourth reactor, the HDN conversion was rather low. This suggests that the basic N-containing compounds poisoned catalyst surface. This was confirmed by the accumulation of nitrogen in MCR isolated from products, as well as in the coke in the spent catalyst. The low HDN conversions in first to third reactors may also be caused by severe deactivation of HYD sites by V deposits as it was observed by Gualda and Kasztelan (100,255). Thus, rather minor amount of V caused the rapid decline in HYD activity. At the same time, the loss of HDS activity was much less evident. The observations made by Nagaishi et al. (254) have some practical implications. Thus, it is evident that a multireactor system may be optimized by selecting different catalysts for each stage rather than using the same catalyst in all reactors. In addition, the H_2S/H_2 ratio may be an important factor as it was pointed out by Bartholdy and Hannerup (114). From kinetics point of view, the decrease in rate constant for asphaltenes conversion with time on stream observed by Nagaishi et al. (254) deserves attention. It confirmed the importance of time on stream at which the estimate of kinetic parameters was made. To certain degree, this factor may be responsible for the different values of kinetic parameters estimated by different authors under otherwise similar experimental conditions.

6.1.3.2 Lumped kinetics

The distribution of the products from Maya crude (10.5 wt% RBC; 12.4 wt% asphaltenes; ~350 ppm of V + Ni) obtained between 653 and 693 K and 7 MPa was determined by Sanchez et al. (256). The lumps of gases, naphtha (IBP-204 °C), distillates (204–343 °C), VGO (343–538 °C) and the unconverted residue (538 °C+) were quantified by simulated distillation. Similar lumps were used by Mosby et al. (257) and Ayasse et al. (258) to study first-order kinetics of hydroprocessing of the VR derived from Athabasca bitumen. Sanchez et al. (256) used the commercial $NiMo/Al_2O_3$ catalyst (surface area of 175 m^2/g, pore volume of 0.56 cm^3/g and 127Å APD) which was crushed to obtain 100 mesh particles. The conversion was calculated as

$$\frac{(538+_{\text{feed}} - 538+_{\text{products}}) \times 100}{538\ \text{C}+_{\text{feed}}}$$

The following set of kinetic equations was developed on the basis of the scheme shown in Figure 6.16 (256):

Residue:	$r_R = -(k_1 + k_2 + k_3 + k_4)y_R$
VGO:	$r_{VGO} = k_1 y_R - (k_5 + k_6 + k_7)y_{VGO}$
Distillates:	$r_D = k_2 y_R + k_5 y_{VGO} - (k_8 + k_9)y_D$
Naphtha:	$r_N = k_N y_R\ k6 y_{VGO} + k_8 y_D - k_{10} y_N$
Gases:	$r_G = k_4 y_R + k_7 y_{VGO} k_9 y_D + k_{10} y_N$

where y_R, y_{VGO}, y_D, y_N and y_G are lumps of residue, VGO, distillates, naphtha and gases, respectively. Thefollowing mass balance equation was used to evaluate the product lumps for a set of kinetic data:

$$r_i = \frac{dy_i}{d(1/\text{LHSV})}$$

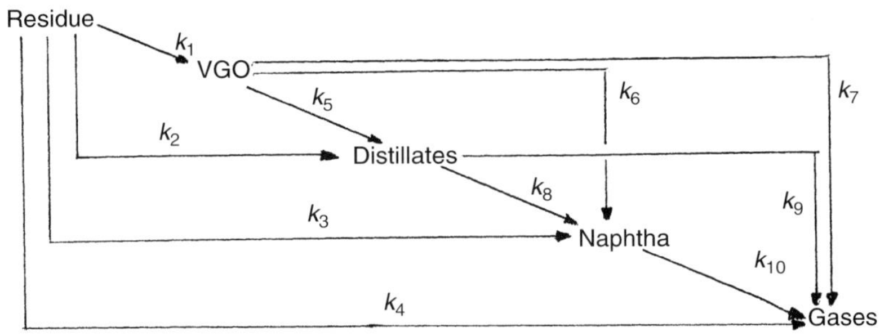

Figure 6.16 Schemes for overall conversion kinetics of residue (256).

The kinetic parameters determined using this model are summarized in Table 6.9 (256). They show that the residue conversion was dominated by reaction 1 involving the formation of VGO, followed by conversion of the latter to distillates. However, the direct conversion of residue to distillates and naphtha was gaining on importance with increasing temperature. Significant increase in the gas production during the residue cracking is supported by more than 20 times increase in the k_4 value between 653 and 673 K. Further temperature increase to 693 K had much less incremental effect. Almost certainly, above 673 K, the direct formation of gaseous products from residue was dominated by thermal reactions. It is believed that the type of catalyst and origin of residue will influence the values of rate constants, although the effect on general trends may be less evident. Moreover, such results can only be obtained in a continuous system enabling the removal of primary products from the reactor. Otherwise, secondary

Table 6.9 Kinetic parameters determined using the scheme in Figure 6.16 (256)

Rate constant ($h^{-1} \times 10^2$)	Temperature (K)			Activation energy (kJ/mol)
	653	673	693	
Residue				
k_1	4.2	14.7	36.7	202
k_2	0.8	2.2	5.7	184
k_3	0.8	2.0	4.3	158
k_4	0.4	9.8	13.7	114
VGO				
k_5	1.8	5.7	10.4	165
k_6	0	0.7	1.6	155
k_7	0	0	0	
Distillates				
k_8	0	0.3	1.0	224
k_9	0	0	0	
Naphtha				
k_{10}	0	0	0	

reactions would play much more important role in the batch reactor. The usefulness of the data generated by Sanchez et al. (256) should be noted. It is, therefore, surprising that so far only few studies on the lumped kinetics of hydroprocessing of VR appeared in the literature.

6.2 MECHANISM OF HYDROPROCESSING REACTIONS

The extensive information on various aspects of the mechanism of hydroprocessing reactions has been published in the literature. Several authoritative reviews were devoted to specific reactions, i.e., HDS (58), HDN (76,259), HDO (260) and HYD (261). Focus has been on both model compounds and real feeds. Usually, the objective of hydroprocessing of the conventional feeds boiling below 350°C has been the removal of heteroatoms and HYD of aromatics to meet specifications of the conventional fuels. The distillate fractions (e.g., naphtha) derived from heavy feeds by carbon-rejecting processes may contain olefins which have to be removed to ensure stability of the final products. Again, the mechanism of reactions occurring during the hydroprocessing of distillate feeds is well documented (61,76,259–261) compared with that for heavy feeds, particularly those containing resins, asphaltenes and metals.

The reactions occurring during hydroprocessing of the feeds boiling below 350°C are common with those for the feeds boiling above 350°C, such as VGO and HGO. However, for the latter, HCR and hydroisomerization reactions may be an important part of the overall mechanism, particularly if the production of middle distillates for transportation fuels and lube base stock is the objective of hydroprocessing. In this case, a high level of dewaxing and HDAr may be necessary to meet specifications of the final products.

For asphaltenes and metals containing feeds, HCR, HDAs and HDM are the most important reactions, as it is documented later in the book. In multistage systems, hydroprocessing will be dominated by different reactions in different stages. The HDM and HDAs are always the main reactions occurring in the first stage. While these reactions may be still important, the conversion of resins may become important in the second stage and stages following after until the overall hydroprocessing is governed by HYD, HDS, HDN and HDO reactions in the final stage. For AR and VR, the reactions occurring during the final stage resemble those occurring during hydroprocessing of DAO and VGO. However, the extent of these reactions in different stages depends also on the origin of heavy feed and the type of catalyst. Therefore, the selection of catalysts for every stage requires attention.

It is generally known that the structural changes of hydrocarbons increase with the increasing acidity of catalysts. This supports involvement of the hydroisomerization and HCR reactions. To certain extent, such reactions proceed via a carbocation mechanism. Because the thermal scission of the C–C bond to form free radicals begins above 600 K, the latter may be formed under typical hydroprocessing conditions. Therefore, both carbocations and free radicals may be part of the overall mechanism of hydroprocessing.

6.2.1 Reactions during hydroprocessing of VGOs and HGOs

The extensive information on hydroprocessing of the AGOs and LCOs forms the basis for elucidation of the mechanism for the heavier feeds such as VGOs and HGOs. For these feeds, the situation is much less complex compared with AR and VR, because only traces of asphaltenes and metals are present. Some differences between the mechanism involving VGOs and HGOs were indicated by the detailed characterization of the feeds and products conducted by Sullivan et al. (262). Thus, HGOs were usually produced under more severe conditions (e.g., at about 800 K) than those applied during vacuum distillation producing VGOs (e.g., ~620 K). This resulted in the presence of the more refractory polyaromatic hydrocarbons and heteroring compounds in the former, although the difference between the AMW of these compounds in VGOs and HGOs may not be so evident. Therefore, to achieve desirable conversion, the HYD reactions followed by HCR reactions are expected to play a more important role during the overall mechanism of hydroprocessing of HGOs than VGOs.

For VGO and HGO, a greater amount of hydrogen and more severe conditions are necessary for achieving a desirable level of hydroprocessing compared with that for the atmospheric distillate feeds. For example, for both VGO and HGO, HYD reactions are important part of the overall hydroprocessing mechanism, particularly if the objective was either the preparation of the feed for FCC or the feed for production of the lube base stock. The large hydrogen consumption was supported by the results published by Jokuty and Gray (263,264) who conducted detailed characterization of the HGO derived from Athabasca bitumen and the corresponding products after hydroprocessing at 9.5–11.0 MPa, 633–673 K over the NiMo/Al_2O_3 catalyst. The product contained about 55% of the fraction boiling above 616 K (343°C). Figure 6.17 (263) shows the reaction network for the HDN of the six-membered N-rings proposed by these authors.

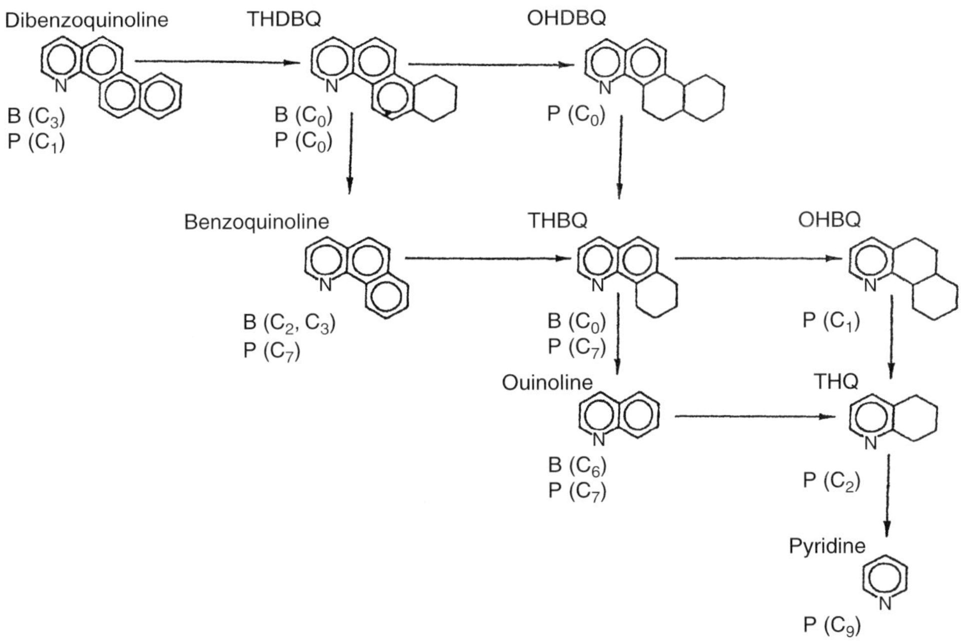

Figure 6.17 Overall network for HDN of dibenzoquinoline; B – bitumen, P – Products, in brackets number of ring substituents (263).

The similar mechanism was also proposed for the five-membered N-heterorings, i.e., benzocarbazoles. The extensive alkyl substitution of the rings should be noted. In order to achieve a desirable HDN conversion, aromatic rings in the benzoquinolines had to be hydrogenated and subsequently hydrocracked until the fully substituted pyridine was obtained as the last N-containing intermediate before the final HDN step. The overall HDN was then dominated by that of the substituted pyridine. This suggests that to achieve high level of the nitrogen removal from such feed, the catalysts possessing high activities for HYD and HCR, as well as for HDN have to be selected.

A similar mechanism as shown in Figure 6.17 (263) may be envisaged for the poly-condensed aromatic structures. For example, the conversion of coronene would begin with the HYD of peripheral aromatic rings followed by HCR of the corresponding naphthenic ring. A stepwise conversion may end up with a highly alkylated aromatic ring with significantly diminished molecular weight, compared with the original reactant.

Sun et al. (265) introduced a concept of the superdelocalizability (S) to study the HYD of aromatic rings, which may be an important reaction in the case that high yields of diesel oil and kerosene are the targets of the hydroprocessing of VGO and HGO. According to this concept, a carbon atom in the ring with the high S value will accept hydrogen more readily. The S values increase in the following order: benzene (0.833 in any position) $\ll$ naphthalene (0.944 in 1-position) $<$ phenanthrene (0.998 in 9-position) $\ll$ anthracene (1.314 in 9-position) $\ll$ 2,3-naphthacene (1.505 in 5-position). This prediction was indeed experimentally confirmed by increasing yield of the HYD products in the same order. Such products are the intermediates in the overall HCR of PAHs. This is consistent with the general observation that high-molecular PAHs are readily converted to lighter fractions providing that the catalyst exhibits an adequate HYD and HCR activities.

The adverse effect of N-compounds on hydroprocessing has been documented (76,259). It is essential that, in the case of VGOs and HGOs, a desirable rate of HDN is maintained in order to achieve high rates of other hydroprocessing reactions. In fact, the ultra-deep HDS using conventional methods may not be achieved unless a substantial part of the N-compounds in the feed was removed. The removal of the most refractory S-compounds such as alkylated DBTs and benzoDBTs is the main objective for achieving the ultra-deep HDS. It has been generally observed that such reactions are poisoned by few ppm of N-compounds present in VGO and HGO (76). The complexity of the HDN mechanism shown in Figure 6.17 (263,264) suggests that for achieving a deep HDS of the VGO and HGO, two-stage hydroprocessing may be necessary. In this case, the first-stage catalyst would have to exhibit a high activity for HYD, HCR and HDN, whereas the second stage could employ a catalyst possessing high HDS activity.

A specific case is the hydroprocessing of VGOs and HGOs to produce feed for the subsequent FCC to obtain transportation fuels. Another case involves hydroprocessing before catalytic dewaxing to produce a lube base stock for production of lubricants. Otherwise, NH_3 and H_2S would poison dewaxing catalysts usually containing noble metals on acidic supports. It is generally known that N-bases have adverse effects on FCC catalysts as well. In addition, the conversion of large aromatic molecules to light fractions during FCC may be significantly enhanced by HYD of the former to naphthenic structures which are subsequently cracked to lighter fractions.

Although to much less extent, the reactions such as HDO, HDAs and HDM also occur during hydroprocessing of VGO and HGO. It is generally known that the content of oxygen increases with increasing boiling range of petroleum fractions. Therefore, for VGO and HGOs, HDO involving mostly alkylated DBFs may also be a part of the overall mechanism (260). The overall hydrogen consumption during the HDO of such heterorings would be somewhere between the HDS of alkylated DBTs and the HDN of alkylated carbazoles and acridines. Few ppm of metals and asphaltenes, which may still be present in VGO and HGO should be removed in order to achieve a desirable performance of the FCC catalysts unless the process was modified to suit RFCC feeds. Therefore, even during the preparation of the FCC feeds from VGO/HGO, HDM and HDAs may be part of the overall mechanism.

The overall mechanism of hydroprocessing of VGO and HGO may be influenced by temperature, as it was observed by Khorasheh et al. (266). In their mechanism, studied over the commercial $NiMo/Al_2O_3$ catalyst, both catalytic and thermal effects were considered. The latter became dominant above 690 K. Below 690 K, the catalytic HCR reactions gained on importance compared with thermal cracking. Above 690 K, HDN was more affected than HDS. This was attributed to the limited availability of hydrogen. Thus, HDN reactions had to compete for hydrogen with the unstable intermediates produced during thermal cracking.

It is believed that for VGO and HGO feeds, HDN is the most important reaction when mutual effects of hydroprocessing reactions are taken into consideration. It is fair to assume that a desirable rate of hydroprocessing reactions cannot be achieved without ensuring a high rate of HDN. The extensive review of HDN of petroleum, addressing all aspects of HDN was published recently (76). It is believed that readers may benefit from the summary of major conclusions of this review with respect to the mechanism of hydroprocessing reactions, particularly that of HDN. Figure 6.17 (263) shows a gradual conversion of benzoquinoline before a highly alkylated piperidine is formed as the last intermediate before HDN, as the final step in the overall conversion is completed. Another possible route (not shown in Figure 6.17) includes the HYD of N-heteroring followed by the opening of the heteroring giving an amine product. The HYD of the alkyl- substituted N-ring is slower than that of the unsubstituted, particularly when alkyl substituents are on carbons adjacent to nitrogen. The HYD of N-heteroring is equilibrium controlled. Thus, with increasing temperature at the same H_2 pressure, the equilibrium will be shifted towards the unhydrogenated reactant. However, for high-molecular weight N-compounds (e.g., alkylated benzoacridines and benzocarbazoles), the HYD of an aromatic ring adjacent to N-heteroring is necessary for opening the latter. Therefore, the product distribution will be dominated by fully or partially hydrogenated products, i.e., in the case of carbazole, phenylcyclohexyl and bicyclohexyl will be in significant excess compared with biphenyl.

The network in Figure 6.18 (76) has been frequently used to study the kinetics of the HDN of quinoline (Q). This rather simplified scheme has been used to identify the reaction stages before the final HDN reaction. Figure 6.18 may be used to illustrate significant complexity of the HDN mechanism in spite of the relatively simple reactant used. In this case, Q can be simultaneously hydrogenated to 1,2,3,4-tetrahydroQ (THQ) and 5,6,7,8-tetrahydroQ (THBzQ). Both intermediates may be converted to decahydroQ (DHQ). The N-rings in THQ and DHQ can be opened to give *o*-propyl aniline (OPA), as well as propylcyclohexane amine (PCHA) and propylcyclohexene amine (PCHEA), respectively. The amine intermediates, not shown in the kinetic network in Figure 6.18, are part of the overall HDN mechanism. These intermediates are finely converted to

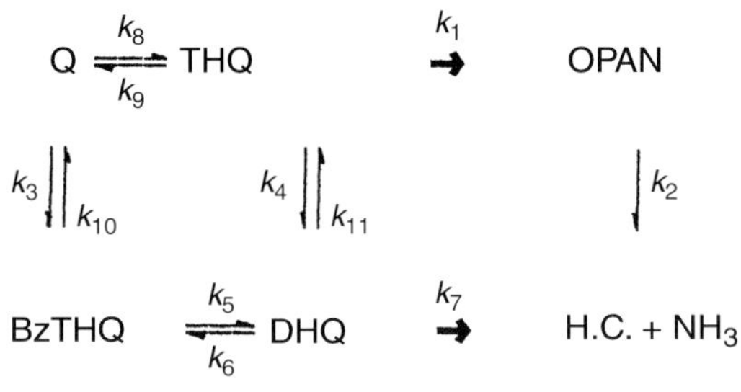

Figure 6.18 Simplified network for HDN of Q.

hydrocarbons. The OPA undergoes HDN either directly giving propyl benzene (PBz) or via HYD to intermediates such as PCHA and PCHEA which are converted to hydrocarbons such as PCH and PCHE. The complexity of the overall HDN mechanism will be further increased for the alkylated Q, as well as for the N-compounds having a greater molecular weight than Q. Moreover, every step in the network shown in Figure 6.18 is affected by self-inhibition by Q and N-intermediates. This suggests that without simplification of the network, kinetics of the overall HDN of Q could not be studied.

Contrary to HDN of carbazoles, the HDS of its sulfur analog such as DBTs does not require the HYD of the adjacent aromatic ring to remove sulfur. Therefore, during the HDS of DBTs, biphenyl can be produced in sizeable yields. However, under H_2 pressures, typically used during hydroprocessing, the HYD of aromatic rings in DBT occurs yielding phenylcyclohexyl or even bicyclohexyl as final products. In this regard, it should be emphasized that slowing down catalyst deactivation and increasing the rate of HDN may be the primary reason for maintaining a high H_2 pressure during hydroprocessing of VGO and HGO. Nevertheless, it should be again stressed that the hydrogen consumption to achieve HDS is lower than that of HDN of the N-containing analogs, whereas that of HDO of similar structures (e.g., dibenzofuran) is between HDS and HDN.

A cursory account of the reactions occurring during hydroprocessing of VGOs and HGOs indicates complexity of the overall mechanism. Yet, resins, asphaltenes and metal porphyrins were not even considered. These components are always part of the overall mechanism of hydroprocessing of DAO, AR and VR. For such feeds, the conversion of resins, asphaltenes and porphyrins have to be included in the mechanism in addition to the reactions occurring during hydroprocessing of VGO and HGO.

6.2.2 Conversion of resins

The temperature applied during vacuum distillation ($\sim 630\,K$) suggests that more volatile portion of resins may end up in VGO. Then, during the hydroprocessing of VGO, both oil and resins components of the colloidal system may be involved. Depending on the type of solvent used for deasphalting, resins may play a key role during hydroprocessing of DAO. Also, they are an important part of the overall mechanism during hydroprocessing of AR and VR.

It was indicated earlier that molecular weight of resins may be smaller than that of asphaltenes by several thousand units. Table 2.5 (33) shows that resins are more polar and more aromatic than oil but less polar and less aromatic than asphaltenes. Then, the hydroprocessing reactivity and severity required for their conversion should be between those required for oil and asphaltenes. It is sometimes suggested that resins are intermediate products of the asphaltenes conversion to oil fractions. This may be depicted using the model in Figure 6.19 (21). Thus, after dissociation of the dimer, the structure and molecular weight of the fragments produced by subsequent cracking in sites 1 and 3 would approach that of resins. The molecular weight of the resins may be greater than that of the reactant molecule shown in Figure 6.17 (263,264). In fact, such molecules may represent a fragment of the resin molecule. Before their formation, such fragments could have been held together by the aliphatic and naphthenic carbon-containing entities, as well as by heteroatoms. Another resin-like structure, or at least a precursor to the formation of one, is shown in Figure 6.20 (274) obtained over catalyst C.

From the hydroprocessing point of view, the content of metals (V + Ni) in resins deserves attention. Thus, Tables 2.4 (32) and 2.5 (33) may represent the upper limits of metals in resins derived from among the heaviest available crudes. It is believed that most of these metals are in the form of porhyrins providing that the amount of asphaltenes in the resin fraction is small. Figure 2.3 shows that the content of metals in resins can be influenced by the choice of solvent used for their separation from heavy crude. For example, a solvent can be selected, to produce DAO containing little of metals, although this can only be achieved at the expense of the decreased yield of DAO. An extreme case represents the DAO containing ~230 ppm of metals used by Reyes et al. (110). In such a case, HDM would be an important reaction occurring during the overall conversion of resins to oil fractions. It is believed that in such DAO, a large portion of porphyrins is either associated with resins or is much less occluded than in the AR and VR containing similar amounts of metals. This suggests that for such DAO, much greater rate of HDM compared with the residues can be anticipated. In this regard, the detailed account of the HDM mechanism is given below.

It is believed that after most of the asphaltenes were converted, the resulting solution of resins in oil should be rather homogeneous. This would ensure a free motion of the resins molecules, particularly under typical hydroprocessing conditions. Under

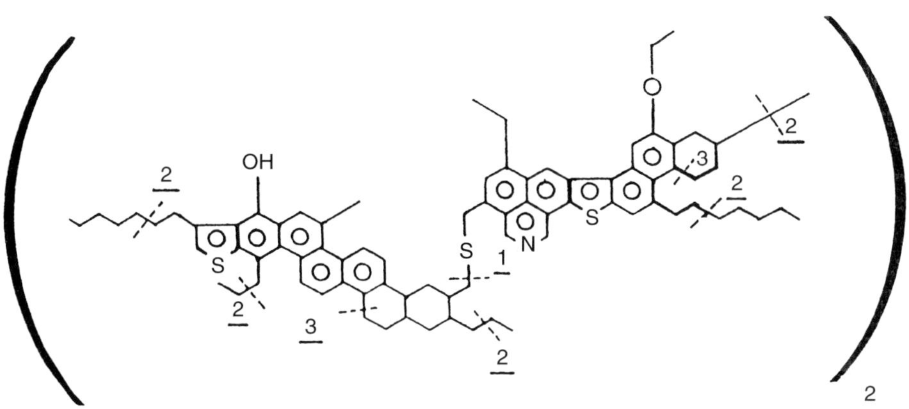

Figure 6.19 Reaction sites for hydrocracking of asphaltenes (21).

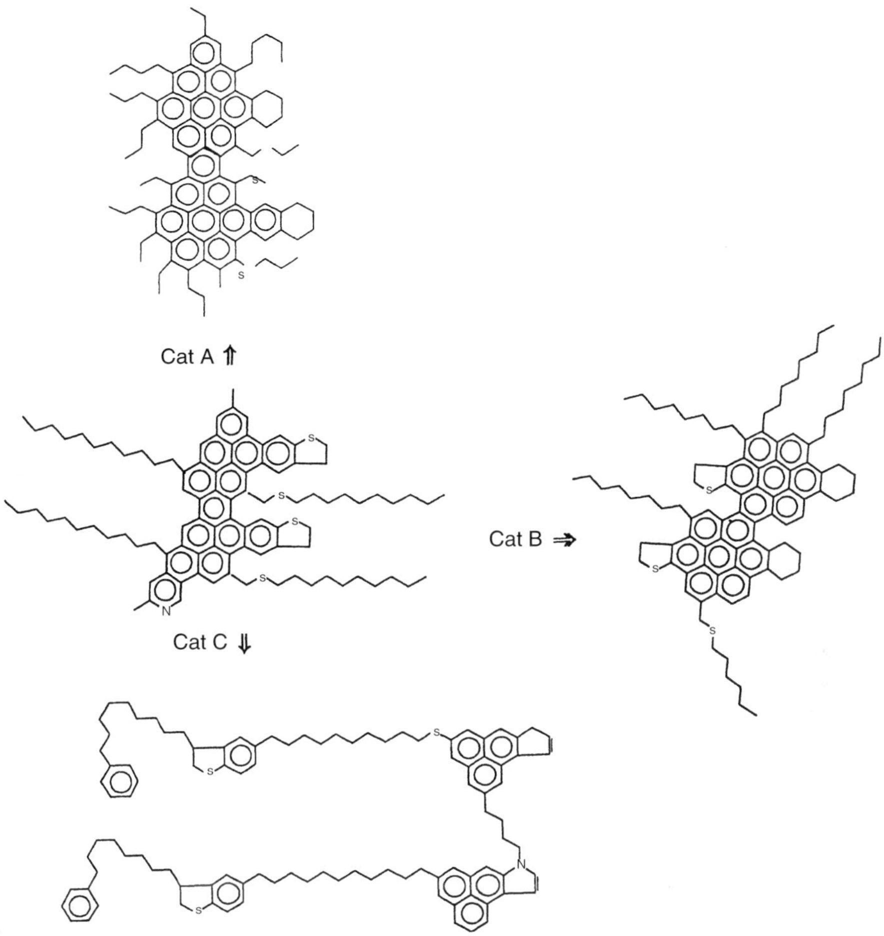

Figure 6.20 Effect of catalyst type on structure of converted asphaltenes (274).

such conditions, active hydrogen required for the conversion of resins can be facilitated more readily than that for asphaltenes, particularly before the aggregates of latter were disintegrated. The structure of resins suggests that HCR of the aliphatic chains and naphthenic rings shall play an important role during the overall conversion to oil fractions. Moreover, the HYD of aromatic rings to naphthenic rings followed by HCR and hydroisomerization of the latter may occur, similarly as it was observed during the conversion of asphaltenes to resins. This is depicted in Figure 6.20 by the conversion of asphaltenes molecule to a resin-like molecule over catalyst C (274). This catalyst had higher HYD activity than that of catalysts A and B. Moreover, it was supported on an acidic support ensuring a higher HCR activity. Therefore, these facts have to be taken into consideration while selecting catalyst for hydroprocessing of high resins and asphaltenes containing feeds.

A "back" reaction of resins to asphaltenes cannot be ruled out. This may occur in later stages on stream, when catalyst lost a large portion of its original activity. Under such conditions, recombination of resinous molecules originally present in heavy feed to asphaltenes-like species may be favored because of a limited active hydrogen supply. An ultimate result of such situation would be a gradual increase in the content of asphaltenes in products with time on stream, particularly when the end of run is being approached.

6.2.3 Conversion of asphaltenes

Because of only small amounts (less than 1%) present in VGOs and HGOs, the involvement of asphaltenes in the overall hydroprocessing mechanism begins with DAO. It was shown earlier that the conditions employed during deasphalting determine the content of asphaltenes in DAO (Figure 2.3). The involvement of asphaltenes in the overall mechanism further increases toward ARs, VRs and topped heavy crudes. Both thermal and catalytic effects play a role during the overall conversion. A sufficient active surface hydrogen is desirable to achieve a high conversion of asphaltenes to lighter fractions and to minimize coke formation. As the consequence of the latter, catalyst deactivation may be slowed down and the life of catalyst extended. These aspects of the asphaltenes conversion are discussed in details later in the book in the Chapter 7 on catalyst deactivation by coke.

6.2.3.1 Thermal effects

A high rate of HDAs is necessary for achieving a high conversion of the heavy asphaltenic species to distillates. Depending on the origin of heavy feed, the molecular weight of some heavy compounds may approach 10 000 (32). Therefore, for heavy feeds, HCR reactions accompanied by depolymerization of the large asphaltenic entities are an important part of the overall mechanism. These reactions, requiring the presence of a catalyst, occur simultaneously with thermal reactions. The relative contribution of catalytic and thermal reactions to asphaltenes conversion, i.e., HCR and thermal cracking, depends on temperature. It was shown that in the absence of catalyst under otherwise typical hydroprocessing conditions, i.e., 633–703 K; 9–18 MPa of H_2, the overall mechanism of asphaltenes conversion started with the disintegration of micelles which resulted in a decrease of the number of unit sheets (267). The V- and Ni-containing porphyrins reacted simultaneously with this step. Depolymerization of the asphaltenes molecules was accompanied by the removal of some sulfur. Most likely, this involved S bridges rather than S-containing heterorings. The $C_{AL}-C_{AL}$ bonds, particularly those in methylene bridges are potential thermal cracking sites as well. This is shown in the mechanism in Figure 6.19. For this purpose, the model of the Athabasca bitumen asphaltenes developed by Suzuki et al. (21) was used. In this model, the potential cracking sites are identified by numbers. It is believed that the weakest site 1 can cleave in the liquid phase even during warming up of the heavy feed. Sites 2 involve $C_{AL}-C_{AL}$ bonds mostly in methylene bridges. Some of these bonds began to cleave at about 600 K without requiring catalyst. Sites 3 are naphthenic rings present in the original feed and/or were produced during the HYD of aromatic rings. The strength of the $C_{AL}-C_{AL}$ bonds in naphthenic rings is greater than that in methylene bridges. Cracking of such bonds may require temperatures which are employed during FCC, i.e., ~800 K. Therefore, after being produced via the HYD of aromatic rings, most of the naphthenic rings should survive hydroprocessing providing that the catalyst comprised a non-acidic support. A little contribution to the overall thermal cracking reactions is expected from the cleavage of $C_{AR}-C_{AL}$ bonds, whereas under typical hydroprocessing conditions the cracking of $C_{AR}-C_{AR}$ bonds may be excluded from the overall mechanism.

Among the important thermal reactions which the asphaltenes undergo is the cracking of long alkyl chains attached to aromatic carbon. Consequently, aromaticity of the remaining asphaltene molecules will increase with time on stream. During this period, little change in the structure of aromatic skeleton is expected. The reactivity of this part of asphaltenes depends on the ratio of the internal aromatic carbon to peripheral aromatic carbon (Ci/Cp). It was difficult to convert asphaltenes having the Ci/Cp ratio greater than three, whereas the ratio of less than two indicated a high reactivity of asphaltenes (268,269). Also the difficulty of converting asphaltenes to lighter fractions increased with increase in their polarity (270). The increased aromaticity of asphaltenes increased the probability for their coagulation and subsequent precipitation from the liquid phase. In industrial units, a situation may be encountered in which the aromaticity of liquid phase decreased because of HYD and that of asphaltenes increased with time on stream. This may enhance coagulation of asphaltenes in spite of their small amounts present. This suggests that with respect to the sediment formation, the structure of asphaltenes may be more important than their amount present in the processing streams. In other words, for heavy feeds containing similar amount of asphaltenes, more sediment and/or coke on catalyst will be formed from the heavy feed containing more polar and more aromatic asphaltenes. The relative importance and mutual effects of hyroprocessing reactions during the overall mechanism will be different as well.

The evolution of the structure of asphaltenes with time on stream was investigated by Ancheyta et al. (271) using Maya crude over the $NiMo/Al_2O_3$ catalyst at 7 MPa in the fixed bed reactor. Initially (at 673 K), the amount of asphaltenes in products abruptly decreased and then gradually increased with time on stream. This was attributed to the conversion of resins to asphaltenic molecules. However, catalyst deactivation affecting the asphaltenes conversion could be another contributor to the increase in the asphaltenes content in products with time on stream. Simultaneously, dealkylation of asphaltenes increased their aromaticity and polarity, while their molecular weight decreased. After 300 h on stream, temperature was increased to 693 K to compensate for catalyst deactivation. From now on, the amount of asphaltenes in liquid streams and their aromaticity continued to increase with time on stream for another 400 h when temperature was increased from 693 to 703 K. This decreased the amount of asphaltenes in products, most likely due to the enhanced thermal decomposition. The removal of V and Ni followed the same trends as that of asphaltenes suggesting that the conversion of asphaltenes was the necessary step before the removal of metals from the feed could take place.

6.2.3.2 Involvement of active hydrogen

The primary products of thermal cracking reactions are radicals. In the case of asphaltenes, most of such radicals are produced in the liquid phase. Because of the low stability, part of the free radicals would recombine to larger molecules unless they are stabilized by gaseous H_2 dissolved in the feed and active hydrogen on the catalyst surface. Thus, it is unlikely that the gas phase H_2 can be a significant contributor to the radical stabilization. The availability of active hydrogen is, therefore, essential for achieving high conversion of asphaltenes without the excessive coke formation. In the absence of catalyst, the consumption of hydrogen by asphaltenic molecules was negligible (272). However, the increased yield of lighter fractions in the presence of H_2

compared with thermal cracking (inert atmosphere) confirmed hydrogen participation in spite of the fact that its consumption was very small.

Another source of the hydrogen required for radical stabilization may be supplied by the hydrogen-donating compounds present in heavy feeds. It is well established that the naphthenic hydrogen can readily participate in various hydroprocessing reactions. With respect to the free radical stabilization, the naphthenic hydrogen can be consumed via abstraction reactions involving radicals. It is believed that for the heavy feeds derived from naphthenic crudes, hydrogen-donating reactions involving naphthenic structures are part of the overall hydroprocessing mechanism.

It was shown earlier (63) that in the presence of a sulfided catalyst (e.g., MoS_2) molecular H_2 is activated to form surface entities such as -SH and MeH (Me = Mo, W, Co and Ni). The reactivity of the surface hydrogen is significantly greater than that of the gaseous H_2 and the H_2 dissolved in the feed. Consequently, the surface hydrogen may be an additional source necessary for stabilizing free radicals formed by decomposition of asphaltenic molecules via the following tentative reactions, e.g.,

$$R^* + \text{-MeH or -SH} = R^*H + (\text{-Me or -S})$$

where R^* is a free radical. Otherwise, the radicals would recombine and/or polymerize to high molecular weight species, e.g.,

$$R^* + R^* = R - R$$

It is believed that with respect to the asphaltenes conversion, an important role of the surface hydrogen may be the stabilization of free radicals. Thus, it is unlikely that the HYD of large asphaltenic molecules can occur before they are converted to lighter fragments such as resins. This was indeed confirmed by Steer et al. (273) who used stable isotopes to find out where the gas phase H_2 ends up during hydroprocessing of the four different VRs. They observed that the hydrogen addition to light products was extensive compared with the unconverted residues. In fact, some residues lost hydrogen. The experiments were conducted in the CSTR system at 703 K and ~14 MPa over the sulfided $NiMo/Al_2O_3$ catalyst. The temperature employed in these experiments suggests that both the catalytic HCR and thermal cracking were contributing to the overall asphaltenes conversion.

Sanford (272) reported that during hydroprocessing of the Athabasca VR, the HYD activity of the $NiMo/Al_2O_3$ catalyst was lost within first few hours on stream, whereas catalyst exhibited a good activity for asphaltenes conversion. This suggests that the active surface hydrogen required for the stabilization of radicals produced by cracking was still available, although the abstraction of paraffinic and naphthenic hydrogens by radicals was also a contributor. The relatively high content of naphthenic structures in Athabasca bitumen compared with other heavy feeds has been noted (1). It is suggested that the catalyst surface in pores, which was not deposited by coke and metals, was an important source of active hydrogen. Thus, the deposits predominantly on the catalyst exterior near the entrance to the pores could not prevent the H_2 diffusion into pores for activation. Once activated, surface hydrogen could migrate to the exterior and as such take part in the radicals stabilization reactions. The ability of surface hydrogen to migrate on the catalyst surface, i.e., from the active phase to support and vice versa, has been documented in the scientific literature (63). It is evident that both HCR and thermal cracking contribute to the conversion of asphaltenes with the importance of the latter increasing with increasing temperature. Therefore, the presence of the radical mechanism during the conversion of asphaltenes to lighter fractions discussed above may be a part of the overall mechanism of asphaltenes conversion.

The overwhelming information suggests that the structural changes of asphaltenes increased with the increasing acidity of catalyst. This was clearly demonstrated in Figure 6.20 (274). Thus, the most extensive change in the structure of asphaltenes was observed over the catalyst C which was the most acidic. This supports the participation of Bronsted acid sites present on the catalyst surface. Such sites can donate proton required for HCR and hydroisomerization reactions and represent another form of the surface hydrogen. Above 670 K, the SH entities on the catalyst surface have the proton-donating ability as well (63). However, because of a suitable acidic sites distribution, supports such as zeolites and amorphous silica–alumina have been used extensively. In this case, molecules possessing a double bond (olefins and aromatics) can readily accept proton and as such form a carbocation. Otherwise, a carbocation could be formed by donating a hydride H^- from hydrocarbon to a Lewis site on the catalyst surface. Therefore, the following tentative reactions may occur on catalyst surface:

$$\text{RH} + \text{Bronsted site} = \text{RH}_2^{\ +} + \text{neutral site}$$

$$\text{RH} + \text{Lewis site} = \text{R}^+ + \text{Lewis site-H}$$

The presence of a double bond and/or an aromatic ring in R would increase the probability of such reactions to occur. The complexity of the asphaltenes structure indicates that rather stable carbocations could be formed. In the case of aliphatic carbon, the tertiary C_{AL} represents a more favorable site for the formation of carbocation than the secondary C_{AL}. The C_{AL} attached to C_{AR} is also a potential site for carbocation formation. In this case, carbocation can be stabilized by resonance of the aromatic ring. It is generally known that carbocations can undergo facile intramolecular migration involving hydride, alkyl or even aryl groups. As the result of this, considerable skeletal isomerizations can be observed. Most likely, a carbocation was precursor to the formation of the product from asphaltenes over catalyst C shown in Figure 6.20. It is believed that similar structural rearrangement could not be possible over the catalysts comprising conventional supports.

The unwanted reactions such as polymerization, condensation and cyclization of carbocation occur as well. Carbocations have to be converted to light fractions to prevent their polymerization. These reactions are considered as the first stage during the formation of coke over catalysts supported on acidic supports. It has been generally observed that for the same feed and the same processing conditions, the amount of coke deposited on the surface of catalyst increased with increasing acidity and cracking activity of the catalyst. It may, therefore, be concluded that for the hydroprocessing catalysts supported on acidic supports, at least three different forms of the surface hydrogen are involved in the overall mechanism of the asphaltenes conversion, i.e., -SH form from which hydrogen can be consumed by either radicals abstraction or as proton, hydride (Me-H) hydrogen and the protonic form provided by acidic supports. To be more precise, the acidic sites distribution suggests the presence of protons varying in protonic strength. As it was indicated earlier, the hydrocarbons with a hydrogen-donating ability usually present in heavy feeds represent another source of hydrogen in addition to several forms of the active surface hydrogen.

It is speculated that for high metals content feeds, the interaction between carbocations and porphyrins may be a part of the overall mechanism. The latter have a tendency to donate electrons from the delocalized π-system of the porphyrin ring. This would result in the neutralization of carbocations followed by an intramolecular rearrangement of the electron acceptor.

It should be noted that the involvement of carbocations in the overall mechanism of hydroprocessing is not limited to asphaltenes. Thus, a carbocation may be formed in every situation in which a proton can be readily supplied from the catalyst surface and reactive molecules (e.g., possessing a double bond) are present in vicinity. Therefore, carbocations may be a part of the mechanism of conversion of resins, as well as of the oil components. Chapter 11 on dewaxing catalysts gives a more detailed account of the typical catalysts used for hydroisomerization. The mechanism of such reactions may always involve the participation of carbocations. This indicates a significant complexity in the mechanism of hydroprocessing even for heavy feeds containing little metals and asphaltenes (e.g., VGO, HGO and DAO).

6.2.3.3 Structural transformations

The determination of structural parameters of asphaltenes in heavy feeds and the corresponding products provides the basis for elucidation of the chemical changes incurred by asphaltenes molecules. Such a study was undertaken by Ali et al. (274). In this case, asphaltenes were fractionated by GPC to various fractions for the determination of aromaticity and molecular weight. The aromaticity of the fractions increased with the increasing molecular weight. After hydroprocessing, the large molecular weight fractions of asphaltenes were found in the unconverted products. It is believed that this portion of asphaltenes was the main contributor to the sediment and/or coke formation. The NMR evaluation revealed that this portion consisted of clusters of four to five condensed aromatic rings held together with aliphatic chains. It is suggested that carboids, as the least reactive part of micelles, located at their center could be the origin of the asphaltenes found in the products.

The removal of sulfur from the S-containing heterorings requires the presence of catalyst (275). Most likely, the HDS of asphaltenes will be delayed compared with the HDS of lighter fractions of the corresponding heavy feed. In the former case, the access of active surface hydrogen to the S-heteroring may be affected because of the steric effects caused by the large size of asphaltenic molecules. One report suggests that at least 50% of the sulfur was removed from heavy feed before the HDS of asphaltenes was noticeable (276). In the other study on hydroprocessing of VR at 685 K, at 70% of the total sulfur removal from VR, the sulfur removal from aphaltenes was 60% (274). At the same time, the removal of nitrogen from asphaltenes was negligible. In fact, in some cases, an increase in the nitrogen content of asphaltenes during hydroprocessing was observed (143,248).

Apparently, the conversion of asphaltenes to lighter fractions may be necessary before the overall HDS of VR became evident. This was supported by the results published by Callejas and Martinez (242) showing the increase in the HDS rate with the increasing conversion of asphaltenes. The HDS leveled off above 50% conversion of asphaltenes. In this case, the AR derived from Maya crude was used between 648 and 688 K at

12.5 MPa in the continuous fixed bed reactor. The asphaltenes conversion increased with increasing temperature. However, the asphaltenes still remaining in the products after high conversions were more refractory than those present in the feed. This resulted from the removal of aliphatic chains attached to aromatic rings followed by recombination of the latter. Simultaneously with these reactions, the recombination of highly aromatic asphaltenic species into larger coke-like molecules could be the initial stage of coke formation on the catalyst surface.

A slight decrease in the content of sulfur and nitrogen in asphaltenes still remaining in the products was observed in the wide range of experimental conditions (e.g., 653–693 K, 7–10 MPa and 0.33–1.5 h^{-1} LHSV) over the NiMo/Al_2O_3 catalyst using Maya crude (277). For example, decreasing the LHSV from 1.5 to 0.33 h^{-1} (increasing contact time) at 673 K and 10 MPa, resulted in the decrease in the content of sulfur and nitrogen from 3.45 to 3.33 wt% and 1.47 to 1.34 wt%, respectively, between the feed and corresponding products. Similar changes were observed by the increasing pressure from 7 to 10 MPa and temperature from 653 to 693 K. In every case, the content of metals in asphaltenes remaining in the products increased with increasing severity. This is the confirmation of the decrease in the reactivity of asphaltenes in the feed with increasing severity. These observations can be partly attributed to the reactions involving the removal of aliphatic carbon, such as dealkylation. Moreover, the contribution of carboids, as the least reactive part of micelle, to the amount of asphaltenes still remaining in the products cannot be ruled out.

Perhaps the most detailed study on the chemistry of asphaltenes conversion during hydroprocessing of heavy feeds was published by Hauser et al. (278). In this case, the objective was to simulate the performance of catalysts during hydroprocessing of the Kuwait AR in the ARDS process consisting of the four trickle bed reactors in series. Thus, the AR was the feed for the first reactor, whereas the product from this reactor (DM-AR) was the feed for the second reactor. The product from the second reactor (DMDS-AR) was the feed for the third reactor. The asphaltenes content in AR, DM-AR and DMDS-AR was 3.8, 2.5 and 0.9 wt%, respectively, whereas the amount of CCR was 12.2, 8.5 and 5.0 wt%, respectively. The asphaltenes separated from every feed and product were subjected to detailed characterizations which included the elemental analysis, NMR to determine aromaticity, AMW and the extent of substitution in the aromatic rings, as well as the XRD. The aim was the elucidation of changes in the structure of cluster of asphaltenes occurring during hydroprocessing. Catalysts A, B and C used for the study were typical HDM (unpromoted Mo/Al_2O_3), HDS and HDS/HDN catalysts, respectively. Catalysts B and C were both of the NiMo/Al_2O_3 formulation, however, the latter contained P and had a higher content of Ni and Mo. Because of the presence of P in catalyst C, its acidity was greater than that in catalyst B. Figure 6.20 (274) shows the AMW structure of asphaltenes in AR and DM-AR. All three catalysts were compared during hydroprocessing of AR to obtain DM-AR. The catalyst structure had a pronounced effect on the conversion of asphaltenes. A more extensive cracking and isomerization was observed over the catalyst C possessing a higher acidity than catalyst B. On the other hand, catalyst A had a low HDS and HYD activity. This was clearly reflected on the average structure of asphaltenes in the DM-AR product obtained over catalyst A. Thus, this structure contained more polycondensed aromatic rings than the original asphaltenes. The dealkylation of the asphaltenes structure obtained over catalyst A may lead to an increase in their aromaticity. Such reactions may be responsible for an increase in aromaticity of asphaltenes still remaining in the products compared with the feed. As it was indicated above, molecular structure of the asphaltenes in the DM-AR obtained over catalyst C (Figure 6.20) approached that of resins. This may suggest that the conversion of asphaltenes proceeded gradually via resin-like species which are subsequently converted to oil and distillates. For catalyst C, significant structural transformations indicate the participation of carbocation during the overall conversion of asphaltene molecule. The change of asphaltenes to resins resulted in the decrease in AMW by few hundreds units before the molecules usually present in distillates were obtained. In the case of the product obtained over catalyst C, such conversion may involve the HCR of methylene bridges which connect aromatic entities.

The study conducted by Seki et al. (279) complemented the observations made by Ali et al. (274) and Hauser et al. (278). In the former case, the Arabian light AR (3.4 wt% asphaltenes) was hydroprocessed in two stages over the Mo/Al_2O_3 (4.2 wt% MoO_3) in the first stage, while over the $Ni/CoMo/Al_2O_3$ (0.7, 1.2 and 10.5 wt% of NiO, CoO and MoO_3, respectively) in the second stage. Asphaltenes defined as the hexane insolubles and toluene solubles were separated from the products after the first and second stages. For runs carried out at 683 K, the amount of asphaltenes after the first and second stages decreased from 3.4 to 0.8 and 0.4, respectively. For every step, Seki et al. (279) proposed the structure of asphaltenes after the determination of AMW and molecular weight distribution of asphaltenes, characterization using NMR techniques, etc. These structures indicated that temperature was the most important parameter for influencing the mechanism of asphaltenes conversion.

6.2.4 Hydrodemetallization

In this case, HDM refers to the removal of metals such as V and Ni from the corresponding porphyrins. Other metals and inorganic solids such as clays, alkali metals salts, etc., can be removed either by physical filtration or dewatering/desalting of heavy feeds. The former involves placing layer of a guard material upstream the first bed of catalyst of the multistage system. Alkali metal salts are usually dissolved in water and present in heavy crudes in the form of finely divided emulsions. The dewatering/desalting processes employing deemulgators can remove most of such emulsions.

The presence of arsenic in some heavy feeds is often overlooked, although it requires attention because of its adverse effects. During hydroprocessing, part of the arsenic is converted to arsine (AsH_3), while the other part deposits on catalyst surface and as such coordinates with the oxygen-containing groups which are present on the catalyst support. Arsine is among the most toxic species listed by health authorities.

The results in Tables 2.1 and 2.3 show that heavy feed such as VGO only contains trace amounts of metals. Moreover, the metal-containing species are highly solubilized in the oil phase. This ensures that during hydroprocessing, they can be removed quite easily. Therefore, the HDM of VGO should not be influenced by asphaltenes depolymerization. At the same time, as Table 2.1 shows, an extensive hydroprocessing may be required to remove metals from the DAO obtained by *n*-pentane deasphalting of the heavy Zuata crude, whereas for the DAO obtained from a sweet crude using the same solvent, the content of metals would be much lower. For AR and VR, HDM may be one of the most important reactions of the overall hydroprocessing mechanism, particularly in the case of AR and VR derived from heavy crudes. For these feeds, a much lower level of solubilization of asphaltenes is expected compared with AR and VR derived from conventional crudes. This suggests that the accessibility of active hydrogen to the metal-containing molecules may be an additional factor to be considered in the mechanism of HDM.

As it has been indicated above, the HDM of heavy feeds occurs simultaneously with HDAs. In fact, a high level of HDAs may be necessary so that this portion of metals which is "hidden" and/or occluded in the micellar entities is released and becomes available for HDM reactions. This was confirmed by Asaoka et al. (280) who observed that the portion of metals, which was not associated with asphaltenes could be readily removed via HDM. A higher HDM reactivity of metals which were associated with resins than those which were associated with asphaltenes is another confirmation of the need to separate porphyrin molecules from the micelles in order to increase the rate of HDM reactions (32). In fact, Rankel and Rollmann (281) concluded that the depolymerization of asphaltenes rather than their intrinsic activity was the rate determining step during the overall HDM. The requirement of the depolymerization is expected to increase from the heavy feeds derived from sweet crudes towards those derived from heavy crudes.

Figure 6.21 is the experimental confirmation of the importance of HDAs on HDM (127). Thus, the removal of V in excess of 30% increased proportionally with the conversion of asphaltenes to lighter fractions. It is believed that below 30%, most of V was removed from resins and oil fractions. The difficulties in removing metals directly from asphaltenes may be attributed to the limited accessibility of porphyrin structures to active sites which supply active hydrogen. Thus, it is unlikely that porphyrins, as part of the asphaltenes aggregate, can undergo an activated adsorption on catalyst surface because of steric effects. Moreover, the solubility of H_2 in asphaltenes phase may be

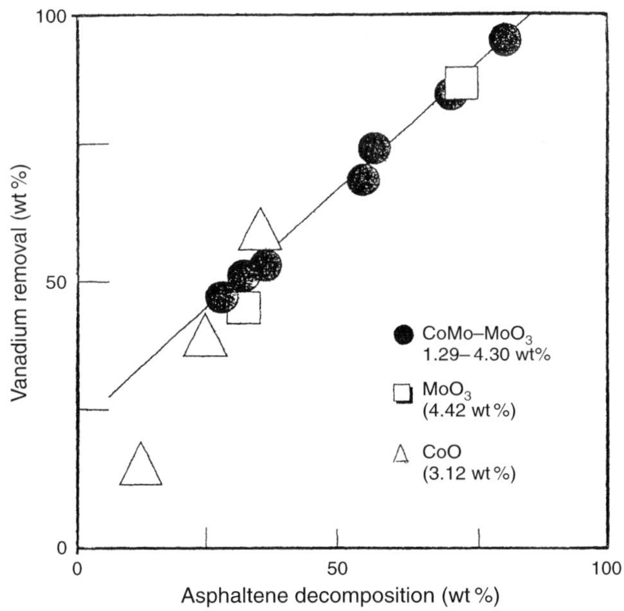

Figure 6.21 Effect of asphaltenes decomposition on vanadium removal (127).

affected by diffusion. This will diminish the availability of H_2 for direct reaction with porphyrins (32). For similar reasons, the direct reaction of porphyrins, still hidden in the asphaltenes entities, with H_2S will be affected as well.

A linear correlation between the metal removal and asphaltenes conversion (Maya crude) was also established by Trejo et al. (277). This study was conducted over the $NiMo/Al_2O_3$ catalyst under varying conditions, i.e., 7–10 MPa, 653–693 K and 0.33–1.5 h^{-1} LHSV. In this case, both the metal removal and asphaltenes conversion increased with increasing severity. However, the content of metals in asphaltenes isolated from the products increased with increasing severity as well. At the same time, the content of asphaltenes in products decreased. The temperature increase from 653 to 693 K at 10 MPa resulted in the increase in the amount of Ni in asphaltenes from 311 to 433 ppm and that of V from 1637 to 1698 ppm. Similar trends were established for the VR derived from a Venezuelan crude (282). This confirmed the importance of asphaltenes conversion before HDM reactions can occur. Thus, it became more difficult to remove metals, after the reactivity of asphaltenes decreased. Then, it may be concluded that the rate of HDAs is the rate limiting step for HDM. This conclusion may be more applicable to the high asphaltenes and metals feed (e.g., VR) compared with the low asphaltenes and metals feed (e.g., DAO). For the latter, a higher dispersion and/or solubilization of asphaltenes in oil phase is more favorable for an efficient transfer of active hydrogen to porphyrin molecules.

It was indicated earlier that the metal porphyrins (M-P; M refers to $V = O$ or Ni) account for most of the metals in heavy feeds, although the conclusions reached during the earlier studies (44,46,49) indicated the presence of non-porphyrinic structures. Thus, the size exclusion chromatography technique used by Reynolds et al. (46) gave a bimodal distribution of V-compounds in the heavy Californian AR. The maximum at the small molecular size was attributed to the porphyrin-like structures, whereas the large molecular size to the non-porphyrinic species. During hydroprocessing, the latter were converted to the small molecular size species (e.g., porphyrins). This was indicated by their increased concentration in the products compared with the feed. It seems more plausible to attribute this observation to the depolymerization of asphaltenes. Consequently, porphyrins were released to the oil phase where their molecular weight could be determined without being obscured by asphaltenes. Indeed, predominance of the porphyrinic structure has been indicated with the introduction of new methods for characterization of heavy feeds (50). Therefore, the studies on HDM, which consider porphyrins as the predominant metal-containing reactants, should be considered as more convincing for elucidation of the overall mechanism of the HDM of heavy feeds.

The information in scientific literature confirmed that porphyrins have been the primary focus of the studies on mechanism of HDM (40). Because of the significant complexity, studies on the HDM mechanism have been dominated by the model porphyrin compounds, whereas for real heavy feeds, the overall removal of metals has been receiving most of the attention. Therefore, the inclusion of the former studies in the present discussion is necessary to obtain a more clear picture on the HDM mechanism. This contrasts with the mechanism of HDAs, in which case only a hypothetical structure of the asphaltenes molecule could be used (274). For HDM, more effort has been devoted to the V-containing porphyrins than to Ni-containing porphyrins. This may be attributed to a much more detrimental effect of the deposited V on the catalyst activity than that of Ni. This results from the preferential accumulation of the former on the inlet of pores thus preventing diffusion or reactant molecules into catalyst interior, as well as from the much higher content of V than Ni in most of the heavy feeds. There are some reports on the autocatalytic behavior of V and Ni deposits. However, this cannot offset detrimental effects such as the restricted diffusion caused by the deposits which are becoming more evident with time on stream (40). It was reported that the V = O entity can interact both with the uncovered support and with the catalytically active metals deposited on the support (283–285). Several potential forms of the interaction of porphyrin molecules with catalyst surface before the HDM reaction were identified (286). The chemical aspects of the metal deposits formation and their interaction with catalyst surface are discussed later in the context with catalyst deactivation by metals.

Earlier studies on HDM mechanism carried out by Wei et al. (287–290) revealed that the first step of the mechanism involved the HYD of one pyrrole ring (in β position) in porphyrin yielding the chlorin intermediate. The formation of chlorin from porphyrin is the reversible reaction which may proceed without catalyst. The conversion to chlorin increased with the increasing H_2 pressure and increasing temperature, as it is shown in Figure 6.22 (53). This seems to be an unexpected observation, i.e., generally observed trends involve the shift of the HYD equilibrium to the left with increasing temperature at the same H_2 pressure. This may be reconciled by assuming a higher aromaticity of chlorin than that of porphyrin. Indeed, hydrogens in methine groups of the chlorin are more acidic (more exchangeable) than that in the pyrrole ring. This may be the reason for higher aromaticity of the former. In the subsequent step, chlorin is converted to another intermediate via HYD of the second pyrrole ring (287,288). These intermediates, i.e., M-PH2 and M-PH4 are stable enough and can be isolated for structural analysis. However, M-PH2 and M-PH4 only accounted for about 20% of the metals. An additional effort by Agrawal and Wei (289), as well as Ware and Wei (290) revealed the presence of a rather complex intermediate (M-PX) formed via HYD of M-PH4. In this intermediate, all pyrrole rings were at least partially hydrogenated. In the mechanism proposed by Wei et al. (287–290) the final products of HDM arose from the fragmentation of the intermediate M-PX. In an attempt to elucidate the HDM mechanism, Reynolds et al. (46) used three different ARs rather than the model porphyrin compounds used by Wei et al. (287–290). The former authors proposed a similar HDM mechanism after extensive evaluations of the heavy feeds and hydroprocessing products using SEC-ICP and UV-V is spectroscopy.

Advancements in the development of analytical methods allowed more detailed characterization of the HDM intermediates. This was evident in the studies on HDM of porphyrin published by Janssen et al. (285,291). They observed the conversion of the tetraphenylporphyrin (M-TPP) to chlorin (M-TPC) followed by conversion of the latter to tetraphenyltetrahydroporphyrin in agreement with Wei et al. (287–290). However, in this step, the formation of the isobacteriochlorin (M-TPiB) was also assumed. The tetraphenylhexahydroporphyrin (M-TPHP) was then formed in the next step. The octahydrogenated species (M-B) arising from M-TPHP was not confirmed but was assumed to be present. It was proposed that this intermediate reacted either via ring cleavage and metal removal or tolyl elimination giving the M-Bil intermediate. All steps discussed by Janssen et al. (291) are part of the mechanism shown in Figure 6.23.

The catalyst composition may influence the HDM mechanism. This was shown in the study published by Garcia-Lopez et al. (292) who compared the NiMo/Al_2O_3

catalyst with the NiMo/Al_2O_3–TiO_2 catalyst during the HDM of Ni 5,10,15,20-tetraphenylporphyrin (Ni-TPP) dissolved in nujol. The experiments were carried out in the batch reactor at 598 K and 8.3 MPa of the total H_2 pressure. The HDM proceeded sequentially via the reversible HYD of Ni-TPP to Ni 5,10,15,20-tetraphenylchlorin followed by the HYD of the latter to Ni 5,10,15,20-tetraphenylisobacteriochlorin, which then reacted via a series of fast reactions ending in demetallation and ring fragmentation. This mechanism was in general agreement with other studies (285,291), however, the rate determining step for the overall HDM was the first HYD of Ni-TPP rather than hydrogenolysis step. The different observation may be attributed to the different acidity of the γ-Al_2O_3–TiO_2 support compared with γ-Al_2O_3.

With respect to the HDM mechanism, the effect of catalyst acidity has received little attention. There are no experimental studies suggesting the formation of porphyrin cations via addition of proton to porphyrin, although in view of the complexity of the structure, this possibility cannot be ruled out completely. Of course, the addition of proton to porphyrin would be more favorable over acidic catalysts. At the same time, the delocalized π-system of porphyrin can donate electron to either Bronsted sites or Lewis sites on the catalyst surface (293,294). Theoretical study of Hirao et al. (295) provides an additional support for the existence of porphyrin cations. However, it is unlikely that electronic effects play any significant role in the overall HDM mechanism of high asphaltenes and metals containing feed because of the additional factors involved compared with the model feeds comprising porphyrin compounds dissolved in a pure hydrocarbon (e.g., hexadecane).

The hydrocarbon medium in which porphyrins are occurring may be a parameter influencing the mechanism of HDM of porphyrins. In this regard, the hydrogen-donating ability of the medium was observed to be the most important parameter (296). Figure 6.24 shows the UV-Vis spectra of an Ni-TP and its intermediates (NiTPH4 and NiTPH2) in a medium A and B possessing a low hydrogen-donating ability and a high hydrogen-donating ability, respectively. The appearance of the intermediates complemented by disappearance of NiTP was more evident in the medium B. The results were obtained over the sulfided CoMo/Al_2O_3 catalyst. This suggests that the rate of HDM of heavy feeds derived from naphthenic crudes, possessing a high hydrogen-donating ability should be higher than that of the heavy feeds derived from aromatic crudes. Therefore, the origin of crudes from which heavy feeds were obtained should be considered as another parameter when analyzing differences in the proposed mechanisms of HDM. For example, a high

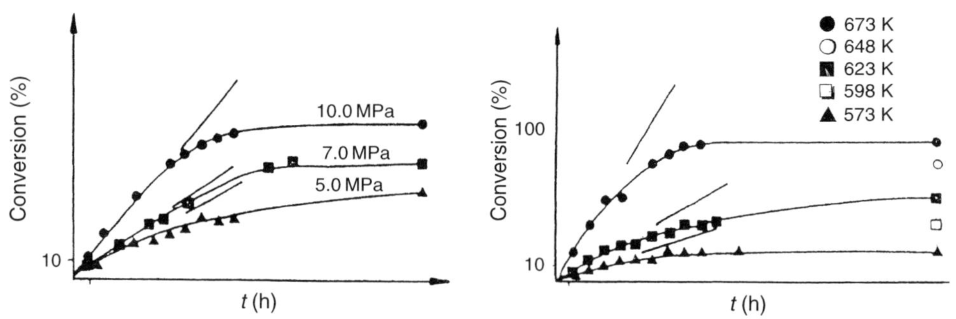

Figure 6.22 Effect of H_2 pressure and temperature on hydrogenation of VO-porphyrin (53).

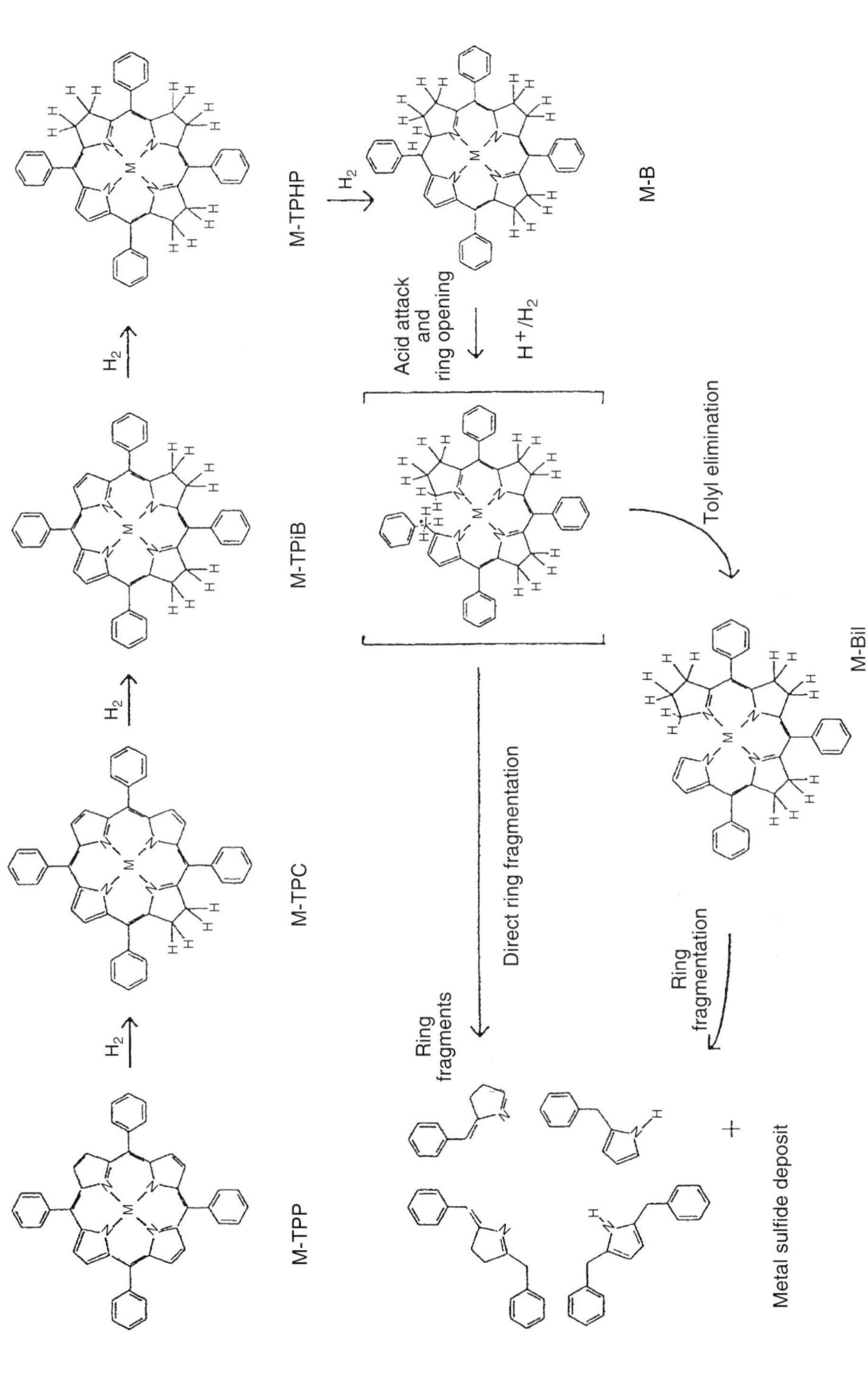

Figure 6.23 A complete reaction mechanism for HDM of metallo-porphyrins; M = Ni or VO (291).

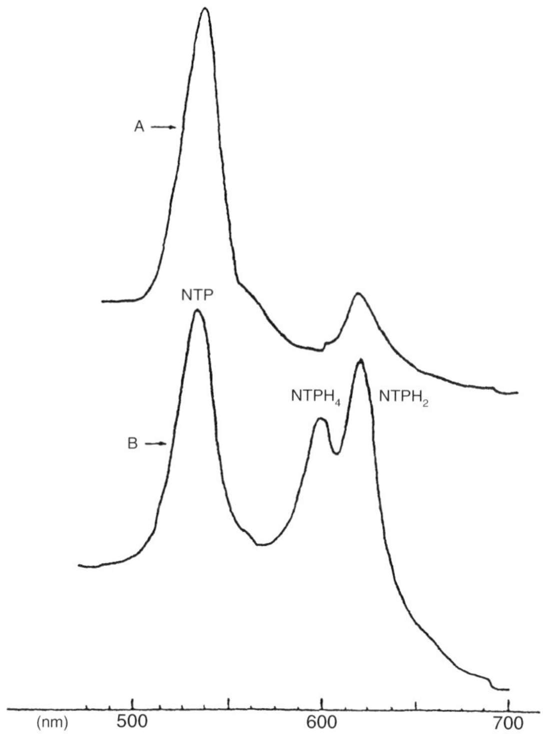

Figure 6.24 UV-VIS spectra of hydrogenation of Ni-porphyrin (A) in mesitylene and (B) in hydrogenated-CDL (296).

content of naphthenic structures in Athabasca bitumen suggests that they may play an important role during the overall mechanism of HDM during hydroprocessing (1).

The studies of Long et al. (297,298) deserve attention, although the focus was on the model compounds, i.e., VO-EP and Ni-EP (E refers to ethyl), rather than on a real heavy feed. During very early stages, mechanism of the metal removal proposed by these authors considered three possible routes such as the thermal decomposition, adsorption on the uncovered γ-Al_2O_3 support and the HYD-hydrogenolysis of the porphyrins. Thermal decomposition became evident above 620 K. It was observed that during very early stages, the removal of metals was associated with their adsorption on the part of γ-Al_2O_3 which was not covered with active metals. In a sulfide form, these metals exhibited some HDM activity, though much lower than that of the active metals. Similar observation was made in the study of Janssen et al (291). In the presence of Fe, the formation of the mixed sulfide such as $(Fe,V)S_4$ which were active for HDM was observed by Embaid et al. (299). In the latter stages, the HYD-hydrogenolysis mechanism dominated overall HDM once the bare γ-Al_2O_3 surface was covered by deposited metals. Contrary to these studies (294,295), Morales and Galiasso (300) attributed the entire HDM activity of the CoMo/Al_2O_3 catalyst to CoMo phase. Thus, little HDM conversion was observed in the presence of deposited metals before the active metals were added to the γ-Al_2O_3 support. These contradictory observations may be attributed to a significantly different H_2 pressure, i.e., 11 and 4 MPa, respectively, used in these studies.

To a small extent, a non-catalytic HDM may also take place in parallel with catalytic HDM (46,54). This depends on the experimental parameters such as temperature, H_2 pressure and, particularly, H_2S concentration. The mechanism of these effects is not well defined, although one may envisage a weakening of the N−Me bonds of porphyrins caused by the interaction with H_2 and H_2S. Moreover, in the case of H_2S, a driving force for these reactions are the stable products such as metal sulfides of Ni and V, which may be produced as the result of such interactions.

Chapter 7

CATALYST DEACTIVATION

The frequent reference to catalyst deactivation is being made in all preceding chapters. This is rather obvious when the type of the feeds evaluated in this book is taken into consideration. In this case, deactivation was evidenced by decline in the rate of hydroprocessing reactions with time on stream (TOS). The primary objective of the development and testing of catalysts was to minimize this activity decline, which could be quantified by kinetic measurements. The phenomena occurring on catalyst surface, which were related to deactivation, were receiving less attention. The aim of this chapter is to describe catalyst deactivation in terms of the modifying effect of components of heavy feeds as well as that of operating parameters on catalyst surface leading to the loss of activity. Among the latter, temperature, H_2 pressure and H_2S/H_2 ratio are the parameters which have pronounced effects on the structure of catalyst surface in relation to deactivation. The properties of catalysts influence deactivation patterns as well.

The preceding chapters clearly indicated a significant difference between the catalyst deactivation during hydroprocessing of heavy feeds and that of light feeds. For the former, continuous attempts have been made to extend the catalyst life by improving properties of the conventional catalysts and/or by developing new catalyst formulations. This requires understanding of the catalyst deactivation phenomena. Part of the activity may be lost because of the change in catalyst structure due to the prolonged exposure to hydroprocessing temperatures. The adverse effect may be further enhanced by the H_2 pressure, provided a sufficient concentration of H_2S is not maintained. Otherwise, in the case of heavy feeds, the deposition of coke and metals is the main reason for the loss of catalyst activity. For VGO feeds, catalyst deactivation is dominated by the poisoning involving N-compounds and coke deposition. The contribution of metals to deactivation becomes evident beginning with DAO and increases toward ARs and VRs. The relative contribution of coke and metals to catalyst deactivation depends on several factors, e.g., the origin of feed, type of catalyst, operating conditions, type of reactor, position of catalyst in the fixed bed (40,301,302).

The N-bases, which are always present in every feed contribute to the catalyst deactivation by poisoning active sites. This results in slowing down the hydrogen activation process (76,303). For the Maya heavy crude ($\sim$290 ppm of V + Ni), a high content of nitrogen found in the coke formed initially on the $NiMo/Al_2O_3$ catalyst confirmed the participation of N-bases during the coke formation (304). With TOS, the nitrogen content in coke decreased, indicating a conversion of the N-containing entities. This suggests that the coke formed initially, i.e., young coke, still possessed some reactivity. However, general trends indicate that during the latter stages, the nitrogen content in coke increased linearly with TOS as coke was becoming more refractory.

It has been established that the relative contribution of N-bases to the overall loss of catalyst activity increased from VR toward VGO/HGO feeds. Furthermore, it was indicated that the conversion of N-compounds to hydrocarbon products may be influenced by the H_2S/H_2 ratio (40). There may be the optimal H_2S/H_2 ratio, for which the conversion (e.g., HDS and HDN) reached a maximum. The optimal ratio may exhibit a continuous change with TOS because of the catalyst deactivation. This further contributes to the complexity of deactivation mechanism. Obviously, this ratio may vary from feed to feed and from catalyst to catalyst. In the fixed bed reactor, this ratio may change between the inlet and outlet of the reactor. The situation becomes even more complex in multilayer and/or multistage catalytic systems.

The recent information showed that during the hydroprocessing of an AR, the catalyst activity was higher and the coke lay down lower in the presence of water (305). It was speculated that water decomposed on the anion vacancy yielding H_2 while oxygen remained attached to the catalyst surface. The latter reacted with carbon on and/or in the vicinity of vacancy. These events may be described by the following tentative reaction:

$$\square + H_2O = \square - O + H_2$$

$$\square - O + C = \square + CO$$

In these reactions, □ represents a vacancy. To certain extent, H_2O cleans and/or protects the vacancy from coke deposition if such reaction indeed occurred. It should be noted that the effect of water on catalyst deactivation during the hydroprocessing of heavy feeds has received little attention, whereas for some model feeds, contradicting results on the effect of water on HDN and HDS were reported (76).

For the asphaltenes and metals containing feeds, the catalyst deactivation by coke and metals occurs simultaneously. For HDM catalysts, few reports suggest that more than 50% of catalyst deactivation is caused by coke (306), whereas the overwhelming information confirmed metals as the main cause of deactivation (40,100,253,307). However, these statements and/or information tend to oversimplify the actual events, particularly in the case of fixed bed reactors. Evidently, for a high metal content feed, front of catalyst bed will be deactivated by metal deposits. The contribution of metals to deactivation will then decrease and that of coke increase toward the outlet of the catalyst bed. However, the front zone of the metals' deactivated catalyst bed will gradually move toward the outlet until the entire bed is deactivated. Before this point, HDM reactions occurring near the end of catalyst bed are affected by the coke deposited during the initial stages of the operation. It is, therefore, obvious that the relative contribution of coke and metals to deactivation will vary between the inlet and outlet of catalyst bed.

The above discussion suggests that relative contribution of coke and metals to catalyst deactivation will also vary with TOS. Thus, during very early contacts of catalyst with a heavy feed, the coke deposition by fouling may dominate catalyst deactivation. At this point, little contribution of metals to the overall loss of catalyst activity may be evident. General observations show that coke deposition reaches a steady state, while the contribution of metals progressively increases in the course of the operation.

For the graded systems comprising several layers of the different catalysts in either one fixed bed or several fixed bed reactors containing different catalyst each connected in a series, e.g., HDM, HDM/HDS and HDS in the first, second and third reactor, respectively, the contribution of metals to deactivation will decrease from the first toward the third reactor (308). At the same time, the contribution of coke to deactivation will increase. It should be noted that N-compounds in the feed are gradually converted to the hydrogenated N-containing intermediates. The basicity of the latter is greater than that of the N-compounds originally present in the feed. This indicates an increased contribution to catalyst poisoning with TOS, before the N-intermediates are completely converted to hydrocarbons. The N-intermediates may be at least partly responsible for the increased coke formation in the downstream reactors.

The study of Al-Nasser et al. (309) gave detailed accounts of the selection of the optimal catalyst bed combinations for the graded system comprising four fixed bed reactors, which are part of the commercial ARDS process. Thus, every stage required a different catalyst. In addition, a control of the H_2S/H_2 ratio between the stages may be necessary to ensure that the concentration of H_2S is not in the inhibition region. To certain extent, this may be achieved by withdrawing a portion of gaseous products between the reactors and replacing them by a make up H_2 to ensure that the optimal H_2S/H_2 ratio is maintained. An option involving scrubbing H_2S from the gaseous effluent of the reactor before entering the subsequent reactor may be less practical.

The study of Rana and Ancheyta (310) indicated the complex deactivation patterns which resulted from different experimental conditions. In this case, the bench scale down-flow reactor and the up-flow microreactor were used to study the Maya heavy crude and 50/50 blend on the Maya crude and diesel oil, respectively. The former reactor could accommodate ten times more catalyst (e.g., 100 mL/85 g). The experiments in microreactor could not be conducted without blending the crude. Moreover, the experimental conditions were different, i.e., 653 K, 5.4 MPa and 10 L/h of H_2 in microreactor compared with 673 K, 7.0 MPa and 100 L/H of H_2 in bench scale reactor. The summary of these results (after 120 h on stream) using the $CoMo/Al_2O_3$ and $NiMo/Al_2O_3.TiO_2$ catalysts is shown in Table 7.1 (310). The experimental conditions had the most pronounced effect on HDAs. Thus, for the $CoMo/Al_2O_3$ catalyst, significant decrease in the HDAs conversion in the bench scale unit compared with the microreactor was observed, whereas the opposite trend was observed for the $NiMo/Al_2O_3.TiO_2$ catalyst. At the same time, for both catalysts, the loss of the HDM activity was more evident in the bench scale unit than in the microreactor. These results may be used to illustrate how the observations and conclusions reached during the catalyst deactivation studies can be influenced by experimental conditions.

Table 7.1 Catalyst activities in microreactor (MR) and Bench scale reactor (BS) (310)

Catalyst	Conversion (%)		
	HDS	HDM	HDAs
$CoMo/Al_2O_3$-MR	56.6	50.0	40.0
$CoMo/Al_2O_3$-BS	56.6	31.2	14.9
$NiMo/Al_2O_3$-MR	37.8	35.4	34.0
$NiMo/Al_2O_3$-BS	35.6	15.7	45.7

7.1 DEACTIVATION DUE TO STRUCTURAL CHANGE OF CATALYST

It is essential that a large number of the active sites on catalyst surface are maintained for a long period of TOS. This requires the suitable H_2S/H_2 ratio to ensure a desirable size of the active site (CUS) comprising a sulfur vacancy and SH group at its proximity (63). It was established that the number of vacancies decreased with increasing H_2S/H_2 ratio. On the other hand, at low H_2S/H_2 ratios, the catalyst over-reduction may occur. Such situation favors the adsorption of N-bases as well as deposition of coke and disfavors hydrogen activation. Furthermore, during the prolonged TOS, the crystal structure of active phase may change. For example, an increase of the MoS_2/WS_2 crystallites in normal direction will increase the number of active sites. This would be an indication of the conversion of the type I active phase to a more active type II active phase. At the same time, the growth in a lateral direction would have an opposite effect. In this regard, the studies of Shimada et al. (311) showed that the lateral growth of MoS_2 crystallites in the $CoMo/Al_2O_3$ catalyst was partly responsible for the loss of activity during the hydroprocessing of VGO at about 660 K and 5.9 MPa.

In an effort to simulate deactivation, Tanaka et al. (312) conducted the accelerated aging in the pilot plant at a higher temperature than that used in commercial units. In the former case, the activity loss due to the lateral crystal growth was more pronounced than that observed in the commercial unit operating at lower temperatures for much longer TOS than that used during the accelerating aging experiments. This agreed with the observations made by Gamez et al. (313) who studied the spent $CoMo/Al_2O_3$ catalyst used for hydroprocessing of the mixture of AGO and VGO. Thus, only a minor change in morphology of the MoS_2 crystallites was observed after 12 months on stream in a commercial unit operating at lower temperature than that used by Tanaka et al. (312) during the accelerating aging. These observations suggest that temperature may be the main parameter influencing the catalyst recrystallization. Therefore, in some cases, the accelerating aging may not properly identify catalyst deactivation patterns.

The chemical composition of the original active phase on catalyst may be gradually modified by the metals deposited from heavy feeds during the operation. The effect of deposited V and Ni on the catalyst activity is rather complex. Moreover, the deactivating patterns will change with progressive deposition with TOS. For example, the deposits had beneficial effects on HDM reaction, as it was demonstrated by a gradual increase in catalyst activity up to maximum attained between 15 and 20 wt% of the deposited metals (314,315). Then, the HDM activity began to decline with further increase in the metal deposition. Almost certainly, the activity decline resulted from the change in pore size distribution which affected the diffusion of reactant molecules into the catalyst pores. Therefore, for an active HDM catalyst, porosity may be at least as important parameter as is its chemical composition. Thus, industrial experience showed that about 5 wt% of MoO_3 in the absence of promoter on the γ-Al_2O_3 support possessing suitable porosity resulted in the active catalyst for HDAs and HDM. Such catalysts have been used industrially. Other catalyst functionalities, e.g., HDS and HDN, were influenced by the metal deposits differently than HDM. This resulted from the transformation of the Co(Ni)MoS phase into the VMoS phase, which was less active than the former (316). Moreover, it was reported that the unsupported V_2S_3 sulfide alone exhibited some activity for HYD and HDS (317,318). However, this was only demonstrated for model compounds rather than for heavy feeds.

7.2 DEACTIVATION BY COKE AND NITROGEN BASES

The studies on the simultaneous deactivation by coke and N-bases have focused on the VGO and HGO feeds as well as on the asphaltenes and metals containing feeds. However, in the number of latter studies, little attention was paid to the contribution of metals to the overall deactivation. Therefore, they are included in this part of the book separately from those works in which attempts were made to decouple the deactivation by coke from that by metals.

To various extent, the deactivation by coke and N-compounds occurs in parallel. For N-compounds, deactivation results from their strong adsorption on the catalytic sites. This slows down the activation of hydrogen, the availability of which is crucial for hydroprocessing reactions to occur as well as for slowing down coke formation (63). Moreover, the prolonged adsorption of N-compounds diminishes the access of other reactant molecules to catalytic sites. Therefore, at least part of the coke is formed as a consequence of the catalyst poisoning by N-compounds. The extensive information on catalyst deactivation by the N-compounds has been recently reviewed in detail (76). This included both the model compound feeds and the real feeds.

7.2.1 VGO and HGO

Among the heavy feeds of interest, the formation of coke and poisoning by N-compounds are the primary reasons for catalyst deactivation only during the hydroprocessing of Heavy feeds such as VGO and HGO. Thus, because of only trace amount present, the contribution of metals and asphaltenes to deactivation is rather small. Furthermore, the adverse effect of coke formed during the hydroprocessing of VGO and HGO on the textural properties of active phase may be less evident than that in the case of heavier feeds because of the less severe conditions employed during the former. This was indeed confirmed by Gamez et al. (313). Therefore, for VGOs and HGOs, the predominant cause of the loss of catalyst activity was the formation of coke and poisoning by N-compounds.

There may be a difference between the poisoning effect of the N-compounds present in VGO compared with that in HGO. This is supported by the presence of the fractions boiling below 350°C in the latter, whereas such fraction is not present in VGO. Thus, depending on the preparation of HGO, this may represent as much as 30% of the HGO. The overwhelming evidence suggests that the poisoning effect of N-compounds increased with decreasing boiling range of the fractions (40,76). This shall be taken into consideration while designing and/or selecting catalysts for the hydroprocessing of HGO compared with that for VGO.

The poisoning effect of N-compounds on catalyst activity was clearly demonstrated in the study published by Kaernbach et al. (319) on HDS of the VGO derived from the Russian crude. In this case, N-compounds were separated from the VGO by the ion exchange chromatography prior to the experiments performed at 633 K and 7 MPa in the continuous fixed bed reactor. As expected, the HDS conversion was much greater in the absence of N-compounds. Similarly, the HDS activity increased by about 60% after the N-compounds were removed from the VGO by adsorption with silica–alumina (241). The poisoning by N-compounds decreased with increasing temperature because of their diminished adsorption on catalytic sites. The important contribution of N-compounds to catalyst deactivation during the hydroprocessing of a VGO was also confirmed by Massoth et al. (320,321). Therefore, VGO and HGO are suitable feeds for decoupling the catalyst deactivation by coke and N-compounds from that by metals and asphaltenes. However, to various degrees, poisoning by N-compounds occurs also during hydroprocessing of the metals and asphaltenes containing feeds, particularly after the large molecules in heavy feed were converted to lighter fractions. For example, when such feeds are processed in the multistage systems, the catalyst deactivation patterns in the downstream reactors approach those observed during the hydroprocessing of VGO and HGO. In fact, in the last stage, the N-compounds poisoning patterns may approach those of an AGO or even lighter fractions.

The catalyst samples taken after 12 months on stream from the different depths of the single fixed bed used for hydroprocessing of a VGO (633–673 K, 8 MPa) had different coke deposition patterns (322). The amount of coke increased with the increasing depth of bed. The graphitic nature of coke increased toward the end of the bed as well. The predominantly amorphous structure of coke on the inlet and graphitic structure on the outlet of catalyst bed observed by Koizumi et al. (323) is in agreement with the results of Anemia et al. (322). It was proposed that the increasing temperature toward the end of fixed bed (because of the increased rate of exothermic reactions) was the main contributor to the difference in coke structure. Almost certainly, the increased rate of poisoning by N-compounds was an important contributor as well. Thus, the HYD of N-heterorings occurring near the front of fixed bed resulted in the formation of N-intermediates possessing a higher basicity than the corresponding N-containing reactants in the feed. An ultimate result of this was the diminished availability of the active surface hydrogen (63). As indicated earlier, such deactivation can be controlled by maintaining an optimal H_2S/H_2 ratio (76). This can be achieved in fixed-bed reactors consisting of several sections of the catalyst, rather than in the stationary fixed beds. As pointed out above, the former may have an option for removing a portion of the gaseous streams to scrub H_2S and/or to introduce the fresh make-up $H_2$and as such to control the H_2S/H_2 ratio.

Trytten et al. (232) investigated a series of the narrow-boiling gas oil cuts (nominal 50°C width), five of which were in a 350+°C range. The HDS, HDN and HYD conversions of the fractions were determined under usual hydroprocessing conditions ($NiMo/Al_2O_3$ catalyst, 13.9 MPa of H_2, 673 K). The study confirmed the presence of diffusion phenomena during hydroprocessing of the VGO feed. This fact has been overlooked in many studies on the hydroprocessing of VGOs and HGOs. These fractions were used to study the effect of N-compounds on the HDN of quinoline (324). In this case, the low-boiling cut (343–393°C) was more inhibitory than the intermediate- (433–483°C) and the high-boiling (524+°C) cuts. This confirmed that the N-compounds in the low boiling fraction are more effective poisons of catalyst than those in heavier fractions. Then, in the case of VGO and HGO feeds, during very early stages of hydroprocessing, i.e., when a rapid coke build-up was occurring, the poisoning by N-compounds may be an important contributor to the overall loss of the initial activity of catalyst by competing with H_2 for the adsorption on active site. The poisoning effect will be more evident for HGO because of the presence of the fraction boiling below 350°C.

In agreement with Trytten et al. (232), Khanda et al. (325) observed that deactivating effect of the N-bases increased with the decreasing boiling range of the feed. Thus, the HDN of quinoline which was added to the 616–666 K, 706–756 K and 797+ K fractions increased in the same order. A conclusion may be drawn from these observations, e.g., between the inlet and outlet of the fixed bed reactor, the inhibiting effect of N-compounds on hydroprocessing reactions may exhibit a maximum before most of the N-compounds were converted to hydrocarbons. Similarly, in a multistage system, the inhibiting effect of N-compounds will increase from the first stage and reach a maximum in one of the downstream reactors.

In the studies of de Jong et al. (326–328) on hydroprocessing of a VGO, the coke formation was observed to be the temperature and catalyst structure dependent. Thus, as Figure 7.1 shows, the coke build-up increased continuously with increasing temperature to a maximum and then suddenly decreased due to the change of the flow patterns, i.e., the predominantly liquid system was gradually changing to gaseous system. According to Figure 7.2, the addition of a small amount of Mo to Al_2O_3 resulted in the significant decrease in coke formation (326,327). The coke build-up increased with the increasing amount of Mo, i.e., with the increasing catalyst activity. Therefore, the formation of this coke, termed as "chemical reaction coke" was associated with hydroprocessing reactions.

The influence of catalyst structure on the coke formation was demonstrated in the study on aging of the $CoMo/Al_2O_3$ and Mo/Al_2O_3 catalysts (328). The aging was conducted at 723 K and 3 MPa in the fixed bed reactor using VGO. For both catalysts, the amount of deposited coke was similar. However, the former catalyst was more deactivated because of the much greater coverage by coke, i.e., about 90 and 50% for the $CoMo/Al_2O_3$ and Mo/Al_2O_3 catalysts, respectively. For the latter catalyst, the islands of coke were present, whereas for $CoMo/Al_2O_3$ the coke was more evenly distributed. The detailed spectroscopic evaluations of the spent catalysts from the hydroprocessing of a VGO conducted by van Dorn et al. (329–331) provided the information on morphology of the

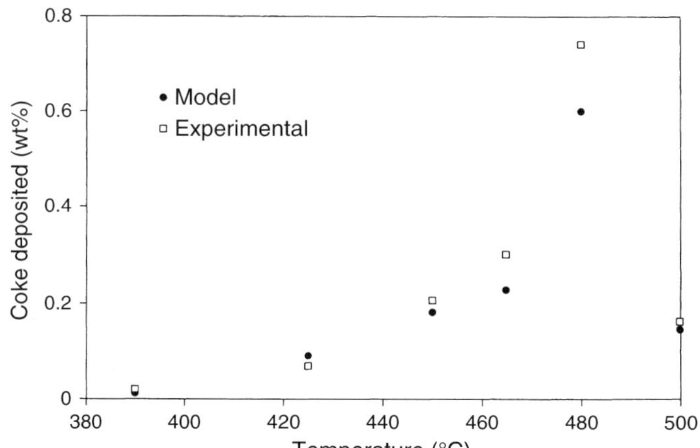

Figure 7.1 Effect of temperature on coke deposition (VGO, $CoMo/Al_2O_3$, 3 MPa) (328).

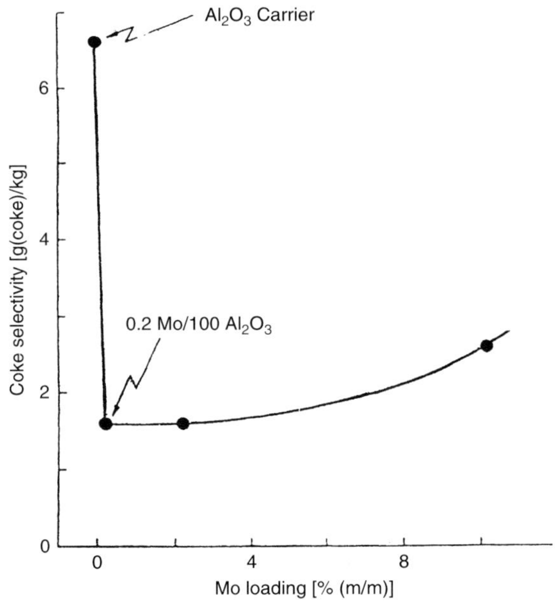

Figure 7.2 Effect of Mo loading on coke selectivity (VGO, 3 MPa) (328).

coke deposited on the catalyst surface. They concluded that most of the coke was located far from the active phase in the form of the irregularly shaped structures covering the entire surface.

7.2.2 Asphaltenes and metals containing feeds

Several studies on catalyst deactivation by coke deposits during the hydroprocessing of AR and VR have appeared in the literature without paying any attention to metals (105,304,332–334). It should be noted that such information may be relevant for the initial stages of coke formation when the metals contribution to deactivation is much less evident than that of coke. Although it may be viewed as incomplete, such information may contribute to the understanding of the coke forming reactions, particularly that of the effect of the feed origin, catalyst type and experimental conditions.

Some information (332) showed that for the model feeds such as carbazole and alkylated carbazoles, as well as PAHs, the initial coke deposition (less than 5 wt%) occurred predominantly on the Al_2O_3 support. Consequently, little deactivation was observed during this initial period. However, when a VGO was used under similar conditions, the catalyst poisoning by N-compounds and PAH present in the feed was observed few minutes after the start of the run (324). Similarly, the initial poisoning by N-compounds was also observed during the hydroprocessing of the Maya heavy crude, although to a lesser extent (304). While using Kuwait AR, Matsushita et al. (333) concluded that the coke formed during the early stages of the operation deposited on the support rather than on active phase. At the same time, N-compounds in the feed adsorbed near and/or on the active sulfide phase and as such contributed to the initial catalyst

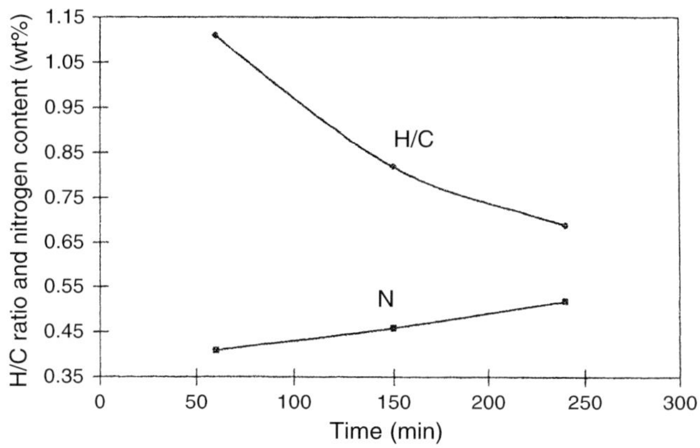

Figure 7.3 Effect of time on stream on H/C ratio and nitrogen content of coke on catalyst (334).

deactivation. Therefore, the catalyst deactivating patterns observed for model compounds (332) were rather different than those observed for the real feeds (304,324,333).

Figure 7.3 shows the decrease in the H/C ratio of coke with TOS (334). At the same time, the N content of coke increased linearly. This suggests that during hydroprocessing of the diluted VR derived from the Chinese heavy crude the accumulation of carbon and nitrogen in the coke deposited on catalyst increased, whereas that of hydrogen decreased with the increasing TOS. An ultimate result of this change was the graphite-like structure of the coke on catalyst. The experiments were performed at 683 K and 8 MPa in an autoclave. During these experiments, no attempt was made to decouple the contribution of metals to deactivation from that by the coke and N-compounds. The duration of these experiments (e.g., max 5 h) suggests that these observations reflect the events occurring during the initial stages of coke formation (334). Contrary to these observations, Callejas et al. (304) reported a decrease in the nitrogen content in coke with TOS for the Maya heavy crude during the early stages of the experiment. This suggests that this coke was rather "young," still possessing some reactivity. It is then evident that the observed trends in coke formation and its structure depend on the origin of crude, type of catalyst and operating conditions. The TOS at which the coke evaluation was conducted is important as well. In the case of ARs and VRs, the metals deposited on the catalyst surface during the operation may be at least partly responsible for the different deactivating patterns by coke compared with the VGO and HGO feeds. Nevertheless, the evidence seems to suggest that the contribution of N-compounds to catalyst deactivation is expected to increase from very heavy feeds (e.g., VRs) to VGO and HGO. However, for VRs and heavy crudes, the contribution of N-compounds to the overall deactivation will increase with the progress of hydroprocessing, i.e., in the fixed bed reactor, from the inlet toward the outlet.

To elucidate the role of asphaltenes during the initial stages of coke formation, Richardson et al. (105) prepared 10 fractions of the variable content of asphaltenes by supercritical extraction of the VR derived from the Athabasca bitumen using different amounts of pentane as the solvent. The solutions consisting of 30% of each fraction in a gas oil were reacted with the $NiMo/Al_2O_3$ catalyst to simulate the initial coke deposition. The amount of carbon on the catalyst increased with the increasing content of asphaltenes in the solution, i.e., from about 7.5 for saturates, aromatic and resin-rich fraction to 24% for the predominantly asphaltenic fraction. When the highest carbon content catalyst was reacted with Athabasca bitumen, the carbon content decreased from 24 to 22%. This was an indication of the equilibration of coke deposit with the surrounding liquid phase. It is believed that such mobility of the coke will diminish as coke becomes more refractory (e.g., with TOS). The HDS activity of the pre-aged catalysts decreased with increasing amount of coke on catalyst. Richardson et al. (105) did not consider the effect of the feed metals on coke formation. However, as indicated above, the effect of metals during very early stages is rather minor but is expected to increase linearly with TOS (40).

7.3 COMBINED EFFECT OF COKE AND METALS ON DEACTIVATION

Even for the feeds such as VGO and HGO, the catalyst deactivation by metals cannot be ignored despite the only trace quantities present. For these feeds, the N-compounds considerably slowed down HDM reactions by preferentially adsorbing on active sites (335). Thus, during the HDM of VO-porphyrin, the highest rate of deactivation was observed in the presence of NH_3 which was the product of HDN. It should be noted that the adsorption constants of the N-bases which are usually present in petroleum feeds are much greater than that of the NH_3 (76), suggesting that they may be much stronger poisons of the HDM reactions than the latter. Therefore, even for the low metal feeds (e.g., ~1 ppm), an adverse effect of the N-compounds and coke formation associated with it on HDM may be still evident.

Most of the nitrogen in coke may be more strongly adsorbed on the catalyst surface than on the other components of coke. This was demonstrated during the TPO of a spent catalyst, i.e., all hydrogen and most of the carbon deposited on catalyst were removed while the evolution of N oxides from the oxidation of the nitrogen in coke was still evident (336). If present, H_2O had the similar deactivation effect on HDM as had NH_3 (334). This was attributed to the coordination of H_2O with the vanadyl group of the porphyrin.

Figure 7.4 identifies major factors causing the catalyst deactivation during hydroprocessing of the asphaltenes and metals containing feeds (337). The results were obtained

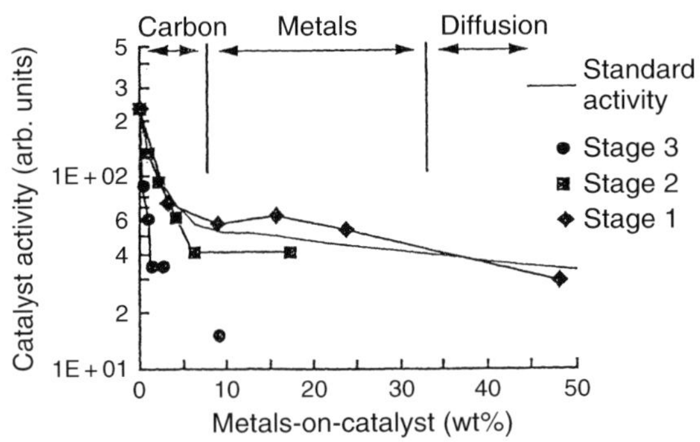

Figure 7.4 Major factors causing catalyst deactivation versus metals-on-catalyst (337).

in the three-stage ebullated bed system processing the heavy feed containing about 400 ppm of V + Ni. For every stage, the last point on the curve was recorded after 110 days on stream. Therefore, for the stage 3 catalyst, deactivation was caused mainly by coke deposition, whereas for stage 1 catalyst, the deposition of metals and restricted diffusion were the predominant modes of deactivation, with the contribution of the latter increasing with TOS until it became the main cause of the loss of activity. Furthermore, the relative contribution of these factors depends on the properties of heavy feeds. It is suggested that deactivation patterns observed in stage 3 may approach those observed during hydrorpocessing of a VGO. Therefore, different types of catalysts have to be selected for every stage.

Figure 7.5 (338) shows that the relative contribution of the above identified factors causing catalyst deactivation will change with TOS. For example, the removal of V from heavy feed was steady since few days on stream until the V deposit reached 50% of the catalyst weight (after more than 7 months on stream). This suggests that the coke deposited initially had little deactivating effect on metal removal since very early TOS. Yet HDS activity exhibited gradual decline before reaching the steady state after more than 2 months on stream compared with the rapid activity decline in Figure 7.4 for stage 3. This was contrary to general observations that the activity levels off after few weeks on stream. Rather slow HDS activity decline shown in Figure 7.5 may be attributed to the compensating effect caused by the deposited metals whose catalytic activity partially offset deactivation by coke. In fact, there are some reports that initially HDS activity increased because of the autocatalysis attributed to the metals which were deposited from the feed (243). In the study on hydroprocessing of the Boscan heavy feed, the abrupt HDS activity decline was observed after about 300 days on stream (338). Most likely, at this point, restricted diffusion was the main cause of the activity loss. This point coincided with the 50 wt% metal accumulation (339). It is, therefore believed that the abrupt decline in HDS activity would also be observed if the experiments in Figure 7.5 were carried out beyond 220 days.

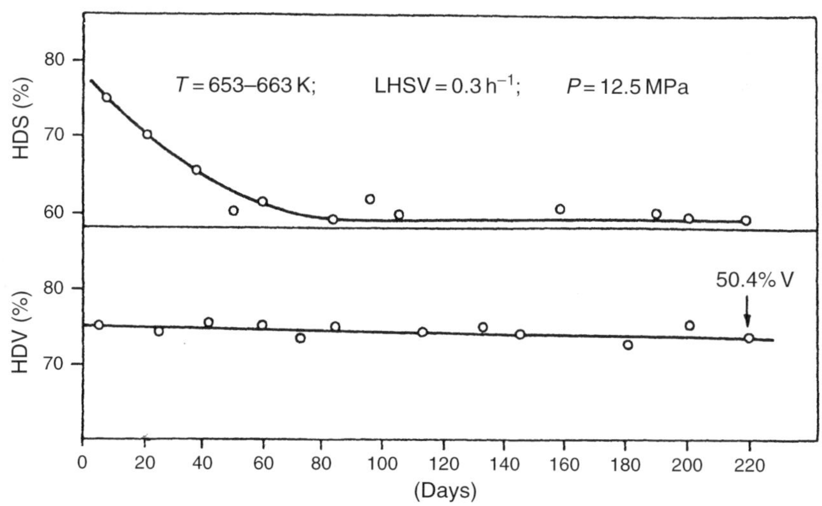

Figure 7.5 HDS and HDV conversions versus time on stream (338).

The catalyst deactivation patterns in Figure 7.6 (340) differ from those in Figure 7.5. The former show continuous decline in activity as indicated by the decreasing HDS conversion and the increasing content of V in the products with the increase in the amount of metals which accumulated on the catalyst. For the same amount of metals deposited on the catalyst, the activity loss increased with the increasing amount of metals in heavy feed. It is believed that all three factors identified in Figure 7.4 are causing catalyst deactivation simultaneously, with the contribution of coke decreasing and metals increasing until most of the desirable catalyst porosity was lost. Again, the deactivating trends will change from catalyst to catalyst, as it was demonstrated by the results in Figure 7.6 (243). Also, these patterns will be different for different feeds. Moreover, it is difficult to quantify the autocatalytic effect of the deposited metals on

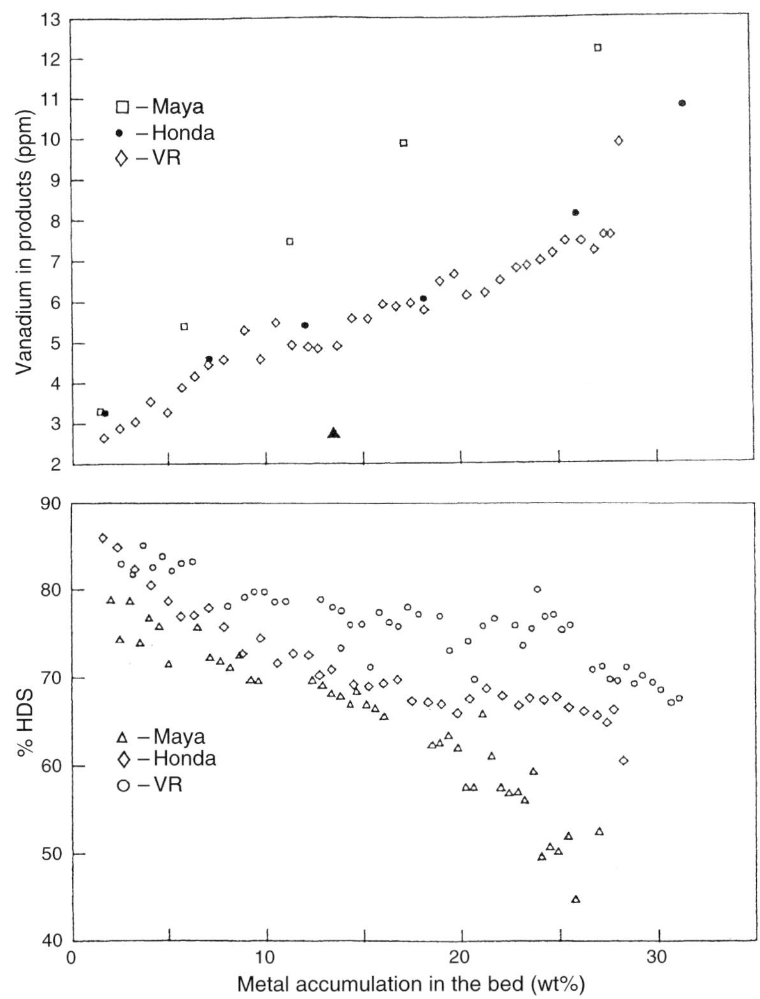

Figure 7.6 Vanadium content in products and HDS levels versus metal accumulation in bed (340).

some catalyst functionalities, particularly on HDM (341). This indicates the significant complexity in describing deactivation phenomena. In the following, catalyst deactivation will be discussed in the order of increasing severity required for hydroprocessing, i.e., increasing content of asphaltenes and metals.

7.3.1 Deasphalted oils

It has been generally observed that the content of metals and asphaltenes in DAO is higher than that in VGO and HGO but lower than than in ARs. However, all this depends on the type of deasphalting solvent, as it was shown in Figure 2.3. The content of metals also depends on the origin of the feed from which the DAO was derived. Thus, it is not unusual to have a DAO which contains more metals than an AR or even VR, particularly when the DAO was obtained from heavy crudes (e.g., Boscan, Maya, Orinoco, Zuata). For example, the DAO studied by Reyes et al. (110) contained ~230 ppm of metals. However, one report suggests that the amount of metals in the DAO obtained from the Boscan crude by hexane deasphalting approached 510 and 60 ppm of V and Ni, respectively (342). For such feeds, the deposition of metals is expected to be the predominant mode of catalyst deactivation from the early stages on stream, particularly when the content of asphaltenes in the DAO was much lower than that in a VR containing similar amount of metals.

In some situations, it was more beneficial to blend VGO with DAO for hydroprocessing, particularly when both were derived from a conventional crude (343). Subsequently, the blend may be hydroprocessed to obtain the feed either for FCC or dewaxing. Correspondingly, the severity employed during the hydroprocessing of VGO/DAO or DAO and catalyst deactivation associated with it would be somewhere between that used during hydroprocessing of VGO and AR.

Figure 7.7 (126) shows the deposition of metals and coke in two trickle bed reactors connected in a series used for hydroprocessing of the DAO containing 27 ppm ov V+Ni and less than 1 wt% of asphaltenes. This DAO was obtained by deasphalting of the VR derived from a conventional crude. The properties of catalysts A and D used for this study are shown in Table 7.2 (126). The catalysts (in baskets) were placed in the central axis of the two fixed bed reactors which were part of a commercial unit. In the first reactor, baskets were placed at the top and middle of the bed, whereas in the second reactor in the middle and bottom of the bed. The objective of the commercial run was to produce feed for the subsequent FCC (112,126). The system operated at the total pressure of about 10 MPa. To compensate for deactivation, the temperature between the start-up and shutdown was increased from 603 to 628 K and from 646 to 658 K in the first and second reactors, respectively. The evaluation of the catalyst was performed after 241 days on stream. For catalyst A, the significant increase in the coke formation toward the end of the second reactor should be noted compared with a little change for catalyst D. It is suggested that in the former case, poisoning of the catalyst by N-bases was the main cause of the catalyst deactivation. Catalyst A exhibited a greater metal storage capacity than catalyst D despite the larger average pore diameter of the latter. Most likely, smaller particle size of catalyst A than that of catalyst D ensured more efficient catalyst utilization. Moreover, the surface area of the former catalyst in the 60–100 Å pore range was nine times greater than that of catalyst D.

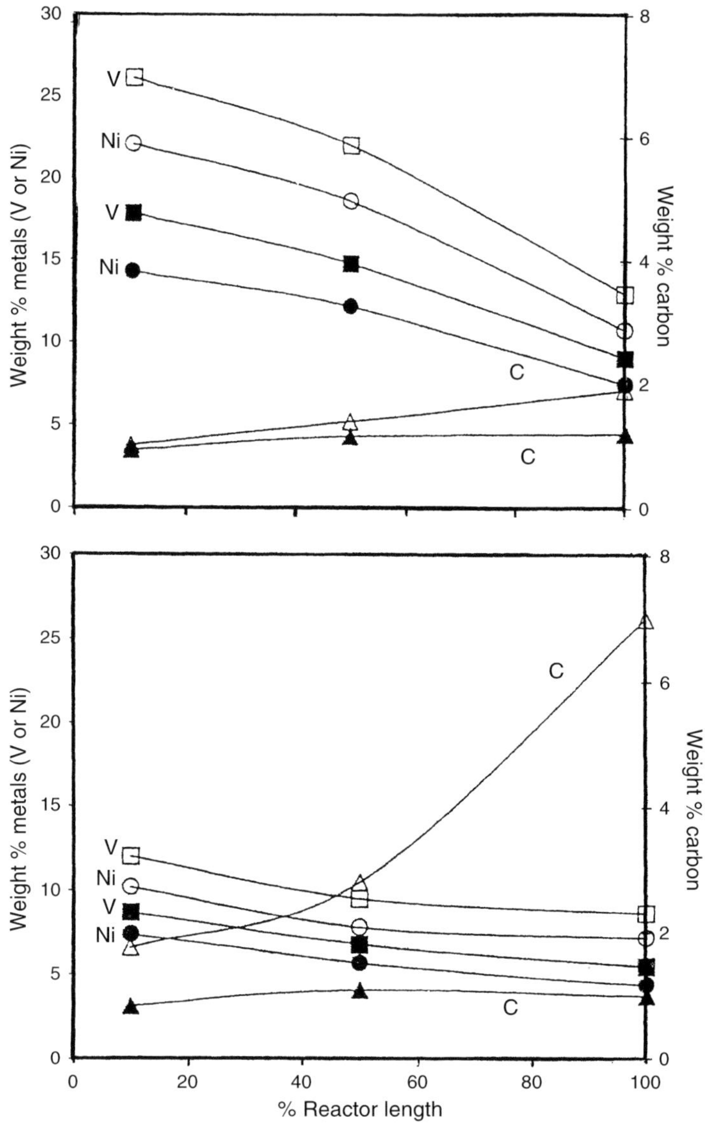

Figure 7.7 Deposition of metals and coke along the length of 1st and 2nd reactors for catalysts in Table 7.2; open symbols catalyst A, full symbols catalyst D (126).

7.3.2 Residues and heavy crudes

For these feeds, asphaltenes and metals are the primary cause of catalyst deactivation. The initial stages of deactivation are dominated by asphaltenes which rapidly deposit on catalyst surface. Part of the asphaltenes which deposit on the support uncovered with active metals is gradually converted to coke. The active hydrogen which spilt on the support from active phase may prevent coke formation in the vicinity of the latter (63).

Table 7.2 Properties of catalysts (126)

	Catalyst A	Catalyst D
Mo (wt%)	8.1	8.0
Ni (wt%)	2.3	2.0
Part. size (mm)	1.0	2.1
Surface area (m^2/g)	320	146
Average pore diameter (Å)	126	233
Surface area (m^2/g) in pore size range (Å)		
0–60	32	4
60–100	80	9
100+	208	133

The deposition of metals and their contribution to deactivation are linear with TOS. Compared with VGO, HGO and DAO, the contribution of N-bases to deactivation is much less pronounced; however, it increases with the progress of hydroprocessing.

The content of metals (V + Ni) in the ARs and VRs derived from conventional crudes may exceed 100 and 300 ppm, respectively. However, for the ARs and VRs derived from heavy crudes, the metals' content may exceed 500 ppm. The content of asphaltenes and CCR increases in the same order. This suggests that the entirely different catalyst formulations and/or catalytic reactor's configurations have to be used for such feeds compared with the low metal content feeds. For conventional ARs and VRs, the systems comprising several fixed bed reactors connected in a series, downstream from the guard reactor, have been used commercially. Significant modifications of the processes employing fixed bed reactors would be required to handle the ARs and VRs derived from heavy crudes. Moreover, for such feeds, a filtration system may have to be installed upstream of the guard chamber to prevent the formation of crust on the front of catalyst bed due to the presence of suspended solids (343). These problems can be overcome by using more advanced catalytic processes, i.e., employing either the moving bed or ebullated bed reactors, allowing a periodic and/or continuous withdrawal and replacement of catalyst without interrupting the operation.

Attempts have been made to develop anti-foulant agents, which can minimize deposit formation due to fouling after the addition to heavy feeds. Xiao et al. (344) studied the P-free and P-containing anti-foulants with the aim to alleviate catalyst deactivation due to coke deposition. They observed that after being added to the AR, the P-containing anti-foulant accelerated deactivation, whereas the P-free anti-foulant had little deactivating effect. The former agent formed an outer layer on the catalyst particles. As a result of this, the diffusion of reactants to active phase present below the layer was slowed down considerably. Moreover, this agent enhanced the aggregation of active metal sulfides. Therefore, the P-containing agents appeared to be unsuitable for the refinery applications.

7.3.2.1 Effect of feed origin and catalyst surface

Hauser et al. (345) used catalysts A, B and C, i.e., Mo/Al_2O_3 (HDM), $NiMo/Al_2O_3$ (HDM/HDS) and $NiMoP/Al_2O_3$ (HDS/HDN), respectively, to study coke formation during the four-stage fixed bed hydroprocessing of the Kuwait AR with the aim to simulate the performance of the ARDS process. Thus, the products from demetallization

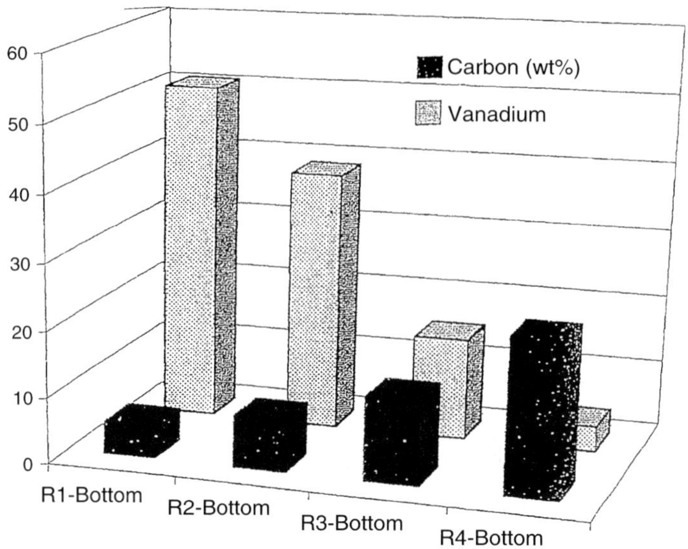

Figure 7.8 Deposition of carbon and vanadium in four reactors connected in series (345).

of AR in the first stage (e.g., DM-AR) were used as the feed for the second stage and so on. As shown in Figure 7.8, the coke build-up increased from reactor 1 toward reactor 4. At the same time, metal deposition decreased. Thus, coke formation increased with the increasing degree of hydroprocessing. This was complemented by the presence of a bulky aromatic structure of asphaltenes in the DM-AR and its products (DMDS-AR) used as the feed for reactor 3. Also, the lower H/C ratio of asphaltenes still remaining in the products compared with that in the corresponding feeds could have increased sediment formation on the catalyst surface because of their lowered solubility in oil fraction. A higher basicity of the N-compounds in the DM-AR and DMDS-AR than in AR (because of the HYD of N-heterorings), i.e., an enhanced poisoning of the catalyst, may be another cause for the increased coke formation.

The results in Figure 7.9 (337) can be interpreted in line with the observations made by Hauser et al. (345) and Al-Dalama and Stanislaus (346), although they were

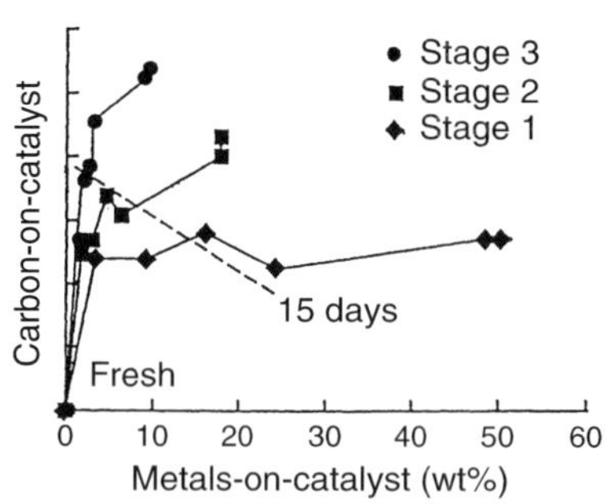

Figure 7.9 Carbon-on-catalyst versus metals-on-catalyst (337).

obtained in the three-stage ebullated bed system using a VR as the feed rather than AR. In this case, for every curve, the last point was recorded after 110 days on stream. It is evident that for stage 1, after about 5 wt% of metals deposited initially, the coke on catalyst exhibited little change, with further metal deposition. Obviously, the catalyst structure should have a pronounced effect on the trends in coke formation shown in Figures 7.8 and 7.9. This is shown in Figure 7.10 on the TPO profiles of CO_2, SO_2 and NO of the spent catalysts from the study of Hauser et al. (345). The properties of the corresponding fresh catalysts are shown in Table 6.6

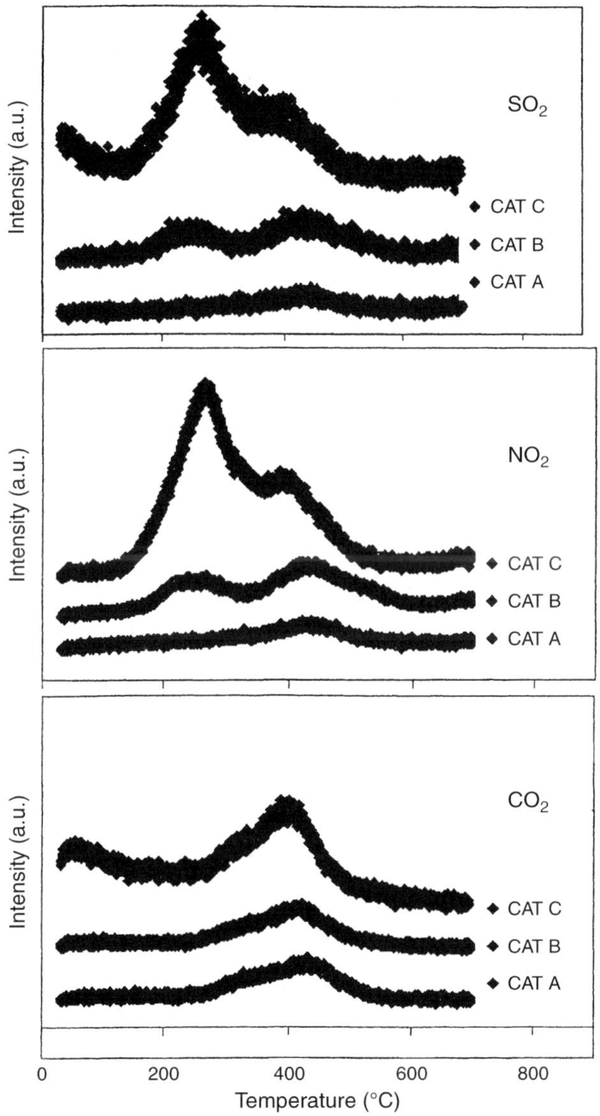

Figure 7.10 Evolution of SO_2, NO_2 and CO_2 during TPO of spent catalysts in Table 6.6 (345).

(243). The largest coke deposits were formed on the NiMoP/Al_2O_3 catalyst (catalyst C). This coincided with the highest acidity of this catalyst compared with the other two catalysts. Apparently, the SO_2 peak which originated from the oxidation of the active phase sulfides coincided with one of the NO peaks. This indicates an association of the N-compounds in coke with active phase. Such N-compounds could include either NH_3 formed during HDN or some unconverted N-bases strongly adsorbed on the catalyst surface (336).

A similar set of catalysts as shown in Figure 7.8 (345) were used by Al-Dalama and Stanislaus (346) to estimate level of the surface area recovery on the oxidative regeneration. For catalysts taken from reactors 1 and 2, the recovery was rather low, suggesting a permanent deactivation by metals. A significantly higher level of the surface area recovery was achieved for the catalysts taken from reactors 3 and 4. For these catalysts, coke deposition was the primary cause of catalyst deactivation.

Figures 7.11 and 7.12 (100,255) confirm the change in the relative contribution of coke and metals to catalyst deactivation during very early stages on stream. It should be noted that the determined trends depend on the type of experimental technique used for generating the data as it was indicated by Figure 4.1 (105). In Figure 7.11, the results identified as P and TS were determined in an autoclave and a continuous reactor, respectively, using the Safania AR (106 ppm of V + Ni, 7.2 wt% of asphaltenes) as the feed. The results in Figure 7.12, obtained in the continuous system, show that during the first contact of the feed with catalyst surface, deactivation was governed by coke, while the adverse effect of N-compounds on catalyst activity was less evident. Thus, within the first 6 h, the carbon content approached about 12 wt%, whereas only about 0.1 wt% of Ni + V was deposited. Within the same period, the relative HDS activity declined from 1.0 to about 0.6. Figure 7.12 shows that the deactivation by coke slowed down because the coke on catalyst reached the steady state. In the context with the results in Figure 7.11 obtained in the autoclave, the loss of the HYD activity was much more pronounced than that of the HDS activity before the steady-state level of coke was attained. From now on, the metals' contribution to deactivation relative to that by coke increased with TOS until the former became the primary cause of the catalyst activity decline.

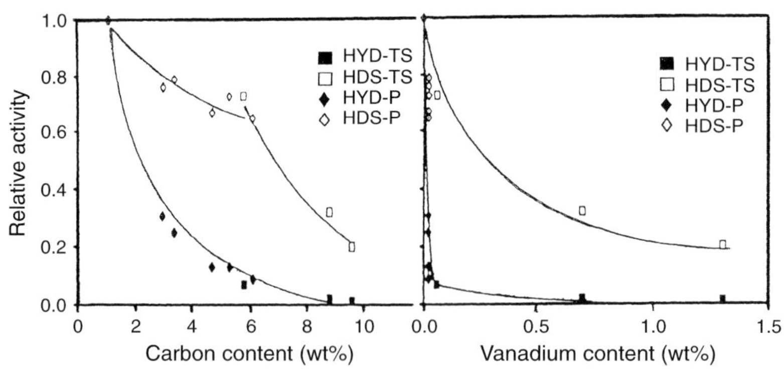

Figure 7.11 Relative HYD and HDS activities versus carbon and vanadium on catalyst (255).

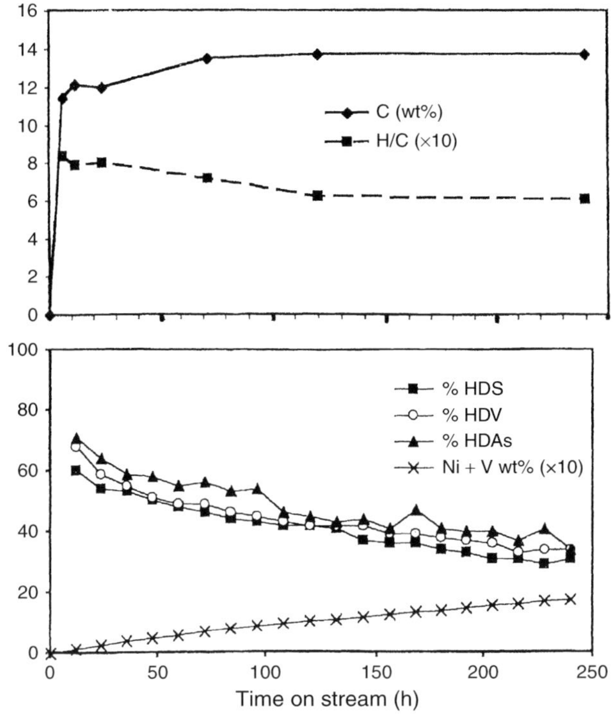

Figure 7.12 Content of carbon and H/C ratio of coke (A) and catalyst activities and metal uptake (B) versus time on stream ($NiMo/Al_2O_3$, AR, 623 K, 6 MPa (255)).

It is expected that the amount of coke deposited initially is governed by the content of asphaltenes and resins in the feed. This is supported by the results in Figure 7.13 (347). In this case, the feeds with different content of resins and asphaltenes were obtained by solvent deasphalting of the two VRs and one AR derived from different crudes each. A close examination of the scatter of data in Figure 7.13 indicates that the amount of deposited coke was influenced by the origin of the asphaltenes and resins. To certain extent, the observations made by Morales and Solari (338) complement the results in Figure 7.13. These authors used several heavy feeds and established the correlation between the content of asphaltenes in the heavy feed and its HDS, HDM and CCR conversions. Thus, the conversions decreased with the increasing content of asphaltenes, but they leveled off when about 20 wt% of asphaltenes in the feed was approached. However, it is unlikely that these observations can be applied to all heavy feeds because the chemical structure of asphaltenes may be another parameter influencing coke deposition. Thus, as indicated earlier, for the heavy feeds having a similar content of asphaltenes but of different chemical structure, the coking propensity increased with the increasing aromaticity of asphaltenes.

After deposition on catalyst surface, V and Ni are gradually converted to sulfides. The overwhelming information showed that V tends to deposit on the external surface of catalyst particles, whereas the radial distribution of Ni is more uniform (8,40,84,347). This is illustrated in Figure 3.12 (8). However, for a macroporous HDM catalyst (pore volume

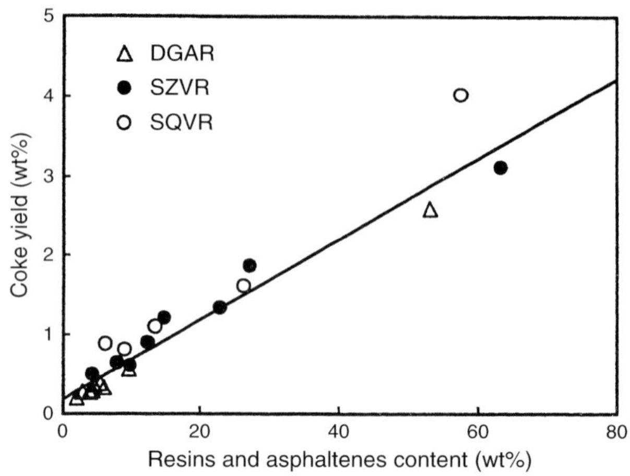

Figure 7.13 Coke on catalyst as function of the content of resins and asphaltenes in fractions from Dagang AR (DGAR), Saudi light VR (SQVR) and Saudi medium VR (SZVR) at 673 K and 8.5 MPa of H_2 over $NiMo/Al_2O_3$ (347).

of 0.95 cm^3/g), difference between the distribution patterns of V and Ni was less pronounced (254). An example of the effect of porosity on metal distribution is shown in Figure 3.14 (87). It was indicated that the presence of V in the vanadyl form is one of the reasons for the enhanced reactivity of V-containing porphyrins compared with Ni-containing porphyrins during the deposit formation on the catalyst surface. As a result of this, V deposited on the surface before it could diffuse into the catalyst interior. It has been observed that a small amount of V may deactivate catalyst because of the blocking active sites, whereas large amount of deposited V deactivates catalyst due to pore mouth plugging (100,225). The pore mouth plugging by Ni deposits is much less evident. In fact, an information suggests that the Ni deposited on the catalyst from the feed may improve catalyst performance, its HYD activity in particular (348,349). Koyama et al. (350) proposed two regions of deactivation by metals, i.e., the initial one involving the poisoning of active sites and the other causing the decrease in effective diffusivity due to pore mouth plugging. It is believed that in both regions the deactivating effect of the V deposits was more pronounced than that of the Ni deposits.

The trends in Figure 7.14 (274) confirmed the initial rapid build-up of coke followed by a much slower build-up, coinciding with the nearly linear deposition of metals with TOS. The results were part of the extensive evaluations conducted by Marafi et al. (129-132,278,351,352) using the Kuwait AR (~80 ppm of V+Ni). The experiments were performed in the continuous fixed bed reactor using either $NiMo/Al_2O_3$ (at 8 MPa) or Mo/Al_2O_3 (~ 4 wt% MoO_3 at 12 MPa) catalysts in the temperature region of 633–693 K. The results of this study confirmed that generally observed trends such as shown in Figure 7.14 may apply to both ARs and VRs irrespective of their origin. This would suggest that during very early stages, the catalyst deactivation by metals using either VR or AR was unimportant, although Gualda and Kasztelan (100,255) observed that a significant loss of the HYD activity due to the very small amount of the deposited

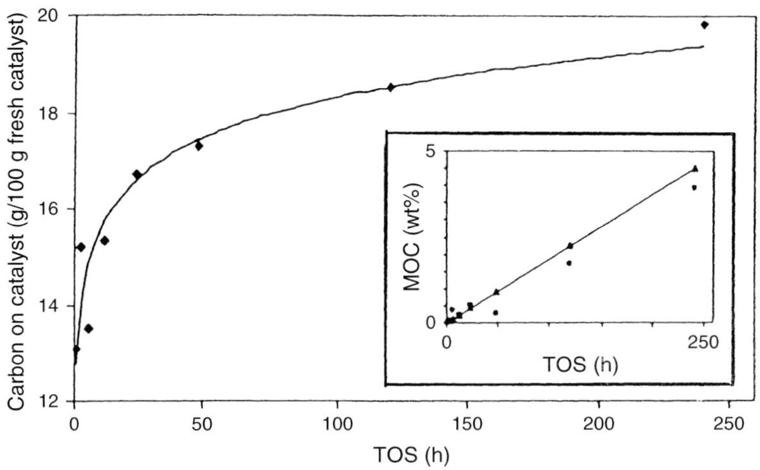

Figure 7.14 Deposition of carbon and metals on Mo/Al_2O_3 catalyst in ARDS process versus time on stream (274).

V was already evident within the first few hours on stream. At the same time, the loss of HDS activity was much less pronounced.

It is believed that during very early stages of the operation, there is little effect of metals on coke formation. On the other hand, the coke formed initially can have a pronounced effect on the rate of the metal deposit formation because of the partial pore plugging by coke. Moreover, this part of the support on which metals could deposit was already occupied by coke. It is, therefore critical that the rate of coke formation is kept at minimum to ensure a high HDM activity of catalysts. In this regard, the results in Figure 7.2 (327,328) can have important implications on the design and preparation of the HDM catalysts, although they were obtained for a VGO feed. Thus, the coke formation may be kept at minimum by selecting an optimal composition of catalyst. At the optimal composition, formation of the "chemical coke" associated with hydroprocessing reactions is slow, thus ensuring a high HDM activity due to the diminished interference by coke. However, this was not confirmed in the study involving the Kuwait AR (90 ppm of V + Ni; 3.6 wt% asphaltenes) conducted by Marafi et al. (129) who compared the Mo/Al_2O_3 (3 wt% Mo) with $NiMo/Al_2O_3$ (8 wt% Mo and 2 wt% Ni) catalysts having pore volume of 0.7 and 0.5 mL/g, respectively. Typically, the catalysts were used for HDM and HDS, respectively. Between 633 and 693 K and at 12 MPa, consistently more coke was deposited on the HDM catalyst. As expected, the H/C ratio of coke on the HDM catalyst was much lower than that on the HDS catalyst because of the higher HYD activity of the latter. The contradictory results reported in the literature underline complexity of the simultaneous deactivation of catalyst by coke and metals, particularly during the initial stages. This may be attributed to the differences in experimental conditions. This is evidenced by the different deactivation patterns for different feeds and different catalysts observed initially. In this case, the method used for catalyst presulfiding may be an important factor for controlling the initial coke deposition.

It is expected that in the case of heavy feeds containing metals, the structure of coke will be progressively influenced by the deposited metals (e.g., V). This was indeed observed by Zeuthen et al. (353,354). In this case, the coke formed in the proximity of the deposited V was more refractory, i.e., it had lower H/C ratio than the coke in the interior of pores. This suggests that in the course of the experiment, V enhanced dehydrogenation of coke. Then, different forms of coke may be present on catalyst surface. The results on the TPO of two spent catalysts, i.e., one used for hydroprocessing of a VGO containing no metals and the other for that of a high metal containing feed published by Bartholdy et al. (355) provide some additional evidence for the existence of the different forms of coke. For the latter spent catalyst, the two CO_2 peaks appearing at different temperatures were observed compared with only one CO_2 peak for the spent catalysts from hydroprocessing of VGO. The temperature of maximum of this peak coincided with that of the low-temperature peak observed during TPO of the catalyst deactivated by both metals and coke. Definitely, the coke giving the CO_2 peak at a higher temperature was more refractory because of the dehydrogenation of coke presumably catalyzed by V. The TPO analysis of coke on catalyst removed from the reactor at different stages of the operation revealed that coke became more refractory with TOS under similar operating conditions (304). In this case, the heavy Maya crude was used as the feed over the $NiMo/Al_2O_3$ catalyst.

The influence of metals on properties of coke was reported by Galiasso Tailleur and Caprioli (282). The study involved the extensive characterization of catalysts removed from the commercial ebullated bed reactor at various stages. They observed that, initially, coke filled pores before depositing on the exterior of the catalyst particles. However, because of its permeability, the liquid phase could reach catalyst surface. The permeability of the coke was gradually decreasing before catalyst was completely deactivated. The permeability decrease was complemented by the increased deposition of metals on catalyst surface. Therefore, it was suggested that metals contributed to the loss in permeability. In this case, the VR derived from Venezuelan crude was studied at 23.6 MPa and between 683 and 703 K over the $NiMo/Al_2O_3$ catalyst.

7.3.2.2 Effect of temperature and H_2 pressure

Temperature is an important parameter which can be used to control coke deposition. This is supported by the results in Figure 7.15 published by Marafi et al. (131) using the same Kuwait AR as above (129–131,278,351,352). For example, decline in the content of asphaltenes and resins in products with the increasing temperature suggested that their contribution to coke formation became less important with increasing temperature. This may be attributed to the enhanced conversion of asphaltenes to light products. Consequently, the HDM rate should be increased as well. If the initial catalyst deactivation was governed by coke deposits, coke had more detrimental effect on HDN than on HDS, as it is shown in Figure 7.16 (351), although the self-inhibition by N-compounds could be another contributor to the decline in the HDN activity. The detailed evaluation of the coke deposited on the Mo/Al_2O_3 catalyst revealed that after 1 h, the H/C ratio and ^{13}C NMR structural parameters of coke on the catalyst approached that of asphaltenes in the feed. This was the confirmation that asphaltenes in the feed were the primary cause of coke formation due to fouling. Subsequently, the H/C ratio decreased rapidly before attaining the region of a very slow decrease.

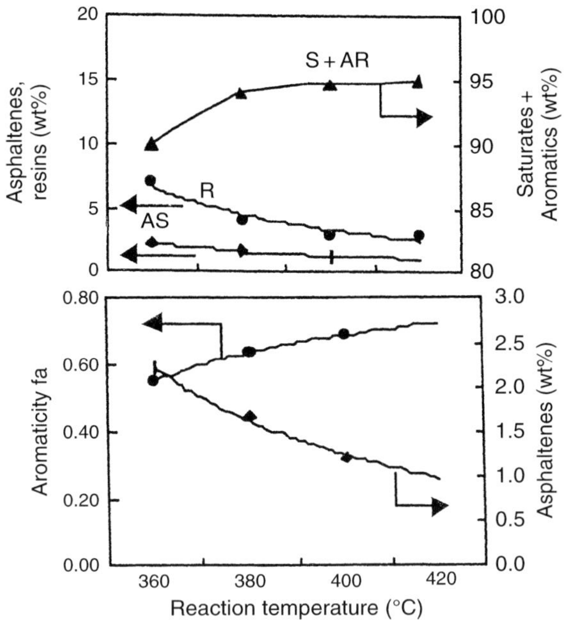

Figure 7.15 Temperature versus (A) content of asphaltenes and resins in products and (B) their aromaticity (NiMo/Al_2O_3, AR, 12 MPa) (131).

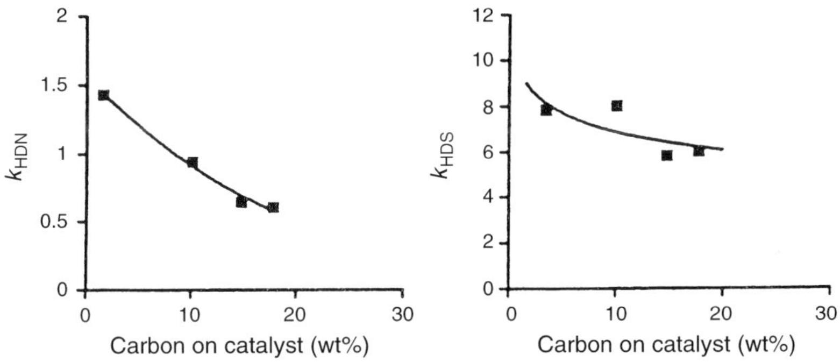

Figure 7.16 Effect of carbon on catalyst on HDN and HDS (NiMo/Al_2O_3, VGO, 648 K, 3 MPa) (351).

The trends in Figure 7.17 (326) may be interpreted in line with the results in Figure 7.15 (131), as well as with the observations made by Trytten et al. (232). The occurrence of the three temperature regions of coke formation such as shown in Figure 7.17 was also observed during hydroprocessing of the heavy feed such as Athabasca bitumen (356). The coke build-up below about 390°C could be attributed to the slow conversion of asphaltenes in the feed. As a sconsequence of this, the asphaltenes deposited on the catalyst surface. A steady conversion attained above 400°C resulted from the slowed down coke formation until about 440°C when the coke build-up increased with further

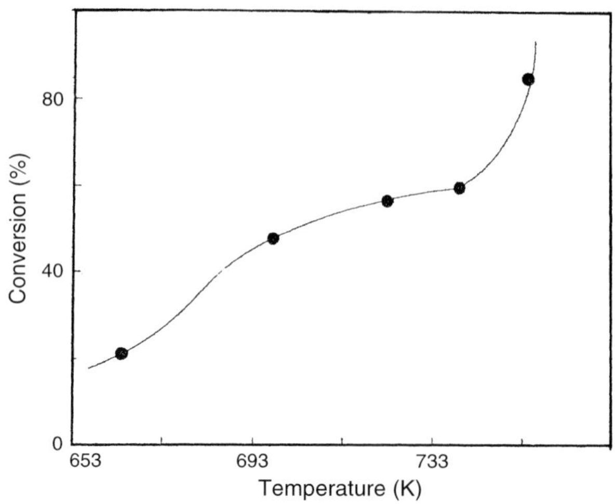

Figure 7.17 Effect of temperature on HCR conversion ($CoMo/Al_2O_3$, VGO, 3 MPa) (326).

temperature increase due to the thermal effects. However, it should be noted that in the study on Athabasca bitumen (356), the effect of metals was not investigated, although their effect on the coke formation may not have been negligible. The temperature effect on coke formation observed by Gualda and Kasztelan (100,255) differed from that in Figure 7.17. In the former case, the coke build-up increased and reached a maximum before further increase in the rate of coke formation with temperature increase was observed. The AR used in their study contained ~110 ppm of V + Ni, i.e., about one-third of that in Athabasca bitumen. On the other hand, in the case of the Kuwait AR, the coke build-up increased linearly with increasing temperature from 633 to 693 K (119). These observations again suggest that trends in the effect of temperature on the coke formation on catalyst surface depend on the origin of feed as well as on the experimental conditions. Therefore, it is not surprising to observe different trends in different studies.

The H_2 pressure is another parameter for controlling coke formation. It is, however, believed that the decreased coke formation caused by an increase in the H_2 pressure would favor the deposition of metals relative to that of coke. Richardson et al. (105) used Athabasca bitumen to study the H_2 pressure effect on the initial coke formation (between 1.5 and 5 h on stream) in the CSTR system and in an autoclave reactor using the commercial $NiMo/Al_2O_3$ catalyst at 703 K. After a rapid coke build-up during the first hour on stream, the coke formation did not change with the increasing ratio of the feed to catalyst. At the same time, increasing H_2 pressure from 7 to more than 15 MPa decreased the amount of coke from about 17 to about 11 wt%. As shown in Figure 7.18, for the AR used by Gualda and Kasztelan (100), the amount of coke decreased from about 10 to 4 wt% by increasing the H_2 pressure from 2 to 15 MPa. Moreover, the H_2 pressure had a pronounced effect on the H/C ration of coke on the catalyst. It is believed that in the case of Athabasca bitumen, large asphaltenic molecules had the predominant role during the initial stages of coke formation. Thus, there was a sufficient amount of asphaltenes to form the same amount of coke even for the low feed/catalyst ratios.

The pilot plant evaluation of an AR conducted by Higashi et al. (357) revealed that the coke deposited on catalyst surface during the very early stages on stream at a low

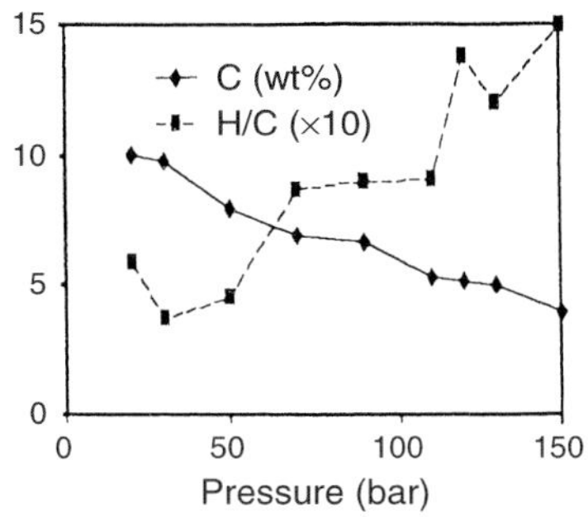

Figure 7.18 Effect of H_2 pressure on H/C ratio and amount of carbon on catalyst ($NiMo/Al_2O_3$, AR, 663 K) (100).

H_2 pressure could not be removed and/or catalyst activity could not be recovered by increasing the H_2 pressure at the same temperature during the later stage on stream. This indicated the permanent deactivation by coke. It is, therefore, essential that the coke deposition control by H_2 pressure begins at the start of the run. In this case, the loss of the HDS activity was noticed in particular. It was observed that the catalyst presulfiding was an important factor in controlling the initial coke deposition.

The heavy Maya feed used by Ancheyta et al. (358) contained more than 350 ppm of V + Ni. This feed was studied in the pilot plant between 673 and 703 K at 7 MPa over the sulfided $NiMo/Al_2O_3$ catalyst. Figure 7.19 shows the change in catalyst activity, expressed in terms of stability, with increasing temperature. In this case, the stability is defined as the ratio of the reaction rate at the end of the run to the initial rate of the corresponding reaction. Apparently, these results complement the results in Figure 7.17 (326). Thus, a slight increase in activity by temperature increase from 673 to 693 K shown in Figure 7.19 coincided with the temperature region (in Figure 7.17) in which

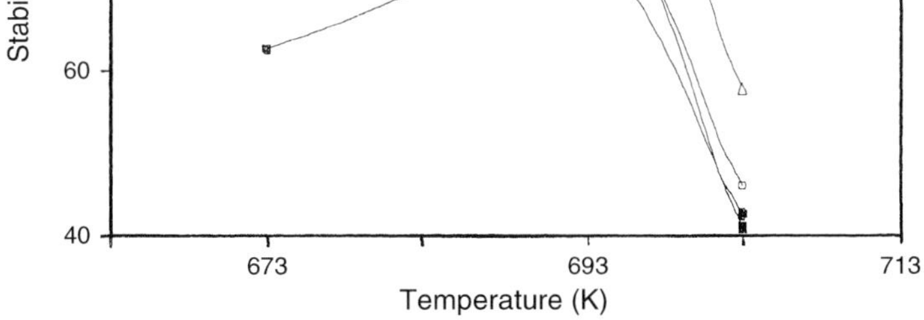

Figure 7.19 Effect of temperature on catalyst stability; (■) HDS, (○) HDAs, (□) HDNi, (△) HDV, (●) RBC conversion ($NiMo/Al_2O_3$, Maya crude, 7 MPa) (358).

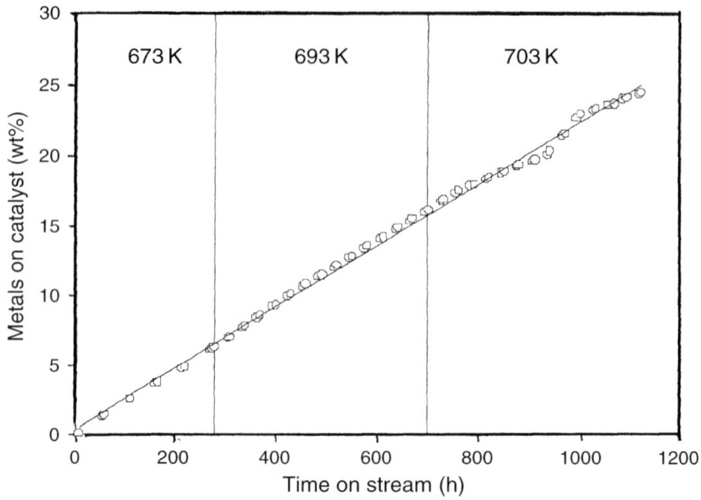

Figure 7.20 Metal deposition on catalyst during hydroprocessing ($NiMo/Al_2O_3$, Maya crude, 7 MPa) (358).

the coke build-up was approaching the steady state. It should be noted that the content of V + Ni in these feeds was similar. The sudden activity decline by temperature increase from 693 to 703 K (Figure 7.19) may then be attributed to the rapid build-up of the thermal coke. At the same time, as Figure 7.20 (358) indicates, during these experiments, the metal deposition increased linearly regardless of the temperature.

The combined effect of temperature and H_2 pressure on the coke build-up was investigated by Kumata et al. (359). As the feed, they used the partially demetallized AR derived from the Heavy Arabian crude. The partial HDM of the AR was conducted over the typical HDM catalyst (4.2 wt% of MoO_3 on Al_2O_3). The experimental system comprised two trickle bed reactors connected in a series. The adjacent reactor was loaded with the typical HDS catalyst of the $Ni/CoMo/Al_2O_3$ formulation. It was observed that at 653 K, coke was rather evenly distributed between the top of the first HDS reactor and the bottom of the second HDS reactor. However, at 683 K, the coke build-up progressively increased in the same direction. When H_2 pressure was increased from 8.0 to 14.0 MPa at 653 K, the amount of deposited coke decreased from about 20 to 14 wt%. This indicates that an optimal combination of temperature and H_2 pressure, for which coke deposition can be minimized, may be established.

Seki and Yoshimoto (360) determined the build-up of the "hard" coke defined as the toluene insolubles on the catalysts ($Ni/CoMo/Al_2O_3$) which were pre-aged during the treatment with Kuwait AR (from 643 to 653 K, 14.0 MPa, 16 h). After pre-aging, the catalysts were washed in situ with the LCO at 623 K for 6 h to remove "soft" coke before being used for aging tests using the demetallized AR as the feed. The tests of 20-h duration were conducted at 643 and 703 K and 8.0 MPa. Under these conditions, the accumulation of the additional "hard" coke decreased with increasing temperature. This may be attributed to the more extensive conversion of asphaltenes to light products with increasing temperature as observed by Seki et al. (279).

The study of Stanislaus et al. (361) may suggest that the conversion of asphaltenes to coke is dominated by thermal effects, while the direct involvement of catalyst is less evident. These authors performed detailed characterization of the sediments, one taken from the stripper bottom of the ebullated (H-Oil) reactor and the other, termed as fresh sediment, was collected from the high-pressure separator of a pilot plant. In both cases, the same VR derived from the Kuwait crude was used as the feed. The sediment obtained from the commercial unit after several months on stream was subjected to much more severe conditions. Therefore, it was less soluble in toluene and THF and its aromaticity was much greater than that of the fresh sediment. The H/C ratio of the asphaltenes isolated from the VR, fresh sediment and sediment obtained from commercial unit was 1.1, 0.98 and 0.59, respectively. The structural parameters determined by NMR revealed that the structure of the sediment was similar to that of the coke deposited on the catalyst surface. For example, the presence of polycondensed aromatic rings, much diminished aliphatic chains and high content of sulfur, nitrogen and metals was quite evident for the deposit obtained from commercial unit. The similar structure was assigned to the coke on catalyst obtained from a commercial operation after almost 9 months on stream (278). The similarity of the structures of sediments and coke confirmed that with TOS, the thermal reactions dominated the conversion of asphaltenes to both sediments and coke deposited on catalyst.

7.4 EFFECT OF MECHANICAL PROPERTIES OF CATALYST ON ACTIVITY LOSS

The mechanical strength of catalyst particles may influence the performance of the bed of catalyst. In a fixed bed, the fine particles formed by attrition may be carried out with liquid streams, thus depleting the original load of catalyst. Moreover, in the fixed bed, the fines may decrease the void space between the catalyst particles. This would affect the flow patterns of the liquid and gaseous streams leading to the development of pressure drops across the bed. Malfunctioning of the fixed bed (e.g., development of channels) ending with the discontinuation of the operation could be an ultimate result of these changes. Although this may not be catalyst deactivation in a true sense, the operating problems caused by fines of catalyst require attention. It was indicated earlier that if not removed from the feed, finely divided mineral matter may cause similar difficulties.

In ebullated bed reactors, the depletion of catalyst material due to the particles' attrition and/or disintegration is much more evident than in the fixed bed reactors. This results from the hydrodynamics of the ebullated bed reactor, which ensures a continuous motion of particles. Also to enhance the active phase utilization, the typical diameter of the catalyst particles is 1 mm ID or less. Without adequate mechanical strength, breaking of such particles could not be avoided. A vigorous mixing in ebullated bed suggests that the fresh particles added periodically may be well mixed with the spent catalyst particles which may need to be withdrawn. Then, a part of the particles are withdrawn, with the spent catalyst without being completely utilized. Little information on these phenomena has been available until the work of Al-Dalama and Stanislaus (346) appeared in the scientific literature. The results from this study are shown in Table 7.3. On the basis of these results, it was estimated that the lightly fouled catalyst accounted for about 30 wt%

Table 7.3 Properties of spent catalyst particles from ebullated bed reactor (346)

Property	Spent mix	Lightly fouled	Heavily fouled
Vanadium (wt%)	10.6	4.4	13.8
Nickel (wt%)	4.0	3.5	5.2
Carbon (wt%)	16.2	15.8	16.3
Surface area (m^2/g)	68	122	55
Bulk density (kg/L)	1.09	0.97	1.21
Side crush. strength (lb/mm)	1.8	2.1	1.2
Pore volume (mL/g)	0.17	0.21	0.11
Particle length distr. (wt%)			
<1.5 mm	25.2	14.4	40.0
1.5–3.0	42.3	23.5	37.0
3.0–6.0	32.5	61.3	23.0
>6.0	0	0.8	0

of the mixture. The catalysts were separated from the mixture by jigging technique using a mineral jig. Of particular importance is the length distribution of particles, which for fresh catalyst was dominated by 3.0–6.0 mm particles. For the heavily fouled catalyst, more than 70% of these particles were broken to less than 3.0 mm length. There was a significant difference in surface area, pore volume and side crushing strength between the lightly and heavily fouled catalysts as well. It was established that fine particles could be carried out from the reactor together with the liquid streams. This represents a loss of activity per unit of the catalyst loaded. The lightly fouled particles withdrawn prematurely represent another source of the activity loss because of their incomplete utilization. These phenomena are physical and/or mechanical and in their nature differ from those occurring during catalyst deactivation. However, because the ultimate result is the loss of catalyst activity, they deserve attention during the catalyst design as well as during the operation.

7.5 KINETICS OF CATALYST DEACTIVATION

The kinetics of hydroprocessing reactions are usually based on the extensive evaluations of the feeds and products. For example, the compositional changes with TOS were followed under variable experimental conditions such as temperature, pressure and LHSV. In the case of kinetics of catalyst deactivation, focus has been on the structural changes incurred by catalyst surface caused by the deposition of coke and metals as well as other factors. For the asphaltenes and metals containing feeds, the kinetics of catalyst deactivation were based on the total amount of coke and metals deposited on the catalyst surface in the course of experiments. These parameters were then related to the decline in catalyst activity for hydroprocessing reactions. The advancements in analytical techniques may allow the detailed characterization and quantification of hydrocarbon types and groups present in heavy feed, as well as in the coke formed on catalyst surface during hydroprocessing. With such information available, the effect of various structural parameters of coke precursors in the feed on kinetics of the deposit formation can be elucidated in more detail.

The structure of coke precursors in VGO and HGO differs markedly from those in the heavy feeds containing asphaltenes and metals. For the former feeds, polycondensed aromatics and N-bases may be the main cause of the coke deposit formation. To a lesser extent, poisoning by N-bases is also present for the metals and asphaltenes containing feeds. For these feeds, the adverse effect of N-bases on catalyst activity is expected to increase with the progress of hydroprocessing. As indicated earlier, in a multistage system, the poisoning by N-bases and coke formation may be the predominant modes of deactivation in the last stage. General observations suggest that coke deposition increased toward the last stage. It is, however, believed that this can be minimized by optimizing the operating parameters.

Attempts have been made to include the amount of contaminants (e.g., CCR, asphaltenes and metals) in heavy feeds in kinetic equations with the aim to develop models predicting the catalyst life. Common approach to study catalyst deactivation has been based on monitoring the decline in hydroprocessing conversions (e.g., HDS, HDN, HDM and HDAs) with TOS. In addition, the operating parameters such as temperature, H_2 pressure, catalyst particle size, pore size and volume distribution have also been incorporated in kinetic equations.

7.5.1 Deactivation by coke

In some studies, focus has been on kinetics of the coke build-up and catalyst deactivation associated with it, paying little attention to metals, although they were present in the feed. Without any doubt, these metals influenced the rate of coke formation and vice versa. In spite of this, the studies dealing only with coke are included in this review though separately from those which include the deactivation by both coke and metals. The primary focus only on coke deactivation may be appropriate for the VGO and HGO feeds. However, for the metals containing feeds, an incomplete information may be obtained when metals are not considered as the cause of catalyst deactivation, particularly during the later stages on stream.

The kinetics of coke formation during the hydroprocessing of a VGO were investigated by de Jong (326–328) who assumed the presence of two types of coke, i.e., catalytic and thermal cokes. The former was the main contributor to the total amount of coke on catalyst during the early stages on stream. The Langmuir type of kinetic expression was used for the rate of catalytic coke formation (R_c), i.e.,

$$R_c = k_c K_{ads} C_q/(1 + K_c C_q)$$

where k_c, K_{ads} and C_q are the rate constant of coke formation, adsorption constant of coke precursor and the concentration of coke precursor in the feed, respectively. The rate constant for coke formation depends on the amount of coke already present on the catalyst surface, i.e.,

$$k_c = k_{c,0}(1 - D/D_{max})$$

where D is the amount of coke deposited on the catalyst and D_{max} the maximum amount of coke determined experimentally. The rate of thermal coke formation (R_t) was expressed as

$$R_t = k_t C_q^2 / P_{H2}$$

The overall rate of coke formation was expressed as the sum of R_c and R_t. The kinetic model could predict quite accurately the effect of temperature and the H_2/feed ratio on the formation of coke.

The kinetic analysis of Richardson et al. (105) focused on the coke build-up over the NiMo/Al_2O_3 catalyst using Athabasca bitumen as the feed. The database was established using an autoclave and CSTR systems. A set of mathematical equations were derived for determining the mass balance of hydrogen in an annular zone surrounding slabs of active phase. Parameters, such as the rate of production of active surface hydrogen, the critical surface flux of hydrogen and the concentration of gas phase hydrogen, were considered. The aim was to simulate the decrease in coke formation with increasing H_2 pressure observed experimentally. The following equation was developed for quantifying the rate of coke deposition (R_{dep}):

$$R_{dep} = d\theta_v / d\omega = K\theta_v$$

where θ_v is the fraction of possible adsorption sites still available, ω is the cumulative feed to catalyst ratio and K is the adsorption constant. It was assumed that the deposit was present in a mono-layer-like form. The actual amount of carbon on the catalyst (in wt%) could then be related to the maximum carbon deposition (C_{max}) and θ_v as

$$C/C_{max} = 1 - \theta_v$$

Substitution of this equation into the former equation gives

$$C = C_{max}(1 - e^{-k\omega})$$

The amount of deposits calculated using this equation was in good agreement with the experimental data.

As part of the same study, Richardson et al. (105) presented the method for estimating the ratio of the effective diffusivity for coked catalyst to that for the fresh catalyst. The ratio varied between 0.065 and 0.105. The estimate was based on the pore size distribution determined experimentally for the fresh and coked catalysts. The values of the empirical constant A for Eqn. (3) (Section 3.2.2) were based on the study of Lee et al. (221) and varied between 3.9 and 5.1. Although present in the heavy feed, the effect of metals on kinetic parameters was not considered.

7.5.2 Simultaneous deactivation by coke and metals

The studies on kinetics of deactivation by only coke provide valuable, although incomplete, information on the overall loss of catalyst activity. It has been indicated that for heavy feeds, the contribution of metals to the overall deactivation may be more important than that of coke, particularly in the latter stages of the operation. To various degrees, the deactivation by coke and metals occurs simultaneously. This adds to the complexity of the deactivation phenomenon which is further complicated by the autocatalysis involving the sulfides of deposited metals. It is difficult, if not impossible, to quantify the compensation effect of the autocatalysis on the overall deactivation. For example, V_3S_4 was more active for HDM than $Fe_{0.85}S$ (299). However, the mixed $(Fe,V)S_4$ sulfide was the most active, suggesting a synergistic effect of Fe on V_3S_4. Apparently, Ni_3S_2 is more active than V_3S_4; however, the presence of the mixed $Ni_xV_yS_z$ sulfide cannot be ruled out. As it was pointed out earlier, the V/Ni and V/Fe ratios will change from the exterior toward the center of the particle. Therefore, because of the significant complexity, only a simplified approach, i.e., lumping variables into single power kinetics, can be afforded.

Correlations in Figures 6.11 and 6.12 established by Bartholdy and Hannerup (114) using the Kuwait AR illustrate the importance of the distribution parameter (Q_V) expressed by Eqn. [10] and Thiele modulus by Eqns [5] and [6] (Section 3.2.2), respectively, on kinetics of the V deposition on catalyst surface. Thus, if the distribution parameter is the ratio of the average V concentration to the maximum concentration between the exterior and center of the catalyst particle, its decrease with TOS would indicate the accumulation of V on the external surface of the catalyst particle. This coincided with the decreasing HDV activity (Figure 6.11) and the decreasing Thiele modulus (Figure 6.12). For the fresh catalyst particle, the value of Thiele modulus for HDNi is expected to be lower than that for HDV. Thus, the rate of HDV was greater than that of HDNi, whereas the rate of diffusion of the Ni-porphyrins was greater than that of the VO-porphyrins. It was pointed out earlier that Thiele modulus is the ratio of the rate of reaction to the rate of diffusion. It is expected that slopes of these correlations will depend on the size of catalyst particles, surface properties of catalyst, properties of feeds and experimental conditions.

The Kuwait AR (~70 ppm of V + Ni and CCR of 10.1 wt%) has been used in several studies on kinetics of catalyst deactivation. Chen and Hsu (155) compared four $CoMoP/Al_2O_3$ catalysts containing different amounts of P, with the commercial $CoMo/Al_2O_3$ in the trickle-bed reactor at 663 K and 7 MPa. The cylindrical form of the catalyst had 1.5 mm ID and 4 mm length. The experimental data were fitted using the modified Voorhies (362) kinetic expression:

$$X_t = X_0 \exp(-bt^n)$$

where X_t is the HDM conversion at time t, X_0 is the HDM conversion at time zero, b is the deactivation constant and n is the kinetic order. Figure 7.21 shows the fit of experimental data for first-order kinetics (155). Two deactivation regions can be identified and deactivation constant determined from the slopes of these correlations. The initial fast deactivation region was dominated by coke deposition. Beneficial effect of the P addition to catalysts was evident from Figure 7.21. In an earlier study conducted by Chen et al. (363) under similar conditions, but focusing on HDS, the best fit of the experimental data was obtained assuming 0.5 order deactivation kinetics. The kinetic order of 0.5 gave the best fit, when the HDS conversion data obtained by Lee et al. (364) were used to follow the catalyst deactivation. In this study, an Iranian AR containing

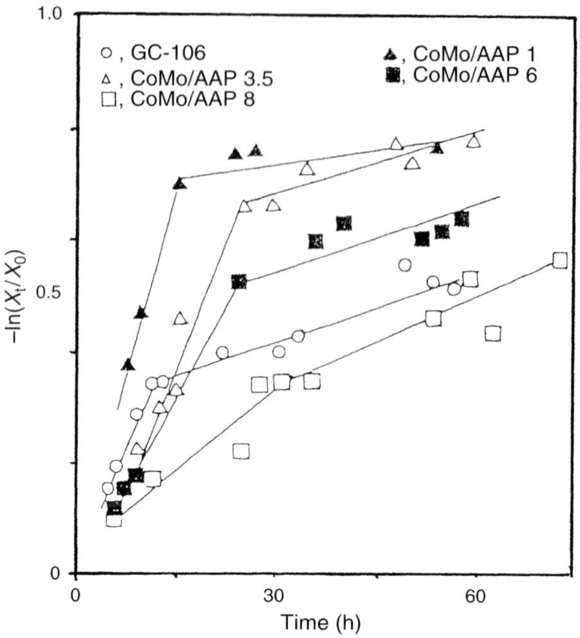

Figure 7.21 Voorhies deactivation correlation (AR, 663 K, 7.6 MPa) (155).

about 30 ppm of V + Ni was used. In this case, the addition of Ni, Ru and W to the $CoMo/Al_2O_3$ catalyst was investigated. The $-\ln(X_t/X_0)$ versus $t^{0.5}$ correlation clearly indicated beneficial effects of Ni and W and little effect of Ru on catalyst performance. This resulted from the diminished catalyst deactivation on the addition of Ni and W compared with that of Ru.

The same kinetic expression which was used by Chen and Hsu (155) in the kinetic study using an AR as the feed was also used by Maity et al. (145) to study the kinetics of the catalyst deactivation during hydroprocessing of the 50/50 blend of heavy Maya crude and naphtha. The study was conducted in the fixed bed up-flow reactor at 653 K and 5.4 MPa over three $NiMo/Al_2O_3$ and one $CoMo/Al_2O_3$ catalysts. The latter authors extended their work on catalyst deactivation to include HDAs and HDN conversions in addition to HDM and HDS conversions studied by Chen and Hsu (155). Figure 7.22 (145) shows the Voorhies' form of correlation, i.e., $-\ln(X_t/X_0)$ versus time, assuming first-order kinetics. For all reactions and catalysts, the presence of two deactivation regions was observed. The observed differences among the catalysts could be attributed to the different properties, such as pore volume and pore size distribution. It was confirmed that the relative effect of deactivation on the catalyst functionalities varied from catalyst to catalyst.

The rate constants for deactivation of the $CoMo/Al_2O_3$ and CoMo/carbon catalysts were determined by Altajan et al. (195) using the VR derived from Athabasca bitumen. The study was conducted in the continuous fixed bed reactor at 698 K and 13.9 MPa. For this purpose, the pseudo-turnover frequency (PTOF) was defined as the number of

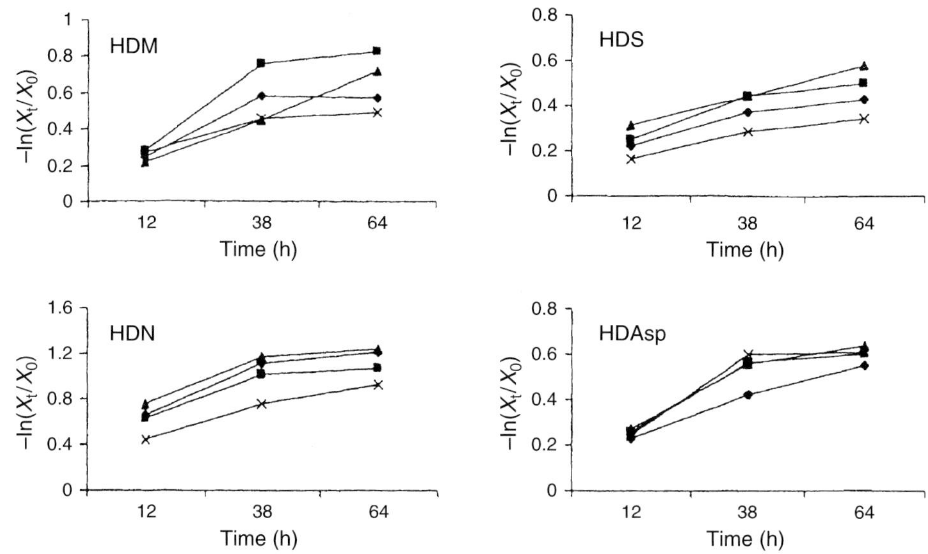

Figure 7.22 Data fit for modified Voorhies deactivation correlation for four different catalysts (AR, 663 K, 7.6 MPa) (145).

reactions per unit time and the unit surface area. The following equations derived by these authors assumed the first-order kinetics for catalyst deactivation:

$$\mathrm{Ln(PTOF)} = -k_D t_{\mathrm{PTOS}} + \ln(k_s C_R/\varrho_{\mathrm{CAT}} S)$$

In this equation, C_R is the concentration of reactant (e.g., S, N, V, Ni and asphaltenes), k_D and k_S are first-order rate constants for deactivation and surface reaction, respectively, t_{PTOF} is the TOS, ϱ_{CAT} is the bulk density and S is the surface area of catalysts. The k_D and k_S values are shown in Table 7.4 (195). These results show that the CoMo/carbon catalyst was more prone to deactivation than the CoMo/Al_2O_3 catalyst. Thus, with the exception of HDNi, all other catalyst functionalities were deactivated at a greater rate over the former catalyst. The difference in HDNi activity resulted from the different mean pore diameter of the CoMo/Al_2O_3 and CoMo/carbon catalysts,

Table 7.4 Pseudo-first-order rate constant for deactivation (k_D) and surface reaction (k_S) (195)

Support	$k_D \times 10^3 (s^{-1})$			$k_S \times 10^5 (s^{-1})$		
	Al_2O_3		Carbon	Al_2O_3		Carbon
Part. diam (mm)	3.2	1.6		3.2	1.6	
HDS	6.8	5.6	13.8	10.3	11.0	5.0
HDAs	12.7	11.6	16.8	7.4	8.1	5.2
HDV	9.4	7.9	14.5	6.9	7.3	6.2
HDNi	9.4	7.4	3.2	9.6	10.0	9.6

i.e., 8.5 and 28.6 nm, respectively. A higher k_D value for HDV over the CoMo/carbon than over the CoMo/Al_2O_3 catalysts may be attributed to a greater interaction of the vanadyl group of the porphyrin with carbon support than with the Al_2O_3 support. Therefore, the deposition of V may be the main cause of the greater deactivation rate observed for the CoMo/carbon catalyst compared with the CoMo/Al_2O_3 catalyst. The following expression was used to compare effective diffusivities of the CoMo/Al_2O_3 $[(D_{eff})_{Al}]$ and CoMo/carbon $[(D_{eff})_C]$ catalysts:

$$(D_{eff})_C/(D_{eff})_{Al} = (1-\lambda)^2_C/(1-\lambda)^2_{Al}$$

where λ is the ratio of the molecular diameter to pore diameter. Assuming the molecular diameter of the former is 2 nm, the value of $(E_f)_C$ was about 1.5 times that of the $(D_{eff})_{Al}$ in line with the greater mean pore diameter of the former.

Takatsuka et al. (365) used the Iranian VR (360 ppm of V + Ni and 8.2 wt% of asphaltenes) and an up-flow fixed bed reactor to study the kinetics of catalyst deactivation based on the loss of HDS and HDV activities with TOS. They used the following equations to determine the rates of HDS (r_s) and HDV:

$$r = \rho_{cat} A E_f k C_{H_2} C^n$$

where C is the concentration of either S or V in the feed, ρ_{cat} is the density of catalyst, A is the surface area of catalyst, C_{H_2} is the concentration of hydrogen in the feed and E_f is the effective diffusivity. It should be noted that the parameter such as C_{H_2} has been rarely included in kinetic equations by other researchers. The coke deposition on the catalyst, including the assumption that part of the deposited coke could be hydrogenated, was expressed in the following form:

$$R = \rho_{cat} A k_{c1} - \rho_{cat} A_c k_{c2} C_{H_2}$$

where k_{c1} and k_{c2} are the rate constants for coke deposition and coke removal by HYD, respectively. The similar set of equations was used by Al-Adwani et al. (366) to study kinetics of the coke deposition during hydroprocessing of the Kuwait AR. As one would expect, C_{H_2} was influenced by temperature, H_2 pressure (P) and solubility parameter. For the H_2 pressure ranging between 5 and 18 MPa and temperature (T) between 573 and 733 K, they used the following expression:

$$C_{H_2} = 8.91 \times 10^{-6} P + 4.16 \times 10^{-6} T(T-273) - 1.40 \times 10^{-3}$$

Effectiveness factor was determined from this equation:

$$E_f = 3\int \eta^2 C_D{}^n \, d\eta$$

In this equation, η is the dimensionless distance from the pellet center, C_D is the dimensionless concentration and n is the reaction order. This enabled the development of the correlation between Thiele modulus and E_f using Eqn. [3.8] discussed earlier (Section 3.2.2). Figure 7.23 (365) shows that sensitivity of the E_f toward the change in Thiele modulus followed similar trends for the different kinetic orders of reaction.

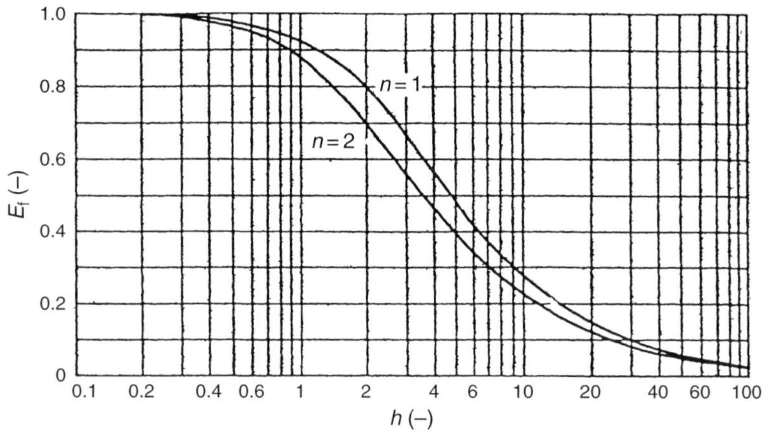

Figure 7.23 Effectiveness factor (E_f) versus Thiele modulus (h) for different reaction orders (n) (365).

7.6 MECHANISM OF CATALYST DEACTIVATION

All evidence suggests that for VGO and HGO, the overall mechanism of catalyst deactivation is much less complex than that for the asphaltenes and metals containing feeds. Similarly, as it was indicated during the kinetic studies, some studies on the mechanism of catalyst deactivation only focus on deactivation either by coke or by metal deposits. In spite of its complexity, the simultaneous deactivation by coke and metals has been attracting attention of researchers as well.

7.6.1 Mechanism of coke formation

The information on the mechanism of coke formation was reviewed in detail elsewhere, with attention being paid predominantly to model compounds and light feeds (40). The presence of resins and asphaltenes in heavy feeds adds to the complexity of the mechanism of coke formation, on catalyst surface. In this case, physical deposition (fouling) of the heavy components may dominate coke formation, particularly during the early exposure of catalyst to heavy feed. The extent of physical deposition may be influenced by the colloidal stability of heavy feeds. This suggests that for the asphaltenes containing feeds, both physical and chemical properties of the feed are important during the coke formation on catalyst surface besides the operating parameters such as H_2 pressure and temperature.

7.6.1.1 Chemical aspects of coke formation

The information on the mechanism of coke formation established for model compounds and light feeds forms the basis for elucidating the mechanism occurring during hydroprocessing of VGO and HGO (40). It is again noted that for these feeds, the interference and/or participation of asphaltenes and metals in the overall mechanism may be negligible. Therefore, coke formation is dominated by factors other than fouling.

7.6.1.1.1 Involvement of free radicals and carbocations

It is generally known that the thermal cracking of C–C bonds begins above 600 K. The primary products of the cracking reactions are free radicals. Unless they are rapidly stabilized, free radicals can combine to large molecules and eventually to coke. The former case was already discussed as part of the mechanism of hydroprocessing reactions. The involvement of free radicals during coke formation was proposed by several authors (367–373). The study published by Kubo et al. (372) provided some support for the involvement of free radical mechanism during coke formation. Thus, the coke formation was suppressed in the presence of a hydrogen-donating liquid which acted as the radical scavenger. The following tentative mechanism of the coke formation via free radicals proposed by de Jong (326,373) accounts for most of these observations:

$$Q \rightarrow P^{\bullet}$$

$$H_2 \leftrightarrows 2H^{\bullet}$$

$$P^{\bullet} + H^{\bullet} \rightarrow PH$$

$$P^{\bullet} + P^{\bullet} \rightarrow P_2$$

$$P_2 \rightarrow \text{coke}$$

In this scheme, Q is the precursor to the radical formation, whereas $P^{\bullet}$ and $H^{\bullet}$ are free radicals. Presumably, the $P^{\bullet}$ radical was formed via scission of a C–C bond. Based on the bond strength, C_{AL}–C_{AL} bonds in methylene bridges are the most reactive, yielding the least stable radicals. Involvement of the C–H bonds scission in radical formation is much less evident unless the tertiary carbon is available. For example, an aromatic structure with the isopropyl substituent attached would yield very stable tertiary radical. Radicals can recombine/polymerize to large species and eventually to coke, unless they are rapidly scavenged. The radical stabilization by hydrogen abstraction from a hydrogen donor may be depicted as

$$RH + P^{\bullet}(H^{\bullet}) = R^{\bullet} + PH(HH)$$

In this case, RH is a large molecule (e.g., resins or asphaltenes), whereas $R^{\bullet}$ is the radical produced by hydrogen abstraction from RH. The $R^{\bullet}$ may be stabilized via an intramolecular rearrangement. The experimental observations made by Nakamura et al. (186) and Kubo et al. (372), i.e., decreased coke formation on the addition of a hydrogen donor agent, were interpreted in terms of the free radical mechanism. Theoretically, the active surface hydrogen in the form of –SH and –MeH entities may stabilize radicals as well. This may proceed according to the following tentative mechanism, i.e.,

$$P^{\bullet} + \text{–SH(MeH)} = PH + \text{–S}^{\bullet}(\text{Me}^{\bullet})$$

However, at later stages on stream, this radical scavenging source may be exhausted due to the diminished hydrogen activation caused by the extensive catalyst deactivation. This is supported by the observations made in commercial units, i.e., a rapid coke build-up during the final stages on stream.

The potential involvement of carbocations was indicated earlier while discussing the overall mechanism of hydroprocessing reactions. In this case, carbocations were identified as the important intermediates of some reactions, i.e., hydroisomerization, polymerization and HCR. If not stabilized, carbocations can combine to higher molecular weight species. The coupling of polynuclear aromatics leading to coke precursors and finally to coke was also proposed (374–379). The rate of such reactions was enhanced in the presence of the Bronsted acid sites. This indicates the involvement of proton (via carbocation) during coke formation. The coke formation was significantly diminished after Bronsted acidity was destroyed by pretreating the catalyst support with basic species. Carbocation mechanism may be part of the overall mechanism of coke formation regardless of the origin of the heavy feed. In the case of such mechanism, the type of support may be the more important factor than the type of feed. It has been generally established that the rate of some reactions of hydrocarbons (cracking, isomerization, polymerization, etc.) was rather low unless the source of protons was available.

The carbocation mechanism of coke formation on a zeolite catalyst shown in Figure 7.24 was proposed by Magnoux et al. (380) after an extensive characterization of the spent catalyst. In this case, the coke was formed between 350 and 450°C in the absence of H_2. Of course, those were not typical hydroprocessing conditions. However, the similar temperature range and the presence of H_2 under typical hydroprocessing conditions are even more favorable for the formation of Bronstedt acidic sites on the conventional catalysts supported on the acidic supports such as zeolites. Thus, it was confirmed that the active phase (e.g., Co(Ni)–Mo(W)–S) could activate hydrogen which subsequently spilled on the support which acted as the active hydrogen reservoir (63). It is proposed that in the case of acidic supports, this hydrogen may be the source of protons required for coupling reactions shown in Figure 7.24 (380). It is, therefore, believed that the mechanism of coke formation involving both free radicals and carbocations may occur simultaneously, with their relative contribution to the overall mechanism of coke formation depending on the acidity of support.

On the basis of the acidity of support, carbon may represent another extreme to zeolites. Thus, it is unlikely that acidic sites are present unless carbons were subjected to

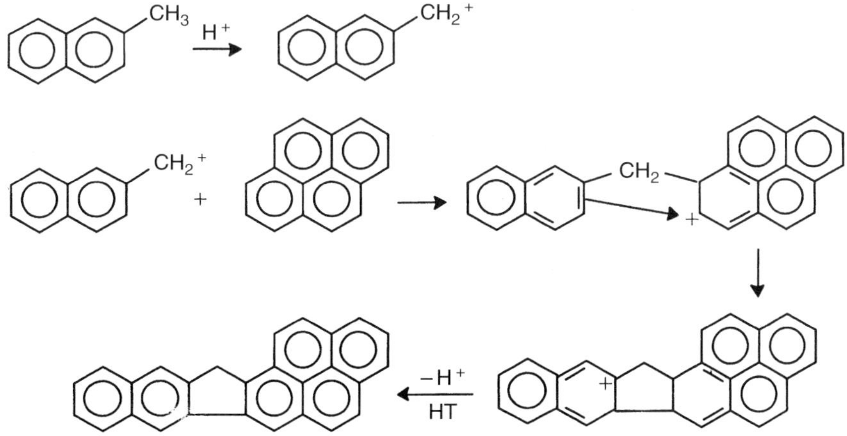

Figure 7.24 Tentative mechanism of polymerization involving carbocations (380).

special pretreatments. The results on hydroprocessing of the Kuwait AR conducted by Nakamura et al. (186) over the carbon-supported catalysts were interpreted in terms of a free radical mechanism. In the case of the Co(Ni)/Mo(W) catalysts supported on carbon, the SH groups could be a source of the hydrogen necessary for quenching radicals unless the heavy feed involved was of a naphthenic origin. Under certain conditions, SH groups may possess a Bronstedt acid character. For example, the Bronstedt acid character of such groups increased with increasing temperature (63). Then, at temperatures approaching 700 K, the – SH groups could donate proton and initiate the formation of carbocations (63). Therefore, even for carbon-supported catalysts, the involvement of carbocations during coke formation cannot be entirely ruled out.

It has been indicated that factors which dominate coke formation during the hydroprocessing of heavy feeds differ from those observed during the hydroprocessing of light feeds and/or model compounds. These facts should be reflected by the different deactivation mechanisms. Potential structures of the radicals formed during the early stages of conversion can be deduced from the models of asphaltenes shown in Figure 2.7. In this case, the weakest chemical bonds need to be taken into consideration. The same model indicates potential sites for accepting proton. Therefore, free radicals and carbocations may be part of the overall mechanism of coke formation regardless of the origin of the heavy feed. It is believed that only their relative contribution to the overall mechanism should be feed dependent.

For the VGO and HGO feeds, as well as some DAO, the poisoning by N-bases may be an important factor contributing to coke formation. It has been indicated that N-bases are the strong competitors for the adsorption on active sites (76). This affects the availability of the active surface hydrogen (63). Consequently, free radicals and other unstable intermediates will polymerize to coke rather than being stabilized and converted to lighter fractions. Therefore, this part of coke was formed as the consequence of the catalyst poisoning by N-compounds. Moreover, because of a prolonged adsorption, N-bases may also be converted to coke. Thus, the trends observed generally indicate the accumulation of nitrogen in coke with TOS, as it was shown in Figure 7.3 (334).

7.6.1.1.2 Characterization of feeds and coke

The early attempts to propose the mechanism of coke formation relied mostly on the elemental analysis of coke and extracts obtained from spent catalysts by successive extractions using various solvents (371,374). The time-consuming Soxhlet extraction can now be replaced by the accelerated solvent extraction (ASE) which can be accomplished within less that 10 min compared with more than 10 h required for the former (381). The advancements in spectroscopic techniques (e.g., ^{1}H NMR, ^{13}C NMR, LD-MS, FTIR) and other analytical methods allowed more detailed analysis of both coke and the corresponding feed and products. This allowed the determination of various structural parameters of the feed, products and the coke which deposited on the catalyst surface. The optical microscopic techniques could characterize coke deposits according to their reflectance, fluorescense and anisotropy. With the availability of such information, the mechanism of coke formation could be defined more accurately and in more detail.

Two spent catalysts from hydroprocessing of VGO were characterized by Sahoo et al. (382), who focused on the CH_2Cl_2 soluble and insoluble parts of the deposit. The structural parameters of the former, termed as a "soft" coke, were similar to those of the heavy components of the VGO feed. At the same time, the "hard" coke was more aromatic but less aromatic than the similar "hard" coke on the spent catalysts from hydroprocessing of ARs and VRs. This is not surprising as the latter requires more severe conditions (e.g., higher temperatures) to attain desirable level of conversions. Also, in the case of VGO, the HYD of some coke components could occur because of the less severe conditions. Then, the factors which dominate coke formation using feeds such as VGO and HGO may differ from those for the heavier feeds such as ARs and VRs.

The extensive characterization of asphaltenes and resins in the products by spectroscopic techniques, during the HDM and HDS of the Kuwait AR over the Mo/A_2O_3 and $NiMo/Al_2O_3$ catalysts, respectively, was carried out by Seki and Kumata (383,384). In this case, the molecular weight of both asphaltenes and resins in products progressively decreased in the course of HDM reactions. The rate of coke build-up significantly increased above 673 K. This was accompanied by the removal of alkyl chains from asphaltene molecules. Therefore, the aromaticity of asphaltenes was increased. Such change facilitated the adsorption of asphaltenes on the catalyst surface and increased deactivation. In the presence of alkyl chains, the adsorption of asphaltenes was diminished because of the steric interference between the coke molecules and catalyst surface provided mainly by aliphatic chains.

The first attempt to characterize coke deposits by the solid state ^{13}C NMR was made by Fonseca et al. (385–387). The $CoMo/Al_2O_3$ catalyst (0.7 wt% CoO; 4.5 wt% MoO_3) used in the three-stage ebullated bed pilot plant was withdrawn after 4, 21 and 120 days on stream from the first and third reactors. The feed was the blend of Khafji VR and a diluent. Less than 69% of the coke carbons could be observed by the NMR technique employed. The invisible part of the carbon was assumed to be of graphitic origin. However, the evolution of this part of carbon with TOS showed a rapid increase during the initial stages followed by a decline with TOS. This suggests that the invisible carbon included quaternary and tertiary carbons in large asphaltenic molecules rather than carbons in graphite-like structure. The decrease in the content of this carbon with TOS may have indicated a cracking and/or rearrangement of asphaltenic molecules, thus making more carbons visible by ^{13}C NMR. The studies confirmed a gradual aromatization of coke caused by dealkylation and aromatic rings condensation with TOS. This was complemented by increasing ^{13}C NMR aromaticity and decreasing H/C ratio of coke.

The mechanism of coke formation developed by Hauser et al. (274,278) was based on the database which, so far, is the only of its kind found in the literature. This included the detailed analysis of the structural parameters of coke using the solid state ^{13}C NMR with the application of the cross polarization with polarization inversion at the low or moderate magic angle spinning. Moreover, the analysis was complemented by using the proton-gated decoupled single-pulse excitation and the results of elemental analysis. With this approach, the limitations of the technique used by Fonseca et al. (385–387) were minimized. The spent Mo/Al_2O_3 (4.3 wt% of MoO_3) catalyst was obtained from the continuous fixed bed unit used for the HDM of the Kuwait AR at 653 K and 12 MPa. The spent catalysts were extracted either by toluene or by THF before their characterization. Figure 7.25 shows the change in the H/C ratio of toluene-insoluble (TIS) coke and THF-insoluble (THFIS) coke with TOS. After 1 h on stream, the H/C ratio of both cokes was similar to that of asphaltenes in the AR, suggesting that the fouling of the catalyst by asphaltenes was the main cause of the coke formation. This would suggest little involvement of the catalyst surface during very early stages. However, the NMR analysis

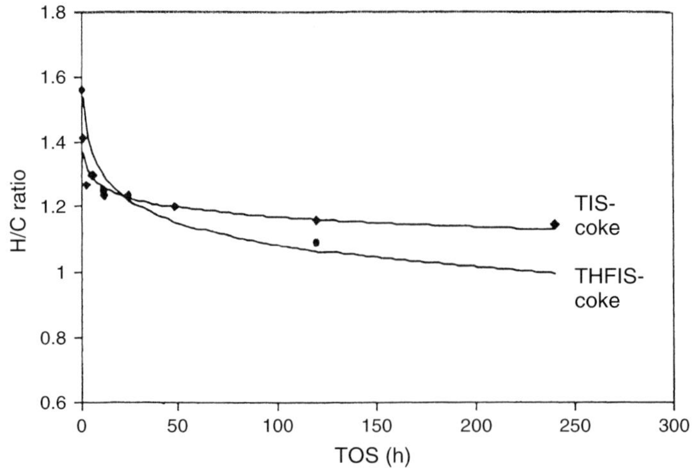

Figure 7.25 The H/C ratio of TIS and THFIS of coke versus time on stream (Mo/Al_2O_3, AR, 653 K, 12 MPa) (274).

of the TIS and THFIS coke indicated that already after 1 h on stream, the structure of asphaltenes changed after being deposited on the catalyst. This included the loss of long chains in particular. As in Figure 7.26, after 1 h, the TIS- and THFIS-coke structures differed from that of asphaltenes. It consisted of the less polycondensed aromatic rings with shorter but heavily branched alkyl substituents attached to them. Between 1 and 12 h, the coke deposition slowed down. In this region, a simultaneous accumulation of aromatic carbon in the coke, some HYD of aromatic rings, isomerization and dealkylation were occurring simultaneously. For the THFIS-coke, some ring condensation occurred as well. The H/C ratio of both TIS- and THFIS-coke was greater than one even after 240 h on stream, although the degree of alkyl substitution decreased significantly. These observations suggest that some of these changes could not occur without the direct involvement of catalyst surface. For example, a strong interaction with catalyst surface could be one reason for a low solubility of the THFIS-coke. Figure 7.27 shows the structure of coke on the same catalyst used for the HDM of the same AR in the industrial ARDS process after 6500 h. In this case, a high degree of the aromaticity of cokes, particularly that of the THFIS-coke, was quite evident. The formation of such structure may be considered as the beginning of the coke graphitization on the catalyst surface.

Several important conclusions can be drawn from Figures 7.25 to 7.27. Thus, the HYD and deHYD of coke on catalyst may occur in parallel together with some HCR, isomerization and condensation reactions. In the steady-state region of coke deposition, these events may be interpreted in terms of the geometry of active sites proposed by Richardson et al. (105) shown in Figure 7.28. Such active sites adsorb and activate hydrogen, part of which can be spilt on the support in the vicinity of active metals and becomes accessible to the deposited coke. Once on the support, active hydrogen protects the catalytically active metals from contamination by coke. Potential involvement of the V and Ni deposited on the catalyst surface during the operation cannot be ruled out. It is well documented that Ni is a good promoter of hydrogen activation and adsorption,

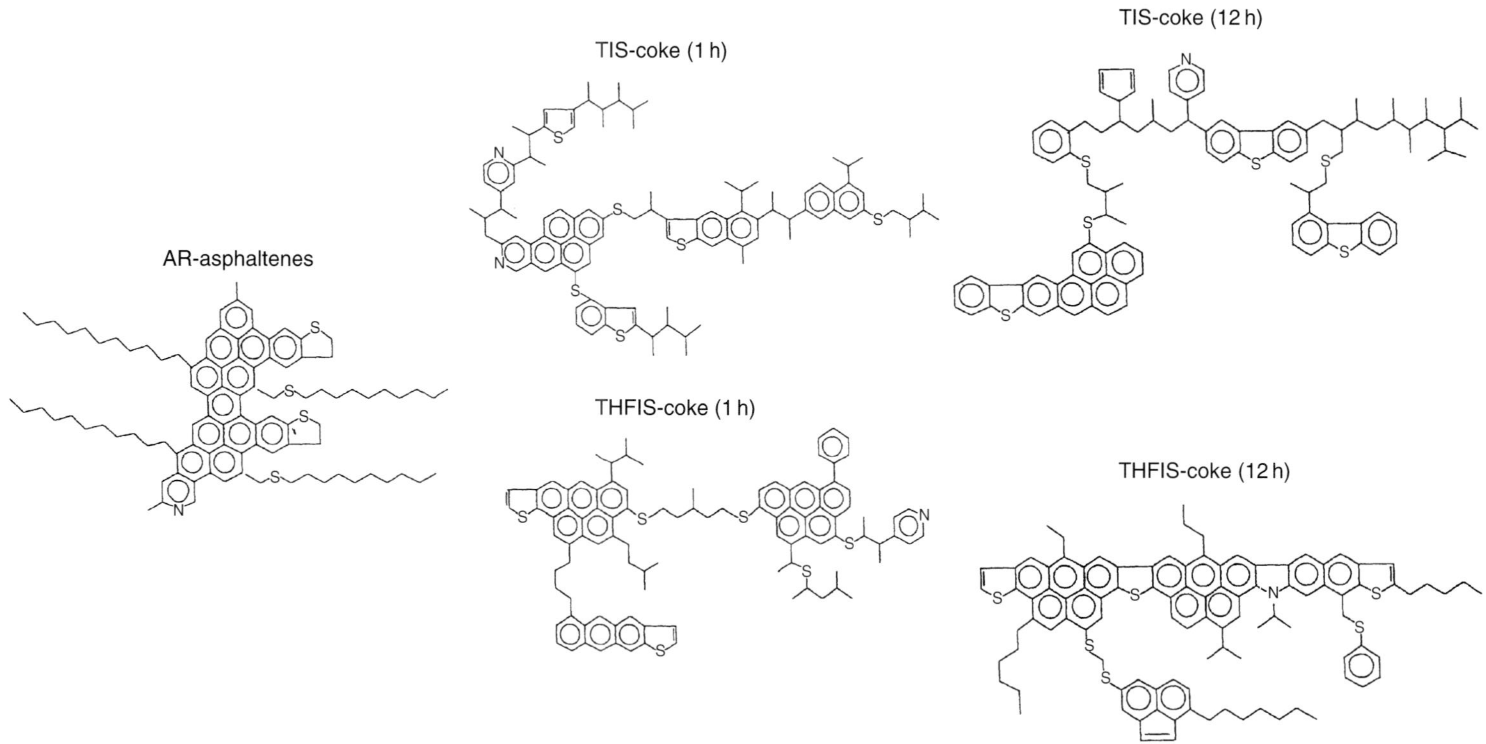

Figure 7.26 Effect of time on stream on structure of TIS and THFIS of coke on catalyst (Mo/Al_2O_3, AR, 653 K, 12 MPa) (274).

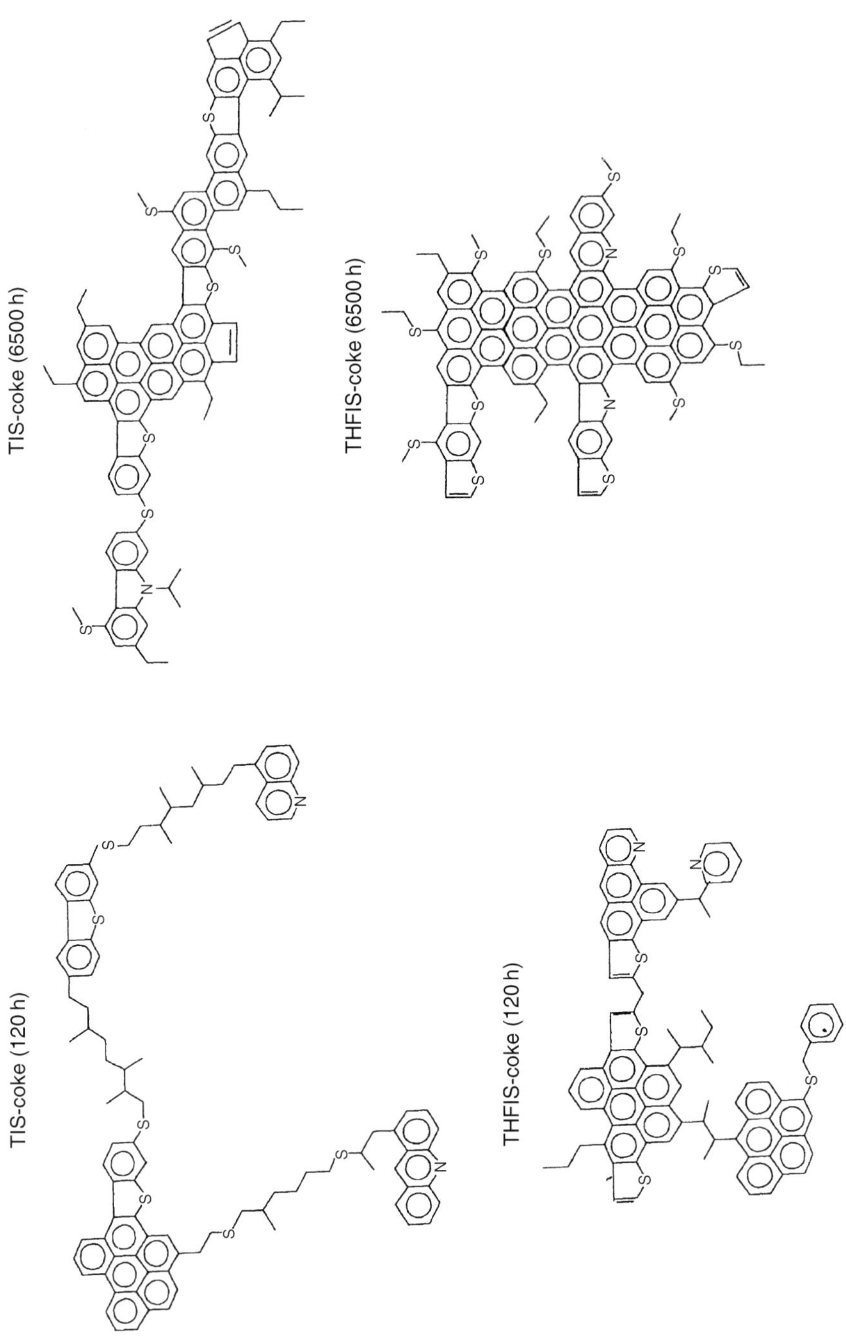

Figure 7.27 Effect of time on stream on structure of TIS and THFIS of coke on catalyst (Mo/Al_2O_3, AR, 653 K, 12 MPa) (274).

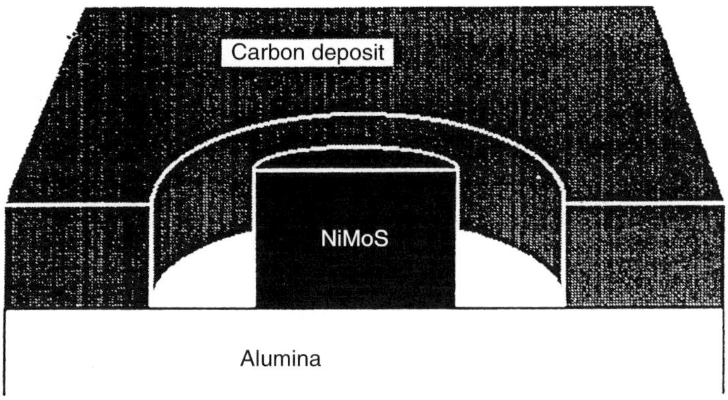

Figure 7.28 Schematic representation of catalyst surface showing cross-section through metal crystallites and coke deposit (105).

whereas V tends to catalyze deHYD of coke, particularly under limited availability of hydrogen, i.e., in the later stages on stream when catalyst was already significantly deactivated (63). This would indicate a higher aromaticity of coke in the vicinity of V deposits.

Matsushita et al. (388) evaluated the spent catalysts from the study of Hauser et al. (274,278) using the TPO, DRIFT and FTIR techniques. During the TPO (Figure 7.29), they observed the two maxima of CO_2 formation, i.e., one at 573 K and the other at 698 K which were formed from the oxidation of a "soft" coke and a "hard" coke, respectively.

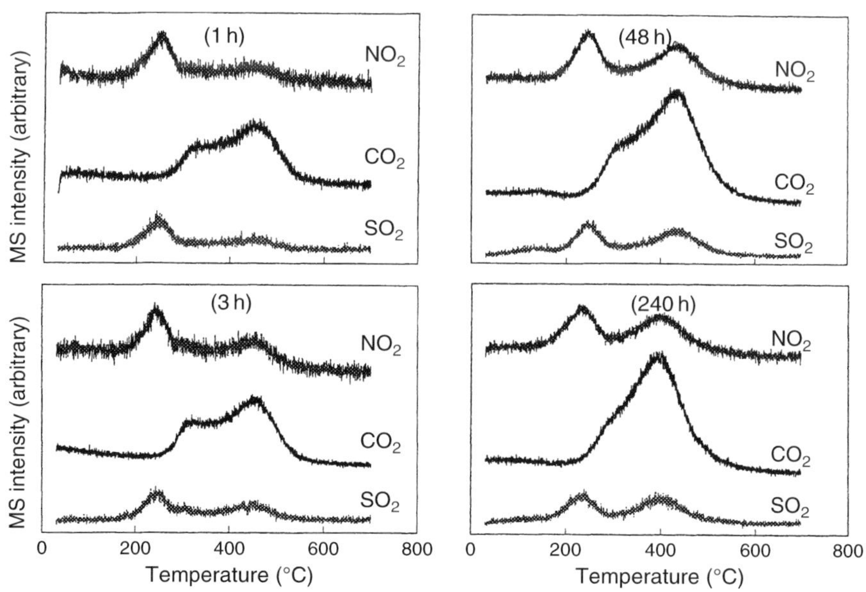

Figure 7.29 Effect of time on stream on TPO profiles of spent catalysts (Mo/Al_2O_3, AR, 653 K, 12 MPa) (388).

It should be noted that in the study of Seki and Yoshimoto (360), the "hard" coke was defined as TIS-coke. The "hard" coke studied by Hauser et al. (278) was much more refractory and was adsorbed strongly on the catalyst surface. In its structure, the "soft" coke may approach the structure of the TIS-coke formed within the first 120 h shown in Figures 7.26 and 7.27, whereas the "hard" coke that of THFIS-coke formed after 6500 h. As expected, the latter coke had very low solubility. It is believed that the "soft" coke was formed predominantly on the uncovered support. With time on stream, the "soft" coke was gradually converted to more refractory coke. This was supported by the decrease of the low-temperature and increase of the high-temperature CO_2 peak between 1 and 240 h shown in Figure 7.29. After 120 h on stream, most of the coke was refractory and was insoluble in THF. More nitrogen and sulfur were concentrated in the refractory coke than in the "soft" coke These observations complement the results in Figure 7.25 and the mechanism shown in Figures 7.26 and 7.27 proposed by Hauser et al. (274,278). The distinction between the "soft" coke and "hard" coke in Figure 7.29 may be affected by the involvement of active metals during TPO. Thus, the transition metals catalyzed oxidation of coke, i.e., oxygen transfer from active metals to carbon has been proposed by several authors (389–391). Then, the coke in the vicinity of active metals would exhibit higher reactivity (e.g., be "softer") than its actual reactivity at the end of hydroprocessing operation.

Rather unique approach to coke characterization was used by Ali et al. (381,392) who compared the ASE with the Soxhlet extraction of spent catalysts. The former was carried out at temperatures higher than the solvent's boiling point at total pressure of about 10 MPa. Under the conditions applied during ASE, a substantial amount of coke was extracted within 5 min. The sequential ASE extractions were performed using heptane, toluene, THF, methanol and dimethylchloride. All spectroscopic parameters indicated the increase in aromaticity in the same sequence. The solubility of coke decreased with time on stream. For example, about 75% of the start-of-run coke was soluble compared with about 10% of the end-of-run coke. This was accompanied by the gradual increase in aromaticity. These observations qualitatively agree with the generally observed trends. However, rapidity of the ASE method compared with the conventional extraction methods provides the opportunity for the generation of a large number of samples required for the analytical and spectroscopic characterizations.

7.6.1.2 Feed compatibility aspects

The discussion on the properties of heavy feeds (Chapter 2) indicated that the colloidal stability of heavy feeds may be affected when resins are converted at a greater rate than asphaltenes. Similar effect would have a high rate of the HYD of oil fraction. To improve pumpability and/or delivery to the reactor, heavy feeds are sometimes blended with lighter fractions. In this case, the choice of the light fraction is crucial for maintaining compatibility of the colloidal system. Compatibility is a non-issue for VGO and HGO because in these systems the oil phase is predominant. The situation may be less clear for the DAO feeds. In this case, the conditions of deasphalting and origin of the feed may have a pronounced effect on the colloidal structure and stability of the feed.

It is evident that during hydroprocessing, the asphaltene entities in the same feed may exhibit a wide range of reactivity. Thus, the most soluble part (the least polar) of asphaltenes may be the most reactive, whereas the insoluble part is the least reactive.

The latter part, sometimes referred to as carboids, can be separated from the asphaltenes by solvent precipitation (1). After most of the reactive portion of asphaltenes was converted to lighter fractions, the remaining carboids may be responsible for the coke deposition during hydroprocessing together with this part of asphaltenes which after deposition on catalyst surface were gradually converted to the high aromatic entities such as shown in Figure 7.27. Because the content of carboids in asphaltenes from different heavy feeds is different, their coke-forming propensity will be different as well. Nevertheless, carboids, as the least reactive component of micelle, may be partly responsible for the increased aromaticity of asphaltenes isolated from the products compared with that in the corresponding feed, sometimes reported in the scientific literature.

The results on solubility of asphaltenes published by Matsushita et al. (393) complement the mechanism proposed by Seki and Kumata (383,384). The former authors introduced the solubility index defined as the ratio of the H/C ratio of asphaltenes to that of the DAO obtained from the same AR using the different solvent/AR ratio. The decreasing solubility index would indicate the loss of alkyl chains in asphaltenes (increase in aromaticity), in agreement with the observations made by Callejas et al. (304). Thus, the paraffinic hydrogen in alkyl chains is an important contributor to the total hydrogen. This would decrease the solubility of asphaltenes in oil and enhance their deposition on the catalyst surface. Also the precipitation of asphaltenes from the products would be enhanced. The onset of the asphaltenes precipitation can be established using the critical solubility parameters, which can be determined by the flocculation onset titration method (394).

Figure 7.30 (393) shows the effect of the solubility index (defined above) on the amount of coke on catalyst and the H/C ratio of the coke. For this purpose, the feeds and products from different runs were separated into the asphaltenes (heptane insolubles–toluene solubles) and maltenes (heptane solubles). The HDM and HDS catalysts were of the Mo/Al_2O_3 and $NiMo/Al_2O_3$ formulation, respectively. These results suggested that the phenomenon known as fouling does not depend on the origin of catalyst, but rather, it is determined by the colloidal stability which depends on the chemical structure of the heavy feed components such as asphaltenes, resins and oil. It was evident from Figure 7.30 that coke deposition was slowing down with the increasing solubility index, i.e., with the decreasing H/C ratio of resins (heptane solubles). This was indeed confirmed by Ali et al. (395) during hydroprocessing of the Kuwait VR in the fixed bed reactor between 678 and 715 K and 13.5 MPa. These authors used three diluents varying in aromaticity and observed the smallest coke formation on catalyst surface for the most aromatic diluent. Using a similar approach, Li et al. (396) developed the colloidal solubility function which predicted coke formation during hydroprocessing of the VR derived from Chinese crude. It showed that any treatment to VR, which decreased its colloidal stability, resulted in the enhancement of coke deposition on the catalyst surface. Then, the coke formation was less pronounced for the VR which possessed a high colloidal stability.

It was pointed out above that the colloidal stability of a heavy feed becomes of crucial importance in the cases that a high viscosity of VR is decreased by blending with lighter fractions to enable feeding into hydroprocessing reactor. Without blending with lighter fractions, some heavy VRs could not be processed. In the study of Yanfei et al. (397), the VR derived from a Chinese crude was blended with the catalytic cracking bottom

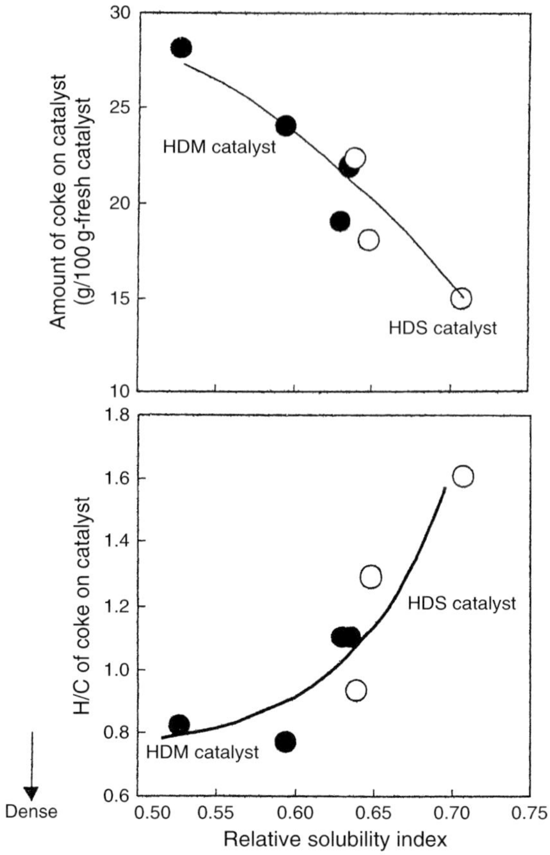

Figure 7.30 Amount of coke on catalyst and H/C ratio of coke versus relative solubility index (Mo/Al_2O_3, AR, 653 K, 12 MPa) (393).

(CCB) oil. Two opposing phenomena were observed, i.e., one which had asphaltenes dissolution effect and the other a dispersion effect. Relative contribution of these effects was CCB/VR ratio dependent. Thus, for high ratios, the flocculation of asphaltenes was quite evident, indicating that the mixture was incompatible. A high rate of coke formation is anticipated during hydroprocessing of such a mixture.

The mixture of decane with the VR derived from Athabasca bitumen showed a phase behavior involving three phases (398,399), i.e., a high-density liquid having a high concentration of asphaltenes and metals, a low-density liquid having a low concentration of asphaltenes and metals and a vapor phase containing predominantly decane. At 653 K, relative amounts of these phases changed with changing the H_2 pressure and VR concentration. Moreover, the phases containing mixtures of these liquids and vapor were also present. The high-density liquid was predominant at the low concentration of VR in decane. Increasing the concentration of VR above 20 wt% resulted in the appearance of the low-density liquid and low-density liquid + high-density liquid phase. This was an indication of sedimentation conditions. The low-density liquid and low-density liquid + vapor phase were predominant at high VR concentrations. During the catalyst testing at 653 K, the most pronounced coke deposition was observed under sedimentation conditions, i.e., the conditions favoring the high-density liquid phase, as well as the high-density liquid + low-density liquid + vapor phase. This region extended from about 20 to almost 50 wt% of the VR in decane. The low-density liquid was predominant at high H_2 pressure and the VR concentrations in decane exceeding 50 wt%. These experiments suggest that the coke deposition may be controlled by adjusting the operating condition to give a low-density liquid system. This may require a determination of an optimal combination of the temperature, H_2 pressure and type of catalyst. An important conclusion can be drawn from these observations, i.e., the amount of coke precursor in the heavy feed may not be the main parameter influencing the rate of coke formation on the catalyst surface. They further show that for graded systems, the relative contribution of asphaltenes to coke deposition in the downstream reactors may be enhanced due to the dilution effect in spite of the decreased concentration of asphaltenes in the process streams. In other words, with respect to coke propensity, the structure of asphaltenes may be a more important factor than their concentration in the feed. For example, for the same quantity in the feed, more aromatic asphaltenes will exhibit a higher coke propensity compared with the less aromatic asphaltenes because of the lower solubility of the former in the oil phase. Then, the catalyst deactivation will be more evident for the former. Similarly, for a heavy feed containing the same amount of asphaltenes having the same aromaticity, but the H/C ratios of their oil fractions are different, the coke deposition may be more pronounced for heavy feed whose oil fraction is less aromatic.

The studies of Mochida et al. (135,400) showed that the compatibility problem in the graded hydroprocessing systems may be alleviated by optimizing operating parameters. Thus, an extensive deposit formation in the one-stage system could be alleviated when a two-stage system was used. In the latter case, the first reactor was operated at 663 K under conditions favoring the HYD and asphaltenes depolymerization. The high rate of asphaltenes conversion at relatively short contact time was achieved in the second reactor which was operated at 693 K. The large pore $NiMo/Al_2O_3$ catalyst was necessary to achieve these results. The studies of Mochida et al. (135,400) focused primarily on the sludge formation in products; however, they are mentioned here because similar factors are involved during the deposit formation on the catalyst surface.

7.6.1.3 Microscopic phenomena

The microscopic evaluations of spent catalysts represent another source of the information to study the mechanism of catalyst deactivation. Such micrographs usually reveal the presence of mesophase, i.e., the spherical domains which exhibit characteristics of liquid crystals. The mesophase is more dense, has a higher surface tension and wets catalyst surface better than the phase from which it was originated. From the structural point of view, this is consistent with the loss of long aliphatic chains from the coke precursors. These chains contributed to the steric hindrance between the catalyst surface and coke precursor, and as such inhibited the wetting of catalyst surface. The mechanism of coke formation involving mesophase as an intermediate phase was proposed by

Beuther et al. (401). With TOS, the liquid crystals could be converted to coke whose structure was changing progressively. This involved ordering and stacking of aromatic sheets. This may be considered as the very early stage of graphitization, the rate of which increases with increasing severity. Figures 7.26 and 7.27 (274,278) offer some support for this mechanism. Thus, the coke after 6500 h represents a sheet which possesses high aromaticity. The stacking of such sheets into platelates may have occurred particularly when the catalyst was approaching the end of its life, i.e., at this point, the active surface hydrogen was very limited. Such conditions favor aromatization, as it was observed by Fetzer (402). Therefore, the rate of change in the coke structure (both chemical and microscopic) at the latest stages of the operation may be rather high.

In the mechanism proposed by Fetzer (402), small aromatic rings were converted to coronene, which was subsequently coupled to form either dicoronylene or even higher molecular weight polycondensed aromatic hydrocarbons. These structures are typical of those present in the sheets of graphite. It is believed that these structures would gradually stack to form the graphite-like crystallites, as it is shown in Figure 2.8. The results published by Sullivan et al. (403) provide some support for the mechanism proposed by Fetzer (402). In the former study, the temperature employed during hydroprocessing was observed to be of primary importance. Thus, the formation of large PAHs was observed as soon as the rate of dehydrocyclization reactions approached and/or exceeded the rate of HCR reactions (404). Under these conditions, the availability of active surface hydrogen was affected. Because of their size, the large PAHs could not attain the activated adsorption on catalytic sites and as such remained unconverted on the catalyst surface before being converted to coke deposits.

The optical microscopy of the polished cross-sections of a series of the spent catalysts used during hydroprocessing of the Athabasca bitumen was investigated by Munoz et al. (405) and Gray et al. (406). The fluorescence due to the presence of the feed components and anisotropy due to the presence of mesophase were observed in addition to the high reflectance which indicated the presence of domains having higher aromaticity than surrounding matrix. This was an indication of the gradual conversion of heavy components in the feed to mesophase which subsequently converted to the high aromaticity species. This was supported by the absence of the feed components and predominance of high aromaticity domains on catalysts after more severe conditions, i.e., higher temperature and longer TOS. The observations made by Munoz et al. (405) are in a good qualitative agreement with the other studies (401–404).

The model of coke proposed by van Dorn et al. (329,330) consisted of the three-dimensional layers of aromatic structures which contained heteroatoms and short alkyl groups. The model further assumed that the coke precursors in the vicinity of active phase were removed, presumably with the aid of active hydrogen which spilt over from the active metal phase (63,69,70). Further studies revealed that the shape of catalyst particles may influence the coke deposition pattern (331). This was indicated by the video image of the cross-section of the NiMo/Al_2O_3 quadro-lobe extrudates which revealed significant differences in the coke amount and structure among the four incisions of the extrudates. This could be attributed to several factors, i.e., the variations in local amount of active metals on the catalyst, occurrence of hot spots and the orientation of the extrudates in the trickle bed used for hydroprocessing of the VGO feed.

7.6.2 Mechanism of metal deposition

It has been established that in heavy feeds metals can be present in both inorganic and organometallic forms. It is believed that when the former are kept undisturbed, a concentration gradient may develop in the case that the viscosity of heavy feed is decreased either by dilution with a solvent or by increasing temperature. As a result of this, the concentration of some inorganic solids in heavy feed will gradually increase toward the bottom. At the same time, little concentration gradient should be present for the organometallic compounds unless colloidal stability of the system was disturbed. These differences indicate that the deposition mechanism involving inorganic solids should differ from that of the organometallic forms of metals. Moreover, for organometallic compounds, the different form of deposits is formed during the non-catalytic demetallization of porphyrins via the reactions with H_2 and H_2S compared with the catalytic HDM of porphyrins.

7.6.2.1 Deposition of inorganic solids

Such solids include minerals and clay-like solids which may have entered heavy feed either in reservoir or during the production of crude oil. For bitumen separated from tar sands, part of the mineral matter originated from the caustic material used in the hot water separation process. In addition, a small amount of the finely divided particles of sand still remained in the bitumen after the separation process. This type of inorganic solids deposits on the external surface of the catalyst particles, if not removed from the heavy feed prior to hydroprocessing. In its nature, the deposition mechanism is physical. This is confirmed by the predominant accumulation of the inorganic solids on the external surface of catalyst particles in a "skin-like" form. Indeed, a high content of Fe and Ca was noted in the "skin" on the spent catalysts from hydroprocessing of Athabasca bitumen (406). A similar form of deposition may undergo the sulfides of V and Ni formed during the non-catalytic reaction with H_2S. However, rather small size of the sulfides of V and Ni particles formed in this way is expected. This suggests that such particles may deposit in large pores of catalyst particles rather than on their exterior.

In the crudes, the solids containing alkali and alkali earth metals are predominantly in an oxidic form. They are usually associated with SiO_2, Al_2O_3 and SiO_2–Al_2O_3 in the form of clays and minerals. Under hydroprocessing conditions, the oxidic form of alkali and alkali earth metals is gradually converted to the corresponding sulfides similar to the Fe oxides. If present, finely divided water emulsions may contain chlorides of alkali metals. The problem with such solids is alleviated by dewatering of the crude oil as soon as it entered petroleum refinery.

The operating problems caused by the deposition of inorganic solids will be more evident in the fixed bed reactors than in the ebullated bed reactors. In the former case, the front of the fixed bed will be most affected. The crust-like layer created by the deposition of such solids on the front of the bed may affect the operation by creating channels and developing pressure drops through the bed. In this regard, attention should be paid to the possible contamination of the feed by metallic particles from the corrosion of upstream equipment. As such, iron scale or fine particles usually do not penetrate deeply the catalyst porous system and do not have any strong deactivating effect. It is more a concern as a contributor to the pressure drop build-up, as these particles may accumulate at the top of the bed or in the interstices between the catalyst granules. Similar problems may be caused by silicon which originates from the anti-foaming agents. Such agents are sometimes added to the feed before coking operations; therefore, they might be present in HGO. For some heavy feeds, a filtration system must be installed upstream of the catalytic reactor to avoid the operational difficulties caused by deposits of inorganic solids. Such function may be fulfilled by a guard chamber filled with a low-cost solid (e.g., alumina, bauxite, clays).

7.6.2.2 Deposits of organometallic origin

From the catalyst deactivation point of view, perhaps the least attention has been paid to arsenic which occurs naturally in many crude oils in an organic form. Under typical hydroprocessing conditions, the organic compounds of As are very reactive. Thus, they are either converted to arsine which ends up in gaseous products or readily decompose. In the latter case, they remain adsorbed on the catalyst surface. In a fixed bed reactor, a very steep arsenic gradient is observed between the inlet and outlet of the reactor. Once on the catalyst, As may be converted to a sulfide. Such conversion is favorable under typical hyroprocessing conditions. Arsenic may be a severe poison, particularly for HDS. This was more evident for regenerated catalysts than for the fresh catalysts (407). In fact, the As in catalyst in excess of 0.3 wt% prevented regeneration of the spent catalyst used during hydroprocessing under moderate conditions.

The organometallic forms of V and Ni in crudes are the main cause of the metal deposit formation on catalyst surface. In crude oils, most of these metals are associated with the asphaltenes entities. The depolymerization of asphaltene aggregates may be considered as a very first stage of the metal deposit formation. As the result of this, porphyrins are released into the oil phase and become available for HDM reactions. The initial stage of the metal deposition may coincide with the final stage of the overall HDM of metal-containing compounds, i.e., the separation of metal from the pyrrole ligand skeleton. It is believed that the sulfides of V and Ni formed during the direct reaction of porphyrins with H_2S remain in the liquid streams in a finely dispersed form. As indicated above, their mode of deposition on catalyst surface will approach that of the inorganic solids, although they are of the organometallic origin.

Predominance of the V-containing compounds compared with the Ni-containing compounds in most of the heavy feeds suggests that the metal deposition patterns are influenced by the former to a much greater extent. Moreover, it was indicated that there is a significant difference between the reactivity of the V- and Ni-containing porphyrins. This was indeed demonstrated by the results shown in Figure 7.31 (408) obtained at 773 K between 13 and 18 MPa. using a Venezuelan VR. In this case, fine particles of a naturally occurring catalyst were slurried with VR containing 512 ppm of V, 77 ppm of Ni, 10.3 wt% of asphaltenes and 17.8 wt% of CCR. The form of catalyst ensured that the restricted diffusion phenomena were not in effect and as such could not contribute to the reactivity difference. As shown in Figure 6.10 (119), the higher reactivity of the V-porphyrins compared with Ni-porphyrins was also confirmed during hydroprocessing of the Khafji AR. Thus, the reactivity difference was maintained for different particle size of the catalyst. Figure 7.31 (408) represents one of its kind comparison of the effect of conversion on simultaneous removal of V and Ni from a heavy feed found in literature because in this case, the different conversions were obtained by using different modes of operation, i.e., once through and recycle at base H_2 pressure of $\sim$13 MPa (130 bar) and increased H_2 pressure. Nevertheless, it is clear that at low conversions, removal of the V compounds would be a much greater contributor to the deposit formation than that of

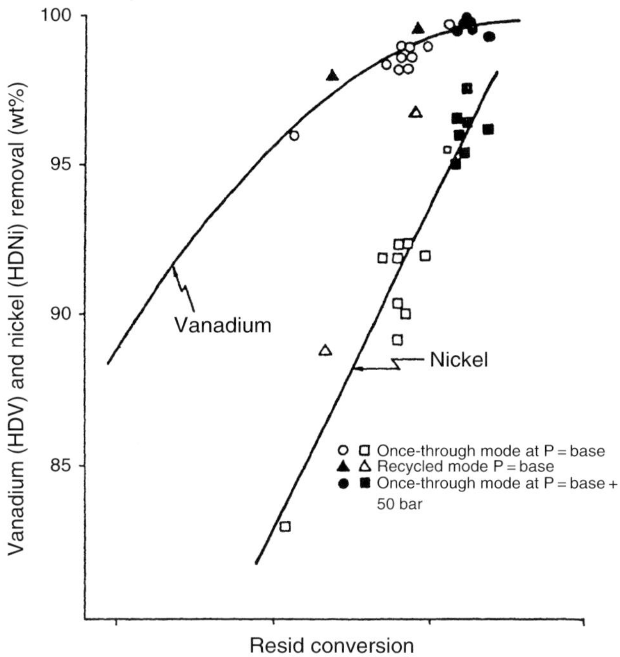

Figure 7.31 Vanadium and nickel removal versus residue conversion (408).

the Ni compounds. There is little reason to believe that under typical hydroprocessing conditions the situation will not be similar. In fact, because of the much stronger interaction of the VO-porphyrins with catalyst surface than that of the Ni-porphyrins, the reactivity difference may be even more evident. At high conversions, e.g., under severe hydroprocessing conditions, the rate of deposition of the Ni compounds will compete more successfully with that of the V compounds. Another way of looking at this issue is the change in conversion between the inlet and outlet of the fixed bed of a catalyst. Thus, it is believed that the conversion will increase toward the outlet of the fixed bed. Then, the contribution of Ni compounds to the deposit formation will increase in the same direction. This was indeed demonstrated by Tamm et al. (8).

7.6.2.2.1 Vanadium-containing deposits

The degradation of the VO-porphyrin structure in H_2 and particularly in H_2+H_2S mixture in the absence of catalyst reported by Rankel (54) may suggest that at least a part of the precursors to the metal deposit formation may not require a direct involvement of the catalyst surface. Apparently, metals can be extracted from porphyrin structures during the non-catalytic reaction with H_2S (281). Dautzemberg and de Dekan (225) suggested that the weakening of the nitrogen–metal bonds in porphyrins may be more pronounced in the $H_2 + H_2S$ mixture compared with pure H_2. They proposed the following mechanism to account for their observation:

$$\left.\begin{matrix}-N\\ ---N\end{matrix}\right\rangle V=O+2H_2S \dashrightarrow \begin{matrix}-N-H\\ -N-H\end{matrix}+VS_2+H_2O$$

Although it was formed without the involvement of catalyst, this V sulfide will most likely deposit on the external surface of catalyst particles. Because of more less physical deposition, this part of the V sulfides shall have much less detrimental effect on activity compared with that caused by the V-containing compounds separated from the porphyrins during the catalytic HDM. However, because of a nearly molecular size in which it is formed, such V sulfides may penetrate large pores of catalyst. It has been noted that with respect to the overall HDM mechanism, the role of H_2S has not yet been investigated in detail in spite of these observations (225,281).

The transformation of the vanadyl group to a V sulfide on the catalyst surface may be affected because of the steric hindrance. Thus, it is generally accepted that the V=O group in the vanadyl-containing porphyrins facilitates a strong adsorption with catalyst surface. This may prevent complete sulfidation of the V=O group after its separation from the porphyrin skeleton. Indeed, there are several reports on spent catalysts and fresh presulfided catalysts confirming only a partial sulfidation, with the V=O entity still being clearly identifiable by spectroscopic techniques (260,386). The V=O entity can interact with both the uncovered support and catalytically active metals deposited on the support (283). According to Loos et al. (284), the V=O moiety was still present, essentially unaltered in the spent $NiMo/Al_2O_3$ catalyst, although additional four sulfur atoms contributed to the average coordination polyhedron of V. This suggests that the sulfidation of V was incomplete compared with that of Mo and Ni/Co. The same was confirmed by Janssen et al. (285). Thus, during the sulfidation of the Al_2O_3 impregnated with the ammonium metavanadate, the complete conversion to V_2S_3 required a temperature of 1273 K, whereas at 673 K, most of the V was still present as an oxy-sulfide.

The presence of the unconverted porphyrins in deposits cannot be ruled out completely. In this regard, several potential forms of their interaction with catalyst surface have been identified. They include a donor–acceptor bonding, in which the π system of the porphyrin ring is the donor and the Bronsted and/or Lewis sites are acceptors (280). However, this information was obtained under low temperature conditions. There is little experimental evidence confirming the presence of unconverted porphyrins in coke. It is believed that because of the complex nature of deposits, a convincing identification of porphyrin structures in coke on the spent catalysts from hydroprocessing of heavy feeds would be a rather challenging task. When a sufficient active hydrogen is available, the adsorbed porphyrin can be hydrogenated, presumably in the mesocarbon position, as the initial stage of the overall HDM. In the final stage, the hydrocarbon products from the HDM of porphyrins end up in process streams, while metal deposits remain on the catalyst surface. This stage of the overall HDM mechanism is depicted in Figure 6.23 (291) by the formation of M-Bil as the last intermediate followed by its fragmentation and simultaneous metal deposition.

Asaoka et al. (280) showed that in the presence of a catalyst, there is a significant difference between the metal deposition patterns in pure H_2 and that in the H_2+H_2S mixture, as it was demonstrated by results in Figure 7.32. In the latter case, the precursor was converted to deposits at the first contact with catalyst surface. Then, the deposits progressively penetrated into the catalyst particle interior. Also, the amount of deposit was decreasing from the inlet toward the outlet of the reactor. The deposition patterns in pure H_2 agree with those frequently reported in the literature (40). It is again emphasized that the role of H_2S during overall HDM and deposit formation requires additional

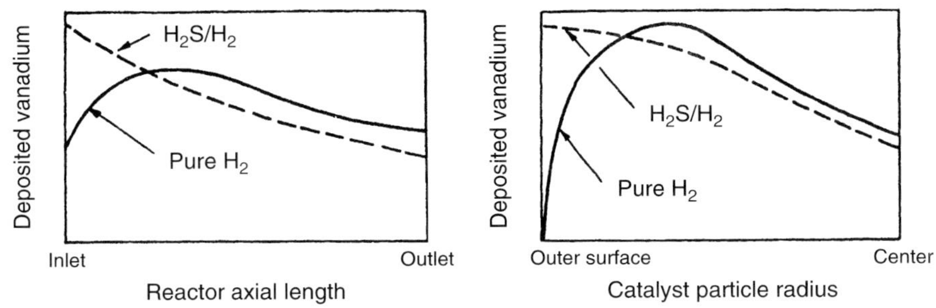

Figure 7.32 Effect of H_2S on distribution of vanadium through reactor length and particle radius (280).

attention. Spectroscopic evaluations of the deposits (formed in H_2+H_2S) carried out by Asaoka et al. (280) identified V_3S_4 as the predominant composition. In V_3S_4, V was present partly as V^{4+} and partly as V^0, with the proportion of the latter increasing toward the catalyst particle exterior. Contrary to the observations made by Asaoka et al. (280), Loos et al. (284) observed the formation of V_2S_3 rather than V_3S_4. However, the latter authors used the model VO-containing porphyrin rather than the heavy feed as used by Asaoka et al. (280), who also confirmed the presence of the unconverted V=O entity.

Kim and Massoth (409) pointed out that the structure of the V deposits formed during hydroprocessing of real feeds may differ from that formed during the treatment with model V-porphyrins. This was indicated by rather different effect of deposits on catalyst functionalities. Thus, the catalyst was much more deactivated by the real deposits than by those formed using model V compounds. The difference between the V/S ratio of the model deposits and the real feed deposits should be noted as well.

7.6.2.2.2 Nickel-containing deposits

Similarly as for the V-porphyrins, the HDM of Ni-porphyrins is influenced by catalyst as well as by the presence of H_2 and H_2S. In fact, H_2S promoted the non-catalytic HDM of Ni-porphyrins by reacting with Ni without requiring the presence of catalyst, similarly as it was observed for the VO-porphyrins (225,281). This was confirmed by Bonne et al. (410), who reported that in the presence of H_2S, the concentration of porphyrinic intermediates decreased. This suggests that the final product of the reaction of porphyrins with H_2S may be at least a partially sulfided Ni, although the main HDM proceeded via hydrogenolysis of the Ni–N bond releasing metallic Ni as the parallel reaction. Then, after deposition on the catalyst surface, Ni will be sulfided via established mechanism. Under typical hydroprocessing conditions, the complete sulfidation of Ni would lead to the formation of Ni_3S_2 sulfide. However, because of so many factors involved, a partially sulfided Ni and/or an oxo/sulfide forms of Ni may be present as well. The formation of Ni carbide during the prolonged exposure of catalyst under hydroprocessing conditions was proposed as well, although as of yet its presence has not been experimentally confirmed (206).

It is believed that the radial distribution of the Ni sulfides formed non-catalytically via reaction with H_2S should differ markedly from that formed catalytically via established HDM mechanism. The former shall deposit physically predominantly on the exterior of catalyst particles in a "skin-like" form, whereas the Ni-containing deposit formed as part of the HDM should be distributed more evenly. It may be rather difficult to distinguish between these two types of deposits on catalyst surface.

7.6.2.2.3 Mixed deposits

It has been generally observed that initially, the metal deposition occurred predominantly on the bare surface of the support (236,250,297,298). The thickness and/or size of the deposit was increasing progressively with TOS. The multilayer deposit would consist of the mixture of V sulfides (e.g., VS_2, V_2S_3 and V_3S_4) and V-oxosulfides as well as Ni sulfides (e.g., Ni_3S_2). The simultaneous deposition of V and Ni supports the formation of mixed sulfides ($Ni_XV_YS_Z$). The formation of a mixed $(Fe,V)S_4$ sulfide was reported by Embaid et al. for the Fe-containing heavy feed (299). The ratio of the V to either Ni or Fe in the mixed sulfide deposit will change from the exterior toward the center of the catalyst particle, i.e., in the case of Ni, the V/Ni ratio will decrease as more Ni-porphyrins than V-porphyrins can penetrate deeper into the catalyst particle interior. At the same time, the V/Fe ratio may increase toward the particle interior because most (if not all) of the Fe deposited on the exterior of catalyst particles. Pore volume and size distribution of the catalyst may play a key role in determining this ratio. These events can be quantified by the distribution parameter discussed in Section 3.2.2 (85,114).

It was suggested that before their separation from porhyrins the V and Ni may coordinate with sulfur of the active metal sulfide already deposited on the catalyst surface (263). Potential coordination with the active phase such as Co(Ni)–Mo–S could lead to the change in activity of the active sites. In this regard, V is expected to have more detrimental effect than Ni. Thus, its interaction may lead to the formation of the V–Mo–S phase which is less active than the Co(Ni)–Mo–S phase. With progressive growth of the metal deposits, the pore diameter becomes less than the molecular diameter of porphyrin molecules. This prevents the access of the reactant molecules to the interior. At this stage, an abrupt loss of the catalyst activity for HDS was observed (339). The TOS at which such point may be approached depends on the metal content in heavy feeds and the catalyst porosity. This was demonstrated by the results in Figure 7.33 (338). In this case, the same HDM ($NiMo/Al_2O_3$) catalyst was used for hydroprocessing of the Gach Saran and Cerro Negro long residues containing 250 and 660 ppm of V + Ni, respectively. As expected, the 100% metal retention by HDM catalyst was approached much faster for the heavier feed. However, to a great extent, the metal retention and/or metal storage capacity depend on the catalyst porosity. For example, for typical HDS catalyst, metal retention before almost total deactivation may approach 20 wt% or even less (Figure 5.6). Although the sudden decline in HDS activity of HDM catalyst was observed at 50% metal retention, its activity for HDM was still retained, suggesting that the deposition of metals could continue beyond this point (28,341). It is believed that the chemical composition and structure of deposits at this level of metal retention are rather complex and will change vertically from the outer surface to the bottom of the deposit layer, which is in contact with the catalyst surface. In this regard, properties of catalyst, its porosity in particular, may play an important role.

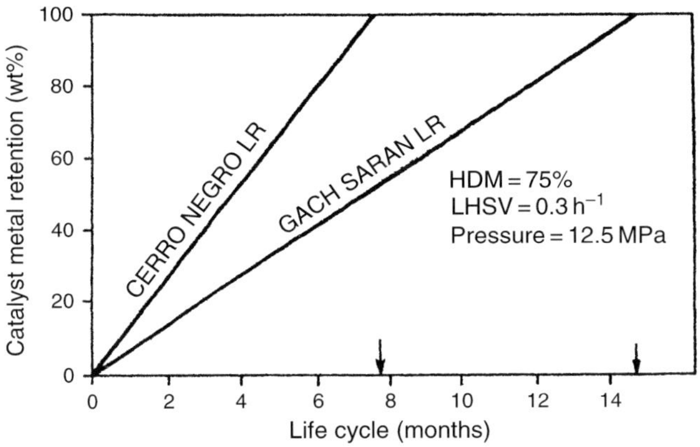

Figure 7.33 Effect of feed origin on metal retention by catalyst (338).

7.7 DEVELOPMENT OF MODELS FOR PREDICTING CATALYST DEACTIVATION

Mathematical models have been developed and used for predicting long-time performance of the catalysts during hydroprocessing of petroleum feeds. In this regard, rather extensive information can be found in the scientific literature. In the following, a summary of these efforts is only given, although the wealth of information suggests that a separate book can be written on this subject.

The first stage of model development involves derivation of the mathematical expressions which may predict the performance of catalyst. Subsequently, the validity of these equations is verified using the experimental data. After modifications, these models can be used to generate the database for selecting catalysts to match the feed of interest with a suitable reactor as well as to predict a long-term performance of the system. The database necessary for the design of catalysts and reactors to suit a particular feed can be readily generated using the models as well.

Models incorporate changes in the interphase, intraphase and interparticulate gradients of temperature and concentration with TOS. The complexity of chemical structure of heavy feeds compared with light feeds suggests that the development of models to simulate hydroprocessing of the former is much more challenging than that for light feeds. It requires the determination of performance parameters such as the change in catalyst activity for hydroprocessing reactions with TOS, the parameters accounting for catalyst deactivation, metal storage capacity of catalyst, pore size distribution, etc. The experimental techniques for determining most of the required parameters are now available. Specific parameters which can be determined by kinetic studies include the intrinsic and apparent rate constants, activation energies, effective diffusivities, efficiency factor, distribution parameter and Thiele modulus. With such parameters available, modeling can be conducted on two levels of scale, i.e., catalyst particle level and the catalyst's active phase level. Modeling on a reactor level requires the information on liquid holdup, height and diameter of reactor, volume of reactor and catalyst bed, superficial liquid velocity, etc. It has been noted that most of the studies were conducted on more than one level of scale.

The feeds ranging widely in properties have been included in modeling studies (40,76). It has been generally observed that the models are feedstock and catalyst structure dependent. Therefore, the model developed for a particular feed may require some modifications in order to predict the catalyst performance using a different feed. Models take into consideration initial and steady-state deactivation by coke which deposited on catalyst surface, as well as a more or less linear and continuous deposition of the metals from heavy feeds. Attempts have been made to simulate deactivation by coke deposition occurring during the very early stages of the contact of feed with catalyst. The validity of models can be verified using the data from the experiments on accelerating aging carried out in bench scale units, pilot plants and from commercial reactors. The conflicting results obtained during the accelerating aging experiments and those obtained in the commercial units have been noted (312). Then, with respect to model development, the former results may have a limited validity. However, the accelerating aging test developed by Alvarez et al. (342) could predict the catalyst performance in a pilot plant unit quite accurately.

The importance of the hydrogen solubility in heavy feeds should be noted, although this issue was receiving little attention during the early attempts to develop models for the predicting catalyst life. It was shown by Ronze et al. (411) that the Henry's constant for dissolution of H_2 in a gas oil decreased by about half by increasing temperature from 573 to 673 K. Then, at higher temperatures, the H_2 pressure has to be increased to maintain a similar solubility. This problem can be overcome by operating under a sufficiently high H_2 pressure in the temperature range of interest to ensure an excess of hydrogen in the liquid phase. The close examination of most of the studies on model development indeed indicates the operation involving the excess of hydrogen (a high H_2 pressure). The excess of hydrogen ensures a high diffusivity of gaseous hydrogen into the liquid phase. Subsequently, the hydrogen dissolved in the liquid phase is transferred to the catalyst surface for hydroprocressing reactions to occur. The H_2 pressure has a little effect on the total amount of active hydrogen on the catalyst surface (63). However, the H_2 pressure influences the rate of achieving the catalyst surface saturation with active hydrogen. Therefore, such state of catalyst surface can be maintained in a wide temperature range if the sufficiently high H_2 pressure is attained.

7.7.1 Modeling on catalyst activity level

In essence, kinetic parameters determined experimentally form a basis for catalyst evaluation on the active phase level. Numerous kinetic models were developed and their summaries were presented in Sections 6.1 and 7.5 for kinetics of hydroprocessing reactions and kinetics of catalyst deactivation, respectively. In most cases, experimental data were compared with kinetic laws, i.e., either power form or LH form equations. Usually, reaction order was chosen to obtain the best fit of the experimental results for a particular kinetic model.

The studies of Long et al. (297,298) deserve attention in spite of their focus on the model compound porphyrins rather than on real feeds. These authors made an attempt to develop models for simulating the change in the catalyst HDM activity during the early stages on stream. The modeling was performed on a catalyst activity level. Figure 7.34 shows a very good fit of the experimental results with those predicted by the model (solid line). However, such a good fit could only be obtained by assuming that during

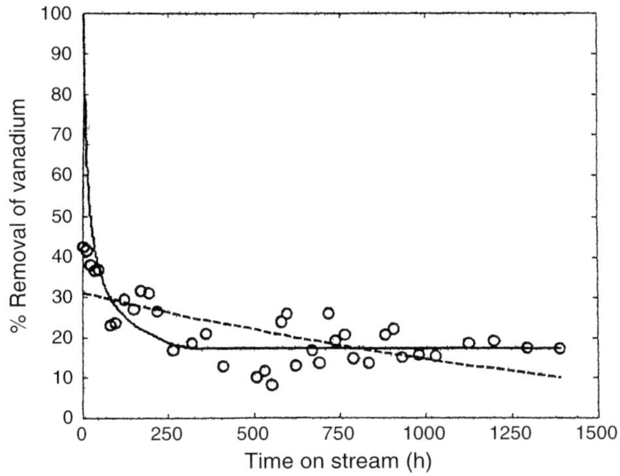

Figure 7.34 Removal of vanadium vs time on stream (CoMo/Al_2O_3, VR, 643 K, 11 MPa) (o) experimental data; ----- model results using simultaneous adsorption; — model results using poisoning mechanism (297).

the very early stages, part of the V was deposited on the uncovered support, thus having little detrimental effect on the active metal phase. It is evident from Figure 7.34 that the model based on an assumption of the classical poisoning of catalyst is less suitable to simulate the early stages on stream.

The results of Long et al. (297,298) do not support the uniform metal layer deposition models proposed by Sato et al. (412) and Newson (413) during very early attempts to develop the models predicting catalyst performance. Based on the latter models, for catalyst having surface area of 200 m^2/g, the complete catalyst deactivation could be attained at 20 wt% of V_2S_5, i.e., after about 4 days on stream (414). After few months, there would be five to six monolayer equivalent of the deposits on the catalyst surface. Therefore, the metal sulfide deposition on the bare catalyst support in a crystallite form appeared to be a more plausible explanation than the deposition in layers.

It has been observed that metal deposits (e.g., V_3S_4) exhibited catalytic activity for HDV, HDAs and HDS (280). This was supported by the results in Figure 7.35. Thus, the activity of bare support increased with the increasing amount of V deposits and approached that of the V/bare support catalyst at about 10 wt% of deposit. As one would expect, the effect of the V deposit on HDS differed from that on HDV and HDAs. Similar trends in HDV and HDAs confirmed the importance of the latter for overall HDM. In order to be incorporated in models, the autocatalysis would have to be quantified. This, of cause is not an easy task. Therefore, the autocatalysis may be at least partly responsible for deviation of the predicted results from those observed experimentally. This may be evident, particularly during the early stages of operation. The model tested by Melkote and Jensen (415) was among few, in which the effect of autocatalysis was considered. Moreover, the non-catalytic HDM may be another source of the uncertainties in applying models to real situations. In this case, the V and Ni sulfides formed via direct reactions of the porphyrins with H_2S, in the absence of catalyst, may deposit predominantly on the exterior of catalyst particles. This may be a sort of

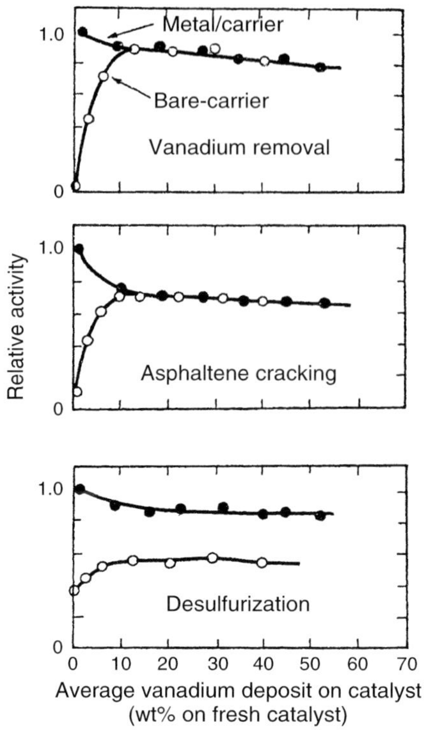

Figure 7.35 Catalytic activities of deposited vanadium (280).

physical deposition. It is believed that such interference may be quantified by performing experiments in the absence of catalytically active metals, under otherwise identical conditions. A correction to the models can be applied when such information is available.

It has been established that deactivating effect of coke on the catalyst functionalities may differ from that of metals. This was clearly demonstrated by Gualda and Kasztelan (100,255) who reported that in spite of a rapid coke build-up during initial stages, the HYD activity was severely poisoned predominantly by very small amounts of V. At the same time, the loss in HDS activity due to the V deposit was much less evident. In accordance with this observation, several studies on the development of models for predicting catalyst life focused primarily on the effect of metals. In this regard, the most detailed account of catalyst deactivation by metals was given in the work of Tamm et al. (8), who used five ARs, the metal content (V + Ni) of which varied from about 40 to almost 500 ppm. Their model confirmed that the metal deposition patterns were feedstock dependent. Both the poisoning of active sites by metals and physical obstruction of pores by metals were the contributors to catalyst deactivation.

The molecular size distributions determined by the size exclusion chromatography with an inductively coupled plasma detector was included in the model developed by Sughrue et al. (416) to predict the life of HDM catalysts. The predictions of catalyst life were more accurate using an average molecular size of the heavy feed compared with its molecular size distribution. This is not surprising when the difference in conditions used for the determination of the size distribution and those used to obtain catalyst deactivation parameters are taken into consideration. Thus, the asphaltene disintegration may have occurred under the latter conditions, whereas the same has been less evident during the molecular size determination. As a result of this, the model predicted a shorter catalyst life at a high HDM conversion and a longer catalyst life at low conversion.

7.7.2 Modeling on catalyst particle level

The surface properties (e.g., surface area, pore volume and pore size distribution), size and shape of catalyst particles are of primary interest for designing catalysts for hydroprocessing of the metals and asphaltenes containing feeds. For this purpose, parameters such as the effective diffusivity, efficiency factor, distribution parameter, Thiele modulus and metal storage capacity are included in the models. The mathematical equations defining some of these parameters were discussed earlier (Section 3.2.2). In addition, the development of models on the particle scale would be incomplete without incorporating data on catalyst activity. This indicates the need for determination of kinetic data, as well as that of the catalyst deactivation pattern. Therefore, it may also be appropriate to refer to this level of model development as the two-scale approach, i.e., an active phase and a single particle scales. Thus, the applicability of the models on particle scale would be somehow limited without including the effects of active phase on the catalyst performance with TOS.

The study of Oyekunle and Ikpekri (417,418) illustrated the usefulness of the particle scale models for designing the catalysts for hydroprocessing of heavy feeds. They performed calculations for the three types of catalysts, i.e., microporous and macroporous, with the predominant portion of pores having APD <100 Å and APD $=100$–250 Å, respectively, as well as the random pore distribution with the predominant APD between the microporous and macroporous catalysts. The following equations had to be solved to determine the catalyst activity decline with TOS:

$$\frac{\partial^2 \alpha}{\partial \beta^2} = \Phi^2 \alpha^2 + \frac{\partial \alpha}{\partial \tau} \quad \text{and}$$

$$\frac{\mathrm{d}\gamma}{\mathrm{d}t} = \frac{\epsilon k (C)^2 M \alpha^2}{\rho r}$$

where α and β are the dimensionless reactant concentration and dimensionless distance along the pore, respectively, whereas Φ is the Thiele modulus and τ is the dimensionless time, respectively. In the second equation, γ is the ratio of the instantaneous pore radius to the initial pore radius, ϵ is the void fraction, k is the rate constant per unit of the surface area, C is the instantaneous reactant concentration, M is molecular weight of the metal sulfides, ρ is the density of metal sulfides and r is the mean pore radius. The incorporation of rate constant in the above equation illustrates the need of modeling on the activity phase level to enable modeling on the particle level. Oyekunle and Ikpekri (417) used the data on hydroprocessing of the heavy Maya crude published by Ancheyta et al. (142,358) to test their models. Figure 7.36 shows a good fit of the process data with those predicted by the models. The catalyst lifetime was then estimated by using the linear regression analysis of the results in Figure 7.36. The total activity loss was predicted to occur after 462, 316 and 150 days for the macropore, random pore and micropore systems, respectively.

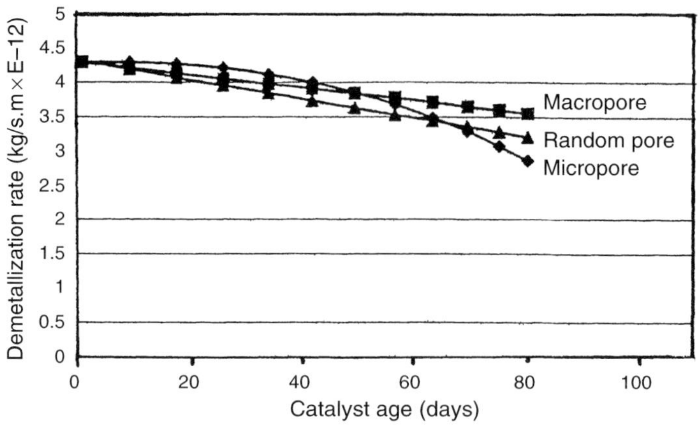

Figure 7.36 Variation of demetallization rate with time on stream for different models (417).

The single particle model developed by Melkote and Jensen (415) considered a uniform and discrete deposition of metals as part of the HDM deactivation mechanism using the Boscan crude and the Heavy Arabian feed. In this case, the effect of both H_2 pressure and temperature on the radial distribution of Ni and V was considered. The best agreement between the experimental and calculated data was obtained for the discrete deposition rather than for the uniform deposition. In this study (415), the effect of autocatalysis on HDM was also addressed, compared with most of the studies in which this fact is usually overlooked.

The model used by Pereira and Beeckman (10) assumed catalyst deactivation due to the build-up of metal deposits on catalyst surface causing the spatially varying decrease in pore dimension. They tested the heavy Boscan feed containing more that 1700 ppm of V + Ni using three bimodal catalysts varying in the shape. The model prediction agreed with the experimental data within the first 10 days of the operation. After 10 days, the model underpredicted the experimentally measured conversions.

The single pellet model proposed by Perreira (419) was suitable for predicting the performance of the bimodal catalyst comprising 0.12 and 0.60 mL/g of micro- and macropore volume, respectively. A set of mathematical equations were developed and their solution obtained for first-order HDM kinetics using the heavy feed containing more than 100 ppm of metals. In this case, the regular perturbation solution approach was applied. A series of correlations involving the change of parameters such as distribution factor, reaction rate, pore radius, Thiele modulus with TOS were developed. In this study, little attention was paid to the effects of temperature and pressure of H_2. Using the models developed by Pereira et al. (419,420), a database on the effect of the feed origin and of the catalyst properties on HDM could be established. The results supported the advantages of bimodal catalysts compared with unimodal catalysts. The model was suitable for selecting and designing the tailor-made catalyst for optimal matching of the type of catalyst with the properties of heavy feed. This includes the size and shape of catalyst particles, as well as the catalyst porosity. Figure 7.37 (419) illustrates the applicability of the model to real situations. Thus, a reasonable agreement between the metal distribution factors determined experimentally and calculated using the model was obtained. For this estimate, performed between 673 and 693 K, the authors used 10 catalysts varying widely

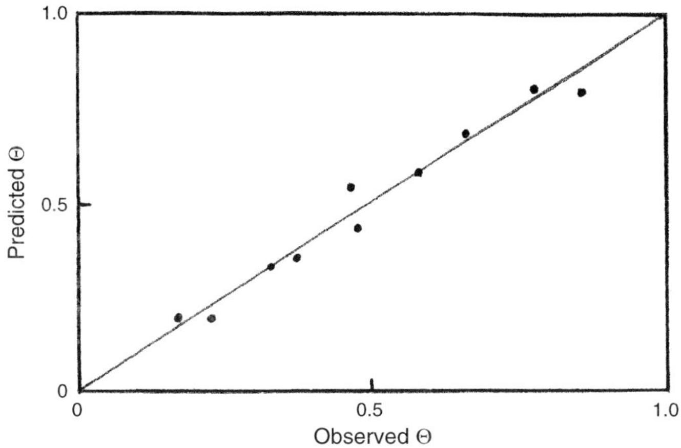

Figure 7.37 Predicted versus observed values of metal capacity factors ($CoMo/Al_2O_3$, VR, 688 K, 14 MPa) (419).

in the physical properties such as pore volume distribution, particle size and shape, as well as the diffusion path length. All catalysts tested were of the similar $CoMo/Al_2O_3$ composition. The estimate was made after more than 10 days on stream.

To a certain extent, the model proposed by Jacobsen et al. (42) involves both pore plugging and pore filling features. In this case, it was assumed that part of the pore having the depth of the particle radius (R) was contaminated by metal deposit, whereas the other part (Rx) was not. Then, the shell of thickness $(1-x)R$ started to build up from the start of the run leaving the uncontaminated core of radius xR, where the reactions could take place. The diffusion of the reactant molecules would be restricted by the shell of metals, whereas in the uncontaminated part of pore, the diffusion would approach that in bulk. Nevertheless, the uncontaminated part of the pore could still be available for hydrogen activation and as such serve as the reservoir of active surface hydrogen, which has a tendency to migrate on the catalyst surface.

In an effort to decouple the deactivation by coke and that by metals, Toulhoat et al. (421,422) introduced the concept of the ultimate storage capacity (USC) of the single catalyst particle of the HDM catalyst. The USC was defined as the mass of metals accumulated in the catalyst pellet, relative to the unit mass of the fresh catalyst, until the catalyst activity became zero. The USC can be expressed in terms of the measurable parameters as follows:

$$\text{USC} = \frac{\Omega\ \rho_{\text{MS}}\ \epsilon_0\ Y}{\rho_{\text{C}}}$$

where ϵ_0 is the initial porosity of catalyst, ρ_C is the initial grain density, ρ_{MS} is the true density of deposited metal sulfides, Y is the metal content and Ω is the fraction of ϵ_0 occupied by the solid deposits in the spent catalyst. For perfect HDM catalyst, $\Omega = 1$, but in general $\Omega < 1$. The actual value of Ω could be estimated from the microprobe determination of the Tamm distribution parameter in the spent catalyst pellet. Figure 7.38 (422) shows the estimate of the metal uptake level at 653 K by extrapolating that part

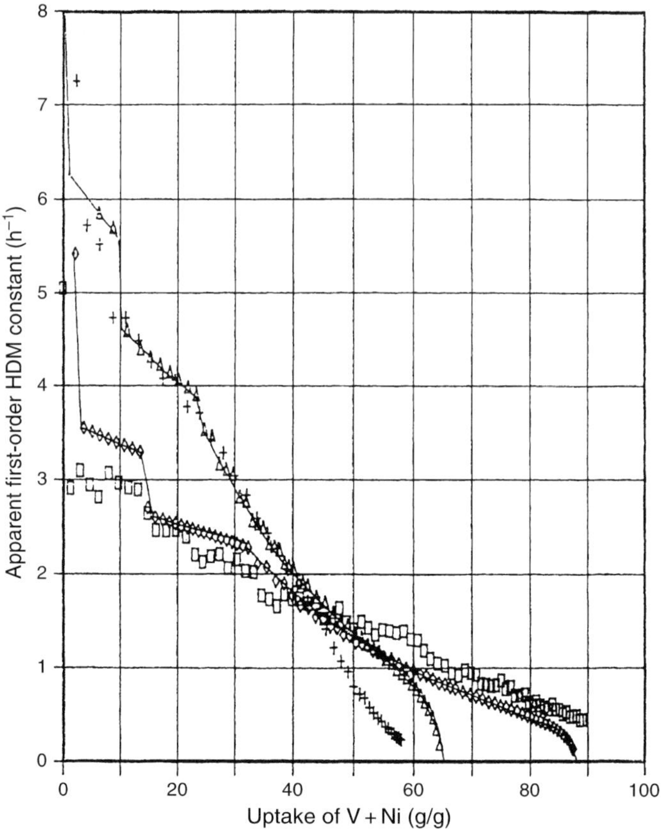

Figure 7.38 Comparison of simulated and experimental data for HDM aging test; (□) experim. at 653 K, (◇) simul., (+) experim. at 673 K, (△) simul. (422).

of the metal decay curve to zero catalyst activity. The USC could then be determined as the ratio of the metal uptake level at zero catalyst activity (X^{final}) and Tamm factor for the spent catalyst (Θ^{end}), i.e., USC = $X^{\text{final}}/\Theta^{\text{end}}$ giving value of 1.325 g/g for X^{final} of 0.64 g/g and Θ^{end} of 0.483. Now it only remains to estimate the value of Ω by assuming the value of 4.26 g/cm^3 as the true density of NiV_3S_4. The properties of the $NiMo/Al_2O_3$ catalyst used for this estimate are shown in Table 7.5 (422).

There are a number of other models, the validity of which was tested using real feeds. For example, the models developed by Dautzenberg et al. (423) and Hannerup and Jacobsen (424) are based on pore-plugging by metals. In this case, the maximum of metal loading and catalyst life associated with it were important parameters for characterizing the deactivation process. Thus, the gradual decline in catalyst activity could be followed and expressed in terms of the change in the relative metal loading with TOS. The pore-plugging model developed by Oyekunle and Hughes (9) predicted the performance of the HDM catalyst lasting at least 1 year, provided the catalyst possessed a suitable pore size distribution. The single particle model proposed by Reyes et al. (110) is noted because of rather unique feed employed. Thus, the DAO obtained from the Orinoco

Table 7.5 Properties of catalysts (422)

Pellet shape	Sphere
Pellet diameter (mm)	1.5
Surface area (m^2/g)	122
Microporosity	0.48
Mesoporosity	0.31
Microporous MPD (nm)	19
Mesoporous MPD (nm)	50
Initial grain density (kg/L)	0.77
Bulk density (kg/L)	0.53
MoO_3 (wt%)	7.0
NiO (wt%)	1.5
Al_2O_3 (wt%)	91.5

crude contained more than 200 ppm of V + Ni and less than 2 wt% of asphaltenes. This suggests that with respect to catalyst deactivation, metals played a dominant role.

7.7.3 Modeling on reactor level

Modeling of the reactors employing the fixed bed of catalyst requires the database on behavior of the active phase of catalyst, as well as the physical parameters determined on the particle scale. Therefore, this may be considered as the modeling on the highest level, requiring much more details compared with the single particle models. Thus, in addition to all parameters required for modeling on a catalyst activity and single particle scale, a new set of parameters have to be determined and incorporated in the models. This includes liquid holdup, bed height, bed volume, catalyst volume, pellet size and superficial liquid velocity. For fixed bed reactors, most of the models assume trickle flow of the liquid and gaseous streams co-currently from the top to the bottom through the void space between the catalyst particles. The wetting of catalyst surface ensures that reactions occur predominantly in the liquid phase. Moreover, the plug flow transport through the reactor, usually assumed for such models, ensures an ideal mixing, in which both radial and axial dispersion or back mixing are neglected. Without such simplification, the development of models would be rather complex and difficult.

The complexity in describing events on a reactor level may be indicated using the results published by Tamm et al. (8). In essence, these results were obtained on a single particle level; however, significant differences in the catalyst deactivating patterns between the inlet and outlet of the reactor were confirmed. On the basis of microscopic evaluations, these authors observed that the V and Ni deposition profiles exhibited either U or M shape (Figures 3.11 and 3.12), i.e, the maximum of metal deposition was located either at the catalyst particle surface or inside the catalyst particle. This was dependent on the location of catalyst particles in the fixed bed. Thus, the M-shape profile was observed in the reactor inlet, whereas the U shape in the reactor outlet. As Figure 7.39 (8) shows, the distribution parameter was influenced by the position of catalyst particles in the fixed bed reactor as well. The higher value of the Ni parameter confirmed more

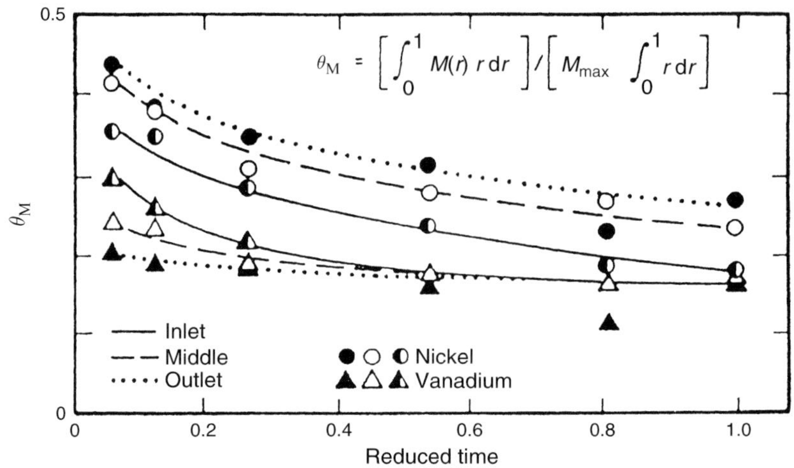

Figure 7.39 Nickel and vanadium distribution parameters as a function of reaction position and time (8).

even radial distribution compared with that of V. The maximum of metal deposition shifted inward with decreasing temperature, decreasing H_2 pressure, decrease in pellet diameter and increase in the average pore diameter. The results if Figures 3.11, 3.12 and 7.39, though obtained on a single particle level, clearly indicate that modeling of the reactor performance requires an additional set of parameters, in addition to those describing the properties of catalyst particles.

The model developed by Khang and Mosby (425) was based on pore-filling by metals, and its suitability was verified using the macroporous catalyst for applications in the trickle bed reactor. Initially, the diffusion of reactants to the pores' interior was unrestricted. A gradual and uniform deposition of metals on the walls of pores restricted the access to active sites in the pores. Deactivation process could be expressed by two adjustable Thiele moduli, i.e., one assuming the bulk diffusivity and kinetics (h_0) and the other effective diffusivity and kinetics in deactivated pores (h_z). These moduli were part of the following mathematical expressions for the normalized HDM and HDS activities (a):

$$a = \frac{(z/h)[(1/h)\tanh\{2h_z(1-z)\}+1]}{[(1/h)+|\tanh\{2h_z(1-z)\}]}$$

$$z = (1 - M/M_o)^{0.5}$$

$$h = h_0^{\,2}/h_z$$

$$h_0 = (k_o/D_z)^{0.5}(R/2)$$

$$h_z = (k_z/D_z)^{0.5}(R/2)$$

where M is the metal on catalyst, M_o is the maximum of metal on catalyst, k_o is the rate constant for original active sites, k_z is the rate constant in the deposited layer, R is the

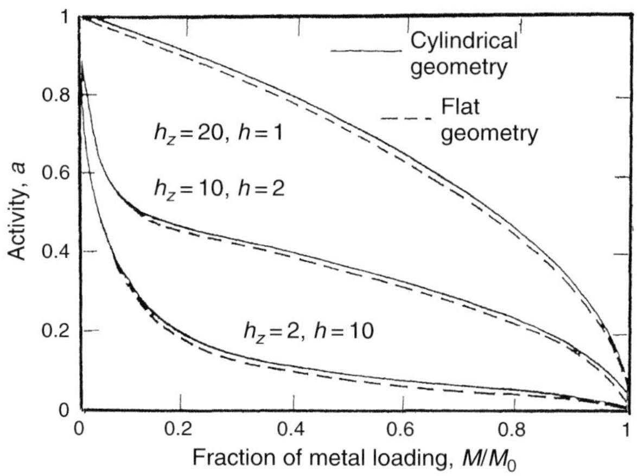

Figure 7.40 Effect of metal loading and Thiele moduli on catalyst activity (425).

initial pore radius and D_z is the diffusivity in the deposited layer. The typical shapes of the activity curves for different values of h and h_z are shown in Figure 7.40 (425).. The model was suitable for predicting deactivation curves for the HDS and HDM reactions in good agreement with the experimental data before less than 50% of pores were filled with metals.

Assuming a perfect plug flow, Khang and Mosby (425) expressed the change in the ~540+°C concentration in the reactor as follows:

$$C_A/C_{AO} = \exp(-gfk_A\ \tau - fk_A u) \quad 0 < u < \tau$$

whereas the inhibition activity over the entire trickle bed reactor is

$$a_{Ave} = \int_0^\tau a(1 - C_A/C_{Ao})^m du \Big/ \int_0^\tau (1 - C_A/C_{AO})^m du$$

In these equations, C_{AO} and C_A are the concentrations of the ~ 540+°C in the feed and products, respectively, at space time τ, f is the ratio of liquid volume/catalyst volume and k_A is the thermal cracking rate constant. In spite of its simplicity, the model data showed a reasonably good agreement with the results obtained in pilot plant such as the HDM activity and metal loading between the inlet and outlet of the fixed bed reactor.

The set of mathematical expressions for normalized HDM and HDS activities (a) derived by Khang and Mosby (425) were used by Togava et al. (426) to analyze five sets of the deactivation data obtained from the commercial operation employing ARDS process using Kuwait AR as the feed. The five evaluated systems comprised various combinations of HDS, HDS/HDM, HDM and HDS/HDN catalysts as shown in Table 7.6 (426). For systems C, D and E, the model was applicable up to 80% of the pores that were filled with metals compared with 50% observed by Khang and Mosby (425). However, for systems A and B, the model was applicable up to about 40% of the metal filling. The simulations of the systems A, B and C performed by Kam et al. (427) are

Table 7.6 Catalyst combination ratio (vol.%) (426)

System	Catalysts			
	HDM	HDM/HDS	HDS	HDS/HDN
A	8	40	42	
B	24	18	58	
C	39	48		13
D	55	45		
E	39	10	47	

shown in Figure 7.41. It was evident that the catalyst performance, particularly during the later stages, could be better simulated after the Amoco model (425) was modified. The deviation for systems A and B was attributed to rather small maximum active metals loading on the catalysts.

Figure 7.41 published by Kam et al. (427) showed that the Amoco model developed by Khang and Mosby (425) can be extended beyond more than 50% of pores filled with metals using a slight modification of h_z, e.g.,

$$h_z = (k_z/D_z)^{0.5} R^{0.5}$$

Moreover, the constant values of h_0 and h_z in Amoco model were updated by the former authors to account for the continuous change in the catalyst surface properties (427). Thus h_0 and h_z are expressed as

$$h = Z + h_0/2$$

$$h_z = -h_0/(k_s/k_{s0})^{0.5}$$

Furthermore, the following equation to determine the metal deposition on catalyst surface (Mi) with TOS was used:

$$M = \frac{100}{W} F_L \sum_{I=1}^{t} \{[\rho_V(C_{V,f} - C_{V,p}) + \rho_{Ni}(C_{Ni,f} - C_{ni,p})] \wedge t\}$$

where F_L is the mass flow of liquid, W is the weight of catalyst, ρ_V and ρ_{Ni} are the densities of the V and Ni sulfide layer, respectively, $C_{V,f}$ and $C_{Ni,f}$ are concentrations of V and Ni in the feed, $C_{V,p}$ and $C_{Ni,p}$ are the concentrations of V and Ni in the product, respectively, and t is the time on stream. The same equation was used for determination of the maximum amount of V + Ni on the catalyst (M_{max}), i.e., for t maximum. Figure 7.42 shows a significant improvement in predicting the change in the k_s/k_{so} ratio, with TOS expressed by M/M_{max} ratio. The Topsoe and Chiyoda results were obtained over different catalysts. Therefore, Figure 7.42 illustrates the ability of models to identify the differences in properties and performance of the investigated catalysts and heavy feeds.

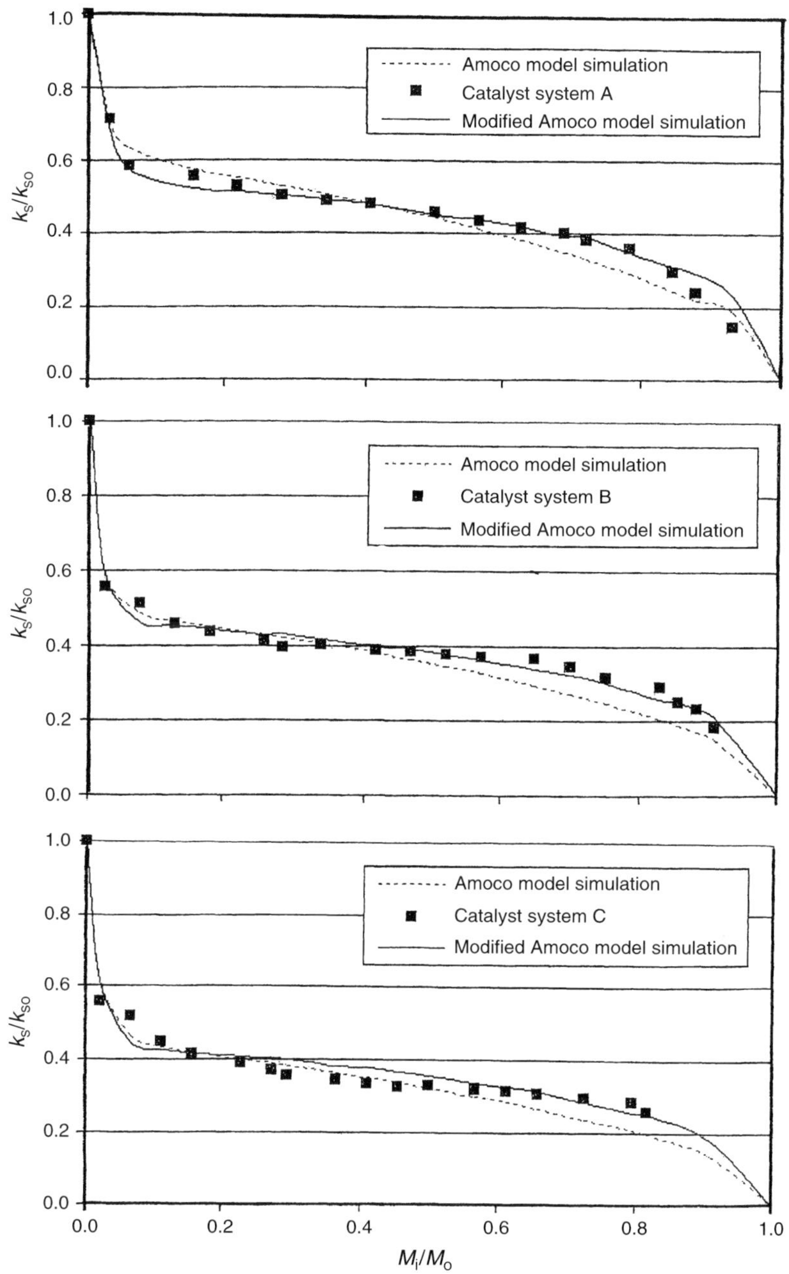

Figure 7.41 Comparison of experimental data (solid squares) with model predictions for catalysts in Table 7.6 (M_i, metal on catalyst at time t; M_o, maximum of metals on catalyst (427).

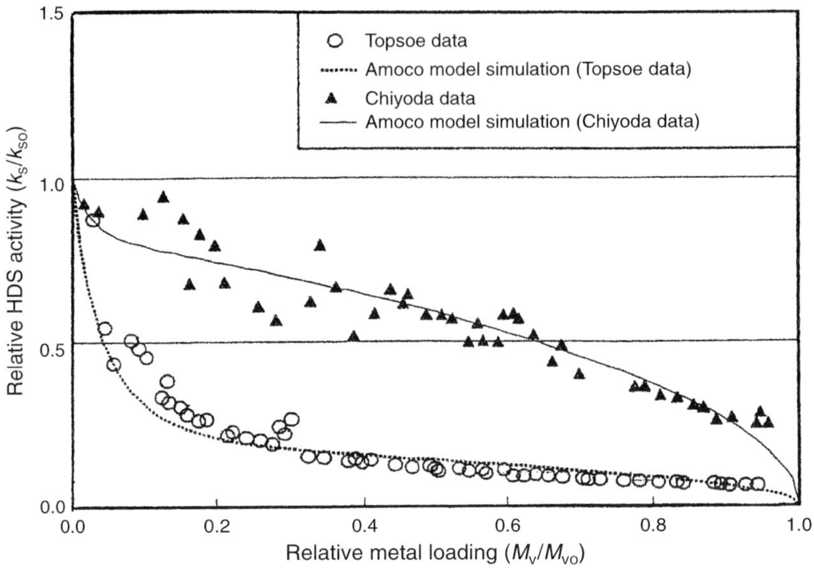

Figure 7.42 Relative HDS activity versus relative metal loading (427).

Kam et al. (428) have continued in their modeling efforts to include the system comprising several fixed bed reactors in a series. The integrated mechanistic reactor model considered the initial rapid catalyst deactivation by coke deposition. In this case, a distinction was made between the "soft" coke formed during very early stages and the "hard" coke formed in the steady state of catalyst deactivation. At the same time, during the start of run period, the deactivation by metals was much less evident, whereas in the middle of run and particularly before the end was approached, the metal deposition was the dominant mode of deactivation. Figure 7.43 compares the simulation data with those obtained in pilot plant during more than 400 days on stream. These results are for the second reactor of the four reactor process. For HDS, a good agreement between the predicted and measured results was obtained for less than 4000 h on stream, whereas the large discrepancies were observed for the HDAs results. Similarly, a good prediction was made for HDV and HDNi. However, the deviation of the HDM data became more evident beyond 4000 h on stream. In an effort to further advance their model, Kam et al. (429) assumed that a part of V and Ni was removed from heavy feed via non-catalytic reactions. After applying this assumption, the prediction of the catalyst performance beyond 4000 h significantly improved.

Based on the extensive kinetic study on the HDS and HDV of an Iranian VR, Takatsuka et al. (365) estimated the material balance for S and V in a plug-flow reactor as follows:

$$U_l \, dC/dz + V_{cat} \, r = 0$$

where U_l is the flow rate of liquid, C is the concentration of either S or V, z is the distance from the reactor inlet, V_{cat} is the volume of catalyst and r is the reaction rate. The solution of this equation was obtained by numerical integration, taking into account the distribution of catalyst deactivation from the inlet toward the outlet of the reactor.

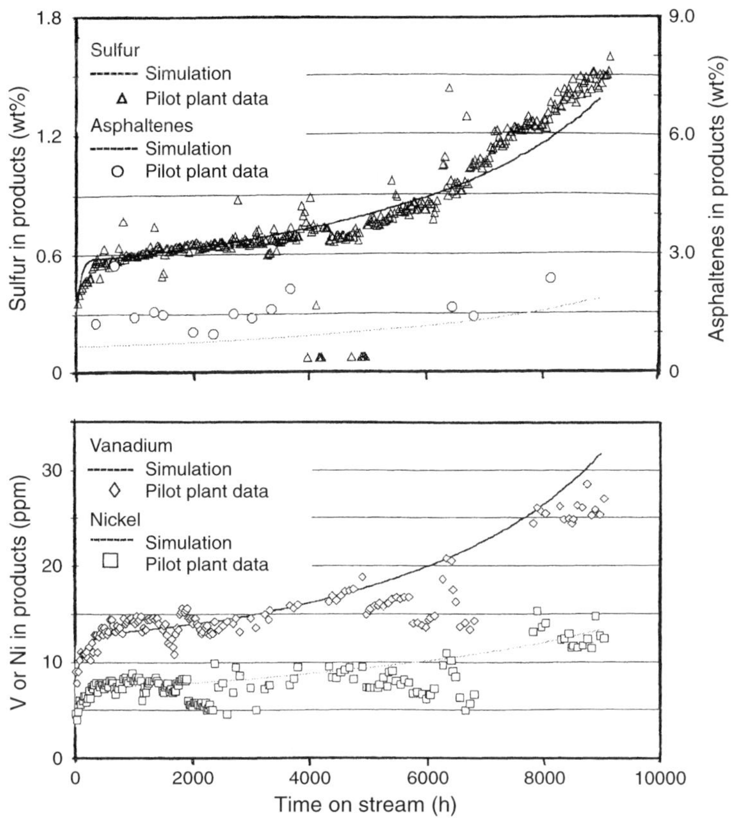

Figure 7.43 Comparison of plant data with simulation data for hydroprocessing Kuwait AR (428).

Figure 7.44 (365) shows the prediction of catalyst performance using this model. The properties of the catalysts are shown in Table 7.7 (365). It was evident that initially catalyst A was more active for HDS than catalyst C, as it was indicated by the lower content of sulfur in the products. However, the activity of the former declined with TOS. The same catalyst was the least active for V removal. The best performance of the catalyst B resulted from the optimal combination of pore volume, particle diameter and surface area.

The percolation model based on the Bethe network was used by Janssens et al. (430,431) to predict the metal deposition patterns during the HDM of heavy feeds. The model consists of the regular network of nodes connected by pores. In this case, there is no need to deal with the change in pore radii and diffusivities. The accessible pore volume and reduced conductivity in the network caused by catalyst deactivation are the only required variables. The fraction of the remaining pores for which the accessibility and conductivity of the network is approaching zero is called the percolation threshold. In this approach, the pellet was assumed to be the binary interdispersion of the void space and the solid deposited material. The total void space ϵ includes the volume fraction of the isolated clusters ϵ^{t} and volume fraction of the accessible void space ϵ^{A}. Transport through the pellet was impossible if ϵ is below the critical value, i.e., percolation threshold ϵ^{C}. The effective transport voidage ϵ^{E} is relevant to the diffusional

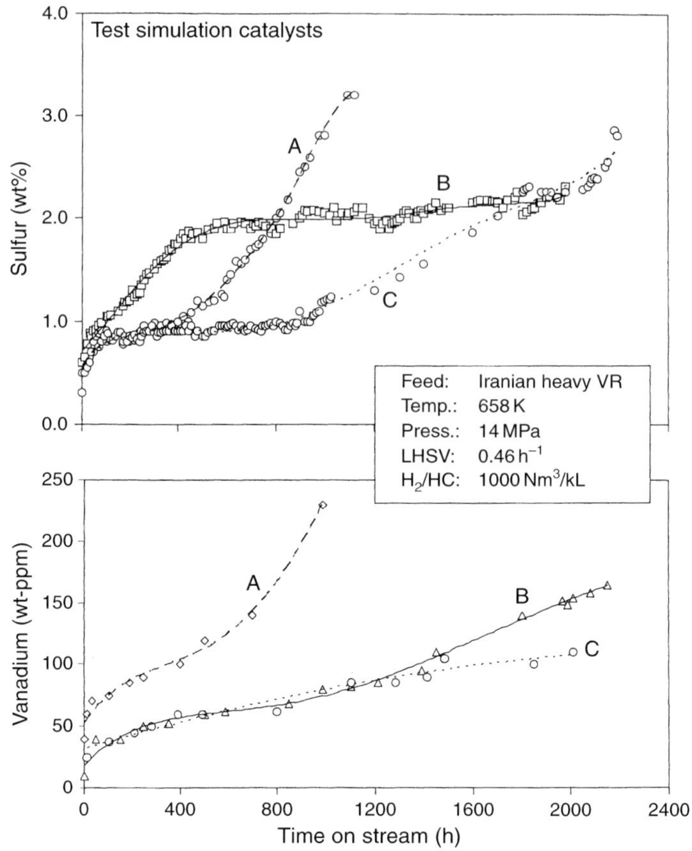

Figure 7.44 Simulation of catalyst aging (pore volume of catalysts A, B and C was 0.44, 0.60 and 0.79 mL/g, respectively) (365).

Table 7.7 Properties of catalysts (365)

	Catalysts		
	A, Ni–Co–Mo	B, Co–Mo	C, Co–Mo
Pore volume (mL/g)	0.44	0.60	0.79
Surface area (m^2/g)	280	155	341
Particle diameter (mm)	1.16	1.34	1.59

transport. For the accessible void space arranged in parallel layers, ϵ^E equals ϵ^A. For disordered interdispersion, the accessible void space for transport may be tortuous and have inactive dead ends. In this case $\epsilon^E \leq \epsilon^A$. The generalized mass balance based on the Fick's law then gives the following:

$$C\frac{\delta\epsilon^A}{\delta t} + \epsilon^A\frac{\delta C}{\delta t} = \frac{1}{R^2}\frac{\delta}{\delta r}\left(r^2\epsilon^E D_{\text{eff}}\frac{\delta C}{\delta r}\right) - S_v R$$

where C, R and S_v are concentration, reaction rate and surface area per unit volume of pellet, respectively; r is the radial length coordinate from the center of pellet and t is the time. Then, the mass balance of the metal deposits could be calculated using the following formula:

$$\delta W_M/\delta t = \alpha M_M S_v R$$

Figure 7.45 (430) compares the V deposition profiles predicted by models for the model compound such as VO-TPP (A) and the real feed such as AR (B) assuming the initial pore radius of 5 and 30 nm.

Most recently, the percolation model involving the Bethe network (430,431) was further expanded by the same research group to include the development of models for the simulation of reactor performance (432). More than 50 mathematical equations were used to describe the events occurring on an active phase level, as well as the transport phenomena taking place on a particle level. The trickle flow reactor in which liquid and gases flow co-currently from the top to the bottom through the void space between catalyst particles was used for modeling on the reactor level. Under such conditions, the wetting of catalyst surface ensured that the reactions occurred in the liquid phase. The variables such as liquid holdup, superficial liquid velocity, bed height and diameter, reactor volume and catalyst volume were incorporated in the mathematical equations in addition to all parameters used on the active phase and single particle level. The plug flow transport through the reactor assumed for the study ensured an ideal mixing in which radial and axial dispersions as well as back mixing could be neglected. To achieve this, the following criteria had to be fulfilled:

$$h_{bed}/d_p > 1000$$

$$d_{bed}/d_p > 10$$

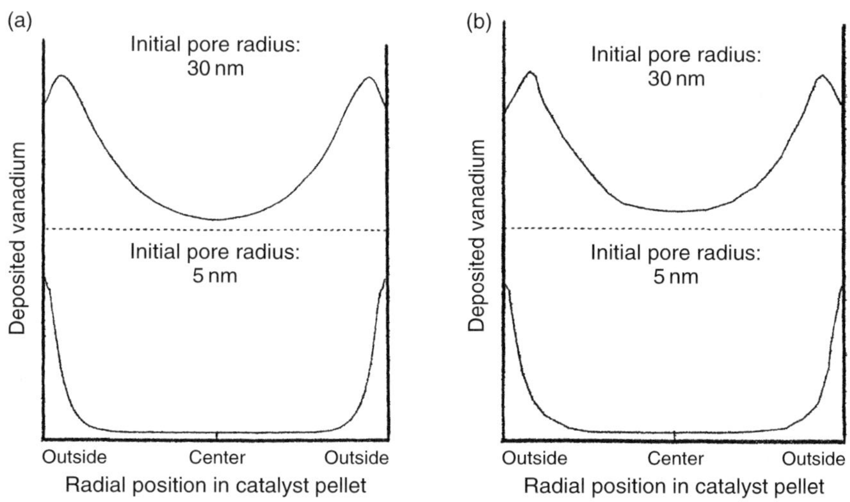

Figure 7.45 Effect of initial pore radius on V deposition profile; (A) – model compound VO-TPP at 673 K and 10 MPa, (B) – AR at 673 K and 15 MPa (430).

where h_{bed}, d_{bed} and d_p are the bed height, bed diameter and particle diameter, respectively. In this case, 90% conversion and second-order kinetics were assumed. All variables used were in a dimensionless form. The experimental results were obtained in a fixed bed divided into four sections. For this purpose, the catalyst (MoS_2/SiO_2) was placed on four trays to determine the V distribution profiles during the HDM of VO-porphyrin after 300 and 700 h on stream. There was a reasonably good agreement between the predicted and experimental data for the top layer of the catalyst. However, a deviation from the latter data was observed for the lower tray's catalysts. Theoretical part of this study includes several correlations in which the values of Thiele modulus were arbitrarily chosen to identify the effect of the presence and absence of diffusion on the V distribution profiles on the particle level. With such correlations available, the meaning of Thiele modulus becomes much more clear. It should be noted that the experimental data in the study of Vogelaar et al. (432) were obtained for a model compound rather than a heavy feed. It is, however, believed that at least a brief account of this study was necessary because of the significant number of details to which attention has been paid.

The THERMIDOR (THERmal Monitoring for Iso-performance Desulfurization of Oil Residues) model was developed by the Institut Francais du Petrole to simulate operation of the HYVAHL process comprising two guard reactors operating in the perturbating mode up-stream from the series of four fixed bed reactors (433). The model considered both the grain scale and the bed scale. The grain scale incorporates a continuous change of the intragranular concentration gradients caused by deposition of coke and metals. A continuous decrease in the effective pore diameter and surface area affects kinetics of hydroprocessing reactions because of the diffusion restrictions. The shape of catalyst particles (sphere, cylinders, polylobe, etc.) was considered at the grain level using the fractional radius of the catalyst particle as space variable, whereas the fractional axial position was used to describe changes between the inlet and outlet of the reactor. More than 40 mathematical equations which were considered represent one of the most comprehensive approaches used for the model development. These equations were beyond the scope of this review. The pseudo-L-H law was used to express the rate of hydroprocessing reactions as first order in disappearing reactant (e.g., sulfur, metals and asphaltenes) using the assumption that about 10% of asphaltenes present in the feed was responsible for coke build-up. The following is a simplified form of this equation for HDS:

$$\rho(r, z, t) = k(r, z, t)\ S\ (r, z, t)\ \exp\ [E^{\pm}/RT] \frac{C^{l}(r, z, t)}{[1 + K_{ads}(T\{z, t)\})C^{l}{}_{asph}(r, z, t)]}$$

where ρ represents the mole of reactant per unit volume of catalyst and unit time at time t, grain radius r and the position between inlet and outlet z; k and S are the corresponding local intrinsic rate constant and specific surface area of catalyst, respectively; C^{l} and $C^{l}{}_{asph}$ are the corresponding liquid concentrations of reactants and the portion of asphaltenes adsorbed on catalyst surface, respectively; and K_{asph} is the adsorption constant of this portion of asphaltenes. The distribution of metals as their homogenous concentration in the solid phase ($C^{s}{}_{m}$) was expressed as

$$\frac{dC^{s}{}_{m}(r, z, t)}{dt} = \frac{M_m}{d^{0}{}_{g}} \sum \rho_m(r, z, t)$$

where M_m is the average molar mass of metals and d^0_g the initial grain density of the catalyst. The following equation was used to express the coke deposition caused by the reversible dehydrogenation of the asphaltenic material deposited on catalyst surface:

$$\frac{dC^s_{coke}(r,z,t)}{dt} = k_{coke}(r,z,t)\,S(r,z,t)\exp[E^{\pm}_{coke}/RT]\{C^{s1}_{coke}(r,z,t) - C^s_{coke}(r,z,t)\}$$

where C^s_{coke} and C^{s1} are the actual and equilibrium local coke concentration in the solid phase, respectively. Figures 7.46 and 7.47 show the experimental and simulated deactivation curves for Boscan heavy feed using the HDS catalyst and the macroporous HDM catalyst, respectively, whereas the coke lay-down and metal deposition are shown in Figure 7.48 (433). Rather remarkable agreement between the experimental data and those predicted by the model should be noted. The THERMIDOR software incorporates Fick law to express the local molar fluxes in the liquid phase. The Stokes–Einstein law and Andrade law were combined to describe the effect of temperature on the feed viscosity and molecular diffusivities. The random sphere, random needle and random

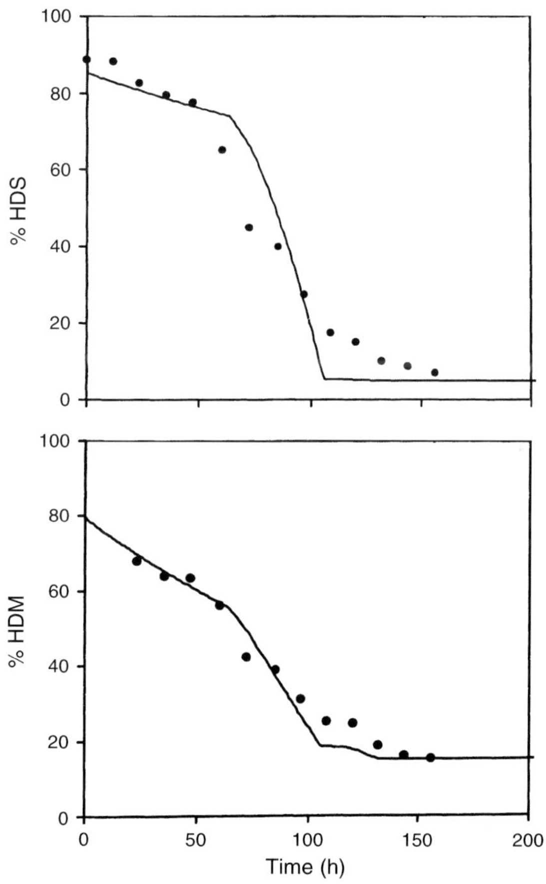

Figure 7.46 Experimental (dots) and simulated (continuous) data for HDS catalyst (Boscan crude) (433).

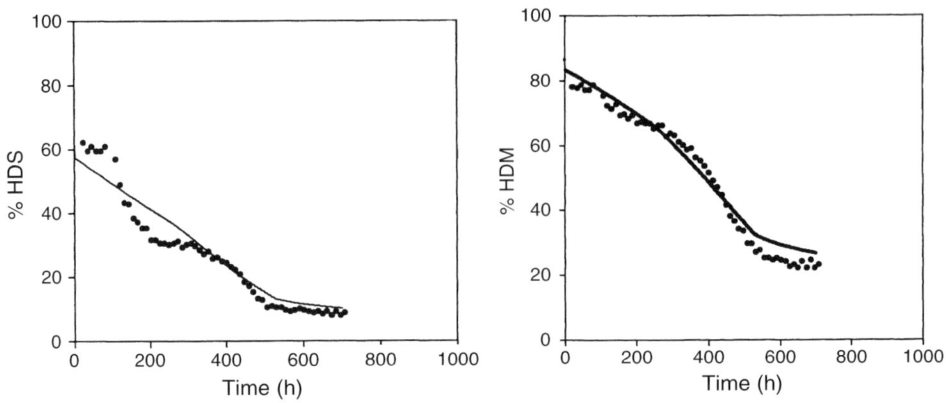

Figure 7.47 Experimental (dots) and simulated (continuous) data for HDM catalyst (Boscan crude) (433).

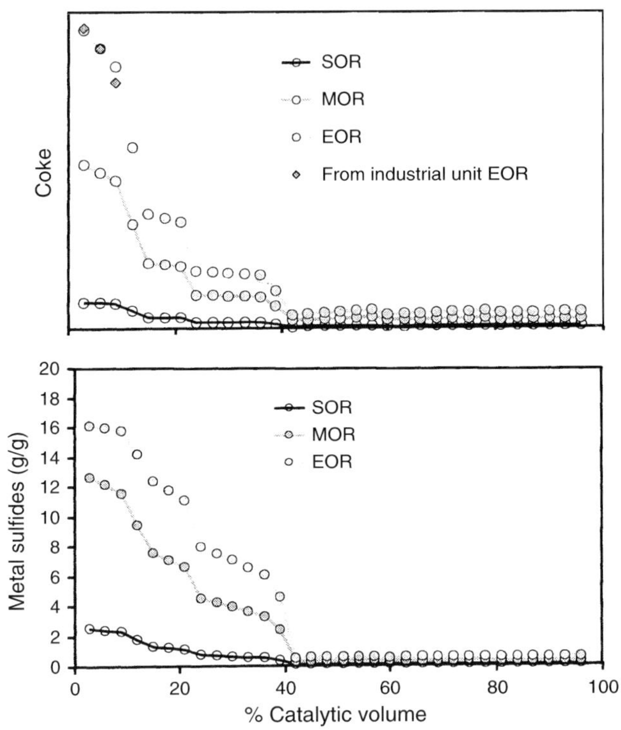

Figure 7.48 Simulated longitudinal profiles of coke and metal sulfides (433).

coin models were used for the representation of the catalyst porous media. Other equations derived as part of the THERMIDOR project enabled simulations of the additional parameters, e.g., the change in the inlet and outlet temperature with TOS, longitudial profiles of temperature, Tamm factor, surface area, porosity, etc., with TOS.

The Chioda model developed by Kodama et al. (434) was used by Al-Adwani et al. (366) to simulate the performance of the four reactors in a series, such as in the ARDS process. Apparently, this model was supposed to be less dependent on the catalyst and feed properties than the other models. The mathematical equations of this model were discussed earlier as part of the kinetics of simultaneous deactivation by the coke and metals (366). The model assumes that HDM reactions interact with the coking reactions. Both these reactions deactivate the catalyst by active site poisoning and plugging. For this study, the HDM catalyst was considered for the first and second reactors and the HDS catalyst for the third and fourth reactors. The focus of the study was on the effect of the LHSV and/or contact time on the catalyst performance. The model could predict the effect of LHSV on the set-point temperature for achieving certain level of HDS. As Figure 7.49 (366) shows, for a given degree of HDS, the set-point temperature increased

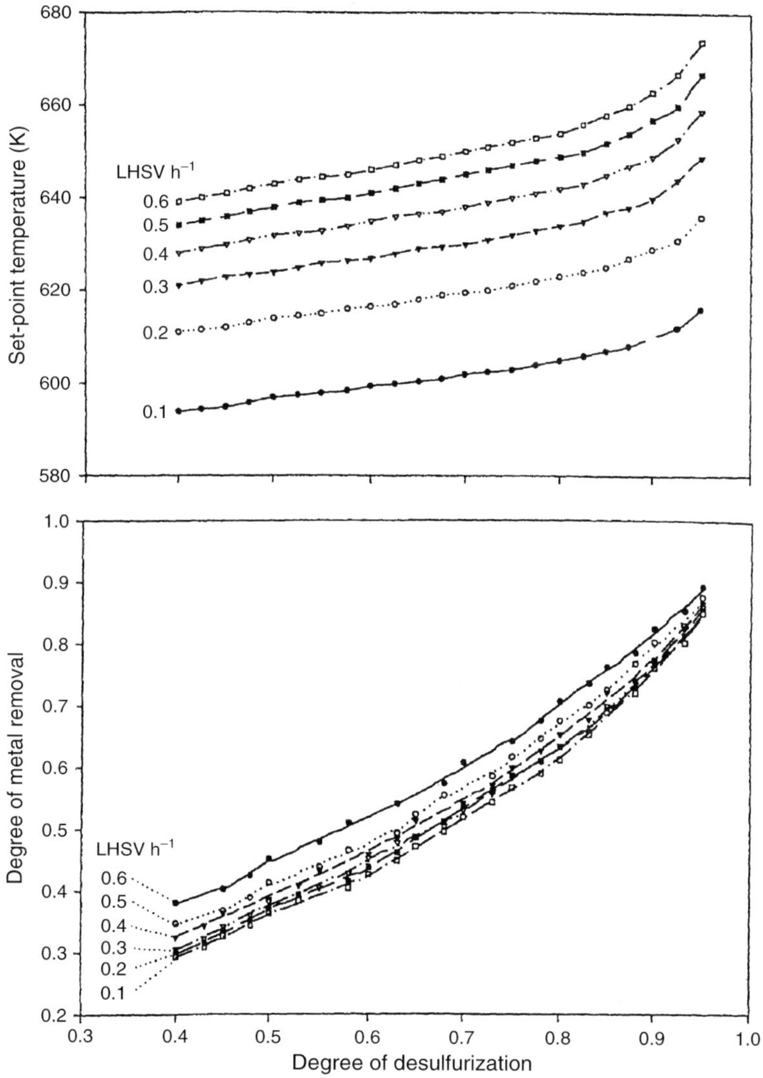

Figure 7.49 Variation of setup temperature and degree of metal removal with degree of desulfurization and LHSV (366).

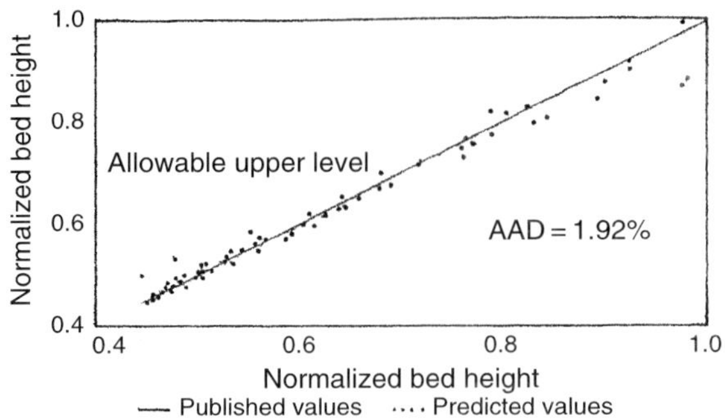

Figure 7.50 Normalized bed height data obtained experimentally versus predicted using artificial neural network (435).

with increasing LHSV, i.e., the decreasing contact time. In general agreement, the model predicted that the degree of HDS parallels that of HDM.

Few attempts have been made to develop the models to simulate the performance of the ebullated bed reactors. In this regard, it has been recognized that the artificial neural network may be a suitable modeling tool (435,436). The neural networks comprise computing systems composed of many simulation elements interacting with bandwidth channels and/or connections to process the information by responding to the external inputs. Initially, it can be built using the historical plant data, as well as the results of laboratory and pilot plant research and those found in the literature. Gradually, the models can be upgraded by incorporating more reliable and advanced information. The neural network architecture developed by Kam et al. (435) consisted of the three neural layers and eight specified input nodes and nine hidden nodes to predict the expanded bed height. Figure 7.50 compares the normalized bed height data obtained from a commercial reactor with those predicted by the artificial neural network. The absolute average deviation (AAD) value indicated the applicability of the model to the expanded bed systems as employed in the commercial H-Oil reactors. Furthermore, the predicted slate of products and their yields were in reasonable agreement with the plant data.

Chapter 8

SELECTION OF CATALYSTS FOR COMMERCIAL HYDROPROCESSING REACTORS

Several detailed reviews of the commercial and emerging processes used for hydroprocessing of heavy feeds were published elsewhere (437–439). Simplified schematics of the petroleum refinery shown in Figure 2.1 indicated the presence of several catalytic units operating in a hydroprocessing mode on the site of the refinery. Among the units, the hydroprocessing of distillates from atmospheric distillation and from FCC, as well as the naphtha fraction from coking are not covered by this review. The properties of heavy feeds and those of the anticipated products after hydroprocessing of the former determine the selection of catalysts and type of reactors. Thus, every unit may require a different type of catalyst to achieve an optimal performance of the reactors. Also, several types of hydroprocessing reactors may be found on the site of the same refinery.

Simplified schematics of the catalytic reactors which have been used commercially are shown in Figure 8.1, whereas the typical operating ranges of these reactors are summarized in Table 8.1 (437). The features of these reactors indicate an importance of the proper selection of catalysts, the size and shape of the catalyst particles in particular, to ensure an efficient and continuous operation. In addition, the properties of heavy feeds must be taken into consideration to achieve the optimal matching of catalysts with reactor. This is the primary objective of this chapter, rather than to give an exhaustive review on catalytic reactors and systems.

There has been decades of experience in the operation of fixed bed reactors, though for the hydroprocessing of light feeds. Progressively, fixed bed reactors have been modified to achieve the steady and prolonged operation using heavy feeds. The degree of modification increased with the increasing amount of asphaltenes and metals in the feed. Therefore, among the heavy feeds considered in this book, the least modifications of reactors would be required for hydroprocessing of VGOs and HGOs. It is believed that if such feeds were derived from conventional crudes, desirable conversions could be achieved with a single fixed bed and/or a fixed bed comprising several layers of different catalysts. Fixed bed reactors consisting of several sections in the same vessel may also be suitable. For metals- and asphaltenes-containing feeds, frequent shutdowns of the operation and catalyst replacement could not be avoided using single fixed bed reactors. This problem can be alleviated by using several fixed bed reactors connected in a series. In this case, the primary function of the first reactor, termed as "guard reactors" is to remove most of the metals with the aim to extent catalyst life in the downstream reactors. A high HDM activity and/or metal storage capacity is the requirement for the catalyst to be used in the guard reactor.

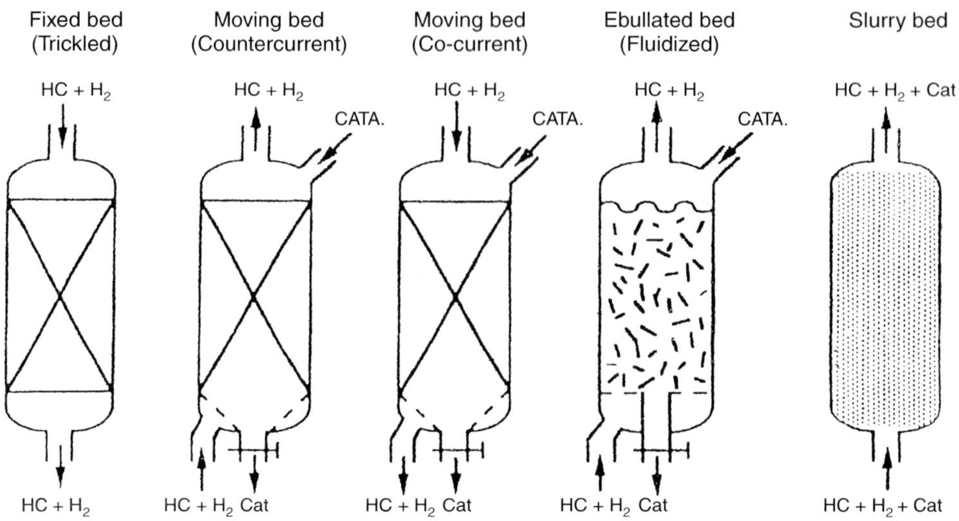

Figure 8.1 Simplified features of catalytic reactors for upgrading heavy feeds (437).

Table 8.1 Operating conditions during hydroprocessing of heavy feeds in different reactors (437)

	Fixed bed	Swing fixed bed	Moving bed	Ebullated bed	Slurry bed
V+Ni max., ppm	120	500	700	>700	>700
Pressure, MPa	10–20	10–20	10–20	10–20	10–30
Temperature, K	655–693	655–693	655–693	655–713	693–753
LHSV, h^{-1}	0.1–0.5	0.1–0.5	0.1–0.5	0.2–1	0.2–1
Max. conv. to 550°C	50–70	60–70	60–70	70–80	80–95
Cycle length, month	6–12	12	CO[1]	CO[1]	CO[1]
Catalyst part, size, mm	~1.2 × 3	~1.2 × 3	~1.2 × 3	~0.8 × 3	~0.002
RCC[2]	1	1	0.55–0.70	1.4–2	

[1] CO, continuous operation
[2] RCC, relative catalyst consumption for the same feed for 1 year

In some cases, a "guard chamber" is placed upstream of the guard reactor which operates mainly in the HDM mode. The function of the former is the removal of inorganic solids dispersed in heavy feeds. Therefore, for the purpose of this book, guard reactor is considered to be the reactor filled primarily with a catalyst possessing a high metal storage capacity. At the same time, guard chamber is filled with the lower value solid materials (e.g., clays, minerals, alumina, etc.) with the aim to filter off the inorganic solids dispersed in heavy feed. Some removal of the V and Ni from heavy feed may be achieved in the case that the guard material includes the γ-Al_2O_3 of a suitable porosity. Most likely, part of these solids was formed during the reaction of V and Ni porphyrins with H_2 and H_2S rather than via catalytic reactions. The number of reactors downstream of the guard reactor increases with increasing content of metals and asphaltenes in the feed. Because of the different properties of the feed (product from the preceding reactor), each reactor may require a different type of catalyst. This depends on the origin of heavy feed and the anticipated slate of the products.

To avoid frequent shutdowns due to catalyst replacement, more advanced hydroprocessing reactors which have provision for either continuous or periodic addition and withdrawal of catalyst during the operation had to be developed. Figure 8.1 (437) shows that one type of the advanced catalytic reactors employs an expanded and/or ebullated bed of catalyst, whereas the other type employs moving beds. In the latter case, the catalyst is added at the top and progressively moves toward the bottom for a periodic withdrawal co-currently with liquid streams. In ebullated bed reactors, the slurry of catalyst with a gas oil is continuously added at the top and spent catalyst withdrawn at the bottom of the reactor. An ebullated bed reactor can be operated without any difficulties even in the presence of inorganic solids dispersed in heavy feed. Thus, difficulties associated with the development of pressure drops, channeling, etc. encountered in fixed bed reactors are not present in the ebullated bed reactors.

Attempts have been made to further advance the existing or develop new catalytic systems for hydroprocessing of heavy feeds. In this regard, the focus has been on the countercurrent reactors compared with co-current reactors which have been used predominantly on a commercial scale. The former reactors employ a co-current flow of the liquid and gaseous streams (426). In countercurrent reactors, a structured catalytic bed in which catalyst particles are enclosed within a packed system are being used.

With the aim of decreasing the cost of catalyst inventory, once through, low-cost solids have been receiving attention. This included throw-away by-products from metallurgical and aluminum industries and fly ash from combustion of petroleum coke and coal, as well as naturally occurring clays and minerals containing catalytically active metals such as iron. In this case, a pulverized form of these solids, slurried with a heavy feed, is being introduced into the reactor operating under more severe conditions than typically employed during the hydroprocessing of heavy crudes and VR. The suitability of this approach for hydroprocessing of heavy feeds containing more than 300 ppm of metals (V + Ni) has been demonstrated on a commercial scale. Definitely, in a pulverized form under otherwise similar conditions, conventional hydroprocessing catalysts would exhibit a much higher activity than the throw-away solids. However, for such system, an economic method for the recovery of metals for reuse has not yet been developed. In this case, metals would have to be isolated from the VR obtained after distillation of the products unless the VR was further converted to liquid products and petcoke in a coking process. If such option was chosen, the catalyst metals together with the metals contained in the heavy feed would end up in the ash providing that the petcoke was utilized via a combustion and/or gasification technology.

8.1 FIXED BED REACTORS SYSTEMS

Several decades of experience in the operation of fixed bed reactors using conventional feeds containing no metals and asphaltenes was the basis for their adaptation and/or modification to suit hydroprocessing of heavy feeds. Many years of experience confirmed that it is easy and simple to operate fixed bed reactors for light feeds, similarly as for heavy feeds such as VGO and HGO. Fixed bed reactors can be operated in the up-flow and down-flow mode (440). The latter, so-called trickle bed mode, has been used predominantly. However, the up-flow reactors ensure a better catalyst wetting at low- and high-mass velocities for both the cylindrical and shaped catalyst particles regardless of the catalyst loading procedure. In trickle bed reactors, the catalyst wetting can be improved by choosing the loading procedure which ensures a minimal horizontal orientation of particles in the reactor.

The fixed bed can comprise either a single stationary bed of the same catalyst of the same particle size and shape or layers of different catalysts. The layers may consist of the catalyst having the same chemical composition but different size and shape of particles, as well as the different pore size and pore volume distribution. For example, the layers may include the HDM catalyst at the reactor inlet, on the top of a HCR catalyst, followed by the HDS/HDN catalyst at the reactor outlet. The choice of catalysts and number of layers depends on the origin of heavy feed, as well as on the anticipated quality of the final products.

There are some advantages of the fixed bed systems consisting of several sections in the same vessel with an empty space between the sections (Figure 8.2). The sections may contain the same or different catalyst each. In any case, with this arrangement the makeup H_2 can be introduced between the sections to quench the heat released by exothermic reactions. Also, some systems have a provision for scrubbing ammonia and H_2S from the gaseous effluent from the first section before it enters into the next section. This enables control of the H_2S/H_2 ratio which is critical for a high conversion of HDN reactions (76). Otherwise, the excessive poisoning of the catalyst by N-bases would affect the operation. Indeed, it has been generally observed that the coke build-up in fixed bed reactors increased from the inlet toward outlet. The potential poisoning effect

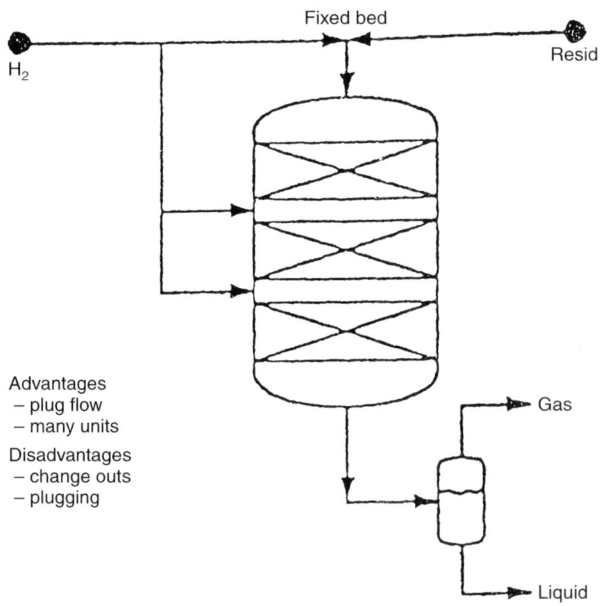

Figure 8.2 Fixed bed three-section catalytic reactor.

of N-bases increased in the same direction. It was indicated earlier that among heavy feeds the adverse effect of N-bases on hydroprocessing would be much more evident for VGO and HGO compared with the heavier feeds.

It has been observed that the performance of fixed bed reactors depends on the method of catalyst loading, i.e., either dense loading or sock loading (441). In the latter case, many catalyst particles will reach the loading surface together, having little time to attain a favorable resting position. Then, particles lay against one another, bridge and maintain random pattern. In this case, large voids are created to hold particles. The bridges may collapse if some forces are exerted on such fixed bed. For example, this may be caused by pressure drop, which may develop during the operation. When catalyst is loaded slowly, particles can settle into place before being inferred by other particles. This prevents bridging and creation of the oversized voids. The bed will have a higher density and shrinkage will be prevented. The advantages of the dense loading compared with the sock loading include the increase in the relative volume activity and decrease in the start of run temperature (441). An increased start of run pressure drop is a negative effect of dense loading.

An optimal combination of the bed void and activity per reactor volume giving the acceptable pressure drops has to be determined to ensure a steady performance of the fixed bed reactors. In this regard, the shape and size of the catalyst particles is important. This is clearly shown in Table 8.2 (11). There is a limit on the maximum pressure drop at which fixed bed can be operated. This depends on the type of the feed. Thus, for light feeds, the particle shape and size may be chosen for dense loading to obtain maximum activity per reactor volume. However, for the high asphaltenes and metal feeds, a small particle size may be needed to achieve a desirable level of catalyst utilization. Then, shape of the catalyst particles must be chosen to obtain the fixed bed with a sufficient level of voidage. For example, this may be achieved by sock loading of the ring and lobe particles giving 35 and 10% higher voidage, respectively, compared with the cylinders (93).

Refinery experience indicates that the heavy feeds containing less than 120 ppm of V + Ni can be successfully hydroprocessed using several fixed bed reactors in a series (442). Under optimized conditions, a high activity and the relatively low metal tolerance catalyst may be suitable for heavy feeds containing less than 25 ppm. A dual catalyst system may be required for feeds containing between 25 and 50 ppm of metals. In this case, the front-end catalyst should possess a high metal tolerance, whereas the tail-end a high catalyst activity. For heavy feeds containing between 50 and 100 ppm of metals, at least a three stage system employing fixed bed reactors may be necessary. In this case, the catalyst in the first reactor should possess a high HDM activity and a high metal storage

Table 8.2 Packed bed properties of catalyst particles (11)

Shape and size of particle	Bed void fraction	Geometric surface area per reactor volume (cm^{-1})
Minilith ($D = 2.54$ mm, $L/D = 1$)	0.52	23
Trilobe (1.3 mm size, $L/D = 3$)	0.41	30
Cylindrical ($D = 1.6$ mm, $L/D = 3$)	0.41	18
Cylindrical ($D = 1.3$ mm, $L/D = 3$)	0.43	23

capacity to ensure the long life of catalysts in the subsequent reactors. It is believed that heavy feeds containing more than 150 ppm of metals can still be hydroprocessed in fixed bed reactor systems providing that some modifications were undertaken. This may include the use of two guard reactors, one in operation and the other on stand-by. The sizing of these guard reactors, i.e., the total metal storage capacity would need to be matched with the content of metals in the heavy feed. An uninterrupted operation could be ensured by switching to the guard reactor with the fresh catalyst as soon as the total metal storage capacity of the reactor onstream was approached (437). The addition of another reactor downstream may also be considered as an option. However, such a step may drive costs of the operation to the unacceptable level.

8.2 COMMERCIAL PROCESSES EMPLOYING FIXED BED REACTORS

These processes have similar features, although they are licensed by different process developers. The number of stages and/or reactors included in the process is determined by the content of asphaltenes and metals in heavy feeds, the projected daily throughput of the heavy feed and the anticipated quality of liquid products. It is unlikely that for any heavy feed, a desirable level of hydroprocessing can be achieved in one stage. Thus, even VGO may require a graded system, e.g., either multilayer bed or multisections reactor, particularly when the objective is to produce the feed for FCC or to increase the yield of middle distillates in the products.

Entirely different configurations of the fixed bed reactors and systems may be necessary when the lube base oil is the targeted product. In this case, catalytic dewaxing reactor may be part of the overall hydroprocessing of VGO and DAO followed by hydrofinishing step performed under milder conditions as usually applied during hydroprocessing. It should be noted that the catalyst formulations required for dewaxing and hydrofinishing may differ from those of the conventional catalysts.

8.2.1 Mild hydrocracking process

The process is a modification of the fixed bed reactor system typically used for the HDS of lighter fractions derived from a conventional crude to suit hydroprocessing of VGOs (443). This may be achieved by merely increasing the reactor temperature. The increased severity resulted in the increased yield of middle distillates. Also the structure of the MHC catalysts may differ from that of the HDS catalyst. Most likely, the former was supported on a more acidic support (e.g., zeolites) to promote HCR reactions. Also some zeolite supported catalysts are active for hydroisomerization of *n*-paraffins. It was indeed confirmed that MHC process employing combination of a more acidic catalyst with the typical HDS catalyst in a stacked fixed bed gave an increased yield of middle distillates with improved cold flow properties because of the enhanced HCR of the long chain *n*-paraffins (444). The desirable level of HDS was achieved as well. Compared with the conventional HDS process, MHC process may require some modifications of downstream units with the aim to recover higher yields of the middle distillates.

8.2.2 Unibon process

Typically, this process has been used downstream of the deasphalting unit. It may also be used for hydroprocessing of either VGOs and HGOs or the blend of VGO with DAO. Depending on the feed, the process can be used as a single-stage or two-stage configuration. For example, the commercial configuration of the Unibon process using DAO as the feed consisted of two single fixed bed reactors; one operating predominantly in HDM mode (guard reactor) and the other in the HDS mode (112,126). The DAO feed contained about 27 ppm of V + Ni and less than 1% of asphaltenes. In the case, that the blend of VGO and DAO is used, the VGO may be produced during vacuum distillation of the AR and DAO produced during deasphalting of the VR obtained during the same vacuum distillation (445). This is depicted by scheme 2 in Figure 2.2. To suit refinery requirements, different configurations of the Unibon process, i.e., BOC Unibon, RCD Unibon, etc., have been licensed (446). For example, the Unicracking RDS version of the Unibon process, shown in Figure 8.3 was designed primarily for the HDS of ARs and VRs derived from the conventional crudes (447). In Figure 8.3, besides guard reactor and two HDS reactors, all necessary downstream and upstream units are shown as well. Most of these units are common for other similar commercial systems employing fixed bed reactors. Furthermore, Unicracking process was modified to suit dewaxing of VGO and DAO to the base oil for production of lubricants. In this case, at least one stage and/or one layer of the fixed bed of catalyst comprised a catalyst supported on an acidic support to ensure a desirable level of the hydroisomerization and HCR of *n*-paraffins.

A number of other commercial processes employing fixed bed reactors have been licensed. For example, the asphaltenic bottom conversion (ABC) process developed in Japan has similar features as the Unibon process (446). A modified version of this process includes the recycling of asphalt from the deasphalting unit to the HDM reactor for further processing, i.e., recycle to the extinction. Apparently, almost complete conversion of the AR could be achieved. The fixed bed reactors which are part of the Gulf RDS process (448) consist of several sections in one reactor vessel, similarly as it is in the Chevron RDS (VRDS) process. Using these processes, a high level of HDS could be achieved with a proper catalyst selection. The Chevron RDS process has also been used

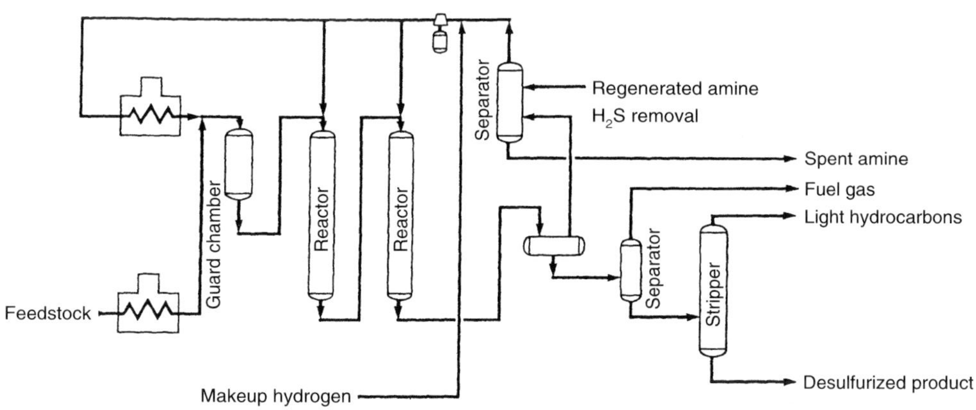

Figure 8.3 Simplified flowsheet of UNIBON process (446).

downstream of deasphalting unit for the upgrading of DAO (449). With the catalyst designed for this process, a high level of HDS and a low H_2 consumption could be achieved. The EXXON Residfining process consists of a guard reactor and the catalytic reactor comprising several sections (450). This process was designed for the HDS of the ARs obtained from conventional crudes with the aim to produce fuel oils meeting all commercial specifications.

With respect to the heavy feeds such as ARs and VRs, the processes discussed in this section were successfully used for the HDS to produce fuel oils. However, this could only be achieved using either a high catalyst/feed ratio or a long contact time. With increasing volume of heavy feeds (e.g., AR and VR), these processes had to be further modified to accommodate higher contents of metals and asphaltenes, as well as greater daily throughputs. The new processes employing fixed bed reactors, comprising various combinations of reactors and catalysts, were developed in response to these new demands. Brief descriptions of the HYVAHL process and ARDS process is given below as an illustration of the efforts to modify fixed bed reactors for hydroprocessing of the asphaltenes and metals containing heavy feeds.

8.2.3 HYVAHL process

Developed and licensed by the Institute France du Petrole (14,451), this process was successfully tested for hydroprocessing of various heavy feeds, i.e., DAOs, ARs and VRs. For example, properties of the VR derived from the Safania crude, which was successfully processed by the HYVAHL process, are shown in Table 8.3 (437). The process consists of the guard reactor placed upstream of the HDM reactors. The guard reactor is sized and optimized to achieve a satisfactory length of the cycle. For this purpose, the HDM catalyst possessing a large metal storage capacity has been developed. To protect catalyst in the HDS section, two more HDM reactors are placed downstream from the guard reactor. As Table 8.1 (437) shows, the swing reactor concept ensured a continuous operation of the process approaching 1 year using heavy feeds, the metal content of which was in the range 500 ppm of V + Ni. In this case, the process included two guard reactors which were switchable during the operation. With this concept, the replacement of catalyst in the guard reactor does not require shutdown of the operation (451). The schematics of the swing reactor concept are shown in Figure 8.4. The guard reactor and two HDM reactors represent about 40% of the total catalyst volume. As Figure 7.48 shows, at this point most of the metals were already removed from the feed (433). This ensured little deactivation of the HDS catalyst by metals in the downstream reactors. All guard and main reactors which are part of the HYVAHL

Table 8.3 Properties of Safania VR (437)

Density, kg/L	1.035
Sulfur, wt%	5.28
Nitrogen, ppm	4600
CCR, wt%	23.0
Asphaltenes (heptane)	11.5
V + Ni, ppm	203

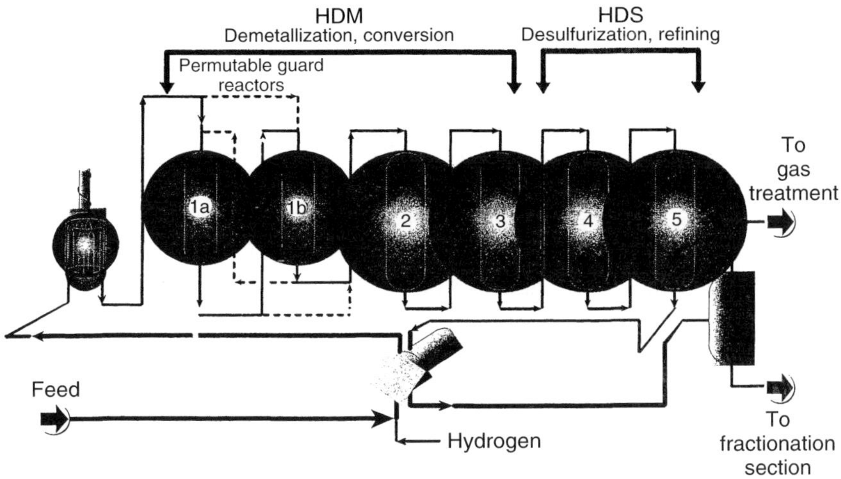

Figure 8.4 Flow diagram of HYVAHL process (451).

process are the single/fixed bed reactors. This simplifies reactor design, loading and unloading of the catalyst, as well as the operation of the reactor.

8.2.4 Atmospheric residue desulfurization (ARDS) process

The ARDS process was developed by Unocal for hydroprocessing of ARs. Apparently, this is an extension of the Unibon process to accommodate more problematic feeds. There are many years of experience in the commercial operation of this process using Kuwait AR, typically containing about 85 ppm of V + Ni and about 12 wt% of CCR. (452). The flow diagram of this process currently in operation in the petroleum refinery in Kuwait is shown in Figure 8.5. The process consists of two trains each having design capacity of 33 000 b/d. Each train comprises one guard reactor and three main reactors with a common fractionation section attached. The guard reactor contains about 7% of the total catalyst inventory and its main function is HDM of the feed. Other three reactors contain 31% of the catalyst inventory each. All three reactors employ a graded bed consisting of either the same catalyst but of different particle sizes and shapes or catalysts of a different composition. The purpose of using the graded bed is to diminish the reactor pressure drop particularly in the front of catalyst system, which is contacted with only partially converted and/or unconverted feed. Because the guard reactor only removes a part of metals, the catalysts in the subsequent reactor must possess an adequate HDM activity. Thus, a relatively large amount of metals was still present in the spent catalysts from all three main reactors (453). However, this problem may be alleviated by an optimal selection of catalyst for the guard reactor and the subsequent reactor. The modification of the Unicracking/HDS process (Figure 8.3) comprises five reactors in a series (454). It has similar features as ARDS process. In this case, the first reactor was in fact a guard reactor containing a high metal storage capacity HDM catalyst. With this arrangement, heavy feeds containing as much as 150 ppm of V + Ni were successfully hydroprocessed.

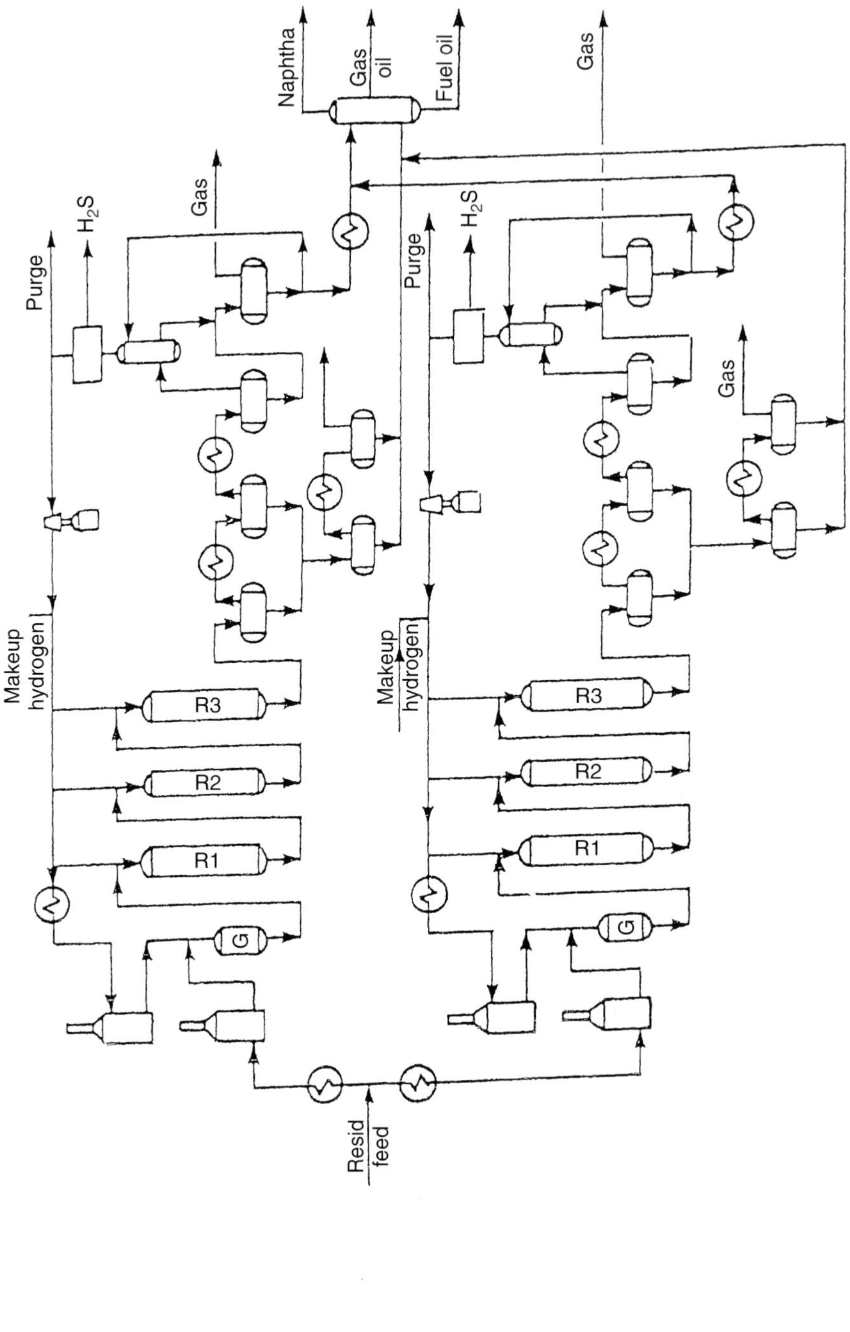

Figure 8.5 Flow diagram of the ARDS process (452).

8.3 MOVING BED REACTORS

It has been evident that for fixed bed reactors, the difficulties in handling heavy feeds could be overcome either by frequent catalyst replacements or by adding more reactors in the series. At certain point, both these options become economically unattractive. Also it is not easy to maintain synchronized operation of so many fixed bed reactors in a series. Because of these problems, reactor design and catalyst development has reached entirely new levels. In this regard, attention has been focusing on the development of a process enabling catalyst replacement during the operation. The bed of catalyst moving vertically through the reactor was one option which had been explored. Several moving bed catalytic reactors reached a commercial scale. Among those, bunker reactor and quick catalyst replacement (QCR) reactor are shown in Figure 8.6a and b, respectively. It should be noted that moving bed reactors require special equipment and procedures for safe transfer of catalyst into and out of the high-pressure and high-temperature vessels and reactors. This may include several high-pressure vessels upstream and downstream of the reactor.

Compared with fixed bed reactors, the problems associated with the development of pressure drops are not present in moving bed reactors. This allows the use of catalyst particles varying widely in the size and shape. Moreover, inorganic solids present in heavy feeds move through the reactor together with the catalyst and exit at the bottom

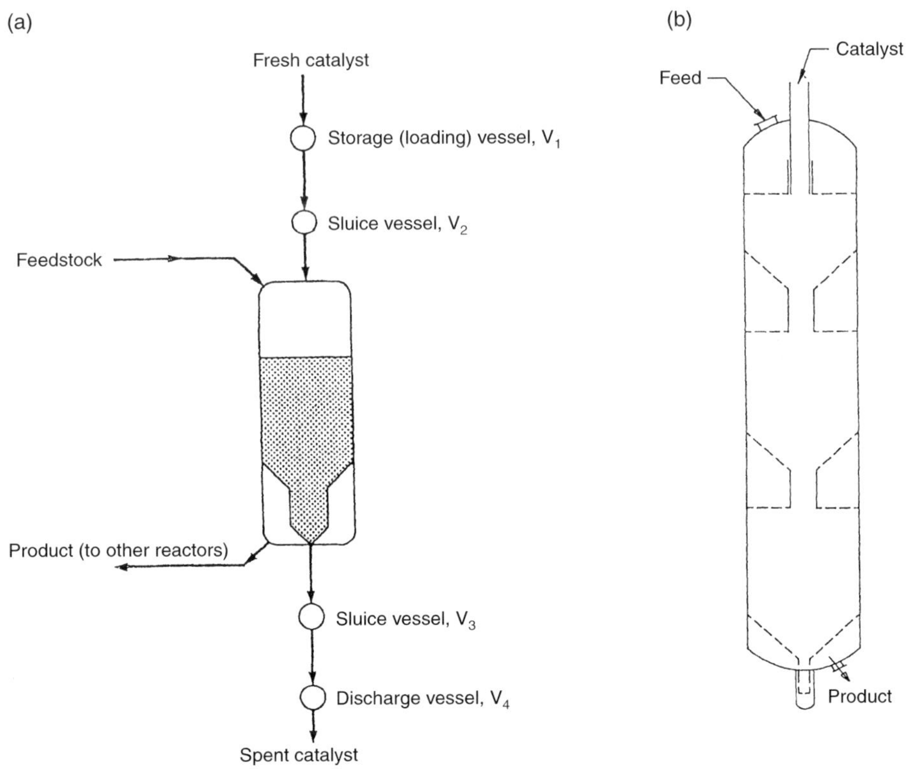

Figure 8.6 Schematics of bunker reactor (a) and QCR reactor (b) (455,456).

of reactor with the spent catalyst. Therefore, the process employing moving bed reactors may not require a guard chamber even for the most problematic feeds.

8.3.1 QCR reactor

This reactor was developed by Shell (455). Its simplified schematics shown in Figure 8.6b indicate that this reactor has conical rather than horizontal support grids. This enables a quick and complete catalyst unloading. Catalyst loading is achieved by hydraulic transport of the catalyst/oil slurry from the storage hopper at the top of the reactor where catalyst and oil are separated. The design of the reactor allows the use of different catalysts at the same time. Thus, for a high metal feeds, a catalyst having a large metal storage capacity may be used at the top section of the reactor in order to protect a high activity catalyst in the lower section. Another commercial process allowing a periodic on stream catalyst replacement (OCR) was developed by Chevron (15,456).

8.3.2 Bunker reactor

Another moving bed reactor developed by Shell is the bunker reactor (455,457). It is the trickle flow reactor in which catalyst is replaced discontinuously. The simplified schematic of this reactor is shown in Figure 8.6a. The main features of the bunker reactor include a special internal construction which allows plug flow of catalyst co-currently with the process stream. The specially designed catalyst holding system allows the separation of catalyst from the process streams. In the reactor, catalyst rests on a conical support with a catalyst outlet duct leading to the sluicing system in the center. High-pressure sluices, rotary star valves and slurry transport loops are used for catalyst loading and unloading. With such system installed on the top and bottom of the reactor, the catalyst loading and unloading could be performed without any difficulties. The fresh catalyst is loaded by hydraulic transport from a storage hopper. Once in the reactor, the catalyst/oil slurry is separated into the oil and catalyst. Furthermore, the reactor internals make the catalyst unloading easier. This arrangement allowed the catalyst replacement onstream without interrupting the operation. The rate of catalyst replacement depends on the content of metals in heavy feed. The top layer of the moving bed always consists of the fresh catalyst. This layer contacts the unconverted heavy feed and as such removes inorganic solids dispersed in heavy feed. This removal is physical causing little detrimental effect to the activity of catalyst.

The current commercial systems employing bunker reactor have been using the VR containing $\sim$250 ppm of V + Ni, although the process can handle much higher metal content feeds (458). The process known as HYCON consists of two parallel trains of five reactors. The first three reactors of each train are bunker reactors operated in series. They were filled with the HDM catalysts possessing a high metal storage capacity. The last two reactors are the conventional fixed bed reactors. Because the metal content of the VR was below the design value, the operation of the third reactor was changed. However, in this mode an accumulation of the fine solids containing iron in the first fixed bed reactor was noted confirming that such solids passed the bunker reactors. This problem was rectified by returning the third bunker reactor which operated in a fixed bed mode to its original function.

Table 8.4 Yields and properties of products from different reactors [437]

	Fixed/moving	Ebullated	Slurry
Naphtha			
Yield/feed, wt%	1–5	5–15	10–15
Density, kg/L	0.71–0.74	0.71–0.72	0.72
Sulfur, wt%	<0.01	0.01–0.2	0.06
Nitrogen, ppm	<20	50–100	200
Gas oil			
Yield/feed, wt%	10–25	20–30	40–45
Density, kg/L	0.850–0.875	0.840–0.860	0.866
Sulfur, wt%	<0.1	0.1–0.5	0.7
Nitrogen, ppm	300–1200	>500	~1800
Vacuum gas oil			
Yield/feed, wt%	20–35	25–35	20–25
Density, kg/L	0.925–0.935	0.925–0.970	1.010
Sulfur, wt%	0.25–0.50	0.5–2.0	2.2
Nitrogen, ppm	1500–2500	1600–4000	4300
Vacuum residue			
Yield/feed, wt%	30–60	15–35	10–20
Density, kg/L	0.990–1.030	1.035–1.100	1.160
Sulfur, wt%	0.7–1.5	1–3	2.7
Nitrogen, ppm	3000–4000	>3300	11 000
Asphaltenes (heptane)	5–10	>20	26

It should be noted that the viability of the process employing moving bed reactors compared with fixed bed reactors may be affected by the capital cost of reactor systems, although this may be offset by the lower relative catalyst consumption. For example, the additional high-pressure vessels upstream and downstream of the bunker reactor and QCR reactor add to the capital cost. Also, the configurations of moving bed reactors are more complex than that of the fixed bed reactors. Therefore, the design of the latter is much more simple. In addition, the different yield and quality of products are obtained from different reactors. This is shown in Table 8.4 (437).

8.4 EBULLATED BED REACTORS

The first process employing ebullated bed reactor was known as the H-Oil process developed jointly by the City Services with Hydrocarbon Research Institute (HRI). The HRI was joined by Texaco and later by IFP to license H-Oil process, whereas City Services jointly with Lummus and Amoco have been licensing similar process known as LC-Fining.

The ebullated bed reactors were designed to handle the most problematic feeds such as VRs and toped heavy crudes having high contents of metals, asphaltenes, sediments as well as dispersed clay and minerals. The flexibility of the operation of the ebullated bed reactors was successfully demonstrated during co-processing using the mixtures of

VRs with coals, as well as VRs and plastics. Table 8.1 (437) shows some operating parameters confirming that the ebullated bed reactors are suitable for hydroprocessing heavy feeds containing more than 700 ppm of metals. This, however, cannot be achieved without significant catalyst inventory. Because of the catalyst being in a continuous motion, particle size approaching 1 mm can be used without any difficulties. This ensures a high level of catalyst utilization. However, for such thin particles, mechanical strength requires an attention to prevent their breaking in the reactor as it was observed by Al-Dalama and Stanislaus (346). To be cost-competitive, this process must produce enough additional liquid products compared with the non-catalytic options, i.e., deasphalting and coking, to compensate for the costs of catalyst inventory. Also, the additional high-pressure vessels and equipment upstream and downstream of the reactor are necessary to ensure safety of the operation similarly as it was noted for moving bed reactors. This adds to the capital cost of the ebullated bed reactors compared with the fixed bed reactors

The most important features of the ebullated bed reactors include their capability to either periodically or continuously add/withdraw catalyst without interrupting the operation. The bed design ensures an ample free space between particles allowing entrained solids to pass through the bed without accumulation and plugging, as well as without increasing pressure drop. Under such conditions, the catalyst particles having a diameter smaller than 1 mm (e.g., 1/32 in extrudates) can be utilized. This results in the considerable increase in reaction rate because of the significantly diminished diffusion limitations. Moreover, under such conditions, the catalyst utilization is significantly enhanced. Depending on the operating strategy of the refinery, the process can operate either in a high conversion mode or in a low conversion mode.

The simplified diagram of the LC-Fining reactor is shown in Figure 8.7 (458–460). It is again noted that the H-Oil reactor has similar features. In ebullated bed reactor, the heavy feed and H_2 enter at the bottom and move upward through the distributor plate at a sufficient velocity to expand the catalyst above the grit into a state of random and turbulent motion. The expanded bed is maintained about 35% above the settled level of catalyst. This can be achieved by controlling the speed of the recycle oil pump. In this regard, the operation is monitored using the density detectors. The suction of the recycle pump is supplied from near the top of the reactor. The recycle pan is used for disengaging the gas before recycling the liquid. The advanced design of the ebullated bed reactor used in the H-Oil process incorporates an improved internal recycle cup enabling a complete separation of gas from the recycled liquid. With this modification, the throughput of heavy feed was increased. On a commercial scale, usually three ebullated bed reactors are used in the series (Figure 8.8). The first reactor serves as a guard reactor, the primary function of which is HDM. The main function of the second and third reactors are HDS, HDN and HCR. In some situations, the ebullated bed reactor can be used as the guard reactor upstream of the fixed bed reactors. However, in the case of a large amount of inorganic solids in heavy feed, part of these solids may not be trapped in the ebullated bed reactor. Such solids may then be carried out with liquid streams to the subsequent fixed bed reactor.

Figure 8.9 (73) shows the simplified diagram of the catalyst handling system consisting of three sections, i.e., fresh catalyst handling, the daily addition/withdrawal of catalyst to and from reactors and spent catalyst handling system. The fresh HDM catalyst is carried as a slurry from the high-pressure vessel to the first reactor. The equilibrium catalyst is withdrawn from the third reactor and transported as a slurry to the second reactor. The

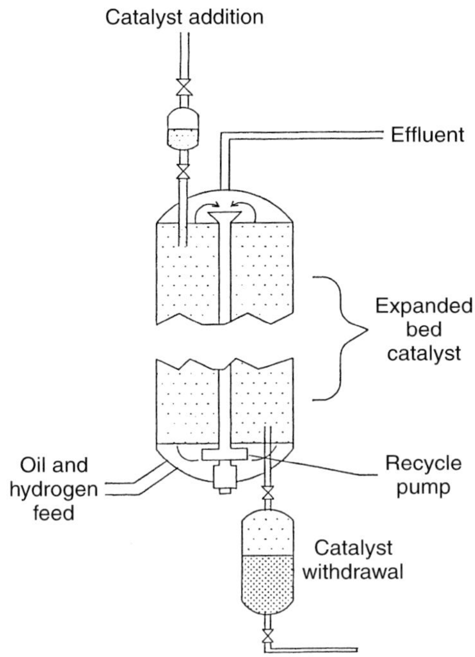

Figure 8.7 Flow diagram of ebbulated bed reactor (458).

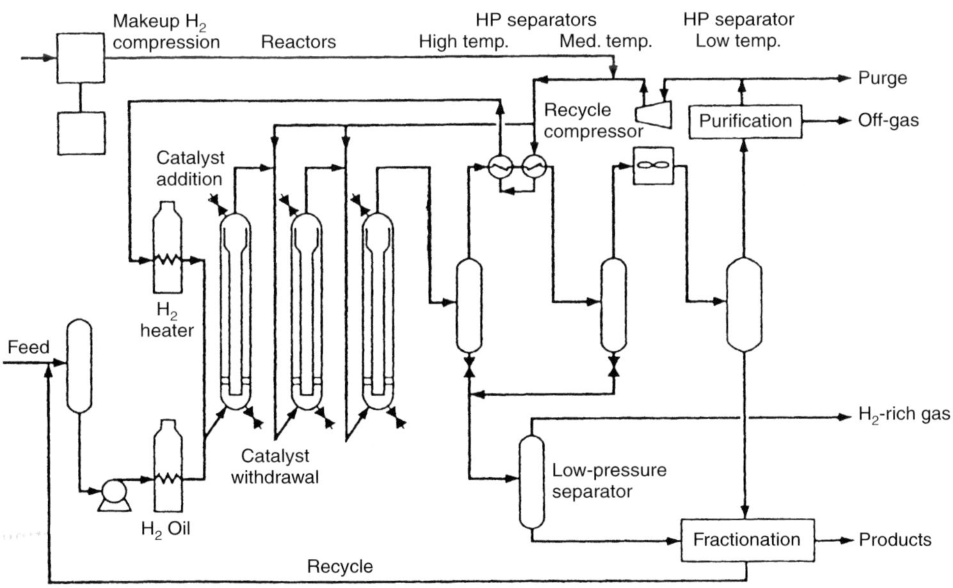

Figure 8.8 Process employing ebullated bed reactors (459).

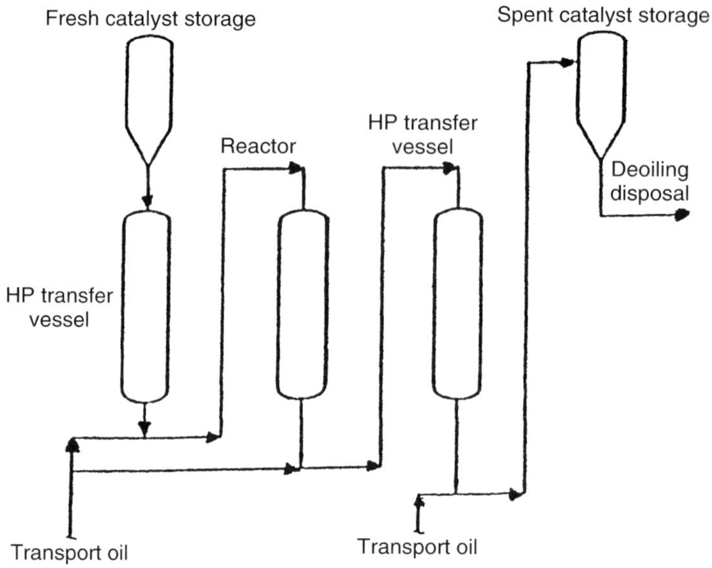

Figure 8.9 Catalyst handling system for ebbulated bed reactors (73).

spent catalysts are withdrawn from the first and second reactors to the transfer vessel. It is then washed, cooled and transferred to the spent catalyst inventory vessel. Further, utilization of spent catalysts from the ebullated bed reactors depends on the level of deactivation, particularly on the amount of deposited metals such as V and Ni.

8.5 SLURRY REACTORS USING LOW-COST SOLIDS

The highest level of catalyst utilization may be achieved using either the micronized size of catalyst particles or a near molecular size of the particles. The former size ensures that less-active solids may exhibit an adequate catalyst activity when slurried with a heavy feed. The molecular size may be approached when the catalyst particles are generated from a precursor in situ. This topic is discussed in detail latter in the book as part of the emerging processes for upgrading heavy petroleum feeds.

Low-cost, "throw-away" materials disposed from various industrial operations may contain some transition metals which are catalytically active. This may include the fly ash from combustion of petroleum coke, "red mud" from aluminum production, coke breeze from the preparation of metallurgical cokes, fines from flexi-coking process, etc. Naturally occurring solids of a suitable composition can also be used. Generally, these solids contain catalytically active metals such as Fe, Ni and V. The solid by-products from various industrial operations are produced in the form of fine particles. Therefore, little preparation may be required before they can be slurried with heavy feeds. Naturally occurring solids may require crushing to a desirable particle size. Besides being catalytically active, fine particles serve as the coke getters, i.e., the seeds of coke which deposit on the catalyst particles are carried out of the reactor with the process streams. Otherwise, coke and/or deposits would be formed on the reactor walls. This could lead to the development of pressure drops and finely cause shutdown of the operation. The beneficial effects of some solids were observed even without H_2 present (461). In this case, fine particles of the solid consisting of kaolinite (75%), illite and montmorilonite suppressed coke formation during a thermal treatment of the Athabasca VR at 703 K. Several processes employing low-cost solids as catalysts have reached a near commercial scale. For the purpose of this book, some of these processes were selected to indicate advantages and disadvantages with respect to commercial applications.

The best-known processes employing throw-away solids include VEBA hydrocracking (30,462,463) process. Simplified schematics of this process are shown in Figure 8.10. In this once-through option, the slurry of heavy feed and finely dispersed solid catalyst enter the reactor at the bottom. The products exit the top and are separated into light ends and heavy ends. The latter enter the vacuum flasher for further separation. Some schematic of VEBA process indicate an option for recycling heavy ends into the slurry before entering the reactor. It is believed that this option may be limited to some specific feeds. Otherwise, this option may affect colloidal stability of the feed leading to an excessive formation of sediments. The VEBA process employs "red mud" as the additive without any option for its recovery. The environmental aspects, as well as the potential options for on-site energy supply during the upgrading Orinoco heavy crude using VEBA process were discussed by Wenzel and Herrera (464). It should be noted that Orinoco crude is among the heaviest crudes found in the world.

There are several other processes employing low-cost solids as catalysts. For example, the HDH process developed by INTEVEP (425,465–467) can operate in both once-through mode and catalyst recovery mode using a naturally occurring solid of a proprietary nature. The naturally occurring limonite (Fe_2O_3) has also been found active under conditions ensuring its conversion to pyrrhotite (468). The Auroban process developed by UOP employs metal oxides of a proprietary nature as well (469). The fines removed from the venturi scrubber of the flexi-coking process were catalytically active at 686 K and 10 MPa in the slurry with VR (470). The activity was attributed to

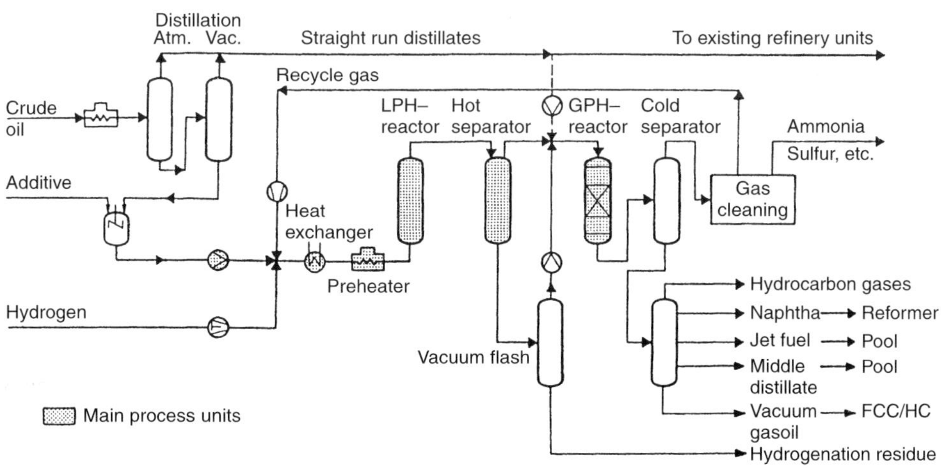

Figure 8.10 Flow diagram of VEBA process employing slurry bed reactor (464).

the presence of metals such as V, Ni and Fe. The pulverized form of theses solids is usually dispersed in heavy feed before processing. All these processes have approached a near commercial scale.

If slurried with a heavy feed, pulverized form of the non-regenerable spent catalysts may exhibit an adequate activity for hydroprocessing conducted in the slurry bed reactor. For example, a spent FCC catalyst was used in the Mild Resid Hydrocracking (MRH) process developed in Japan (471). However, an excessive coke formation could not be prevented in spite of the mild conditions applied. The problem was overcome with a new catalyst consisting of a mixture of the spent FCC catalyst and a proprietary catalyst. Apparently, the FCC portion of the mixture played an important role as the coke scavenger. The crushed spent HDS catalyst was also tested. This catalyst exhibited much higher activity. Unfortunately, the limited availability prevented its utilization in commercial units. Without any doubt, fresh conventional catalysts in a pulverized form would be much more active than once-through solids. However, the recovery of these catalysts from the residue obtained after the distillation of hydroprocessed products would be rather complex and costly. This prevents utilization of the conventional catalysts in the slurry bed hydroprocessing units.

8.6 COMPARISON OF HYDROPROCESSING REACTORS

Morel et al. (437) estimated ranges of the yields and of the properties of the products from hydroprocessing of the Safania VR in different types of reactors. The properties of the VR are shown in Table 8.3, whereas those of the products together with their yields in Table 8.4. With respect to the content of contaminants (S, N and CCR) in products, fixed/moving bed reactors were the most efficient followed by ebullated bed reactor. Because of the higher temperature employed, the latter reactor gave the larger yields of naphtha and gas oil. In the slurry bed reactors employing throw-away solids, the conversion to liquid products have exceeded 80%. This resulted from the temperatures which were higher than those typically used during conventional hydroprocessing. The residence time was usually longer as well.

The quality of products (Table 8.4) from different reactors reflects the difference in operating conditions. The lower quality for ebullated bed reactor compared with fixed/moving bed reactor is attributed to a higher temperature used in the former. This may be offset by a lower yield of VR in ebullated bed reactor. The lowest quality products are obtained in slurry bed reactors, most likely, because of the highest temperature used compared with the other reactors. This suggests that a significant hydroprocessing of the liquid products from the slurry bed reactors would be required to achieve specifications of the commercial fuels. Moreover, feasibility of the slurry bed reactors may be affected by availability of the catalytically active solids. Thus, the plant processing 10 000 t/d of heavy feed, requiring about 0.5 wt% of catalyst to achieve acceptable conversion, would consume about 50 t/d of the catalytically active solid. Therefore, the integration with an industrial process (e.g., aluminum production) generating low-cost solids would enhance the viability of slurry bed reactors.

The safety aspects of hydroprocessing operations deserve attention. Decades of the experience using heavy feeds varying widely in properties shows that it is quite easy and safe to operate fixed bed reactors. The additional high-pressure equipment upstream and downstream of the moving bed and ebullated bed reactors adds to the complexity of the operation. More severe conditions, i.e., higher temperatures and pressures, than in fixed bed reactors indicate that ebullated bed reactors may require special materials for the construction of equipment, similarly as it is for slurry bed reactors. However, the simple features suggest that the design of the slurry bed reactor may be less challenging compared with that of the moving and ebullated bed reactors.

Chapter 9

PATENT LITERATURE ON HYDROPROCESSING CATALYSTS AND REACTORS

The intention of this patent summary is to identify the areas relevant to the development of catalysts and processes for hydroprocessing of heavy petroleum feeds, which are covered by patents, rather than to give an exhaustive review of the subject. For this purpose, the focus has only been on the patents issued by the US Patent office during the last two decades. In this case, a large representation by the non-US inventors has been noted. Therefore, it is believed that this database is the largest and the most comprehensive collection of patents in the world. Because of its extent, this database may adequately reflect the activities related to the development of the catalysts and catalytic systems for upgrading heavy petroleum feeds with emphasis on the industrial applications.

The patent literature indicates the efforts to improve performance of the conventional catalysts and to develop novel catalysts, as well as to optimize existing and/or to design the new process configurations, suitable for hydroprocessing of the problematic feeds. It has been noted that the performance of conventional catalysts could be improved by the addition of modifiers and by using different supports, as well as by varying conditions, during the catalyst preparation. A number of the non-conventional catalysts have been patented as well. From the process development point of view, the attention has been paid to guard beds, mixed and layer beds, multibed and multistage systems, as well as to novel systems, and reactors. It has been observed that, in the patent literature, the information on the development of catalysts and catalytic processes for hydroprocessing of heavy feeds is in line with that published in the scientific and technical literature.

9.1 CATALYST DEVELOPMENT

The patent literature covers all aspects of catalyst development, i.e., preparation of the conventional catalysts, modified catalysts and novel catalysts, as well as the modified conventional supports, and entirely novel supports. The procedures for the laboratory evaluation of catalysts, as well as their utilization in the commercial units, have also been part of the patent development efforts.

9.1.1 Conventional catalysts

There is an extensive information on conventional catalysts, i.e., the Co(Ni)Mo(W) catalysts supported on γ-Al_2O_3, in the patent literature. For the great part, the aim has been an improved catalyst performance. In this regard, the emphasis was on the suitable surface properties (e.g., porosity) of catalysts with a specific heavy feed in mind. This could be achieved by optimizing the conditions used during preparation and designing new catalyst formulations. The procedures for reuse and/or cascading of the spent hydroprocessing catalysts have been receiving attention as well.

The preparation of the catalyst composition consisting of W(Mo) and Ni(Co) metals in the presence of an organic oxygen-containing agent was described by Plantenga et al. (472). Two preparation routes were reported, i.e., one in which the catalyst was dried under such conditions that at least part of the additive remained in the catalyst and the other procedure, in which the catalyst was subjected to a calcination step at such conditions that the additive was removed from the catalyst. The reason behind the increase in catalyst activity was the influence of the remaining additive on the interaction between the metal components and the alumina support. The process of the invention allowed one to prepare a catalyst, while using the combined volatile content reduction-sulfurizing step (473). In this method, a porous support was combined with Group VI and Group VIII metals containing precursors, followed by temperature increase below the calcining temperature. The latter step was performed in the presence of the sulfur-containing agents. The conventional $NiMo/Al_2O_3$ catalyst suitable for hydroprocessing of VRs, prepared according to the invention of Sherwood et al. (474,475), had the total surface area of 150–210 m^2/g, a total pore volume of 0.50–0.75 cm^3/g and the pore size distribution such that the pores having the diameters of less than 100 Å constituted less than 25%, the pores having diameters of 100–160 Å constituted 70–85% and pores having diameters greater than 250 Å constituted 1.0–15% of the total pore volume of the catalyst.

A special preparation procedure developed by IFP resulted in the macroporous "chestnut bur"-like γ-Al_2O_3 ideally suited as the support for HDM catalysts (476). The acidity of this support was lower than that of the conventional γ-Al_2O_3. Therefore, the adverse effect of coke deposition during HDM was minimized. This was demonstrated during the extended runs using the high metals and asphaltenes VRs. In this case, Ni and Mo were used as the active metals. The metal (V + Ni) storage capacity of this catalyst has approached the weight of the fresh catalyst.

The catalyst consisting of the porous alumina support bearing the metals of Groups VIII and VIB and optionally phosphorus had the surface area of 240–310 m^2/g, pore volume of 0.5–0.75 cm^3/g, and a 63–78% of the total pore volume was present in the micropores of diameter 55–115 Å and 11–18% in the macropores of diameter greater than 250 Å (477). The conversion during hydroprocessing of VRs was greatly improved by using this catalyst. Another catalyst prepared by impregnation of the alumina support with the metals of Groups VIII and VIB and optionally phosphorus had the surface area of 165–230 m^2/g, pore volume of 0.5–0.8 cm^3/g. Less than 5% of the pore volume of this catalyst was present in micropores with the diameter less than 80 Å, about 65% of pore volume was in 100–135 Å range and 22–29% in more than 250 Å range. This catalyst was particularly effective for achieving high levels of HDM, HDS and HCR of VRs (478). The Kemp's (479) invention gives the detailed description of the method for using hydrogels for preparation of the conventional Ni(Co)/Mo(W) catalysts for the HDM of heavy feeds. The final calcined catalysts had the surface area about 300 m^2/g, at least 20% of its pore volume was in the pores having diameters greater than 350 Å and at least 20% of the pore volume in the pores having diameters less than 70 Å.

The de Boer's (480) invention relates to the process for preparing catalyst from a spent hydroprocessing catalyst. It involves the removal of part of the coke and sulfur by burn-off, grinding the obtained spent catalyst and mixing it with additives. The mixture was shaped to obtain new particles which were subjected to a high-temperature calcination. The catalyst had good properties for applications in the ebullated bed reactor. The invention of Artes and Kukes (481) describes another catalyst for the ebullated bed reactor. It comprised a porous alumina support and had an incremental pore volume maximum at the pore radius greater than 50 Å and at least 0.4 cm^3/g of pores having the radius less than 125 Å. The typical HDM catalyst for ebullated bed process consisting of the Mo(W) oxides (3.5–5.0 wt%) and CoO (0.4 to 0.8 wt%) was patented by Beaton et al. (482–484). It had surface area of about 150 m^2/g to about 220 m^2/g, and the total pore volume within the range of about 0.85 cm^3/g to about 1.5 cm^3/g. Furthermore, the pore volume of the pores possessing diameters greater than about 1200 Å ranged from about 0.15 cm^3/g to about 0.4 cm^3/g. Another catalyst to be used in ebullated bed reactors developed by Peck et al. (485) comprised a porous support having pore volume of the pores with diameter greater than 1 200 Å of 0.1–0.3 cm^3/g and not more than 0.15 cm^3/g pore volumes in the pores with diameter greater than about 4 000 Å.

The method of preparation and composition of the catalyst suitable for the simultaneous HDS and HDM of heavy feeds, having a long life was disclosed (486). The catalyst consisting of the alumina support and the Groups VIII and VIB metals had the pore volume from 0.5 to 1.0 cm^3/g, and the average pore diameter varied between 80 and 300 Å. Another catalyst, which was active for HDS, HDM and HDAs had 5–11% of its pore volume in the form of macropores, and the surface area greater than 75 m^2/g (487,488). The catalyst had the peak mesopore diameter greater than 165 Å and the average mesopore diameter greater than 160 Å. Several formulations of the catalysts for hydroprocessing of ARs and VRs were disclosed by Boon et al. (489). The first comprised the Group VI and VIII metals on the support having at least 40% of its pore volume in the pores with diameters in the range from 170 to 250 Å and the surface area from 100 to 160 m^2/g. The second catalyst contained the same active metals and had at least 40% of its pore volume in the pores with diameters in the range from 30 to 170 Å and the surface area in the range of 160 m^2/g to 350 m^2/g. The support used for the third catalyst had at least 40% of its pore volume in the pores with diameters in the range from 170 to 250 Å and the surface area from 100 to 160 m^2/g. The last catalyst was at least 1.5 times more active for HDM than the first catalyst.

A method for preparation of the bidisperse support to be used for the preparation of conventional hydroprocessing catalysts was patented by Pereira and Cheng (490). The support had in one region small micropores with the average pore diameter of less than 100 Å, whereas in the other region large micropores with the average pore diameter between 100 and 600 Å. By adding active metals, the catalyst was suitable for hydroprocessing of the high metals- and asphaltenes-containing feeds. Apparently, the metals catalyzing HDS were placed in the small micropores and the HDM metals in the larger pores (491).

Gibson (492) disclosed a process for hydroprocessing of heavy feeds using shaped catalysts. When the shaped catalysts were used for diffusion limited reactions, i.e., for HDM in particular, the lifetime of the catalyst was extended and its utilization enhanced. The preferred shaped particles included the oval and elliptical shapes with and without bumps.

Rather unique catalyst preparation procedure was patented by Duddy et al. (493). According to this invention, used catalyst was withdrawn from the second stage reactor and treated using a suitable catalyst rejuvenation procedures to remove about 10–50 wt% of the contaminant metals and usually at least about 80 wt% of the carbon deposits. Using the additional treatment by suitable solvent and acid washing, almost 85% of the original catalyst pore volume, surface area and catalytic activity were recovered. The rejuvenated catalyst particles were then cascaded to the first stage reactor, together with any fresh makeup catalyst as needed therein. A sufficient amount of the fresh catalyst was added to the second stage reactor to replace the used catalyst.

9.1.2 Conventional modified catalysts

This group includes catalysts consisting of the active metals which are part of the conventional catalysts, i.e., Co(Ni)Mo(W) and the γ-Al_2O_3 support, which were modified by various additives. In addition, the catalyst properties can be modified by using different inorganic oxides as supports and/or a combination of several inorganic oxides. The acidity of supports has been receiving attention, whenever a high conversion of large molecules (e.g., resins and asphaltenes) in the feed to distillates was the objective. In this case, the origin of support may play the decisive role.

9.1.2.1 Effect of additives

Phosphorus has been most frequently used as the additive for modification of the conventional hydroprocessing catalysts. Other additives included fluoride, boride, alkali and alkali earth metals. An improvement in the catalyst performance could also be achieved by combining Al_2O_3 with SiO_2.

The properties and the method for preparing the NiMoP/Al_2O_3 catalyst for hydroprocessing of heavy feeds were disclosed by Abbo et al. (494). The total pore volume of the alumina support, as measured by the conventional mercury porosimetry methods, was usually between 0.3 and 0.9 cm^3/g and surface area of about 125 m^2/g. The preferred support had the mean pore diameter from 60 to about 100 Å. The support had less than 20% of the total pore volume in the pores having diameter less than 60 Å and less than 5% of the total pore volume in the pores having diameter less than 50 Å. In addition, 15% of the total pore volume was in the pores with diameter greater than 90 Å.

It has been evident that the methods for preparing the catalyst support having the controlled pore size distribution have been receiving attention. They are usually based on the incorporation of phosphorus into the support. Thus, Angevine et al. (495) described the method for preparation of the support containing the amorphous magnesia–alumina–aluminum phosphate. This involved the admixture of an organic cation having the size equal to or greater than 2 Å, the organic cation preferably being the tertiary or the tetraalkylammonium or phosphonium cation. A porous refractory support containing the under-bedded phosphorus-containing component was prepared by incorporation and calcination of the phosphorus with the support, followed by further incorporation of the catalytic promoter components, such as the Groups VIB and VIII metals (496). The catalyst was especially active when the under-bedded phosphorus components were incorporated into the support at calcination temperatures lower than 760 K. The catalyst was active during the simultaneous HDN and HDS of VGOs.

Details of the preparation of the Mo(W) catalysts supported on γ-Al_2O_3 and promoted with Ni(Co) (0.1–2.5 wt%), designed for the HDM and HDAs of heavy feeds, were given by Fujita et al. (497). The additional active components included phosphorus or boron. The catalyst of this invention had 10–30% of its pore volume in the pores with the diameter above 1000 Å, the so-called macropores. The preferred pore volume of the catalyst was in the range of 0.6–1.0 cm^3/g, surface area between 110 and 140 m^2/g and the particle size of 2–7 mm.

The NiMo catalyst having a specified pore distribution required for the conversion of VRs was disclosed by Sherwood (498,499). The catalyst support was γ-Al_2O_3–SiO_2 (2.5 wt%) combined with 2.2–6 wt% of the Group VIII metal oxide, 7–24 wt% of the Group VIB metal oxide and 0.3–2 wt% of the loaded phosphorus oxide. Phosphorous was loaded onto the catalyst as the aqueous phosphoric acid. The catalyst had surface area of 175–205 m^2/g and pore volume of 0.82–0.98 cm^3/g. The pore diameter distribution included 29.6–33.0% of the macropores having diameter greater than 250 Å, 67.0–70.4% of the micropores with diameter less than 250 Å of which 65% had pore diameter between 110 and 130 Å. The conventional catalyst supported on the alumina containing between 2 and 8 wt% of silica was prepared by impregnating the support particles having a narrow pore size distribution and a median pore diameter greater than about 120 Å, with the solution containing active metal precursors, followed by drying and calcining. The catalyst was active for a number of hydroprocessing reactions (500).

The preparation of Mo(W) catalyst promoted with Ni(Co) and supported on the Al_2O_3–SiO_2 (3.5 wt% SiO_2) and containing 0.1–2 wt% of alkali metal component was disclosed by Abe (501). The catalyst had the surface area between 185 and 250 m^2/g, the total pore volume of 0.6–0.9 cm^3/g, 40–65% of its pore volume present in pores with diameter of 100–200 Å, 8–25% with diameter of above 1000 Å and less than 20% of its pore volume in the pores with diameter of less than 100 Å. The catalyst was suitable for hydroprocessing of ARs.

9.1.2.2 Effect of supports

The objective of the catalyst development for the mild hydroconversion of VRs at Texaco was the increased yield of middle distillates (502–506). These disclosures include combinations of the conventional γ-Al_2O_3 with more acidic supports in comparison with the additives discussed above. The improved catalysts consisted of ~2.0 to ~6.0 wt% of an oxide of the Group VIII metals, ~12.0 to ~25.0 wt% of the oxide of molybdenum and 0 to ~3.0 wt% of the phosphorus oxide supported on the porous alumina support containing about 4.0 to ~30 wt% of silica (502,503). A modification of the same catalyst comprised the support containing 0.1–10 wt% of lithium oxide (504), as well as the precipitated alumina, silica or silica–alumina and dealuminated Y-zeolite (505,506).

The zeolitic alumino-silicate was the support for the catalyst patented by Best et al. (507). The catalyst was designed for hydroprocessing VGOs to give high yields of middle distillates, similarly as the invention of Habib and Dahlberg (508). In this case, the catalyst comprised the ultra stable Y-zeolite base of the unit cell size greater than ~24.55 Å and the crystal size less than ~2.8 μ. Moreover, it contained an amorphous cracking component, a binder and a catalytic amount of hydrogenation component consisting of a Group VI metal, a Group VIII metal and/or their mixtures. A novel support invented by Occelli (509) comprised β-zeolite and the layered magnesium silicate such as sepiolite. The latter was especially suited for the use in combination with one or more hydrogenation components as the catalyst for conversion of VGOs to middle distillates. The HCR catalyst suitable for hydroprocessing of VGOs contained the mesoporous crystalline material together with the molecular sieve component of the relatively smaller pore size as the support (510). The latter consisted of either Y-zeolite or an intermediate pore size zeolite such as ZSM-5. This component provided a higher level of the acidic functionality than the mesoporous component and permitted the metals loading and acidic activities to be optimized for the enhanced catalyst selectivity to middle distillates.

The catalyst of the Nakaoka's (511) invention comprised the Groups VI and VIII metals supported on various supports, e.g., alumina alone and alumina containing at least one of silica, magnesia and calcium oxide, Y type zeolite having a unit lattice constant of 2.425–2.445 nm and zinc oxide. A boron compound was also added to improve cracking and HDS activities, as well as to prolong the catalyst life.

9.1.3 Novel supports and catalysts

The catalytic phases and supports, which differ from those found in the conventional catalysts have been receiving attention. For example, the catalyst support of the Quann's (512) invention consisted of the γ-Al_2O_3 and α-Al_2O_3. The catalyst had at least 25% of its pore volume in the pores having diameter 300–1000 Å, whereas no more than 10% of its pore volume in the pores with diameter greater than 1000 Å and the surface area of less than 100 m^2/g.

The novel catalyst for the hydroprocessing of VRs and heavy crudes was patented by Fukuyama et al. (513). The catalyst contained iron and activated carbon and had the specific surface area of 600–1000 m^2/g and pore volume of 0.5–1.4 cm^3/g. It also had at least 60% of its pore volume in the mesopore range with an average pore diameter varying between 30 and 60 Å. The amount of iron added to the activated carbon varied between 1 and 20 wt% of the weight of activated carbon.

The preparation of the unconventional catalyst comprising the silicoaluminophosphate molecular sieve such as SAPO-11 and SAPO-41, as well as platinum and/or palladium, was described by Miller (514). The catalyst was suitable for simultaneous HCR and isomerization of VGO to produce high yields of middle distillates. The products had good low temperature fluid characteristics, especially reduced pour point and viscosity.

The layered metal oxide components were used as the supports by Angevine et al. (515). They could be prepared from the layered oxides of the Group IVB metals, such as titanium, zirconium and hafnium. In this case, the layered titanates, e.g., $Na_2Ti_3O_7$ were suitable starting materials in particular. Such materials comprised the interspathic cationic species between their layers. The layered metal oxide contained stable polymeric oxide such as silica between adjoining layers. The resulting silico-titanates were the heat-resistant materials. Such materials were suitable for the operation under high severity conditions.

The invention Audeh and Yan (516) described the method for preparation of an improved sepiolite-based hydroprocessing catalyst by controlling and modifying the acidity of the catalyst. For this purpose, the catalyst was treated to back-exchange magnesium ions for acidity control. Optionally, sepiolite could be mixed with the low acidity zeolites. Initially, sepiolite was contacted with an aqueous solution of the salt of metal such as cobalt to form the metal-exchanged sepiolite product. Subsequently, the dried powdered product was contacted with the aqueous solution of a molybdenum compound and finally was either contacted or ion-exchanged with the aqueous magnesium salt solution to effect the ion-exchange of magnesium ions. In the pulverized form, this catalyst composition could be admixed with zeolite of the ZSM-5 type having a high silica/alumina mole ratio (e.g., greater than 10). The resulting catalyst exhibited the enhanced hydroprocessing activity when heavy feeds were used.

9.2 CONFIGURATIONS OF CATALYTIC REACTORS AND SYSTEMS

The efficient utilization of catalysts and the extension of operations can be achieved by modifying the configurations of catalytic reactors and processes to match hydroprocessing of a particular heavy feed. This may involve the different features and combinations of guard chambers, guard reactors and fixed bed reactors. As it was indicated earlier, for the fixed bed reactors, the modifications may involve either multilayered single fixed bed or a single reactor comprising several sections of the catalyst separated by an empty space in the same reactor vessel. For the reactor comprising several sections, the methods of the feed introduction into the each section can play an important role as well. Several fixed bed reactors connected in a series may be needed to achieve desirable conversions of heavy feeds. The advanced reactors employing either moving bed or ebullated bed have been developed for the most problematic feeds. As documented by the patent literature, the efforts in response to these requirements are in line with the information published in the technical and scientific journals.

9.2.1 Guard chambers and materials

In this case, guard chambers are considered to be the fixed bed systems filled with the low value inorganic solids (e.g., bauxite, alumina, magnesium silicate, etc.) rather than with hydroprocessing catalysts. The main function of guard chambers is the removal of inorganic solids dispersed in heavy feeds. If present in heavy feed, the finely dispersed sulfides of V and Ni formed (in feeding lines) via non-catalytic reaction with H_2S will also be removed in guard chambers. According to the system disclosed by Wolk et al. (517,518), installing the guard bed filled with the alumina having pore diameter greater than 125 Å upstream of the catalyst fixed bed reactor increased hydroprocessing conversion of the AR containing about 100 ppm of metals in the subsequent reactor. Similarly, alumina was used as the guard bed material in the inventions patented by Hilbert et al. (519) and Howell (520). Another two-step process comprising the guard bed filled with magnesium silicate, upstream of the fixed bed filled with the conventional catalyst, was patented by Fukui (521).

The performance of the guard chambers can be improved by dividing the feed stream into a number of portions, each of which is passing through the different annular guard bed (522). After exiting guard chamber at the bottom, the feed streams entered the catalytic reactor placed below the guard bed system. In the similar arrangement, an improvement of the process performance was achieved by withdrawing gaseous products from the top of the catalytic reactors and introducing them to the bottom of the guard bed placed on the top of the former (523).

In some cases, the guard bed material was placed on the top of catalyst in the same fixed bed reactor. Thus, the spherical alumina was used in the system claimed by Kukes et al. (524–527). The performance of alumina was improved by pretreating either with ammonium or magnesium salts, followed by heating at above 700 K. The alumina layer was efficient for the removal of V and Ni from the feed. The performance of the fixed bed, the removal of metals from the feed in particular, was improved by mixing the catalyst particles with the particles of the activated bauxite and alumina (527). Similarly, the guard catalyst bed was part of the reactor comprising the multistage fixed catalyst beds, contained within the single on stream vessel, with the separate catalyst addition and the separate catalyst withdrawal system. The guard catalyst bed was on the top for removing metals to extend the life of one or more fixed catalyst beds serially disposed underneath (528). Catalyst particles could be withdrawn from the fixed catalyst beds in the form of a slurry. The multistage fixed bed reactor vessel, with separate catalyst addition and separate catalyst withdrawal system for the top fixed bed functioning as the guard bed for removing major contaminants from heavy feed, was designed with the aim to extend the life of one or more fixed beds of catalyst underneath the top guard bed of a low-cost solid. The reactor could be retrofitted to accommodate two or more fixed beds of catalyst downstream from the guard bed.

The graded system used for removing Ca and Na from heavy feeds comprised two sections of the catalyst having decreasing porosity and increasing activity in the direction of heavy feed flow through the system under conditions typically applied during HDM (529). A method for selecting catalyst and/or matching its properties with the amount of Ca and Na in the feed was disclosed as well.

9.2.2 Mixed layer and multiple bed systems

The systems discussed in this part consist only of one reactor vessel. In most cases, they have been designed for the conversion of VGOs either to commercial products (e.g., naphtha, diesel oil, lube base oil, etc.) or to the feed for FCC. The single fixed bed systems comprising either catalysts of the different chemical composition or the same composition but different particle size and shape have been patented (530–536). Several advantages were observed using the two-layer fixed bed, i.e., one layer contained the NiMoP/Al_2O_3 catalyst and the other the Ni(W)Mo/SiO_2–Al_2O_3–zeolite catalyst (530–532). In this case, the conversion of VGO to middle distillates was significantly enhanced.

The mixed bed consisting of two catalysts of a different chemical composition was part of the invention disclosed by Ho (533). This invention consisted of the layered catalyst system for HDN of VGOs. The first layer was of the macroporous NiMoP/Al_2O_3 formulation, whereas the second layer catalyst was the NiW/SiO_2–Al_2O_3–zeolite (534). Operating time of the fixed bed reactor comprising the top and bottom layer of a catalyst could be extended according to the invention of Gupta et al. (535). In this process a portion of heavy feed could bypass the top layer, as soon as the development of pressure drop due to fouling became evident. This portion of heavy feed was then introduced into the bottom layer of catalyst.

A two-phase process, where the need to recirculate hydrogen through the catalyst was eliminated, was invented by Ackerson et al. (536). In this process, a heavy feed was mixed with a diluent in which the solubility of hydrogen was high. The type and amount of diluent, as well as the reactor conditions, could be adjusted to ensure that all hydrogen required for hydroprocessing reactions is in the solution. The oil/diluent/hydrogen solution can then be fed to the plug flow reactor packed with catalyst. With this setup, hydrogen recirculation and trickle bed operation of the reactor was avoided. The intention of this invention was the replacement of large trickle bed reactors with much smaller tubular reactors.

An enhanced performance was achieved using the fixed bed of the non-uniformly sized catalyst particles of the same chemical composition (537). The largest size particles were placed on the front of the fixed bed. However, a physical mixture of the high-void and low-void catalyst particles mixed in different amounts and different layers of the fixed bed gave a better performance than the fixed bed consisting of the uniformly sized particles. In the fixed bed patented by Chou et al. (538), the graded particles were arranged with the largest particles in either upstream or downstream portion of the bed. Compared with the conventional bed of the uniformly sized particles, the graded bed exhibited the enhanced activity for hydrocarbon conversion of VGOs. Such catalyst bed was particularly useful in the moderate HCR operating at less than 7 MPa pressure.

The graded configurations of reactors can be used for the low VGO conversions processes (539–542). Thus, it has been generally known that the enhanced product selectivity can be achieved at lower conversion (60–90% conversion of fresh VGO) per pass through the catalytic HCR zone. The low conversion per pass was generally more expensive; however, the present invention greatly improved the economic benefits of the process with the low conversion per pass. At the same time, the product quality was preserved. Other benefits of the low conversion per pass operation included the minimization of the need for interbed hydrogen quench and the minimization of the fresh feed preheat. Also, the higher flow rate of recycle liquid could provide the additional process heat to initiate catalytic reactions and an additional heat sink to absorb the heat of reaction. An overall reduction in fuel gas and hydrogen consumption, as well as the reduced light ends production, could also be achieved. Moreover, the low conversion per pass operation resulted in the lower catalyst consumption.

The mixture of VR and the light cycle oil was successfully processed using the multiple catalyst bed which consisted of the $CoMo/Al_2O_3$ in the first section and the $NiMo/Al_2O_3$ in the second section of the reactor (543). An example of the multiple bed system comprising three sections in the same reactor vessel is shown in Figure 8.2. In this case, the processing conditions such as temperature and H_2 pressure were similar in all three sections. Several disclosures of dual bed systems can be found in the patent literature. For example, the dual bed reactor patented by Banta (544) consisted of the first bed of large pores and the second bed of small pores of the $CoMo/Al_2O_3$ catalyst of the same composition. In the similar process, VR mixed with the recycle oil was contacted with the first bed containing the large pore quadrulobal shape catalyst, before entering the second bed of the small pore quadrulobal catalyst of the same chemical composition (545). The dual bed reactor disclosed by Gillespie (546) had 1.5 times more of the active components on the catalyst in the second bed. Also, the effective diameter of the catalyst particles was 0.75 times or less than that of the particles in the first bed. It was evident that the chemical composition, as well as the size and shape, of the catalyst particles represented the main difference among the dual bed reactors discussed above. It appears that these systems were developed for hydroprocessing of the particular feed to produce specific final products. In the system patented by Buchanan (547), the feed was introduced at the top of the lower reaction zone for downward flow through. A partially reacted liquid effluent from the lower reaction zone was pumped to the top of the upper reaction zone for downward flow through with the feed. The H_2-containing gas was introduced at the top of the upper reaction zone for flow downwardly and co-currently through catalyst beds in the upper and lower reaction zones. The liquid products were recovered as the effluent from the upper reaction zone.

Multibed reactors (the same vessel) may have a provision for the withdrawal of vapor between the sections and at the same time, for the addition of a makeup H_2 (548). Such withdrawal and replacement reduce the concentration of NH_3 in the gas stream and enables control of the H_2S/H_2 ratio. Furthermore, an improvement in operation can be achieved by removal of light fractions between the sections. This ensured that the colloidal stability of the heavy feed was maintained in the subsequent bed of catalyst (549). The reactor comprising two beds of the HCR catalysts was designed for hydroprocessing of VGOs with the aim to enhance conversion to middle distillates (550). In this case, the product after the first bed was fractionated to obtain light fractions and the heavy fraction which was subsequently processed in the second bed. The product from the second bed was fractionated as well. Thus, the heavy fraction from this fractionation was then recycled to the first bed. For more difficult VGOs, three beds may be required to meet the target specifications of the products (551,552).

Another multibed process was utilizing the HDM, HYD and HDS/HDN catalysts (553). The first bed comprised the $Co(Ni)Mo/Al_2O_3$ catalyst containing small amounts of active metals. The second portion of the multibed was filled with the catalyst with large amount of active metals on the Al_2O_3–TiO_2 support. The third portion of the multibed contained the $NiMo/Al_2O_3$ catalyst. This combination of catalyst ensured lower operating process temperatures. Moreover, the duration of the operation was extended.

The invention consisting of the multiple bed down-flow reactor comprising the vertically spaced apart reaction beds of the different catalyst particles and a mixing device for mixing fluids between the adjacent beds was patented by Ven der Meer and Zonnevyle (554). The mixing device consisted of the horizontal collection tray and the swirl chamber for mixing liquid, arranged below the collection tray. In addition, the horizontal distribution tray was located below the swirl chamber. The former had a plurality of openings for downward flow of gaseous and liquid streams.

9.2.3 Countercurrent systems

Currently, there is no commercial hydroprocessing system employing the countercurrent flow of liquids and H_2-containing gas through catalyst bed, although the patent literature indicates interests in such systems. The invention of Kalnes et al. (555) describes the continuous process consisting of the reactor comprising several parallel countercurrent vapor–liquid flow reaction zones. This arrangement allowed internal recycling of the unconverted VGO to the inlet of each reaction zone. Furthermore, the continuous operation was ensured by regenerating and/or replacing the catalyst in one reaction zone, while other zones were in operation. The countercurrent process disclosed by Laccino et al. (556,557) involved several steps, such as the mechanical filtration of the feed to remove particulates and/or the precursors to foulant; introducing treated feed into the reaction vessel upstream from at least one reaction zone and passing the feed stream through one or more reaction zones. Each reaction zone contained the bed of hydroprocessing catalyst. The H_2-containing gas was introduced at the bottom of reaction vessel and passed upward countercurrent to the flow of liquid stream. The non-reaction zone would typically be a void (with respect to catalyst) horizontal cross-section of the vessel of suitable height, although it may contain an inert packing material. Suitable catalysts for this process included the conventional hydroprocessing catalysts containing Mo(W) and Ni(Co) metals on Al_2O_3 and other supports.

The invention of Stangeland et al. (558) makes possible countercurrent flow of the uniformly distributed hydrogen and liquid feed across the densely packed catalyst bed ensuring that the entire volume of the reactor vessel was filled without ebullating the catalyst bed. Catalysts were selected by density, shape and size at the design feed rate of liquids and gas to prevent ebullation of the packed bed. Catalysts were selected by measuring bed expansion, such as in a large pilot plant run. At the desired flow rate, such catalysts continually flowed in the plug-like manner downwardly through the reactor vessel by introducing fresh catalyst at the top of the bed. Catalyst could be removed as the slurry of catalyst particles in liquid stream at the bottom of the bed.

The processes employing the countercurrent flow in combination with the co-current flow were disclosed by Gupta et al. (559–561). The reaction vessel used for this purpose contained vapor and optionally liquid passageway means to bypass one or more catalyst beds. This permitted a more stable and efficient reaction vessel operation. Another process based on the combination of countercurrent and co-current flows systems consisted of the reaction vessel comprising one or more vertically disposed reaction zones. Each zone was loaded with catalyst and was immediately preceded and followed by the non-reaction zone. At least one of the latter zones employed the vapor passageway means so that a portion of the up-flowing vapor could bypass a portion of the vertical section of reaction zone. At least one reaction zone allowed the flow of liquid countercurrent to the up-flowing H_2-containing stream in the presence of catalyst. The co-current stage liquid effluent may be the feed for the countercurrent stage, and the countercurrent stage liquid effluent was the hydroprocessed product liquid. If necessary, the uncondensed vapor rich in H_2 could be scrubbed and recycled back into the co-current stage as H_2-containing treat gas. Fresh H_2 could be introduced into the countercurrent stage. The effluent from this stage contained sufficient and/or all of the H_2 required for the vapor stage reaction.

9.2.4 Multistage systems

These systems have been developed to accommodate hydroprocessing of the metals- and asphaltenes-containing feeds, although similar configurations comprising a smaller number of the stages were designed and used for VGOs and HGOs. Therefore, they have to employ more than one reactor in a series. The reactors, which are part of the series of the reactors, may operate either in the single bed mode or consist of several layers and/or sections of different catalysts. It appears that the patent development has been focusing on the process configurations with a particular feed and properties of products in mind. The pre-1980 patent literature was dominated by the two-reactor and three-reactor systems. This reflected the properties of heavy feeds which dominated the market at that time. It should be noted that similar processes discussed in the preceding Chapter 8 on commercial processes for hydroprocessing have been also covered by the patent literature. However, in the former case, the reference was only made to the technical and scientific literature without paying any attention to the patent literature.

Two reactors processes have been designed for the conversion of VGOs, HGOs and DAOs to middle distillates. The process patented by Wolk et al. (562) has the provision for removing light fractions from the effluent of the first stage to prevent precipitation of heavy components in the second stage. Various modifications of the two stage processes were disclosed by Dahlberg and Myer (563), Nelson and Wray (564) and Hamner et al. (565–568). The two-reactor process employing the large-pore and small-pore catalysts was designed to handle +550°C and −550°C fractions of a heavy feed in separate reactors each (568). The former process had the provision for removing gas oil from the effluent of the first stage, similarly as in the process developed by Wolk et al. (562).

The two-stage process suitable for hydroprocessing of VGOs and DAO employed the catalysts which possessed a high HCR activity in addition to usual catalyst functionalities (563). The second stage was limited to HCR. Conversion in the second stage could be improved by the addition of multiple reaction zones (in the same vessel) for HCR with flash separation zones between the stages. The yield of middle distillates was thereby increased and the volume of the recycle stream was reduced. This invention reduced the need for equipment which would normally be required in the case of the large recycle stream.

The process consisting of three fixed bed reactors claimed by Angevine (569) was designed for the production of diesel fuel from heavy feeds such as DAO, as well as from the low metals and asphaltenes ARs. A high level of HDM and HDS was achieved in the first reactor over the conventional catalyst upstream of the HCR reactor filled with the zeolite-based catalyst. The last stage was used to attain the required product specifications. According to the process described by Henke and McKinley (570), VGOs, DAO and ARs could be upgraded in the three-reactor process. The macroporous $NiMo/Al_2O_3$ catalyst used in the first reactor contained relatively small amount of active metals. The amount of active metals on catalyst increased, and H_2 pressure increased toward the third reactor. Another invention was the integrated hydrotreating process for HDS of VRs comprising three or more reactors (571). It encompassed the method of revamping the existing unit to increase the capacity and to increase the production of middle distillates. This improvement was achieved by removing the process stream before passage into the last reactor and then passing it through the carbon-rejection zone. The gas oil recovered there was passed through the last reactor.

The process allowing the on stream catalyst replacement (OCR) contained two or more catalysts designed for different functions (e.g., HDM and HDN). The different size and density catalysts could be employed (572). This ensured that one type of catalyst could be removed from the OCR reactor faster than the other type of catalyst. Two and/or more different (i.e., different in physical and/or catalytic properties) and distinct catalysts were all employed in the single on stream reactor. The system allowed for the partial spent catalyst particles withdrawal from the single OCR and the fresh catalyst addition without interrupting the operation, similarly as it can be achieved in the process disclosed by Scheuerman (573). In this case, reactor comprised the first moving bed of catalyst and the second moving bed of catalyst supported by the first and second bed support, respectively. The properties of the first moving bed catalyst differed from those of the second moving bed catalyst. The support for the first catalyst bed allowed the untreated feed stream to pass through, while preventing catalyst particles from the first catalyst moving bed to exit. The support for the second catalyst bed allowed the initially treated feed stream from the first catalyst moving bed to pass through, while preventing catalyst particles from the second catalyst moving bed to exit.

The most problematic heavy feeds cannot be processed using fixed bed reactors. For this purpose, several processes were disclosed employing either several ebullated bed reactors in a series (562,574) or one ebullated bed reactor in a series with several fixed bed reactors (575). The latter approach is the basis of the process described by Kunesh (576). In this process, a heavy feed was hydrocracked in the ebullated bed reactor before being additionally treated in the fixed bed reactor under more severe conditions. In one case, the thermal HCR step performed in the presence of the coke-suppressing agent preceded the ebullated bed reactor. The selection of catalyst for the second stage of the ebullated bed reactor was addressed by the patent of Gibson et al. (577). The catalyst suitable for this purpose had a smaller APD, a lower density but exhibited a high HDS activity, although its HDM activity was low. However, this was offset by the high HDM activity of the catalyst used in the first stage.

Details of the operation of the ebullated bed reactor with emphasis on the heat exchange and the fired heater fuel consumption were given in the patent of Steinberg et al. (578). Another patent, describing the operation of the ebullated bed reactor, was awarded to Mosby et al. (579). The invention of Artes and Kukes (580) described the catalyst suitable for the ebullated bed reactor. It comprised the porous alumina support and had the incremental pore volume maximum at the pore radius greater than 50 Å and at least $0.4\,cm^3/g$ of the pores having radius less than 125 Å. There is a number of different configurations of the process (581–583) for converting heavy feeds in at least one ebullated bed reactor equipped with the system for catalyst withdrawal at the bottom and catalyst addition at the top. At least a portion of the effluent from the ebullated bed reactors was subsequently treated in the fixed bed reactors. The products from the latter were fractionated to recover middle distillates and the residue being sent for catalytic cracking. Another version of the process includes atmospheric distillation followed by vacuum distillation. The VR can be deasphalted and DAO sent to hydroprocessing reactor (584,585). Catalysts claimed by Morel et al. (586) to be used in their process employing ebullated bed reactors consisted of the Mo(W) and Ni(Co) metals and supports such as alumina, silica, silica–aluminas, magnesia, clays and mixtures of at least two of these minerals. The supports could also comprise other compounds, i.e., boron oxide, zirconia, titanium oxide or phosphorous pentoxide. In the form of extrudate, these catalysts were used in the ebullated bed reactors for hydroprocessing of the great variety of heavy feeds, i.e., VGO, DAO, AR, VR and diluted heavy crude. The configuration of multistage system employing ebullated bed reactors patented by Rockwell (587) has the provision for catalyst cascading from the last reactor to one of the preceding reactors.

Chapter 10

SPENT HYDROPROCESSING CATALYSTS

At the end of the utilization cycle, spent hydroprocessing catalysts are removed from the reactor and replaced by either fresh or regenerated catalysts. Depending on the type of reactor, the replacement is performed either periodically on stream or during the shutdown of the operation. In order to ensure safety, a specially designed equipment has been used for catalyst withdrawal. Moreover, after the withdrawal from reactor, spent catalysts must be handled according to the procedures prescribed by the environmental and regulatory authorities who have been classifying spent hydroprocessing catalysts as the hazardous toxic wastes (41). De-oiling may be the only treatment performed on the refinery site. However, loading the catalyst for transportation and shipment to the off-site companies deserve attention as well. In most cases, additional treatments of spent catalysts are carried out off site. This may involve regeneration and/or rejuvenation for reuse, metal reclamation and other applications. The storage facilities approved by environmental and regulatory authorities must be used for the final disposal of spent catalysts to prevent release of the toxic species to ground water and the atmosphere.

10.1 REGENERATION

Because of the extensive information available in the literature (68,546,588), only a brief account of regeneration is given with emphasis on the spent catalysts used for hydroprocessing of heavy feeds. For the purpose of this review, regeneration is considered to be a process for only removal of the coke deposited on catalyst surface with the aim to restore as much as possible of the original catalyst activity. The removal of metals with the aim to restore activity is discussed as part of the catalyst rejuvenation. At least 80% recovery of the original activity is desirable to make regeneration attractive. It is difficult to reach this level of the activity recovery when both coke and metals, which were deposited from the feed, are present on the catalyst surface. Apparently, among the evaluated heavy feeds, only the spent catalysts from hydroprocessing of VGO and HGO could be regenerated. It is, however, believed that catalysts used in the downstream layers and/or reactors of the graded systems can also be suitable for regeneration, particularly after using DAO as the feed. However, for such spent catalysts, the number of regeneration-utilization cycles will be less than that for the spent catalysts from hydroprocessing of light feeds derived from conventional crudes. Similarly, in the case of AR and VR, the regeneration of catalysts from the fourth- or even third-stage reactors of the multistage processes, i.e., ARDS and HYVAHL, can be performed without difficulties. This was confirmed by the results published by Al-Dalama and Stanislaus (346) showing a significant recovery of surface area and porosity of catalysts taken from the third and fourth reactors of the ARDS process, compared with a little recovery for the catalysts taken from the first and second reactors. The coke removal may be achieved using either the oxidative or the reductive regeneration. The former technique has been used on commercial scale for several decades, whereas the reductive regeneration has not yet reached a commercial stage.

10.1.1 Oxidative regeneration

Before 1980, most of the regenerations of spent hydroprocessing catalysts were conducted in situ, using air diluted with either steam or nitrogen. Since that time, the off-site regeneration has been gradually replacing the in situ regeneration. The former became preferred for several reasons, e.g., corrosion issues, safety and environment, time considerations, availability of the experienced staff and better activity recovery. The temperature control was one of the drawbacks of the in situ regeneration. Thus, temperature runaways damaging the catalyst structure and the reactor material could not be entirely avoided.

In the case of in situ regeneration, at the end of the hydroprocessing operation, flows of the feed and H_2 were gradually replaced with the flow of an inert gas (e.g., N_2) and/or steam, while spent catalyst remained in the reactor. After displacing combustible gases and liquid streams from the reactor, inert gas was gradually replaced by the diluted air and air. There are reports on significant problems, such as temperature runaways and mal-distribution of oxidizing gas during the in situ regeneration (68). Also fines which are usually formed during regeneration complicate the subsequent hydroprocessing operation due to the development of pressure drops and mal-distribution of liquid and gaseous stream through the bed. In some cases, this required unloading the catalyst from reactor for the removal of fines and subsequent reloading. All these problems were encountered during the in situ regeneration of catalysts used for hydroprocessing of light feeds. Of course, even more complications could be envisaged for heavier feeds. Therefore, it is not surprising that most of the petroleum refiners prefer to have their catalysts regenerated offsite. In fact, there were no published reports, indicating that a petroleum refinery would make an attempt to perform in situ regeneration of the catalyst fixed bed after hydroprocessing of heavy feeds, which are considered in this book.

Normally, regeneration involves oxidative removal (by burn-off) of the carbonaceous deposits, generally termed as coke, formed during the operation. The diluted air has been used as the oxidation medium. A safe contact of spent catalysts with air can only be maintained at temperatures lower than 250°C after most of the soluble and/or reactive material deposited on the catalyst surface was already removed. Otherwise, an uncontrolled temperature runaway would lead to recrystallization and/or to sintering of the catalytically active phases. In this form, the catalytic functionalities could not be restored. Figure 10.1 shows the effect of temperature on the surface area and dynamic O_2 chemisorption (DOC) of the regenerated catalysts (589,590). The latter is usually performed on the regenerated-presulfided catalysts and is an indication of the catalyst activity recovery. An abrupt surface area decline above 600°C is rather evident. At the same time, the decline in the DOC is more gradual. Nevertheless, only regeneration below 500°C ensured a desirable level of the activity recovery.

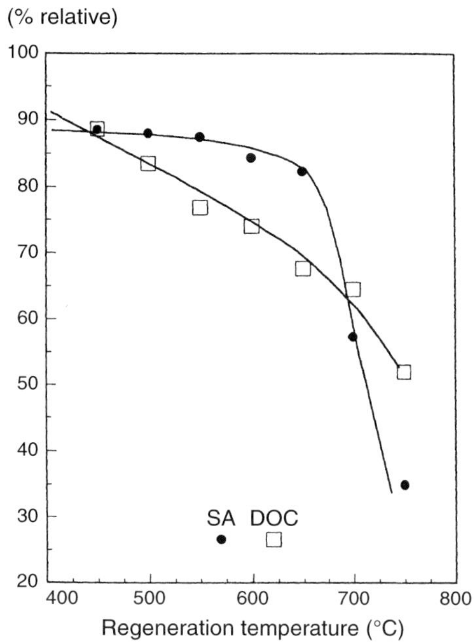

Figure 10.1 Effect of regeneration temperature on recovery of surface area (SA) and on dynamic O_2 chemisorption (DOC) (590).

The oxidative gases such as O_3 and N_2O were also tested as the reactive gases for coke oxidation (68). With these gases, coke could be removed at temperatures lower than 250°C. However, a commercial use of these gases for catalyst regeneration has been prevented because of their limited availability in desirable concentrations and quantities. Another potential low-temperature option for the oxidative removal of coke from catalyst involves the use of the solution of H_2O_2. Because of the limit on temperature, the coke removal via H_2O_2 oxidation is expected to be rather slow. It is also believed that the liquid phase oxidation of spent catalysts to remove coke could affect mechanical properties of the catalysts.

Figure 10.2 shows general trends observed during the regeneration (decoking) of spent catalysts via oxidative burn-off (68). It is evident that hydrogen was the most reactive component of the coke. Thus, more than 90% of hydrogen was already removed, while about half of the carbon was still on the catalyst. Looking at the model of coke (THFIS after 6500 h) in Figures 7.26 and 7.27 (274,278), this would involve the removal of most of aliphatic carbon and all hydrogen associated with it, leaving behind the highly aromatic and non-reactive residue.

The nitrogen associated with the bulk coke is released simultaneously with carbon. This is shown in Figures 7.10 (345) and 7.29 (388). Part of the nitrogen is the least reactive component of the coke. This results from its strong interaction with catalyst surface because of the basic nature of N-compounds (76). It is believed that during the later stages of the burn-off, the removal of this part of nitrogen may be delayed compared with that of carbon. This is indicated by a broken line in Figure 10.2. The origin of this nitrogen may be the N-bases in the feed which strongly adsorb on the bare support of

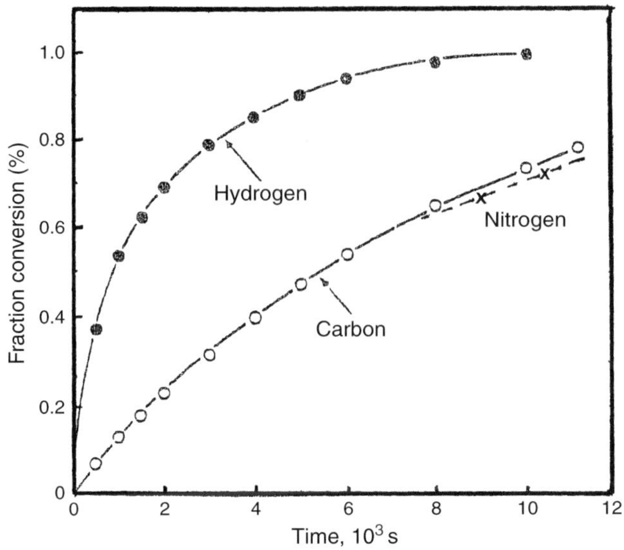

Figure 10.2 Burn-off profiles of hydrogen, carbon and nitrogen during regeneration of spent catalyst (68).

catalyst (76,336). Their presence in spent catalysts cannot be ruled out, although their molecular weight may be significantly lower than that of the coke molecules. A part of the nitrogen is removed simultaneously with inorganic sulfur, as indicated by an overlap of the SO_2 and low-temperature NO regions (346,388). Most likely, this part of the NO originated from N-compounds adsorbed on and/or near catalytically active phase, rather than on the bare support. It was suggested that the last part of carbon and nitrogen may be gradually converted to metal carbides and nitrides (206). These phases are catalytically active for hydroprocessing reactions. Therefore, the last amount of carbon and nitrogen left in catalyst may not be detrimental, although their removal requires prolonged exposure to the oxidation medium. This was supported by the results published by Noguchi et al. (591) who reported that the activity of the spent catalyst was almost completely restored as soon as the amount of carbon on the catalyst was less than 3 wt%.

As shown in Figures 7.10 (345) and 7.29 (388), during regeneration, the sulfur removal occurred simultaneously with that of carbon and nitrogen. Two maxima of SO_2 shown in Figures 7.10 and 7.29 include the inorganic sulfur, predominant part of which was released at lower temperatures compared with the organic sulfur, with the maximum occurring at higher temperatures. Then, the latter maximum coincided with the CO_2 maximum, confirming that this part of SO_2 was originated from organic matter. Two maxima in CO_2 formation indicate difference in reactivity of coke components, i.e., more reactive, predominantly aliphatic carbon and less reactive carbon, which is part of the refractory and highly aromatic structures.

Catalyst regeneration is usually influenced by diffusion phenomena. This may be seen in Figure 10.2 (68) which indicates the presence of two burn-off regions, i.e., chemically controlled occurring during the early stages of burn-off and predominantly diffusion controlled occurring during the final stages of burn-off. The former involved the coke deposited on the exterior of catalyst particles, whereas the diffusion controlled burn-off involved the coke deposited in pores. The slow burn-off during the later stages resulted from the diminished accessibility of O_2 in the pores. Thus, the diffusion of O_2 into the catalyst interior was obstructed by the burn-off products (CO, CO_2, SO_2 and NO_X) exiting from the pores.

Information on the regeneration of catalysts used for hydroprocessing of heavy feeds and at the same time deactivated only by coke is limited. The results shown in Figure 10.3 were obtained during the regeneration of a VGO (589,590). In the same study, several catalysts used for hydroprocessing of lighter feeds were also included. The main differences included the higher temperatures required for the removal of sulfur from the spent catalyst used for hydroprocessing of the VGO than that for the lighter fraction. Moreover, the regeneration lasted longer because of the greater amount of the deposited coke. A lower reactivity of coke, because of more severe conditions applied during hydroprocessing of VGO, may be a contributing factor as well. Nevertheless, as Figure 10.1 shows, more than 80% of the activity recovery could be achieved for this catalyst. It was reported that the regenerability of catalyst may be influenced by the conditions to which the catalyst was exposed during hydroprocessing (592). In one case, during hydroprocessing of a VGO, catalyst was exposed to a temperature runaway under low H_2 pressure. For this catalyst, more severe regeneration conditions were necessary for achieving a desirable level of the coke removal.

The regeneration of spent catalysts shown in Table 10.1 was used to illustrate the other extreme of the catalysts deactivation due to both coke and metal deposits (593). The spent catalysts were obtained from the two different HDM operations using the

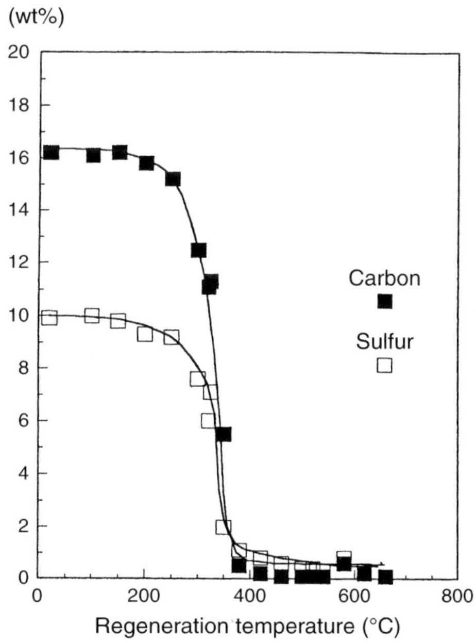

Figure 10.3 Burn-off profiles for removal of carbon and sulfur during regeneration of spent catalyst (590).

Table 10.1 Analysis of toluene-extracted spent catalysts (593)

Element (wt%)	Catalyst A	Catalyst B
Carbon	15.0	13.6
Sulfur	5.9	13.5
Vanadium	3.3	10.1
Nickel	2.7	4.6
Iron	0	7.7

Table 10.2 Properties of fresh and regenerated catalysts (593)

Catalyst	Treatment	Surface area (m^2/g)	Conversion (wt%)	
			HDS	HYD
Fresh	None	140	46.7	27.5
Spent A	None	17		
Spent B	None	11		
Spent A	2% O_2	127	15.0	10.7
Spent B	2% O_2	60	15.7	12.9
Spent A	Air	54	12.3	9.3
Spent B	Air	17	6.4	5.9

HYVAHL process (14,450,451). In the fresh form, the macroporous catalyst consisted of 2.7 wt% of NiO and 14.0 wt% of MoO_3 and had a surface area of 140 m^2/g The results of regeneration shown in Table 10.2 (593) were obtained on either fresh or regenerated catalysts which were subsequently presulfided according to the established procedures. They clearly demonstrate that neither controlled regeneration in 2% O_2 nor that in air could ensure an acceptable activity recovery in spite of the fact that the HDS and HYD activities were determined using small molecules such as thiophene and cyclohexene, respectively. It is believed that if a heavy feed (comprising much larger size molecules) would be used for the activity determination, the difference would be even more evident. It is, however, believed that in a pulverized form, such solids could be suitable catalysts for hydroprocessing conducted in the slurry bed reactors.

The effect of the amount of deposited metals (V + Ni) on catalyst activity was investigated by Clark et al. (594). The coke in spent catalysts was removed by oxidative burn-off. The results in Figure 10.4 show that the activity of the decoked catalyst containing about 8 wt% of metals was similar to that of the fresh catalyst. This amount of metals represented the coverage of one-half of mono-layer. Apparently, most of these metals were deposited on bare support. During hydroprocessing of a heavy feed, the activity of the decoked catalyst rapidly declined with the further addition of metals and approached that of the fresh catalyst containing the same amount of metals. This suggests that the metals in excess of 8 wt% deposited near or on the active sites and as such caused permanent deactivation. Then, such catalysts may be too heavily contaminated with Ni and V to be regenerated, particularly when the intention was the reuse of the

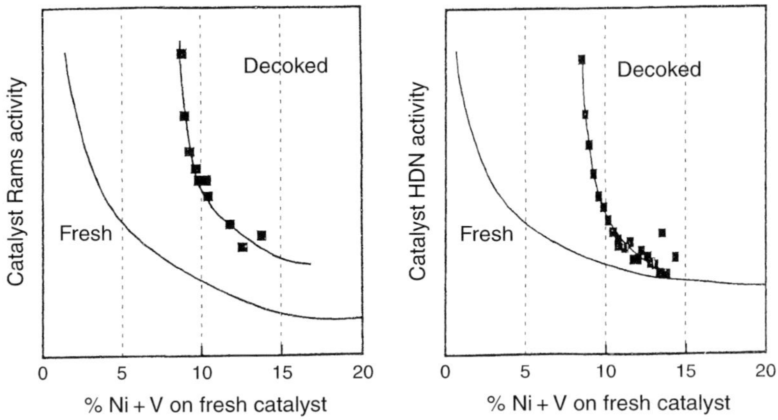

Figure 10.4 Effect of metals on catalyst on HDN and RBC conversions (594).

regenerated catalyst for hydroprocessing of the same heavy feed. It is uncertain, how this catalyst would perform for a feed free of metals and asphaltenes.

It was indicated that there may be some exceptions where regeneration of the catalyst used for hydroprocessing of the metals and asphaltenes containing feeds can be justified, i.e., the catalysts from the third and fourth reactors either of the ARDS process or of the HYVAHL process. It is believed that in some cases, even the second reactor catalyst may be at least partly utilized. For example, the catalyst unloaded from the bottom of the bed can be (after regeneration) cascaded to the top bed for the next utilization cycle.

For both light and heavy feeds, arsenic may accumulate on the catalyst surface during the operation. It has been established that under hydroprocessing conditions organo-arsenic compounds in the feed readily decompose and deposit on the catalyst surface (595). Because of the severe poisoning of the HDS active sites, the oxidative regeneration of catalysts may be prevented when the arsenic content approached about 0.4 wt%.

The choice of conditions used for testing regenerated catalysts requires attention for obtaining reliable information. Thus, as it was pointed out by Dufresne (595), the difference between the activity of the fresh and regenerated/contaminated catalyst was usually greater at start of the run than after several weeks/months of operation. An example of this specific behavior had been shown on the residue catalyst contaminated with the overall contaminants content as high as 21 wt% (596). After regeneration, it was less active for HDM and HDS than a fresh catalyst on start of run, but, after 1 month of the operation, the activity difference was negligible for HDM and still existing but reduced for HDS. So catalyst reuse has to be carefully examined on a case-by-case basis, either for the same application or for cascading in less severe applications.

10.1.2 Reductive regeneration

In an ideal case, catalyst could be regenerated at the end of the operation by discontinuing the feed supply while continuing the H_2 supply. In spite of this, there is little information on the reductive regeneration of catalysts used for hydroprocessing of any petroleum feed. This may result from the fact that a desirable level of coke removal could not be achieved even at much higher temperatures than those applied during the oxidative regeneration (68,388). For example, during the reductive regeneration of the $NiMo/Al_2O_3$ catalyst used for hydroprocessing of a VGO, the evolution of most of CH_4 formed via hydrogasification of coke required temperature of more than 1000 K. As a result of this, sintering of the MoS_2 particles was observed. Almost certainly, hydrogasification of coke catalyzed by active metals was the source of the small CH_4 maximum at about 800 K.

A brief comparison of the reductive regeneration with the oxidative regeneration was made by Noguchi et al. (591). The spent catalyst was from the hydroprocessing of the Arabian heavy VR conducted at 713 K and 8.5 Mpa. The comparison revealed that the recovery of surface area and activity during the reductive regeneration was higher than that during the oxidative regeneration. However, the latter was conducted in air at 923 K, whereas the reductive regeneration at 713 K. Apparently, at 713 K, part of the coke in the vicinity of active metals could be removed presumably via hydrogasification catalyzed by active metals without having an adverse effect on catalyst compared with the oxidative regeneration at 923 K. In the latter case, sintering of the catalytically active phase could not be prevented. Moreover, at this temperature, transformation of the γ-Al_2O_3 support to other forms of the Al_2O_3 could have occurred.

The coke on catalyst surface can also be removed via its reaction with H_2O and CO_2, i.e., carbon gasification (68). The products of these reactions include CO and H_2. This ensures the presence of reducing conditions in the regeneration system. Again, also for these reactants, much higher temperatures were necessary for achieving desirable rates of coke removal. For example, the study on the CO_2 regeneration of catalyst used for hydroprocessing of the VR derived from Athabasca bitumen revealed that the formation of CO via Boudart reaction, i.e.,

$$CO_2 + C = 2CO$$

just began above 800 K (597). More than 1000 K was necessary for removing most of the carbon in pure CO_2. Similar trends may also be observed when CO_2 was replaced by H_2O as the gasification reactant.

The structures in Figures 7.26 and 7.27 (274,278) may be used to illustrate difficulties which may be encountered during the reductive regeneration in H_2. It is unlikely that such structures (e.g., THFIS after 6500 h in Figure 7.27) could be hydrogasified at a near atmospheric pressure of H_2 and temperatures at which the catalyst stability can be still maintained. The increase in temperature and H_2 pressure to increase the rate of hydrogasification of coke could have adverse effects on catalyst structure. Thus, at a temperature suitable for coke removal, the catalyst over reduction could not be avoided. This would also be accompanied by recrystallization of the catalytically active phases. Therefore, the in situ regeneration using reductive agents may not be feasible, particularly for the catalysts used in severe hydroprocessing operations.

10.1.3 Regeneration by attrition/abrasion

The objective of this method is the removal of either the "skin" of inorganic solids or this part of metals, which deposited predominantly on or near the external surface of catalyst particles. For V and Ni, the sulfides formed via the non-catalytic reaction of porphyrins with H_2S could be easily removed from spent catalysts by the attrition. Under dry conditions, such an attrition test could be conducted in a fluidized bed reactor. The spent catalyst has to be kept in motion when the attrition test is conducted under wet conditions.

The method used by Gray et al. (406) involved rolling the mixture of the metal deposited catalyst particles with the particles of α-Al_2O_3 in water. After drying, catalyst was separated from the α-Al_2O_3 by sieving. Some removal of the "skin" metals (e.g., Fe and Ca) was observed, but it was less efficient than the extraction with the diluted HCl. The experiments could also be conducted under dry conditions. Nevertheless, this method can only remove metals from the external surface of catalyst. The removal of metals from pores would require a prolonged attrition which could affect the size and mechanical properties of catalyst particles. It is believed that the attrition/abrasion methods for regeneration of hydroprocessing catalysts have limited practical applications. Apparently, such methods may be more suitable for removing metals from the spent FCC catalysts.

10.2 REJUVENATION

The primary objective of rejuvenation is the removal of metals deposited on the catalyst surface during the operation, while leaving the active metals intact. It was indicated earlier that in the conventional hydroprocessing catalysts, the active metals exist in the Co(Ni)-S-Mo(W) phase. At the early stages of operation, the active phase accounts for most of the metals on the support. With time on stream, inorganic solids in the feed, as well as the V and Ni produced during HDM reactions, are deposited on catalyst surface. The active phase metals together with these deposits represent rather complex mixture, particularly in the later stages of the operation. This may include sulfides and mixed sulfides of V, Ni, Fe and other metals. To a certain extent, Co and Ni promoters in the active phase may be replaced by V and Fe and as such decrease the catalyst activity. The objective of rejuvenation is to liberate the active phase from the unwanted deposits in the mixture. Moreover, because of the deposition in pores, the metals have to be removed to restore the original porosity of catalyst. This appears to be rather challenging task. It should be noted that several methods for the removal of contaminant metals have been tested with a varying degree of success.

Besides removal of the unwanted metals, catalyst has to be decoked by oxidative burn-off. The former may be performed either prior to decoking or on the decoked catalyst. In the latter case, the temperature of decoking must be carefully controlled to prevent sintering which would affect the removal of the unwanted metals, e.g., by extraction. Clark et al. (594) showed that a desirable level of the metal removal can be achieved prior to catalyst decoking. Then, after subsequent decoking, the rejuvenated catalyst had acceptable activity. However, the extraction affected mechanical properties of the rejuvenated catalyst.

Numerous solvents have been used for metal extraction from spent hydroprocessing catalysts. Both organic and inorganic solvents have been evaluated. The former ensures the non-corrosive environment requiring much less safety precautions than that which has been present during leaching using inorganic agents. Moreover, their leaching efficiency can be increased and/or modified by various additives with the aim to selectively remove the unwanted metals.

10.2.1 Organic agents

This method of metal extraction is based on the pioneering work of Beuther and Flynn (598), who recognized that some organic agents were capable of forming complexes with the transition metals which are part of the hydroprocessing catalysts. Rather small concentration of acids, i.e., ~ 1 wt%, was sufficient for removing substantial amount of the metals from spent catalysts. The tested organic agents included acids such as oxalic, lactic, citric, glycolic, phthalic, malonic, succinic and salicylic as well as acetylacetone, ethylenediamine, *o*-aminophenol and salicylaldehyde. Among these agents, acids were much more efficient leaching agents than non-acid compounds. This method has been significantly advanced by the researchers at Kuwait Institute for Science and Research (KISR) who established the following order of leaching efficiency of the most efficient agents: oxalic > malonic > acetic acid. In addition, combination of these acids with various agents was evaluated as well (599–601).

As the most efficient agent, oxalic acid was used for rejuvenation of the spent catalyst obtained from the first stage of the ARDS process used for hydroprocessing of the Kuwait AR (601). The composition and surface properties of the fresh and corresponding spent catalysts are shown in Table 10.3. Leaching experiments were performed in the continuous up-flow reactor at 298 K using both the decoked and the spent as-received catalysts. The selectivity for the V removal was much higher for non-decoked catalyst. The results of the experiments are shown in Table 10.4. The trends in the recovery of surface area and pore volume with the amount of V leached out are shown in Figure 10.5 (602). They indicate that these parameters, surface area in particular, improve before about 35% of V was removed. The HDS activity recovery followed the same trends as that of the surface area. The HDS activity data of the rejuvenated catalysts were obtained in the continuous fixed bed reactor system at 623 K and 4.0 Mpa. For the activity estimate, the AGO containing about 2 wt% of sulfur was used as the feed. On the basis of the extensive testing program which has been conducted at KISR, one may conclude that the catalyst rejuvenation using organic agents is approaching a commercial stage (599–605). As a consequence, the rejuvenated catalyst may be cascaded for less severe applications and/or at least used as the solid for guard beds. In both cases, economics are a decisive factor.

Table 10.3 Chemical composition and physical properties of the fresh and spent catalysts (601)

Property	Fresh catalyst	Spent catalyst
Surface area (m^2/g)	240	52
Pore volume (mL/g)	0.48	0.12
Bulk density (kg/L)	0.73	1.18
Mo (wt%)	8.80	5.40
Co (wt%)	3.20	1.90
Ni (wt%)	0	3.07
V (wt%)	0	14.90
Fe (wt%)	0	8.00
Carbon (wt%)	0	15.60
Sulfur (wt%)	0	5.30

Table 10.4 Surface area, pore volume and HDS activity of fresh, spent and rejuvenated catalysts (601)

Catalyst	Agent	Surface area (m^2/g)	Pore volume (mL/g)	HDS activity	
				%	% recovery
Fresh		240	0.40	61	100
Spent		52	0.12	17	28
Method A	Oxalic	63	0.15	22	34
Method A	Oxalic + H_2O_2	141	0.27	52	85
Method B	Oxalic + H_2O_2	181	0.43	59	97
Method C	Oxalic + H_2O_2	197	0.46	36	59
Method C	Oxalic	180	0.42	37	61

Method A – leached but not decoked; Method B – leached and decoked; Method C – decoked and leached.

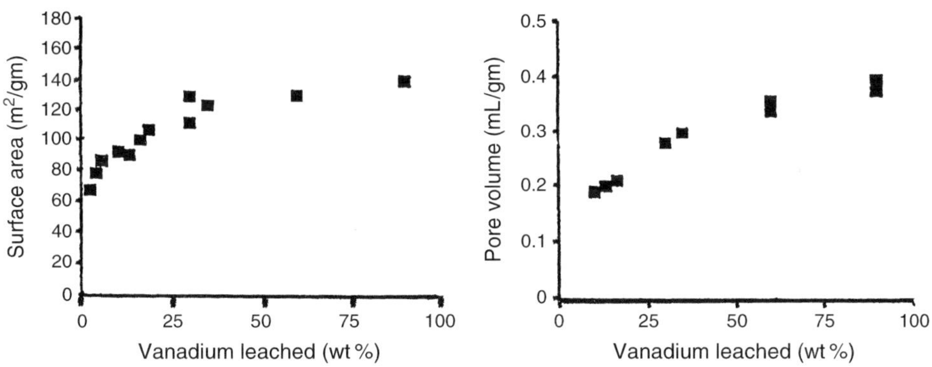

Figure 10.5 Effect of vanadium leached out on surface area and pore volume (602).

Several combinations of the organic acids with different agents were also evaluated. Table 10.4 (601) shows that the leaching efficiency can be enhanced in the presence of H_2O_2. This was attributed to the oxidation of the metal sulfides in spent catalysts to corresponding oxides (603). The presence of the latter is more favorable for the formation of the water-soluble complexes with organic agents. Furthermore, the surface area and pore volume improvement, as well as the increased recovery of activity, were observed in the presence of $Al(NO_3)_3$ (604). Compared with H_2O_2, $Al(NO_3)_3$ was a more efficient additive for the removal of Ni, whereas for V removal both additives had a similar effect (605). However, as Figure 10.6 shows, for both V and Ni, the most efficient system comprised a mixture of $Fe(NO_3)_3$ and oxalic acid. However, with this system, the unwanted removal of Mo occurred as well. The $Fe(NO_3)_3$ was more efficient than $Fe_3(SO_4)_2$, presumably because of the higher oxidation strength of the nitrate group (606). The pH of the solution had a pronounced effect on leaching efficiency.

The $Fe(NO_3)_3$ + oxalic acid system was evaluated in the fixed bed and ebullated bed pilot plants (605). The latter system was specially designed for leaching experiments. The

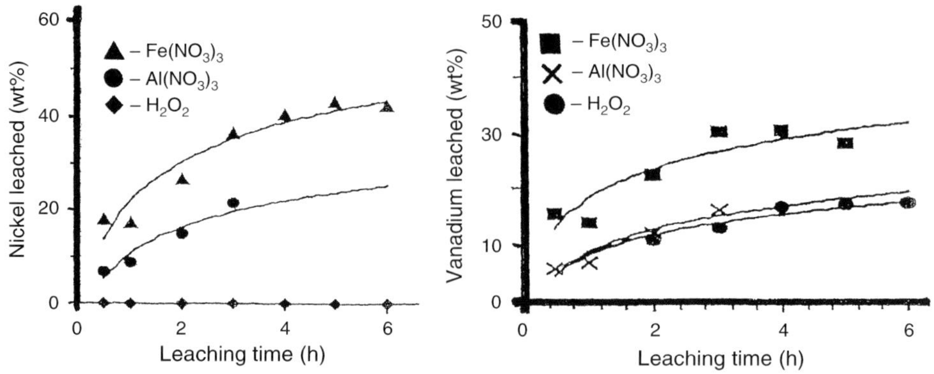

Figure 10.6 Effect of additives on leaching efficiency of oxalic acid (605).

amount of V leached out was significantly greater in the ebullated bed. Moreover, three modes of the $Fe(NO_3)_3$ addition to oxalic acid, i.e., continuous, successive and batch additions, were tested (607). With respect to the V removal, the continuous addition was the most efficient. In contrast to the report by Clark et al. (594), the rejuvenated catalyst only lost about 5% of the side crushing strength compared with the fresh catalyst (608). Some evidence suggests that there are other Fe ions containing systems, i.e., the redox couple (Fe^{2+}/Fe^{3+}), which when added to an organic acid may enhance both leaching efficiency and particularly the selectivity for V (609).

10.2.2 Inorganic agents

The strong acids such as HCl and H_2SO_4 and/or their solutions with metal ions have been used for rejuvenation of the spent catalysts as well. Corrosive nature of these solutions suggests that special precautions have to be taken to ensure safety of the operation and environment, although many years experience with similar methods gained in hydrometallurgy can be applied during rejuvenation as well.

Perhaps, the most simple case of the catalyst rejuvenation is the removal of the "skin" of inorganic solids which deposited on the exterior of catalyst particles. The "skin" may comprise clays, alkali and alkali earth metals compounds, V and Ni sulfides formed via the reaction of porphyrins with H_2S, etc. The study of Gray et al. (406) showed, that such "skin" could be easily removed by treatment with diluted HCl. Much more severe conditions are necessary to remove metals deposited in pores.

The studies were selected to illustrate the use of inorganic agents for selectively leaching the contaminant metals from spent hydroprocessing catalysts (610–612). For example, the spent $CoMo/Al_2O_3$ catalyst used for hydroprocessing of a VR was decoked by the oxidative burn-off before being extracted with either Fe^{3+}+HCl or $Fe^{3+}+H_2SO_4$ solutions (610). The spent catalyst contained about 15 wt% of coke, 14 wt% of V and 3.5 wt% of Ni. The solutions removed contaminant metals, as well as the active metals. However, after presulfiding of the decoked catalyst, selectivity for the removal of the former metals significantly improved. The activity of the rejuvenated catalyst was measured in an autoclave using a residual feed. In this case, the ratio of rate constants for the rejuvenated catalyst to that of the fresh catalyst was determined. The values of 0.7,

0.7, 3.9 and 2.5 were obtained for HDS, HDN, HDV and HDNi, respectively. Although only 70% of the original HDS and HDN activity was recovered, the HDM activity of the rejuvenated catalyst was superior to that of the fresh catalyst, suggesting that the former may be suitable for cascading, e.g., in a guard reactor. In another treatment, the spent catalyst was presulfided at 813 K in 5 vol.% H_2S in He and then extracted with the $Fe^{3+}+H_2SO_4$ solution before being decoked (611). This treatment resulted in the significant removal of both V and Ni. The efficient leaching of the metals (V+Ni) from spent catalyst with H_2SO_4 could be achieved by optimizing the concentration of the latter (612). For example, for 15% H_2SO_4, the removal of V and Ni was significantly greater than that of Mo and Al. Decoking of the catalyst increased the removal of Mo relative to that of V and Ni.

10.2.3 Supercritical extraction

Potential applications of this method for catalyst activity recovery were discussed by Seapan and Guohui (613) with focus on solvents such as CO_2, SO_2 and pyridine. Under supercritical conditions, these agents exhibit lower viscosity and surface tension and higher diffusivity compared with normal conditions. Under such conditions, even highly aromatic materials can be displaced from catalyst surface and solubilized. The level of coke removal was influenced by temperature, pressure and duration of extraction. However, even for the best case, only about 50% of the coke removal was achieved. This increased pore volume from 0.17 to 0.22 mL/g and surface area from 106 to 137 m^2/g for the spent and extracted catalysts, respectively. There is little information available on the removal of metals under similar conditions, although one study indicated that less than 50% of Co could be removed from the spent $CoMo/Al_2O_3$ catalyst using the supercritical and subcritical solution of ammonia (614). However, with respect to catalyst rejuvenation, the removal of Co is an unwanted reaction.

10.3 METAL RECLAMATION

For some spent hydroprocessing catalysts, the recovery of catalyst activity via either regeneration or rejuvenation may be so low that other utilization options have to be considered. A number of commercial processes have been developed for the recovery of metals from spent catalysts (41). In this case, both active metals and the metals deposited during the operation are of interest. Ideally, the recovery of all metals, as well as the support, would be the objective; however, this is not always achievable. One group of the metal reclamation processes is based on leaching the metals with the same organic and inorganic agents which have been tested for catalyst rejuvenation (615). Metals could be removed from spent catalysts using the solutions of inorganic acids (H_2SO_4 and HCl) as well as the hydroxide and carbonates of alkali metals and ammonia. Roasting of the spent catalysts with inorganic acids and/or alkali carbonates followed by dissolution has also been used. Once in solution, the metals can be recovered in a pure form either by precipitation or by extraction using organic solvents.

10.3.1 Leaching of metals

The selective leaching of metals from various solid materials has been conducted on an industrial scale for decades. In this regard, spent hydroprocessing catalysts have been receiving attention as well (615–627). The organic acids used for rejuvenation have also been evaluated for the metal reclamation (598–606). In this case, the experimental parameters such as temperature, duration of extraction and the solvent/catalyst ratio, as well as the spent catalyst pretreatment, may be adjusted to maximize the recovery of metals (615). Figure 10.7 shows the effect of the glyoxylic acid concentration on

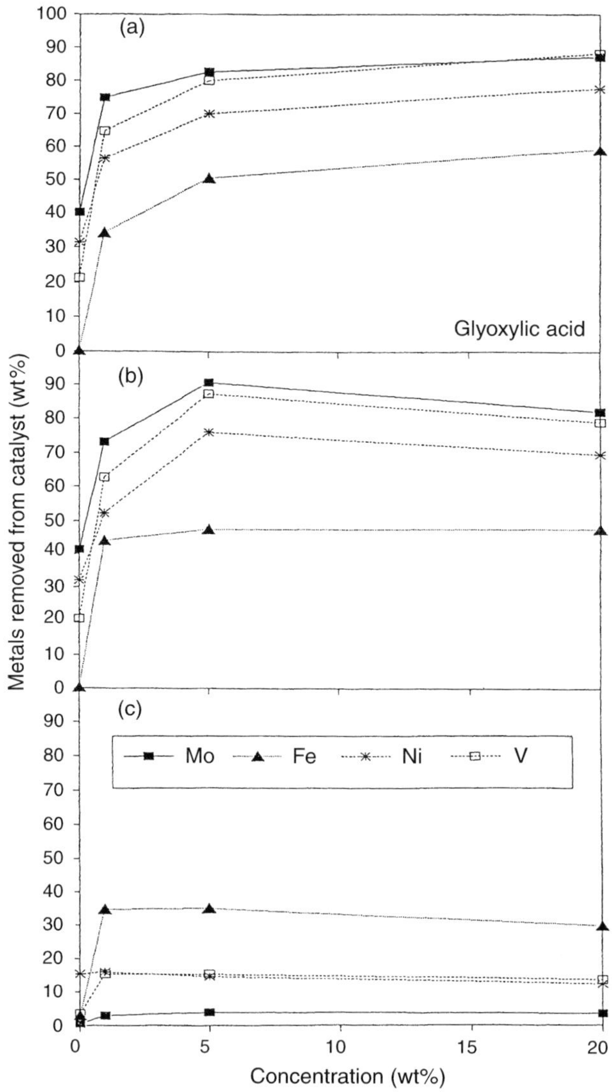

Figure 10.7 Effect of acid concentration on metal removal (615).

Table 10.5 Analysis of spent catalyst (wt%)[1] (615)

Carbon	25.0
Hydrogen	3.2
Nitrogen	0.14
Sulfur	12.9
Mo	4.4
Ni	5.9
V	11.6
Fe	2.5

[1] Metal analysis for decoked catalyst

the amount of metals leached out from the catalyst shown in Table 10.5. It can be estimated by extrapolation that the optimal concentration is about 2 wt% of glyoxylic acid. Similar trends were established for other acids. Decoking of the catalyst resulted in a significant increase in the metal dissolution. Thus, the results in Figure 10.7a and 10.7b are for decoked catalyst in a powder form and in an operating form (2-mm-diameter sphere), respectively, whereas in Figure 10.7c for the catalyst which was de-oiled but not decoked. The rate of leaching was the greatest during the first contact of the solution with the catalyst (Figure 10.8). The rate levelled off after about 1 h of leaching. For leaching experiments, 2.5 g of the catalyst was mixed with 50 mL of the solution. The mixture was kept in the suspension either by magnetic stirrer or by ultrasonic treatment. The former was consistently more efficient. The experiments were of 4-h duration and were performed at 50°C. The extent of leaching was determined from the concentration of metals in leachates.

A high level of metal leaching could be achieved using H_2SO_4 either alone or in the combination with other agents (616–618). In METREX process, metals can be leached out from decoked catalyst with H_2SO_4 (613). After removing silica and alumina, metals, i.e., Mo, Co, Ni and V, were separated from filtrate by solvent extraction carried out in several stages. Almost complete dissolution of Ni, Mo and Al in H_2SO_4 was reported by Siemens et al. (617). However, only a trace of Al and about 90% of Ni and Mo were dissolved with the mixture of $NH_4OH + (NH_4)_2SO_4$. In another case, H_2SO_4 was used to leach out Ni after Mo or W was removed with the solution of NaOH (617). Similar methods were used by Olazabal et al. (619,620) and Angelidis et al. (621) who selectively removed Mo and V using the solution of NaOH before treating the same catalyst with H_2SO_4 to leach out Co and Ni. Another process using the mixture of HNO_3 (0–6%) and H_2SO_4 (2–4%) as leaching agents was tested by Kelebek and Distin (622).

The high efficiency of the NaOH solution for leaching Mo/W and V from spent catalysts was indicated. The leaching efficiency could be further enhanced by applying an electric current (623). Among the K-containing compounds, KOH was much more efficient than K_2CO_3 and CH_3COOK for the selective removal of V (624). The aqueous solutions of $(NH_4)_2OH$, $(NH_4)_2CO_3$, NaOH and Na_2CO_3 were used either alone or in the mixture with H_2O_2 (625,626). The best recovery of Mo and V was achieved using the mixture of $Na_2CO_3 + H_2O_2$. The metal dissolution could also be achieved using H_2O_2 alone (627). Thus, the dissolution of metals increased with the increasing concentration from 0 to 6%.

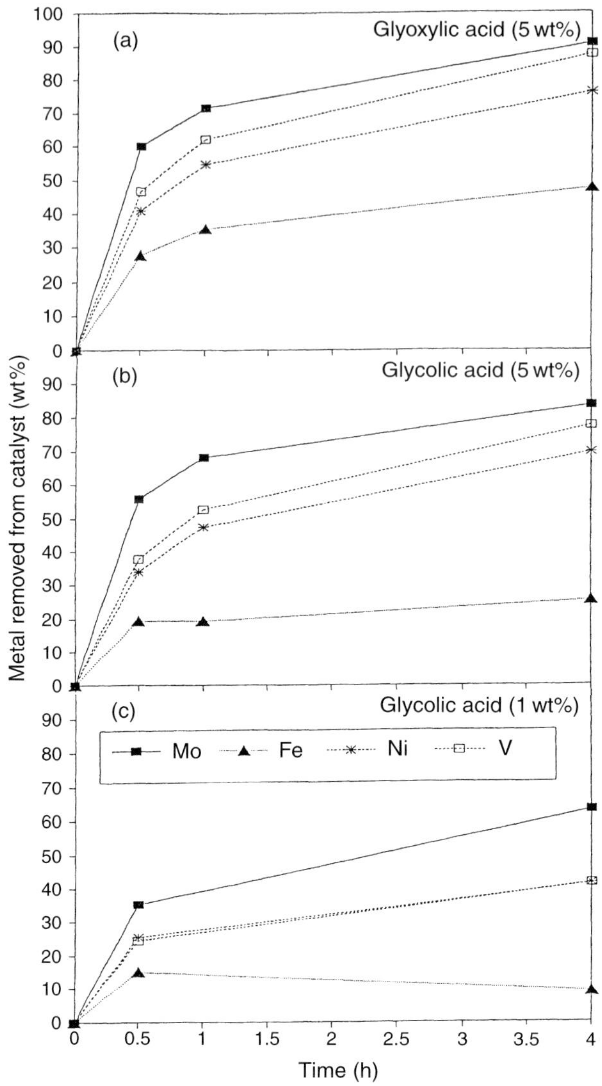

Figure 10.8 Effect of time on metal removal (615).

10.3.2 Roasting treatments

Roasting with a suitable agent may convert the metals in spent catalysts into the water-soluble compounds. For example, roasting the spent HDS catalyst with the Na-containing compounds such as NaOH, Na_2CO_3 and Na_2SO_4 resulted in the dissolution of Mo and V (628). The best roasting agent, giving about 96% recovery of Mo and V, was Na_2CO_3. In the similar study, the roasting with Na_2CO_3 produced the water-soluble product such as Na_2MoO_4, whereas Co and Al remained in the filter cake (629). The roasting of the spent $CoMo/Al_2O_3.SiO_2$ catalyst with NaCl converted most of the Mo into the water-soluble Na molybdate (630). Pure MoO_3 could then be recovered by adjusting pH of the solution. The EURECAT process based on the caustic treatment of spent catalysts can also convert most of the Mo, W , V, As and P into the water-soluble compounds while leaving Ni, Co, Fe and most of the Al in the filtrate cake (631). Other roasting agents used for the dissolution of Mo/W and V included H_2SO_4 (632) and NaCl + water vapor.

10.3.3 Chlorination

This method is based on the difference in volatility of the metal chlorides and/or oxy-chlorides (41). For example, the oxy-chlorides of Mo, W and V (e.g., MoO_2Cl_2, WO_2Cl_2 and $VOCl_3$) are much more volatile than the chlorides of $NiCl_2$, $CoCl_2$ and $AlCl_3$. The mechanism of chlorination of the Co(Ni)Mo(W)/Al_2O_3 catalysts was discussed in detail by Jong et al. (618). Gaballah and Djona (633,634) confirmed that experimental conditions can be optimized to maximize the formation of the volatile oxy-chlorides. In this regard, in the temperature range of 523–873 K, the composition of the chlorination mixture (Cl_2, air, N_2, CO and CO_2) was the key parameter. After chlorination, the volatile Mo-, W- and V-containing oxy-chlorides were washed out from the receiver and the lines. The obtained slurry was filtered and cake hydrolyzed to isolate Mo, W and V. The Ni and Co chlorides remaining in the spent charge were water leached at 100°C. As the results in Table 10.6 (634) show, the catalyst decoking prior chlorination had an adverse effect on the metal recovery confirming the involvement of carbon in the overall chlorination mechanism (618). Apparently, chlorination is a more efficient method than two-step leaching, i.e., the isolation of Mo, W and V using the solution of NaOH followed by leaching out Ni/Co using the solution of H_2SO_4.

10.3.4 Other methods

Among novel methods, bio-leaching is based on the observation that Mo and/or W can be isolated from spent catalysts using bacteria (635,636). However, the bacteria with a high tolerance to some metals have to be used. For example, some bacteria were suitable for isolating Ni; however, their efficiency was diminished when Mo was present. Apparently, the electrochemical dissolution may be in a more advanced stage than bio-leaching. In the former case, the solution of an oxidizing ion (e.g., Ce^{4+}) with spent catalyst added is on the anode side of the electrochemical cell (637). Here, coke is oxidized to CO_2 and H_2O, whereas metal sulfides are converted to the corresponding

Table 10.6 Effect of roasting on metal recovery from spent catalysts (634)

Catalyst	Process	Temperature (K)	Recovery (wt%)				
			Co	Ni	Mo	V	Al
Unroasted	Chlorination Cl_2-O_2-N_2 N_2 = 0–67%	655–773	91.8	96.5	95.3	70.9	9.6
Roasted	Chlorination Cl_2-CO-N_2 N_2 = 33–67%	673–873	80.9	76.3	90.5	51.4	9.7

metal oxides. This significantly increased solubility of the metal species in water. Further enhancement in solubility could be achieved by adding HNO_3 to the solution. Potential of the radiation thermal treatment has also been explored (638). It was observed that such treatment decreased sublimation temperature of MoO_3 by about 573 K. This enhanced the separation of Mo from the spent decoked catalysts.

10.3.5 Separation of metals from solution

An extensive information on the separation of metals from solutions can be found in the literature dealing with various aspects of hydrometallurgy. This experience can be applied to spent catalyst. In these cases, more than one metal is usually present in the solution after leaching and/or dissolution of spent catalysts. Additional treatments are required to isolate metals in their pure form. For these purposes, numerous organic solvents with a high selectivity for a metal of interest have been available for commercial applications. The selectivity can be further enhanced by optimizing the conditions applied during extraction. As the final step, a pure metal compound can be obtained from the solution by precipitation under controlled conditions (e.g., pH) using a suitable precipitant. In some cases, only minor modifications to the methods used commercially in hydrometallurgical industry are required before they can be applied for metal reclamation from the spent hydroprocessing catalysts. Some typical examples were selected to describe these events in more detail.

An important contribution to the development and testing of the organic agents used for the extraction of metals from various solutions was made by Inoui et al. (639,640). Some of the agents evaluated by these authors are shown in Figure 10.9. With respect to the purity of the isolated metals, pH of the solution may be the most critical parameter. This was demonstrated by the results shown in Figure 10.10. In this case, before the extraction, the solution contained 2.7, 0.75, 0.03, 13.5, 1.0 and 0.17 g/L of Mo, V, Fe, Al, Co and Ni, respectively. An excellent separation of the Mo from the other metals could be achieved at pH approaching zero using 20 vol.% the Cyanex 272 in EXXSOL D 80 as diluent. Subsequently, the aqueous solution of ammonia could be used to strip Mo from the extract. After the Mo separation, pH of the scrub solution was increased to 1.5 by the addition of $Ca(OH)_2$ powder. At this pH, V was separated with Cyanex 272 and subsequently stripped from the extract with the aqueous solution of ammonia. After the separation of Mo and V, the efficient separation of Ni and Co from Al could be achieved using the mixture of Lix 63 and SYNEX DN-052 in the EXXSOL D 80. Conventional methods have been available for the separation of Ni from Co (641). The METREX process can be used as another example illustrating the use of organic agents for the extraction of metals from leachates. The order of metal separation was identical as that used by Inoui et al. (639,640). The optimization of conditions (e.g., type of agents, pH of extraction, ratio of agent/diluent) for the separation of metals from solutions using organic agents was carried out by Olazabal et al. (619,620). Another study involving numerous extractants is that of Sato et al. (642).

In the leachates containing Mo and V, the latter can be isolated by precipitation with $(NH_4)_2SO_4$ at pH of 8.6 (632). The Mo was separated from the remaining V in 0.05 M sulphite ion medium by extracting with tri-*n*-octylamine, stripping with aqueous ammonia and precipitating by acidification of the stripped solution. The method based on the use of ion-exchange resins was also suitable for the separation of Mo from V in

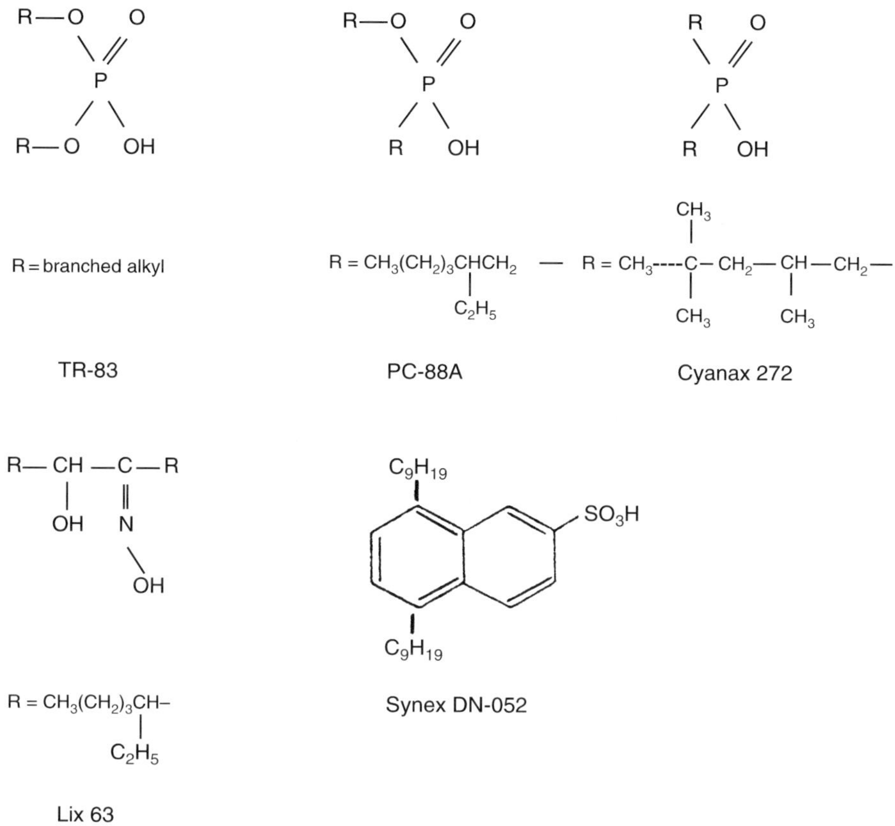

Figure 10.9 Structures of agents used for extraction of metals from solution (639).

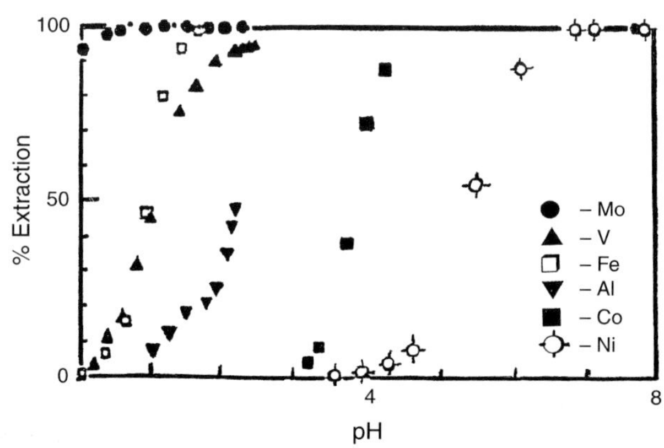

Figure 10.10 Effect of pH on extraction of metals from solution (639).

the same solution (630). Among several resins, the polystyrene resin cross-linked with bis-(22-hydroxyethyl) amino group was the most efficient for the separation of Mo from V (643). In this case, the solution containing Mo and V was passed through the column of the water-swollen resin and the effluent analyzed.

10.4 OTHER POTENTIAL USES OF SPENT HYDROPROCESSING CATALYSTS

Metals which were isolated from spent catalysts may be returned to the catalyst manufacturing companies and used for catalyst preparation. In this regard, other than hydroprocessing, catalysts can also be prepared. Of course, after isolation, there are numerous applications for the pure compounds containing metals such as Mo, W, Ni, Co and V. In fact, the extracts containing Mo, V and Ni obtained by leaching spent catalysts were used directly for catalyst preparation by impregnating amorphous SiO_2 support (644). The processes, which can carry metal reclamation to the "extinction", produce pure Al_2O_3 which can be reused as the support for catalyst preparation. The production of refractories is another potential application for the reclaimed Al_2O_3. There is a report on the use of the crushed spent HDS catalyst for the HCR of VR in a slurry bed reactor (645). However, its commercial application was not possible because of the limited availability.

A direct utilization of spent catalysts has also been receiving attention. They were among waste solids identified by Schreiber and Yonley (646) as having a potential to be utilized for the cement and/or concrete production. Their suitability was indeed confirmed by Stanislaus et al. (602,647). Thus, the mechanical strength of the concrete made by mixing cement with synthetic aggregate containing the pulverized spent hydroprocessing catalyst was comparable to that of the commercially produced concrete. However, firing of the synthetic aggregate (spent catalyst + gatch + clay + sand) at about 1180°C was required to ensure a low leachability of the regulatory metals. In this case, the spent catalyst was used in the ARDS process and contained about 10 wt% of V + Ni.

Spent decoked hydroprocessing catalysts may still possess adequate activity for some other reactions, although most of their hydroprocessing activity was lost. Thus, there is little potential for "cascading" of such catalysts in petroleum refineries. However, there are reports that spent decoked catalysts used in typical HDS operation could still be used for hydroprocessing of heavy feeds (648,649). Almost certainly, decoked spent catalyst could be used as the guard material for removing inorganic solids from heavy feeds in guard chambers. Large amount of V deposited predominantly on the exterior of catalyst particles may indicate on the suitability of spent catalysts for the preparation of oxidation catalysts, which comprise V oxides as catalytically active phase. However, there are no reports to support the same.

It was reported that after presulfiding, spent decoked catalysts were active for the decomposition of H_2S into H_2 and elemental sulfur (41). A high affinity of the metals in spent decoked catalysts for H_2S makes them suitable as the potential hot gas clean-up sorbents for the removal of sulfur from the coal gasification products, coke oven gas, etc. (650). The spent decoked $CoMo/Al_2O_3$ catalyst was also active for the CO_2-aided decomposition of H_2S giving H_2, CO, COS and elemental sulfur (651).

10.5 DISPOSAL AND STORAGE

After all options for utilization of spent catalysts were exhausted, petroleum refiner must ensure that further handling of spent catalysts is carried out in accordance with environmental regulations. Thus, spent hydroprocessing catalysts have been classified as hazardous solid wastes (41). If mishandled, such a waste poses substantial or potential hazard to human health and environment by releasing toxic species while in contact with air or water. A pyrophoric nature of some spent catalysts has also been reported. Therefore, regulations have been in place to deal with all aspects relevant to handling of the spent hydroprocessing catalysts (e.g., storage, disposal, transportation).

In the USA, besides the Environmental Protection Agency (EPA), there are additional acts and laws giving detailed accounts of the methods used for handling spent catalysts. Detailed description of the methods to be used for long-term storage such as land-filling as the final option is also available. It is inevitable that the specially designed land-fills are used to prevent the leaching of hazardous metals to ground water. It should be noted that 20 years' liability which was in effect previously has been changed to an unlimited liability. Other acts and laws have been continuously changing toward more stringent regulatory limits. There are at least two regulatory levels in Europe, i.e., one national and the other established by the European Commission. The latter is based on the Basel Convention signed in 1989. Apparently, continuously evolving new directives will supersede the less stringent national legislation (652). Similar activities have been conducted by the United Nations, with focus on countries which are part of the Organization for Economic Cooperation and Development.

The results in Table 10.7 are used as an example to illustrate the hazardous nature of spent hydroprocessing catalysts (653). The regulatory levels of trace elements (last column in Table 10.7) and volatile organics are reported in Federal Register, which has been issued and periodically updated by the US EPA. Catalysts 1, 2 and 3 were

Table 10.7 Analysis of TCLP leachates (ppm) of spent catalysts and EPA regulatory levels (653)

Metal	Catalyst 1	Catalyst 2	Catalyst 3	EPA regulatory levels
As	53	<0.1	<0.1	5
Ba	<0.1	<0.1	nd	100
Cd	nd	<0.1	0.1	1
Co	1	2	293	–
Cr	<0.1	<0.1	nd	5
Fe	3	83	15	–
Pb	0.3	0.3	0.4	5
Hg	nd	nd	nd	0.2
Mo	249	9	310	–
Ni	657	455	4	–
Ag	nd	nd	nd	5
Se	<0.1	<0.1	nd	1
V	<0.1	205	<0.1	–
Zn	0.2	<0.1	0.3	–

nd-not determined

from the commercial operations in which synthetic distillate, AR and middle distillate, respectively, were used as feeds. For the analysis, catalysts were used without decoking. The leachability of the catalysts was determined according to the Toxicity Characteristics Leaching Procedure (TCLP) developed by EPA for the evaluation of hazardous solid wastes. This involved shaking catalyst particles in the solution of a buffer for 18 h. Subsequently, solid particles and suspended fines were filtered off to obtain leachates for analysis. The test simulates potential release of metals to the environment during contact with water. Catalyst 1 must be classified as hazardous toxic waste because the amount of arsenic leached out was almost 10 times that allowed by regulations. For this catalyst, the feed/catalyst ratio was very high because it was used in continuous operation for more than 2 years. Large contents of Mo and Ni in the leachate from catalyst 1 should also be of concern. Apparently, for catalysts 2 and 3, the content of regulated elements in leachates indicates that the solids are non-hazardous. However, high contents of V and Ni for catalyst 2 as well as Co and Mo for catalyst 3 in leachates should not be ignored. Indeed, because of the unlimited liability currently in effect, most of the refiners treat spent catalyst as hazardous wastes to avoid potential litigations in future. Potential release of PAHs deserves attention as well. In this case, the same leachate which was used for the determination of trace elements is being used. The set of analytical methods used for PAHs analysis is described in detail in the EPA's Federal Register as well.

There is also potential for the release of toxic gases during the contact of spent catalysts with water and air. For example, after a prolonged exposure/contact with water, H_2S may be released because of the hydrolysis of the sulfidic form of metals which are present in the spent hydroprocessing catalysts. At elevated temperatures, toxic gases such as HCN and NH_3 can be formed even in the absence of air, whereas SO_X and NO_X may be formed in the presence of air (654). In the latter case, precautions need to be taken because of the pyrophoric nature of spent catalysts.

Chapter 11

HYDROPROCESSING OF VGO AND DAO FOR PRODUCTION OF LUBRICANTS

Lubricants can be distinguished from other products obtained from petroleum crude by a high-viscosity and a high-boiling range, i.e., >400°C (673 K). Therefore, heavy feeds suitable for the production of lubricating oils include VGO and DAO because the substantial portion of these feeds boils above 400°C. The crudes from which VGO and DAO were obtained influence the overall processing scheme. For example, VGO and DAO derived from a naphthenic crude are more suitable for the production of lubricants than those derived from an aromatic crude. In this case, propane would be the solvent of choice used for the production of DAO. This would almost completely eliminate metals and asphaltenes from DAO. At the same time, this would ensure that oil and to a lesser extent also resin phases of the colloidal system are the main components of the feed for the production of lube base oil. Subsequently, the lube base oil is blended with various additives to attain performance characteristics of the lubricating oils.

Although of non-petroleum origin, the wax by-product of the FT synthesis has been identified as an important feed for the production of lubricating oils. Rather high content of *n*-paraffins makes this wax a suitable feed for investigating the catalysts required for an extreme case of dewaxing. Today, the advances in catalyst development resulted in the process which can readily hydroisomerize the wax by-product to produce the lube base oil possessing similar properties as that derived from petroleum feeds. Middle distillates and AGO are other feeds which may require dewaxing to meet the final products specifications. For this purpose, the same or slightly modified catalytic processes which are suitable for dewaxing of the VGO and DAO feeds can be used. The middle distillates and AGO feeds are out of the scope of this book.

Conventional dewaxing involved the removal of aromatics by extraction using solvents such as furfural and phenols. In a concentrated form, aromatics were separated from the solvent by distillation. The extraction of aromatics was usually followed by the removal of long-chain paraffins. The latter involved mixing the feed with a solvent, e.g., acetone–benzene, barisol (ethylenedichloride-benzene), trichloroethylene, and cooling the mixture until most of the paraffins solidified in a crystalline form. Subsequently, the solid paraffins were filtered off in rotary drum filters and solvent distilled off for reuse. There have been several outlets for industrial utilization of the solid paraffins from conventional dewaxing The removal of aromatics was necessary to improve viscosity behavior, whereas the removal of paraffins to improve cold flow properties of lubricating oils. In some cases, these treatments are followed by contacting the base oil with the activated clay to remove polar compounds. Otherwise, the required stability of the resultant lubricating oil could not be maintained.

Catalytic dewaxing has been gradually replacing conventional dewaxing. In many cases, hydroprocessing of the VGO and DAO had to be performed as the first step before catalytic dewaxing to produce lube base oil. The former topic, as it applies to VGO and DAO, was covered by the previous chapters. It was evident that, in this case, the main focus was on the removal of contaminants (sulfur, nitrogen, metals, resins and asphaltenes) and on the increasing yields of desirable products. The objectives of catalytic dewaxing for the production of lube base oil are rather different. Thus, they include the removal/conversion of *n*-paraffins and the HYD of aromatic structures. Therefore, the catalyst formulations suitable for these applications may differ from those of the conventional hydroprocessing catalysts, although processing conditions and systems may be similar.

The catalytic dewaxing of hydroprocessed VGO and DAO has been conducted to ensure that lubricants meet the performance specifications. As was indicated, the high-performance lubricants must possess desirable cold flow properties as defined by parameters such as pour point, cloud point and freezing point. This can be achieved by the removal of *n*-paraffins from the feed via either their HCR or hydroisomerization. The latter may be the preferred route, if the lube base oil is the anticipated product. Thus, HCR tends to increase yields of the fractions which are below the boiling range of the lube base oil. Among hydrocarbon groups, *n*-paraffins possess the highest melting point. For example, the melting point of the C_{20} *n*-paraffin approaches about +35°C compared with about −15°C for the C_{20} iso-paraffin containing 5-methyl substituent. Moreover, the lube base oil used for the lubricant preparation must exhibit as little change in viscosity with temperature as possible. This can be achieved either by extracting aromatic fractions from the feed or by their HYD to naphthenic structures. It has been noted that most of the information on the development of catalysts and processes for these applications can be found in the patent literature, whereas its cursory account was given in the monographs published recently (655,656).

Besides suitable cold flow properties, the enhancement in the viscosity index (VI) is necessary to meet the specification of the lube base oil. A high VI, indicating high-quality lubricants, ensures a little change in viscosity with temperature change. Among hydrocarbon groups, VI increases in the following order: aromatics $<$ naphthenes $<$ iso-paraffins $<$ *n*-paraffins (e.g., ~50, ~110, ~155 and ~175, respectively). Unfortunately, the highest VI *n*-paraffins have to be removed because of the undesirable cold flow properties which result from their high melting point. To a certain degree, the adverse effect of the removal of *n*-paraffins on VI can be offset by converting polycondensed aromatics to the corresponding naphthenes. The loss of *n*-paraffins can be prevented by their isomerization to iso-paraffins. Both iso-paraffins and napthenes have a high VI, particularly when the large branched chains remained attached to the napththenic rings. The HYD of aromatic rings to naphthenic rings can be achieved at temperatures and H_2 pressures typically used during hydroprocessing of heavy feeds in the presence of catalysts possessing a high HYD activity. In the case that dewaxing catalysts contain noble metals (e.g., Pt and Pd), the HYD of aromatics for the preparation of lube base oil may be conducted at lower temperatures than typically used during conventional hydroprocessing. This suggests that an ideal dewaxing catalyst would exhibit a high activity for HYD as well as for isomerization.

Middle distillates used for the preparation of fuels require a similar dewaxing as it is used for the preparation of lubricants. For example, for diesel oil and heating oil, cold flow properties which are requested by the product quality standards could not be established without dewaxing. In this case, *n*-paraffins have to be removed to avoid their precipitation at low temperatures similarly as it was in the case of lube base oil. Without dewaxing, some difficulties during the delivery of fuel to the combustion chamber could not be avoided. Moreover, the HYD of middle distillates is necessary to ensure that the content of aromatics (in diesel fuel) was below the levels prescribed by fuel specifications. In addition, the improvement in cetane number may be necessary in the case that these products are intended for use as diesel oil. Some products from the FCC/RFCC of VGO and DAO may also require catalytic dewaxing to attain desirable cold flow properties of fuels.

11.1 CATALYTIC DEWAXING

As was pointed out, HCR and hydroisomerization are the main reactions occurring during catalytic dewaxing. The relative contribution of these reactions to the overall dewaxing depends on the catalyst structure and the origin of the feed. A high activity of the catalyst for hydroisomerization is desirable to maximize the conversion of VGO and DAO to the lube base oil fractions by converting *n*-paraffins to iso-paraffins. However, in some cases, the co-production of transportation fuels (gasoline, jet fuels and diesel) with lubricants may be part of the refinery strategy (657). The catalyst possessing a good HCR activity besides the adequate hydroisomerization activity, would be more suitable for such applications.

Compared with conventional dewaxing, which involves several extraction steps, the catalytic dewaxing applied to VGO and particularly to DAO for the production of lube base oil extends the end boiling point of the products by allowing the heaviest fractions of these feeds to be included in the lube base oil fraction. If a suitable catalyst is used, this may also result in the increased yield of middle distillates per unit of the feed. In the case of lubricants, this improves both the yield and the quality. Thus, hydrocarbon groups in the heaviest fractions of VGO and DAO exhibit relatively good lubricating properties. However, because of their high melting points, the long-chain unbranched paraffins contained in these fractions would have the most adverse effect on the cold flow properties of middle distillates and lubricants. As a general rule, the content of *n*-paraffins increases with increasing boiling range of the fractions. Therefore, they must be removed from the heaviest portions of VGO and DAO. This can be achieved catalytically via either conversion of the long-chain paraffins into iso-paraffins or removing the former by cracking. Therefore, for this purpose, catalysts with either a high isomerization activity or HCR activity are necessary. An ideal catalyst would exhibit a high activity for both these functionalities. Otherwise, if the conventional dewaxing would be used instead of catalytic dewaxing, the heaviest fractions of the VGO and DAO feeds would end up in the unwanted residues rather than being included in the lube base oil product.

Although it has been used on a commercial scale, catalytic dewaxing became part of the petroleum refineries only since the mid-1980s (658). Catalytic dewaxing is conducted under conditions which only slightly differ from those employed during conventional hydroprocessing (e.g., 260–430°C and 2–5 MPa of H_2, 250–450 m^3 of H_2 per m^3 of the feed). Typically, catalytic dewaxing is carried out either in the trickle bed reactor over a bi-functional, zeolite catalyst or in the fixed bed comprising several layers/sections of different catalysts. For VGO and DAO feeds, the structure of dewaxing catalysts used in such operations differs from that of the conventional hydroprocessing catalysts.

It was indicated above that VGO and DAO are the most common feeds used for the production of lubricants. In some cases, prior to catalytic dewaxing, these feeds require hydroprocessing to remove sulfur, nitrogen, resins as well as the traces of metals and asphaltenes. This step needs to be conducted using conventional catalysts with the aim to protect dewaxing catalyst in the subsequent step, particularly if the latter catalyst contains Pt and Pd metals. Thus, a dewaxing catalyst, which contains noble metals, may be sensitive to NH_3 and H_2S. In the case of lubricants, the primary product of dewaxing may require additional hydrofinishing to ensure that the properties of base oil are in line specifications. The latter is then blended with various additives to produce lubricants.

For the production of lubricants, a process in which both the HYD of aromatics to increase VI and isomerization/cracking of *n*-paraffins (dewaxing) to improve cold flow properties could be accomplished in one stage would be an ideal process. In view of the low limits of aromatics set by fuel specifications, such process could also be used for the production of diesel oil. This would require a catalyst with high activity for both HYD and isomerization/cracking. However, practical experience suggests that catalytic dewaxing and HYD conducted in either two stages or using a fixed bed comprising several layers and/or sections of different catalyst have been used commercially. The processes in which the HYD precedes dewaxing step and vice versa have been noted as well. Combinations including the extraction methods (e.g., of aromatics) with catalytic dewaxing have also been practiced. The choice of the process configurations depends on the origin of the VGO and DAO feeds. The choice becomes more challenging in the case that these feeds were derived from heavy petroleum crudes.

Apparently, the conventional multilayer or multistage fixed bed reactors such as those discussed in Chapters 8 and 9 have been used for dewaxing. In this regard, the Unicracking process, which has been used for the hydroprocessing of petroleum fractions, was modified for dewaxing applications (657). Figure 11.1 shows a version of the Unicracking process comprising two stages including the separators and fractionator (444,658). According to this flowsheet, the process incorporates the conventional hydroprocessing in the first stage followed by dewaxing in the second stage. Two reactors and one reactor of the first and second stages, respectively, ensure versatility to suit a wide range of feeds. Although there might be other commercial dewaxing processes, the modified Unicracking process in Figure 11.1 (444,658) is used as an example to illustrate commercial dewaxing. For this purpose, the DAO obtained from an unconventional crude is assumed to be the feed. In this case, the catalyst for the first reactor of the first stage has to be active for HDM and HDAs to remove last amounts of the metals and asphaltenes from the feed. The four sections of the second reactor of the first stage offer flexibility to use different catalysts in the same vessel, i.e., from the top toward the bottom the HDN/HDS, HDS, HCR and HYD catalysts could be used. The product separator and fractionator after the first stage ensure that most of the H_2S and NH_3 are removed from the process streams. Therefore, the process stream comprising the fraction taken from the bottom the of fractionator mixed with fresh H_2 is free of H_2S and NH_3 before entering the second stage for dewaxing to produce lube base oil. Thus, it was indicated above that H_2S and NH_3 poison dewaxing catalysts, unless the catalyst which is resistant to poisoning by these compounds was used. In this case, the catalysts from the top toward the bottom section could exhibit a high activity for HCR, hydroisomerization and hydrofinishing.

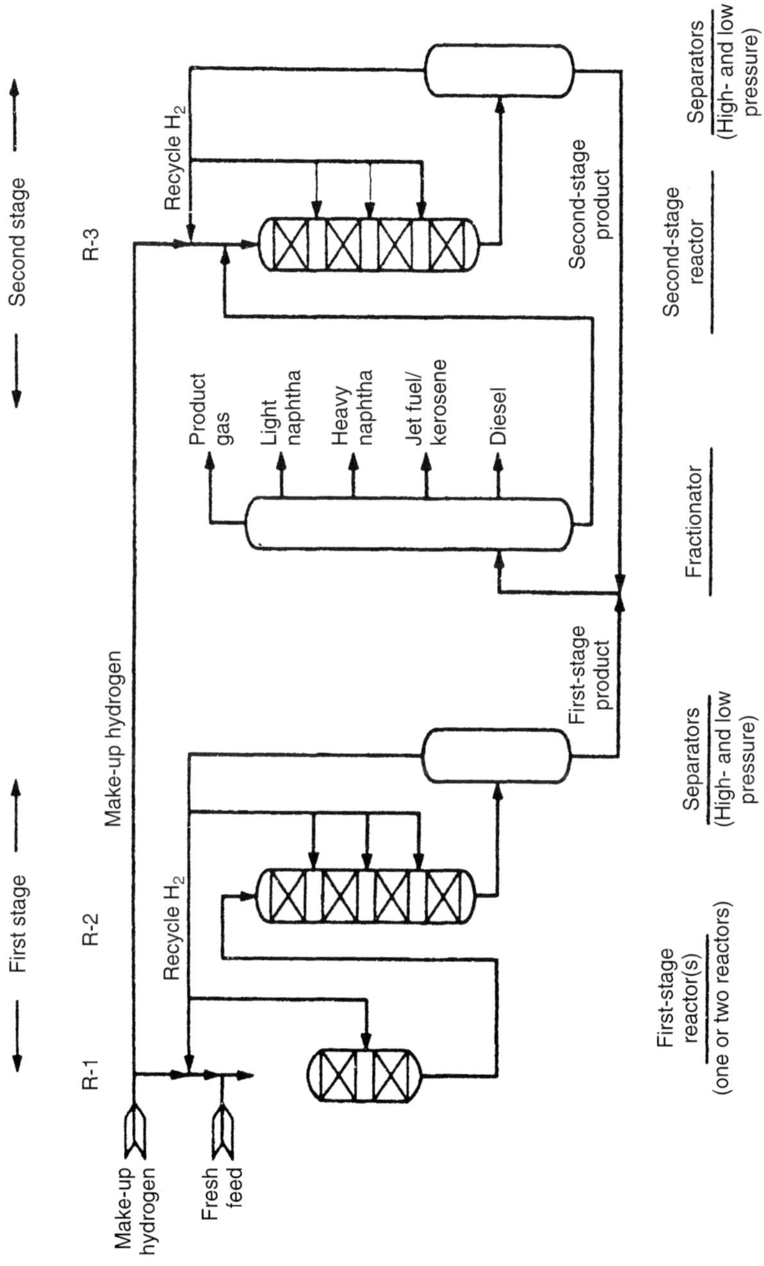

Figure 11.1 Modification of Unicracking process for dewaxing petroleum feed (657).

The co-production of the lube base oil with other products, i.e., naphtha, kerosene and diesel oil, during dewaxing is indicated in Figure 11.1. The operating conditions (e.g., type of catalyst and feed) may be chosen to maximize the yield of these products. This depends on the overall strategy of the refinery. The origin of the feed and the slate of the anticipated products determine the degree of modification of the Unicracking process. The extent of hydroprocessing required in the first stage (Figure 11.1) depends on the origin of the feed as well. For example, less severe conditions may be necessary when these feeds were obtained from sweet crudes compared with those from heavy crudes and/or synthetic crude. The extensive information presented in previous chapters suggests that conventional hydroprocessing catalysts may be suitable for these applications. A suitable porosity of the catalyst for the upper section of the first reactor (first stage) in Figure 11.1 would be required to remove the last amount of metals from the feed. General trends indicate that NiMo(W) catalysts have a higher HDN and HYD activity, whereas the CoMo(W) catalysts are suitable for HDS. However, as a result of various modifications of the catalyst preparation methods, these trends may be reversed.

The first stage shown in Figure 11.1 may not be necessary in the case of the VGO derived from a sweet crude, particularly if the latter was of naphthenic and/or paraffinic origin. In the absence of asphaltenes and metals, such VGO may be dewaxed in one reactor comprising several sections containing a different catalyst each, i.e., HDN/HDS catalyst in the top section, followed by HCR/hydroisomerization catalyst and a HYD/HDAr catalyst in the bottom section to achieve a final hydrofinishing of the product. The reactors in Figure 11.1 show a provision for adding a make-up H_2. It is believed that in such reactor, the gaseous products could be treated between the sections to remove NH_3 and H_2S before entering the next section, unless the catalyst in this section is resistant to poisoning by these compounds. Therefore, such configuration of the reactor may be suitable for achieving desirable levels of dewaxing and HDAr in the same vessel. This is supported by the results in Table 11.1 (659) obtained in the Unicracking unit. Thus, a higher VI was obtained at higher conversions. A slight decrease in VI after dewaxing was caused by the removal of the high VI *n*-paraffins. However, a significant decrease in pour point of the resulting lube base oil was the main benefit of dewaxing. Before dewaxing, the pour point of the VGO exceeded 25°C. This confirmed the need for the dewaxing step.

Some information suggests that rather low yields of the lube base oil obtained during traditional dewaxing involving several extraction steps were perhaps the main reason for its gradual replacement by catalytic dewaxing (660). This was indeed confirmed in the

Table 11.1 Properties of dewaxed VGO (659)

Property	Conversion	
	Low	High
Yield of 315°C + (wt%)	79	68
460°C + (wt%)	39	34
Density (Kg/L)	0.8524	0.8519
VI not dewaxed	100	120
VI dewaxed	95	115

study of Taylor and McCormack (661), in which catalytic dewaxing was compared with solvent dewaxing. However, the former was more suitable for the heavy components of VGO and DAO, whereas the solvent dewaxing gave higher yields of lube base oil of good quality, when lighter fractions were used as the feed. Moreover, higher yields of base oil and higher VI were obtained when the isomerization step preceded dewaxing step. This suggests that the integration of dewaxing with other processes is dependent on the origin of the feed. For example, the optimal process configurations such as isomerization/dewaxing/hydroprocessing, hydroprocessing/dewaxing and FCC/dewaxing may be identified to ensure a high quality and yield of the products. This was indeed confirmed by Hennico et al. (662) who showed that the combination of processes depends on the properties of VGO and DAO. In this regard, the origin of crude oil from which these feeds were derived was an important factor. For example, a different process configuration is anticipated for the HGO obtained from primary products after thermal upgrading of a heavy crude, compared with a virgin VGO obtained from a sweet crude Therefore, the choice of an optimal configuration may require some testing prior to process design. In this regard, the configuration of the second-stage reactor in Figure 11.1 provides an opportunity for arranging different catalysts between the top and bottom sections in order to suit the origin of the feed and to maximize the yield of the targeted products.

11.2 HYDROGENATION OF VGO/DAO FOR LUBE BASE OIL

In the overall schemes of the lube base oil production, HYD reactions may be as important as HCR and hydroisomerization reactions. This was already indicated above by emphasizing the need for removal of aromatics to improve the VI of lube base oil. The degree of HYD required depends on the origin of the VGO and DAO feeds. For example, the catalyst possessing a high HYD activity will be more desirable in the case that these feeds were obtained from aromatic crudes than from naphthenic crudes. In some cases, hydrofinishing is used as the last step during the lube base oil production. Usually, a high HYD activity catalyst is used for this purpose. For the dewaxed product, HYD may be the most important reaction, particularly when the simultaneous production of diesel oil and lube base stock from VGO and DAO obtained from aromatic crude is considered. Thus, for both products, the removal of aromatics is necessary to attain the performance standards. Prior to this step, the light fractions are removed from the dewaxed product in the fractionator. The former fractions enter fuel pool. A high hydroisomerization activity of the catalyst used for dewaxing suggests that these fractions may be a valuable blending component with transportation fuels because of the high concentration of iso-paraffins. A bifunctional catalyst, exhibiting a high activity for both HYD and hydroisomerization reactions during dewaxing, would be desirable in the case that a high yield of the fuel blending components was the objective.

For diesel oil and lubricants, low levels of aromatics have to be maintained to meet performance standards, e.g., cetane number and freezing point as well as VI and cloud points and pour point, respectively, in addition to a number of other important parameters. For lubricants, hydroprocessing under conditions ensuring the conversion of the low VI aromatics to high VI naphthenic structures is always necessary. Therefore, a catalyst selected for these purposes must exhibit a high HYD activity. Among novel catalysts, the carbides and nitrides of transition metals exhibit a high HYD activity (206). So far, this was demonstrated for model compounds rather than for real feeds. The noble metals (Rh, Ru, Pt and Pd) containing catalysts possess a high HYD activity at relatively low severity. The sulfides of these metals were successfully tested as the catalysts for hydroprocessing of the high aromatic feeds such as coal-derived liquids (40). Other forms of noble metal catalysts, e.g., metallic form, may be sensitive to poisoning by H_2S and NH_3. If necessary, hydrofinishing is conducted as the last step before producing final products such as diesel oil and lube base stock. This step is conducted under conditions favoring HYD equilibrium of aromatics, i.e., at a high H_2 pressure. At the same time, the temperatures employed are at least 100°C lower than those typically used during conventional hydroprocressing. The noble metals, such as Pt and Pd, are usually part of the hydrofinishing catalysts. In some cases, dehazing must follow hydrofinishing as the last step. The purpose of dehazing is to remove discoloration/turbidity of the lube base oil, which may still be present, in spite of several stages of treatment. It is believed that conventional hydroprocessing catalysts supported on the acidic supports may be suitable for this purpose; however, the catalysts containing Pt and Pd may be suitable as well. In every case, these processes employ simple trickle bed reactors containing several sections.

With respect to HYD/HDAr, there is no process configuration which could be universally used for a wide variety of VGO/DAO. Thus, for highly paraffinic and naphthenic VGO and DAO feeds, HYD may be of much lesser importance compared with hydroisomerization and HCR. For some VGO and DAO, hydroprocessing must be performed prior to dewaxing, whereas for others, e.g., those derived from sweet crude, FT wax, this step may not be necessary. Similarly, for many feeds, hydrofinishing and dehazing may not be required. This suggests that the involvement of HYD reactions during the overall production of lube base stock may vary from feed to feed. This again confirms that the process configuration required for production of diesel oil and lube base stock depends on the origin of VGO/DAO as well. Experimental testing is necessary before the optimal configuration of the process units and a suitable catalyst can be identified.

11.3 DESIGN OF DEWAXING CATALYSTS

The above discussion showed that HCR and hydroizomerization are the main reactions occurring during catalytic dewaxing. It has been established that such reactions proceed via a carbocation mechanism. Therefore, an active catalyst for isomerization and cracking must possess an adequate surface acidity. Such acidity is provided by acidic supports such as zeolites, silica–alumina and silico-alumino phosphates (SAPO). The information on the development of catalysts for dewaxing applications in scientific literature is limited compared with the wealth of information in patent literature.

The design of dewaxing catalyst takes into consideration the shape selectivity of some supports for long unbranched *n*-paraffins, which can enter pores of a medium size. Such shape selectivity is exhibited by some zeolites. This ensures the access of *n*-paraffins to catalytic sites inside the pores. At the same time, the diffusion rate of the branched paraffins into similar size pores is low. Therefore, their conversion will be low as well. Because of the high VI and low pour point, it is desirable that most of the iso-paraffins remain unconverted during dewaxing.

Table 11.2 Effect of catalyst type on properties of lube base oil (660)

Property	Feed	ZSM-5		Pt/mordenite	
Conversion (wt%)	0	20	23	27	36
Pour point (°C)	+35	−12	−26	−12	−26
Viscosity index	–	94	89	90	78

The information published in scientific journals shows that both noble and non-noble metal (e.g., Ni) catalysts supported on zeolites and other acidic supports have been receiving attention. To prepare the Ni-containing dewaxing catalyst, zeolite is usually mixed with binder (e.g., γ-Al_2O_3) to obtain a composite, which is subsequently calcined before being exchanged with the ammonium salt solution and treated with the Ni-nitrate solution to obtain the Ni-exchanged catalyst (657).

Among zeolites, a high selectivity was achieved using the medium pore size ZSM-5 zeolite. Bendoraitis et al. (660) observed that the ZSM-5 zeolite and Pt/mordenite exhibited the shape selective properties. The latter had larger pores than ZSM-5 zeolite. The pour point and VI of the lube base oil obtained over these catalysts are compared in Table 11.2 (660). A modification of the ZSM-5 zeolite was tested by Chen and Garwood (663) for dewaxing a middle distillate fuel. In this case, the zeolite was exchanged with Zn/CH_4Cl solution to obtain 0.9 wt% of Zn. The activity of this catalyst was in the range of that of the Ni/kieselguhr catalyst (664). In view of the similarity of the process, the observations made during dewaxing of the middle distillate feeds may also be relevant for dewaxing of VGO and DAO.

An ideal case of dewaxing would be a once-through, single-stage process in which hydroprocessing, dewaxing and HYD could be performed in the same reactor. An attempt to develop such a process was described by O'Rear (665). A multifunctionality of the catalyst suitable for such process appears to be an essential requirement. However, the fixed bed consisting of several layers of different catalysts in the same vessel may be another alternative to the multistage processes. The dewaxing catalyst suitable for this concept consisted of 65% H-ZSM-5 (67/1 molar ratio SiO_2/Al_2O_3) bound with 35% Al_2O_3 (666). The ZSM-5 was prepared by hydrothermal synthesis, filtration, washing, calcination and ion exchange to remove sodium. This was followed by binding with alumina and extrudation. Conventional Co(Ni)Mo(W) catalysts were suitable for hydroprocessing step.

11.4 CATALYSTS AND CATALYTIC SYSTEMS IN PATENT LITERATURE

The difference between the properties of the conventional hydroprocessing catalysts and those used for dewaxing, HYD, hydrofinishing and dehazing to produce the lube base stock may not be properly identified without at least cursory account of the relevant information available in the patent literature. For this purpose, a selected number of the patents issued since 2000 as well as few earlier patents were reviewed in detail. It should be noted that this is not an exhaustive review. Thus, it was felt that a more detailed review of the patent literature on the production of lubricants was beyond the scope of this book. For comparison, few patents on dewaxing of the waxy by-product from FT are also included because they represent anextreme requirement for dewaxing involving HCR and hydroisomerization, compared with the VGO and DAO derived from sweet crude.

Apparently, some similarities between the structure of the catalysts discussed in Chapter 9 on patent literature and those of the dewaxing catalysts may be evident. However, the upgrading of heavy feeds via catalytic hydroprocessing of heavy feeds was the primary objective of the development catalysts and processes discussed in Chapter 9. At the same time, little attention was paid to the parameters which are essential for achieving performance standards of the lube base oil and diesel oil.

11.4.1 Catalysts for controlling cold flow properties and VI

The conversion of *n*-paraffins to iso-paraffins is the preferred route of catalytic dewaxing to control cold flow properties compared with their removal by HCR. The former method gives larger yields of the lube base oil compared with the latter, which produces more gaseous as well as light and middle distillate fractions. However, some HCR activity may be necessary if the co-production of lube base oil and fuels is the objective. With respect to the production of lube base oil, the increasing number of patents on the development of hydroisomerization catalysts has been noted. It was observed that efforts have been made to develop multifunctional catalysts with the aim to decrease the number of stages used for production of the lube base oil. The catalysts possessing both HYD and hydroisomerization activities were of the primary interest.

11.4.1.1 Zeolite-containing catalysts

The disclosure of LaPierre and Partridge (667) represents an attempt to develop a multifunctional catalyst. Based on this patent, a VGO may be simultaneously hydrocracked and hydrodewaxed to produce liquid product of the satisfactory pour point and viscosity. Such promising result was obtained by the use of the catalyst which contained zeolite beta as an acidic component to induce the cracking reactions. In addition, the catalyst included a HYD component to promote HYD reactions. This component may include either a noble metal (Pt and Pd) or a non-noble metal (Ni, Co, Mo and W).

A family of the crystalline molecular sieves with unique properties, referred to as the zeolite SSZ-54, were used as the catalysts for hydroisomerization/dewaxing processes (668). The hydrogen form of the zeolite SSZ-54 was obtained in the form of silicate, aluminosilicate, titanosilicate, vanadosilicate or borosilicate. In this case, “silicate” refers to a zeolite having a high SiO_2/Al_2O_3 mole ratio (preferably greater than 100), including zeolites composed entirely of silicon oxide. The term alumino-silicate refers to the zeolite containing both alumina and silica, whereas boro-silicate refers to zeolite containing oxides of both boron and silicon. In addition, the catalyst contained at least one Group VIII metal.

The catalytic dewaxing process based on passing VGO through the fixed bed of catalyst comprising the zeolite ZSM-22 and a noble metal, e.g., Pt, was disclosed by Chester et al. (669). In this case, the VI of the produced dewaxed oil was higher than that obtained by methods employing other catalyst compositions containing zeolites. Also, the activity and selectivity, as measured by VI and the lube base oil yield at a given pour point of this catalyst composition, were superior. The invention of Angevine et al. (670) involved the HCR of VGO or DAO followed by dewaxing. The former employed sequential beds of the catalyst containing β-zeolite. In this case, the content of zeolite in the catalysts bed was increasing in the sequence in which the oil passed over the catalysts. With this arrangement, the degree of dewaxing in the next step increased compared with the catalyst containing equivalent amount of zeolite. The patent literature also indicates that for some heavy feeds and catalysts, the hydroprocessing step may be necessary to protect HCR and hydroisomerization functionalities. However, the sulfur-resistant dewaxing catalyst patented by Barre et al. (671) could tolerate as much as 500 ppm of sulfur in VGO. This catalyst comprised a hydrogenation component, the dealuminated aluminosilicate zeolite crystallites and the low-acidity refractory oxide binder material which was free of alumina.

Heavy paraffinic feeds (e.g., petroleum wax, FT wax, deoiled waxes) with the end boiling point exceeding 650°C could be converted to a good quality base oil by hydroisomerization (672). Compared with the VGO and DAO obtained from the sweet crude oils, such feeds may represent another extreme of catalytic dewaxing, i.e., the latter feeds require much more extensive dewaxing. If necessary, for such feeds, hydroisomerization may be preceded by hydroprocessing to achieve desirable levels of HDS and HDN. Among hydroisomerization catalysts, zeolites of the ZSM type and aluminum phosphates (e.g., SAPO and MAPO) exhibited good activity in such applications. With a suitable catalyst, lube base oil may be an attractive outlet for the wax by-product of FT. Indeed, one patent (673) describes the isomerization of the FT wax to obtain a blending component with a petroleum-derived base oil. The resulting blend had lower pour point and cloud point, as well as higher VI compared with the individual blending components.

The process claimed by Chou et al. (674) was based on the use of the ammonia vapor to control the operation of the hydrocracker or catalytic dewaxer. Catalyst activity and selectivity may be controlled by the addition of the base to the feed with the aim to maintain an optimal balance between isomerization and HCR in an operation using the zeolite beta catalyst. The temperature runaway conditions could be avoided by the addition of basic compounds to regulate the temperature profile within the reactor.

In another patent, Benazzi et al. (675) gave a detailed account of the preparation of the zeolite ZSM-48 as the catalyst for improving the cold flow properties of VGO. The zeolite was mixed with 40% of alumina gel. The paste was extrudated, dried and calcined. The extrudates were impregnated with the Pt solution to obtain 0.45% of Pt. The catalyst exhibited a high hydroisomerization activity. The hydroisomerization catalysts suitable for highly paraffinic feeds were patented by Marchal-George (676). One catalyst was silica–alumina (29.3% SiO_2 and 70.7% Al_2O_3) in the form of extrudates impregnated with Pt (0.60 wt%). Using similar materials, another catalyst was prepared with the much smaller particle size of Pt on the support. Details of the method of preparation to obtain catalyst with precisely defined surface and chemical properties were given in the patent.

In the process patented by Carroll et al. (677) VGO was hydrocracked and products distilled to obtain a diesel oil fraction and the feed for dewaxing. Both these fractions were subsequently catalytically dewaxed. Hydrofinishing was used as the last step before producing the final lube oil stock. In the dewaxing step, the catalysts that contained ZSM-5 were more active than those containing SAPO-11 and ZSM-23. This resulted in the lower start-of-run temperature. The catalysts also contained Pt and Pd. The advantage of the ZSM-5-containing catalyst was that it could be reactivated by H_2 at the end of the cycle. The similar reactivation could not be accomplished with the other catalysts. These catalysts had to be removed from the reactor and regenerated by oxidative burn-off. It was noted that this is one of the few patents where the catalyst reactivation/regeneration issue is being addressed. The lubricating oil base stock having a low pour point and a high viscosity index can be produced by using the catalyst comprising zeolite ZSM-5, which has been subjected to controlled acidity reduction, before being deposited with a finely dispersed noble metal component (678). Moreover, the more open pore structure of the ZSM-5 permitted dewaxing of a wider range of feeds than the highly shape-selective zeolites (i.e., ZSM-23). This catalyst was less susceptible to deactivation by pore blockage. Forbus and Kyan (679) disclosed that the improvements in cycle duration both in the first and in the subsequent cycles may be achieved by using the zeolite in the dewaxing catalyst in its hydrogen or "decationized" form. In addition, the use of the hydrogen form zeolite leads to improvements in the quality of the lube base oil, especially its oxidative stability. Also, very low aging rate was achieved during the later portions of the cycle, i.e., the aging rate falls to less than 0.5°C/day.

Rather detailed description of the method for catalyst preparation was given by Benazzi et al. (680,681). In this case, the fraction boiling above 340°C was dewaxed over acidic catalysts. The acid function was provided by a zeolite selected from the group which included ZSM-48, EU-2, EU-11 and ZBM-30. The matrices that are used for carrying out the shaping of the catalyst were alumina gels, aluminas, magnesia, amorphous silica–aluminas and mixtures thereof. The techniques such as extrusion, pelletizing or sugar-coating can be used for carrying out the shaping operation. The HYD function of the catalyst was provided by Pt and Pd (~1.0%). The hydrofinishing step may be conducted under milder conditions than the dewaxing step. The HYD function of the catalyst for this step was provided by noble metals (Pt or Pd). Moreover, this catalyst contained either chlorine or fluorine.

The process for simultaneous production of the lube base stock and middle distillates disclosed by Benazzi et al. (682) comprised hydroisomerization over the catalyst containing noble metals (Pt or Pd) dispersed on amorphous acidic support. Subsequent dewaxing was conducted over the catalyst containing Pt or Pd supported on molecular sieves such as NU-10, EU-1, EU-13, zeolite and ferrierite. Another invention of Benazzi et al. (683) concerned the NU-85 zeolite from which a portion of the elements such as Al, Fe, Ga or B has been removed, e.g., by dealumination. The catalyst containing this zeolite combined with a noble metal, (e.g., Pt and Pd) exhibited a good activity for the reduction of the pour point of the lube base oil.

The rare earth (RE) exchanged ferrierite exhibited better overall selectivity for producing lube oil fractions having a low pour point and a high VI than the hydrogen form of either ferrierite or mordenite (684). RE metals include lanthanide elements such as La, Ce, Pr, Nd, Sm, Eu, Gd, Tb, Dy, Ho, Er, Tm, Yb, Lu, Y and their mixtures. The "RE-ferrierite" may include either natural or synthetic ferrierite. To obtain an active dewaxing catalyst, at least one noble metal (e.g., Pt or Pd) is added to RE-ferrierite.

The method for selectivation of dewaxing catalysts was disclosed by Cody et al. (685). The method involved contacting catalyst with the liquid stream containing about 100 ppm of oxygen provided by oxygenates such as alcohol, carboxylic acid, ester, ketone, aldehyde or ether between 400 and 670 K and at H_2 pressure ranging from atmospheric up to 20 MPa. The dewaxing catalysts included either molecular sieve containing at least one 10- or 12-ring channel or the ZSM-48 containing Pt and Pd. Before dewaxing, the feed contained more than 50% of *n*-paraffins. The dewaxed product was finely hydrofinished over the catalyst from the M41S family. After this treatment, the VI of the base oil product approached 125.

11.4.1.2 Non-zeolitic catalysts

The inventions of Gillespie et al. (686–688) were based on the observation that the catalyst containing lanthanide series and platinum-group components provides a superior performance and stability in the isomerization of the full boiling range of *n*-paraffins to iso-paraffins. In this case, the catalyst comprised the tungstated support zirconium oxide, a lanthanide element and/or yttrium as the first component and a second component being a metal from the platinum group. The first component consisted of a single lanthanide series element or yttrium, while the second component of a single platinum-group metal. Preferably, the first component was ytterbium, holmium, yttrium, cerium, europium or their mixtures, whereas the second component was Pt. The catalyst optionally contained the inorganic-oxide binder, such as alumina. Another solid-acid hydroisomerization catalyst patented by Gillespie (687) consisted of sulfated ZrO_2 as well as a Group IIIA component and Pt. The catalyst was active for the conversion of a paraffinic feed to an iso-paraffin-rich product with significantly enhanced cold flow properties. The catalyst comprising the support of the tungstated zirconia, at least one lanthanide element, preferably ytterbium or holmium and Pt, was active for the selective upgrading of a paraffinic feed to iso-paraffin-rich product (688). Other suitable supports included HfO_2, TiO_2 and SnO_2. Another isomerization catalyst consisted of a combination of the sulfated ZrO_2 support with one lanthanide element or yttrium component and the Pt-deposited Al_2O_3. This was very active catalyst for hydroisomerization as well. In this case, the Al_2O_3 was used as the binder (689).

The method for the preparation of dewaxing SAPO catalysts was disclosed by Rodriguez et al. (690). The SAPO catalyst having the silicon content between about 7 to about 40 mol%, aluminum content from 30 to 50 mol%, and the total phosphorus amount from 10 to 50 mol% had unique silicon distribution. The SAPO catalyst was iso-structural with a SAPO-11. It was disclosed by Miller (691) that the crystalline SAPO-11 catalyst gave the improved lube base oil yield and VI because it reduced pour point by a different mechanism than the ZSM-5 catalyst. The crystalline SAPO-11 catalyst was shape selective in that it appeared to isomerize normal and slightly branched chain paraffins and cyclo-paraffins without much cracking of the highly branched paraffins. Then, the yield of gas was reduced. Further improvement in lubricating oil yields was achieved by the processing scheme consisting of hydrocracking followed by dewaxing using the Pt and Pd SAPO-11. This catalyst was, however, very sensitive to N-compounds, which had to be removed from the feed to less than 10 ppm of the total nitrogen.

The novel hydroisomerization catalysts invented by Benazzi et al. (675,692) were successfully tested for decrease in pour point and increase in VI of VGO. In this case, a highly paraffinic VGO was contacted with the catalyst comprising at least one dioctahedral 2:1 phyllosilicate which was synthetized in the fluoride medium in the presence of HF (692). The space between the phyllosilicate sheets comprised pillars based on at least one oxide of elements from the groups IVB, VB, VIB, VIII, IB, IIB, IIA, IVA or any combination of these oxides, preferably selected from the group comprising SiO_2, Al_2O_3, TiO_2, ZrO_2 and V_2O_5, or any combination of the latter oxides.

The catalyst contained at least one hydrodehydrogenation metals, e.g., Fe, Co, Ni, Ru, Rh, Pd, Os, Ir and Pt.

11.4.2 Process configurations

The patent literature suggests that various configurations of catalytic systems using a wide range of catalysts have been considered for the production of middle distillates and lubricants. Various combinations of methods involving extraction methods with catalytic systems were noted. Catalytic processes are dominated by multistage systems always employing fixed bed reactors. For example, a multistage process may consist of hydroprocessing to remove sulfur and nitrogen to protect the noble metals containing catalyst used in the subsequent dewaxing stage, followed by hydrofinishing and dehazing. The properties of VGO and DAO determine the optimal process configurations.

11.4.2.1 Combinations of extraction and catalytic processes

Several patented concepts are based on the combination of conventional solvent extraction with catalytic systems. The concept disclosed by Fujihara and Mosishima (693) was designed in response to the continuous depletion of the naphthenic crude oil which traditionally has been used for the production of lube base oil. The VGO and DAO derived from such crude are known to have a low wax content. The paraffinic crude oils have been replacing the naphthenic crude oil. The VGO and DAO obtained from the former require dewaxing. Using solvent dewaxing, the wax precipitated by diluting the paraffin crude oil with a solvent and cooling was removed by filtration. Subsequently, the wax was subjected to hydrodewaxing over the form-selective zeolite catalyst. The hydrodewaxed product was sent to the next step for hydrofinishing conducted below 350°C at typical hydroprocessing pressure of H_2. The method patented by Cody et al. (694) involved solvent extraction of the feed in the first step. The aim was to produce the underextracted product which was subjected to conventional hydroprocessing, followed by hydrofinishing conducted at lower temperatures than the former. It is believed that such a process may be suitable for the feeds derived from some unique crude oils. In another method disclosed by Cody et al. (695), hydroprocessing was conducted in two stages before the hydrofinishing step. This arrangement appeared to be more feed flexible.

O'Rear and Biscardi (666) disclosed the process, in which the fraction boiling below 620°C was catalytically dewaxed, whereas the fraction which boiled above 620°C was subjected to a solvent dewaxing. The products could be blended to obtain the lube base oil for lubricant production. If necessary, the solvent dewaxed fraction may by either hydroisomerized or hydroprocessed to remove olefins and impurities which would affect the lubricant stability. A similar process configuration was disclosed by Briot et al. (696). In their process, the first catalytic hydroprocessing was conducted over the conventional $NiMo/Al_2O_3$ catalyst. The second-stage catalyst comprised the HY zeolite characterized by 13.6 wt% of SiO_2, 13.49 wt% of MoO_3, 2.93 wt% of NiO, 5.09 wt% of P_2O_5 on the support consisting of 64.89 wt% Al_2O_3.

In one process, the feed was extracted by *N*-methyl-2-pyrrolydine to remove aromatics and heterocyclics (697). The raffinate was then passed through hydroprocessing reactors before entering the dewaxing reactor. The process consisting of solvent (furfural, phenol, etc.) extraction in the first step, followed by hydroconversion under typical hydroprocessing conditions and hydrofinishing conducted under milder conditions produced the lube base oil having at least 90 wt% saturates and the VI higher than 105 (698). The hydrofinishing catalyst contained either Pt or Pd. Also, the combination of both Pt and Pd was very active. With this catalyst, the hydrofinishing could be conducted at temperatures which were about 100°C lower than that typically applied during conventional hydroprocessing.

The invention disclosed by Garwood et al. (699) showed that further improvements in the quality of the lube base oil by the additional non-catalytic treatment could be achieved. In this case, the lube base oil produced by catalytic dewaxing was additionally treated using an organic peroxide. The treatment increased the product viscosity without any significant adverse affect on viscosity index or pour point. In fact, with higher peroxide dosage treatment, there was an improvement in the lube base oil properties, i.e., the VI was increased and the pour point and cloud point were lowered. The process is, therefore, capable of making more viscous grades of lubricants of improved properties from the high wax containing feeds.

11.4.2.2 Catalytic processes

In every case, catalytic processes employed fixed beds, i.e., the bed consisting of several layers of different catalysts, several sections of different catalysts in the same vessel and several fixed bed reactors operating in series. Therefore, the reactors which are part of the Unicracking process shown in Figures 8.2 and 11.1 seem to be representative of those used traditionally for dewaxing.

A two-stage process, comprising the first stage operating at high H_2 pressure and the second stage operating under less severe conditions, was suitable for the conversion of VGO to the lube base oil (700). Light products together with NH_3 and H_2S were separated from the effluent of the first stage which was used as the feed for the second stage. The fixed bed reactor of the second stage had one or more beds containing different hydroprocessing catalysts and was maintained at typical hydroprocessing conditions. In the process disclosed by Leta et al. (680), the isomerization catalyst and the dewaxing catalysts were contained either in a single reactor in a layered configuration or in two separate reactors. The former catalyst was of an amorphous SiO_2–Al_2O_3 (35–55 wt% or Al_2O_3) origin, whereas the dewaxing catalyst was a crystalline molecular sieve. In both cases, the feed made the first contact with the isomerization catalyst. According to the process claimed by Riley et al. (701), waxy feeds can be hydroisomerized to produce good yields of an isomerate of high VI using a silica–alumina-based catalyst. The silica-alumina may be modified with a RE oxide of yttria, boria and/or magnesia. The lube fraction of the isomerate was dewaxed to provide the lube base stock having a high VI.

An example of another two-stage process consisting of hydroprocessing to remove sulfur and nitrogen to protect the noble metals containing catalyst, followed by dewaxing stage, is the invention disclosed by Albinson et al. (702). In the two-bed process patented by Kyan and Oswald (703), the feed was divided into the low-boiling stream and the high-boiling stream. The latter was hydroprocessed at the top bed. The effluent from this bed was mixed with the lighter fraction in the interbed distributor before entering the dewaxing bed at the bottom. Another process claimed by Cody et al. (704) consisted of the conventional hydroprocessing of a waxy feed followed by dewaxing over a ZSM-48 zeolite doped with either Pt or Pd, as well as a mixture of both. In the last step, the feed was hydrofinished under milder conditions than applied during the preceding step. The last step was conducted using the MCM-41 catalyst which contained at least one noble metal, i.e., Pt or Pd.

Several patents were disclosed involving processes in which the dewaxing step preceded the hydroprocessing step (705–707). The process described by Chen and Degnan (708) comprised two trickle bed reactors in series containing a dewaxing catalyst and a hydroprocessing catalyst in the first and the second reactors, respectively. The dewaxing catalyst was supported on the vertically tiered and staggered trays, while H_2 was concurrently introduced into the catalyst on each of the trays. A similar arrangement in the second reactor allowed trickling of the liquid, while H_2 introduced into each tray flowed upwardly together with the light fractions separated from the liquid streams.

In the process configuration disclosed by Carrol et al. (709), VGO was subjected to conventional hydroprocessing prior to hydrocracking. The fractions boiling below the IBP of VGO and diesel oil were separated from the hydrocracked product, while the heavier fraction was catalytically dewaxed before being introduced into the hydrofinishing reactor to obtain the stable lube base stock. In a similar process (710), VGO was hydrocracked in the first step. The products from this step were distilled to remove gaseous and lighter products. The heavier fraction was catalytically dewaxed before entering the hydrofinishing step.

Murphy et al. (711) invented the process for the lube base stock production which consisted of several stages. In the first stage, the removal of sulfur and nitrogen was accomplished under the conventional hydroprocessing conditions. The hydroisomerization of this product was conducted at about 250°C over the catalyst containing the active metals on zeolite together with dopants such as yttrium, halogen and boron. The product from the second step may be either dewaxed by solvents or catalytically.

The patent claimed by O'Rear (712) was suitable for the production of lube base oil from the feed containing olefins and aromatics. Such feed was directed to the alkylation unit to produce alkyl aromatics with boiling point in the range of the lube base oil. Subsequently, this product was the feed to the HYD unit where the aromatics were converted to naphthenic structures possessing a high VI.

Several patents (713–716) claim the process in which the isomerization of the feed preceded the dewaxing step. In this case, the *n*-paraffins, which would be otherwise removed during dewaxing, were isomerized in order to decrease the freezing point/pour point of the product. This results in the increase in the yield of the latter. The invention of Berlowitz et al. (717) shows that the waxy feed such as obtained during FT and/or derived from paraffinic crudes may not require the hydroprocessing step and as such may be sent directly to hydroisomerization. After distilling off the products boiling below 350°C, the hydroisomerate was subjected to catalytic dewaxing.

In some cases, the products from dewaxing of the highly paraffinic feeds lack clarity. Johnson et al. (718) disclosed process and catalyst suitable for dehazing of dewaxed products. Dehazing of such products requires catalyst and is conducted at relatively low temperatures, whereas H_2 pressure may vary between 3 and 15 MPa. A conventional hydroprocessing catalyst may be used for this purpose. However, preferred catalysts are those which contain Pt and Pd. The dehazing method patented by Biscardi and Fong (719) involved contacting the catalytically dewaxed base oil with the solid adsorbent to remove the haze precursors, thereby reducing the haze-forming tendency.

11.5 SPENT CATALYSTS FROM DEWAXING

It is believed that the life of catalysts used for dewaxing as well as those for HYD, hydrofinishing and dehazing should approach at least 2 years, provided no unexpected events occurred during the operation. In most cases, the feed for dewaxing was free of contaminates because it was already pretreated by hydroprocessing. Therefore, the coke deposition and poisoning by N-compounds could be the main causes of catalyst deactivation. However, N-bases would have to be removed down to few ppm of nitrogen in the case that dewaxing catalysts contain noble metals. Such catalysts can also be poisoned by H_2S. Therefore, the deep HDS of the feed may be necessary before dewaxing. To a certain extent, recrystallization of the catalytically active phases during prolonged exposure to hydroprocessing conditions could occur as well.

In view of the above facts, regenerability of the catalysts from all stages of catalytic dewaxing (e.g., HYD/HCR, hydroisomerization, hydrofinishing and dehazing) for reuse should be rather high. Because no contaminant metals were present in the feed, the catalyst regeneration would only require decoking using the oxidative burn-off. This would involve the same procedures which were discussed in Section 10.1.1 for the conventional hydroprocessing catalysts. Hydrofinishing and dehazing steps, which may be part of the overall catalytic dewaxing schemes, are usually conducted under mild conditions. This indicates a high regenerability of the catalysts used in these operations. Thus, under such conditions, the duration of the uninterrupted operation could approach 5 years.

The presence of noble metals (e.g., Pt and Pd) in some dewaxing catalysts suggests that the conditions of their regeneration may approach those applied during regeneration of the reforming catalysts (720,721). For the latter, more than 100 and in one case almost 230 regeneration-utilization cycles could be performed before catalyst replacement was necessary almost after 5 years (722). The large number of the utilization-regeneration cycles attained for the reforming catalyst may be partly attributed to rather clean feed (e.g., hydroprocessed naphtha) used for reforming compared with the hydroprocessed VGO and DAO. Therefore, it is believed that for dewaxing catalysts, the number of the regeneration-utilization cycles will be less than that for reforming catalysts but greater than that for hydroprocessing catalysts. The regenerability of the Pt/Pd containing catalysts used for hydrofinishing and dehazing as part of the overall middle distillates and lubricants production is expected to be much greater than that of the typical dewaxing catalysts. In fact, the duration of these catalysts in the operation may approach that of the reforming catalysts. Thus, the temperatures employed during the hydrofinishing and dehazing operations are usually about 100°C lower than that used during dewaxing. With respect to the HYD equilibrium of aromatics, such conditions are favorable for maintaining a high rate of HDAr.

After being replaced, spent non-regenerable catalysts have to be handled according to the procedures prescribed by environmental and health authorities, similarly as it was indicated for hydroprocessing catalysts (Section 10.5). Entirely different approach is used for the non-regenerable catalysts containing noble metals. Thus, because of the high value on the noble metals, refineries must pay special attention to the inventory and handling of such catalysts. This involves an accurate chemical analysis determining the content of noble metal, which has to be performed before shipping spent catalysts for metal reclamation. This was confirmed by Rosso (723,724). For example, in the case of Pt, a tolerance of $\pm 0.5\%$ may represent a significant monetary value. The accuracy of the analysis can be significantly improved by the removing contaminants (e.g., coke, sulfur, moisture,) using preburning (725–727). Both, on-site and off-site preburning has been practiced, although with respect to the final settlement with the metal reclaiming company, the former appears to be more favorable for the refinery.

There are a number of processes operating on the commercial scale for the recovery of precious metals from various types of spent catalysts (41). These methods can be applied to spent dewaxing catalysts without any modification. One method is based on the dissolution of the γ-Al_2O_3 support, while leaving precious metals in a solid form. The dissolution can be achieved either by NaOH or by H_2SO_4 (724,728). On the other hand, the dissolution of metals, while leaving the support undissolved, can be achieved by the mixture of HCl and an oxidant, i.e., HNO_3, H_2O_2, $NaClO_3$, NaOCl (729,730). Once in solution, precious metals can be isolated in the pure form by precipitation (731,732), solvent extraction (733) and membrane separation (734). Other processes include the gas phase volatilization of precious metals using the Cl-containing agents and pyrometallurgical processes such as plasma arc smelting (729).

Chapter 12

NON-CONVENTIONAL CATALYTIC UPGRADING OF HEAVY FEEDS

The scientific literature has been indicating the attempts to develop more advanced hydrogen addition processes for upgrading the most problematic feeds such as VRs derived from heavy crudes, as well as the direct upgrading of the latter. However, lighter feeds such as VGO were also investigated using methods which are still in the early stages of development. Although the process configurations may be similar, the type and form of catalysts being developed for the emerging processes differ markedly from those used during the conventional hydroprocessing. It is felt that a brief review of this information is necessary to put emerging processes in prospective with the currently used commercial processes for upgrading of heavy feeds.

For heavy crudes, one of the objective has been the improvement in pumpability for transportation by pipeline. In this regard, the down-hole upgrading and/or upgrading on the site of the production well has been investigated. The hydrogen addition processes accounted for most of the novel methods which have been investigated for the upgrading heavy feeds to distillate fractions. These processes utilize catalysts either in a dissolved form or in a finely dispersed form. In their configurations, the latter processes resemble those utilizing once-through dispersed low-cost catalytically active solids. It has been generally known that the average molecular weight of heavy feeds can be decreased with the aid of a living organism, i.e., so-called bio-catalysis and/or bio-upgrading. It was hoped that the severity of the upgrading of heavy feeds could be decreased using these novel methods. Apparently, the methods such as down-hole upgrading, the bio-catalysis and the use of the dispersed catalyst are still in various stages of development. Thus, additional efforts are necessary before these catalysts and processes can be used commercially on the site of refineries. In this regard, novel processes must be cost-effective in comparison with several carbon-rejection processes (coking and thermal cracking) and hydrocracking processes, which have been used commercially for upgrading most problematic feeds for decades. They must also be competitive with the conventional hydroprocessing. It is believed that the scale of operation may be an important factor influencing commercial viability of the novel processes.

12.1 DOWN-HOLE UPGRADING

The upgrading of heavy crude either in the reservoir or inside the producing well represents the most attractive and also most ideal case of the heavy feed upgrading.

However, this is also one of the most challenging task. The information on non-catalytic processes used for these applications is predominantly in the patent literature (735). The mechanism, the state of the art and complexity associated with the down-hole upgrading using microbes, i.e., microbes enhanced oil recovery (MEOR) was discussed in detail by Fujimara et al. (736) who indicated that a new knowledge is required how to achieve a balance between the beneficial and detrimental effects of the microbial growth in reservoirs. The downstream upgrading of petroleum feeds using microbes is discussed in the latter section 12.3.

The relevant studies on the down-hole upgrading using catalysts were reviewed by Weissman et al. (737). Potential use of the conventional hydroprocessing catalysts for down-hole upgrading has received attention as well (738). Unusual conditions, such as the presence of water and/or brine, as well as the heat and reactant gas requirements, impose significant challenges compared with the conventional hydroprocessing methods. For example, a viable process should consist of the catalyst bed placed in the producing well-bore. Heavy crude would then move through the bed of a catalyst co-currently with reactants (e.g., hydrogen, hydrogen donor, water, CO, etc.). The heat requirements could be supplied either by the partial combustion of heavy crude or by steam injection. It is believed that under such conditions, the rapid catalyst deactivation would occur, particularly in the presence of clay-like mineral matter and brine. The method described by Weissman et al. (737) involved placing a fixed bed of catalyst before the entrance into the production well. The fixed bed was placed using the conventional gravel pack method. The in situ partial combustion was then used to generate reactive gases and to drive oil over the catalyst bed to production well.

In the early stage of development seems to be called "toe-to-heel" air-injection process (739). In this case, the perforated horizontal producer well is positioned along the whole length of the reservoir. The vertical planar combustion zone is created by injecting air and steam. A catalyst is emplaced on the external surface of the perforated producer well. The mobile heavy crude zone, created by the heat of combustion, is moving through the layer of catalyst and entering the producer well. The tests conducted with and without catalyst indicated significantly improved quality of the products for the former, i.e., a markedly lower viscosity and specific gravity. For this purpose, the conventional $CoMo/Al_2O_3$ catalyst was used. However, so far, there are no indications of such process being used commercially.

The in situ upgrading using the injection of catalyst dispersed in the hydrocarbon donor liquid may be another potential method, although so far, there are no reports to indicate any activities in this direction. One may conclude that the down-hole upgrading using a catalyst is still in very early stages of development. The upgrading of a heavy crude near the producing well (on the surface) is sometimes referred to as the down-hole upgrading. This, of course, is not the case because the conventional processes based either on hydrogen addition or on carbon rejection can be used for these applications.

Apparently, there might be an option to inject a catalyst during the enhanced oil recovery (e.g., together with CO_2), although this approach has never been explored. In this regard, one may only speculate about the type and form of catalyst to be employed. Nevertheless, it is fair to conclude that the down-hole upgrading using a solid form of catalyst is still in its infancy, although the use of bio-catalysts has already been noted.

12.2 PROCESSES USING DISSOLVED/DISPERSED CATALYSTS

Except for low-cost, throw-away materials, so far there is no process utilizing dispersed/dissolved catalysts which would operate on a near-commercial scale, although the brief information may suggest the potential start-up of such process by 2010 (740). The comparison of the flow sheet of this process shown in Figure 12.1 (740) with the commercial VEBA process shown in Figure 8.10 (462–464) indicates some similarities. The catalyst used in the former consists of the nano-dispersion of active metals, such as present in the conventional hydroprocessing catalysts, whereas VEBA process utilizes once-through low-cost throw-away solids (e.g., red mud). In the latter case, the catalyst recovery for reuse is a non-issue.

A provision for recycling at least part of the VR from vacuum tower to the feed entering the reactor shown in Figure 12.1 (740) should be noted. In this case, a part of the original catalyst would be also recycled with the VR. There is little information on the activity (if any) of this part of the catalyst. In an ideal case, the viability of the process could be enhanced if the activity of this catalyst could be somehow restored. Because of the limited information given, the process in Figure 12.1 (740) may still

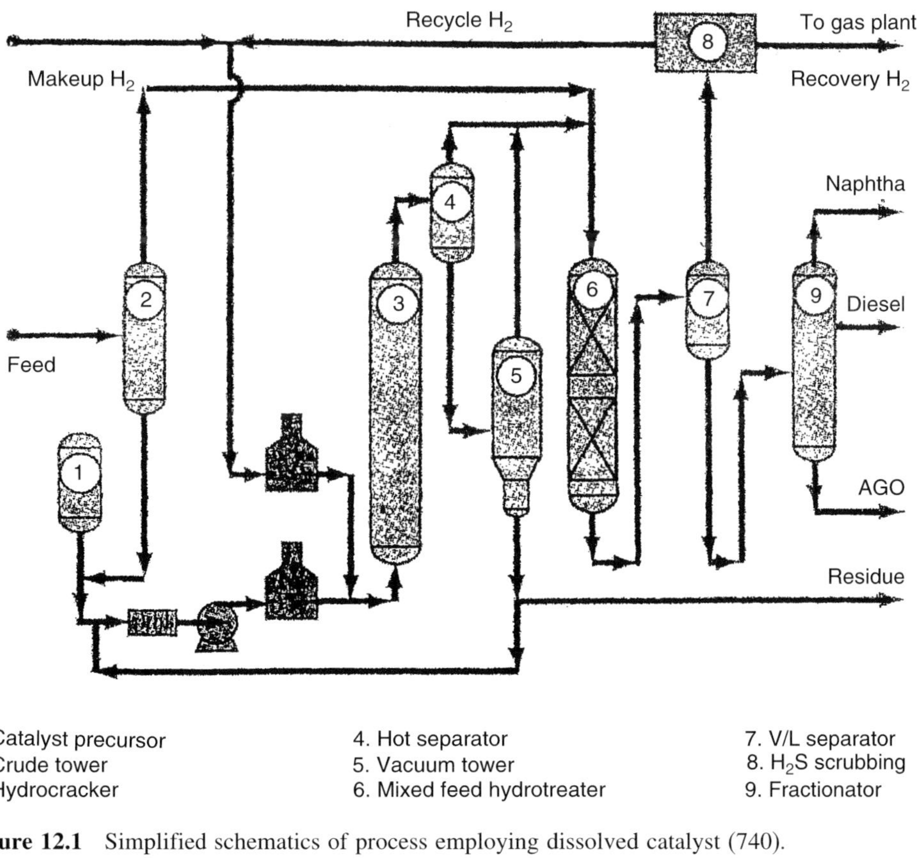

Figure 12.1 Simplified schematics of process employing dissolved catalyst (740).

be considered as only an emerging rather than a near-commercial process until more technical data is provided.

A brief account of the methods employing dispersed/dissolved catalysts will only be given for comparison with the commercial processes using fixed bed, moving bed and ebullated bed reactors, as well as the slurry bed reactors using once-through finely dispersed solids. The former methods utilize the catalysts either finely dispersed in heavy feed or dissolved in a liquid phase which is subsequently blended with heavy feed. Most of the published studies indicate that the operating parameters such as temperature and H_2 pressure approach those used during the hydroprocessing of heavy feeds using conventional methods (741), although one of the main objective for developing the processes employing dispersed/dissolved catalysts has been the decrease in processing severity.

The first review on hydroprocessing using the dispersed/dissolved catalysts was published by Del Bianco et al. (741) in 1993. In this case, the attention was paid to the patent literature, as well as to the basic research. Table 12.1 summarizes some examples of heavy feeds and the oil soluble catalyst precursors which have been tested in the batch systems (60 min duration). Apparently, the temperature and pressure used during these experiments approach the upper range of those usually employed during the conventional hydroprocessing of heavy feeds. High levels of the metal removal at a low coke yield should be noted. Generally, the active metal concentration, H_2 pressure and temperature ranged from 300 to 2000 wppm of the metal based on the feed, 7 to 16 MPa and 693 to 723 K, respectively, have been tested.

The proposed mechanism of hydroprocessing involving dispersed catalysts was based on the formation of a coke-like deposit and/or coke precursor from heavy feed simultaneously with the hydrogen activation on the catalytically active metal (742). The presence of active hydrogen ensured that the coke precursor was converted to liquid products rather than to coke. A high H_2 pressure may be favorable for maintaining a desirable rate of hydrogen activation required for this reaction. Because of a high H_2 pressure, part of the catalyst leaving the reactor with the process streams may still exhibit some activity.

Table 12.1 Catalyst precursors and operating conditions (741)

Precursor	Feed	Temp. (K)	Pressure (MPa)	Metal in feed (ppm)	Coke (wt%)	HDM (%)
Fe-naphthenate	JB[2]	711	17	800	2.6	56
Ni-octoate	JB	711	17	800	0.7	91
Cr-resinate	CL[3]	711	17	820	0.5	95
Mo-resinate	CL	711	17	820	0.5	93
PMA[1]	CL	711	17	382	0.3	87
$Mo(CO)_6$	CL	711	17	390	0.4	88
Mo-naphthenate	CL	716	14	108	2.6	64
Mo-naphthenate	CL	711	17	700	0.7	84
Mo-sulfonate	HH[4]	711	14	100	1.9	–

[1] Phosphomolybdic acid
[2] Jobo crude
[3] Cold Lake
[4] Hondo heavy crude

This suggests that recycling the VR obtained from the products to the feed entering the reactor for reuse of the catalyst dispersed in the former, may have some merit, although little information indicating this could be found in the scientific literature.

An efficient contact of the active phase with the feed makes dispersed catalysts suitable for hydroprocessing of the high metals and asphaltenes content heavy feeds. Metals in heavy feeds, i.e., V, Ni, Fe and Ti, may deposit on the fine catalyst particles and as such may be catalytically active rather than deactivate catalyst. The latter case has been usually observed during the hydroprocessing in commercial units employing fixed bed, moving bed and ebullated bed reactors. Because of the absence of diffusion limitations, fine particles ensure a complete catalyst utilization. The submicron size particles of carbon (e.g., carbon black) are among potential solids to be used for dispersion. The metals in heavy feed would then deposit on the carbon particles. Moreover, fine carbon particles may act as the nucleation site for coke formation. In this form, coke is carried out with the process streams rather than being deposited on the reactor walls. This extends the operation because plugging of the reactor is avoided.

12.2.1 Soluble catalysts

Once dissolved in a liquid, the solution of the catalyst precursor can be blended with heavy feed. Depending on the type of catalyst, both aqueous and hydrocarbon phases can be used for catalyst dissolution. However, some precautions have to be taken while injecting an aqueous solution into the hot heavy feed. Under hydroprocessing conditions, the catalyst precursors are decomposed and subsequently converted to catalytically active phase via the reactions with H_2S which was released during HDS. The catalyst made in situ is in a nearly molecular size. This ensures a high dispersion and contact with the reactant molecules in heavy feed. Moreover, in such a state, a nearly ideal catalyst utilization may be approached.

12.2.1.1 Oil soluble precursors

There is a number of the organometallic compounds being sufficiently soluble in hydrocarbon liquids, e.g., metal salts of organic acids (naphtenic, acetic, oxalic, octoic, etc.), organic amines, organometallics, metal containing quaternary ammonium compounds, etc. The carbonyl compounds of transition metals (e.g., $Fe[CO]_5$) and ferocenes are soluble in oil and as such are the potential catalyst precursors as well (743). Several examples found in the scientific literature may be used to assess the potential of the oil soluble catalysts for hydroprocessing of heavy feeds with the aim to make the distinction between the non-conventional and conventional hydroprocessing.

Sato et al. (744) compared the ultra fine MoS_2 with the MoS_2 made in situ from the oil soluble Mo-dithiocarbamate during hydroprocessing of the Kuwait AR. An extensive micronization of the former was required in order to achieve activity, which would approach that of the in situ made MoS_2. Catalytic activity of the dissolved Mo-catalyst prepared from the Mo-dithiocarbamate during upgrading of VR was significantly enhanced in the presence of hydrogen donor such as tetralin (745). Apparently, this resulted from the special synergy between the catalyst and tetralin. In concentration lower than 800 ppm of Mo, the MoS_2 formed in situ from the dissolved Mo-naphtenate suppressed coke formation during the hydroprocessing of Athabasca bitumen (746). However, when a basket with the conventional $CoMo/Al_2O_3$ catalyst was present in the same system to achieve an additional upgrading, the beneficial effect of the dispersed MoS_2 was not present. Moreover, activity of the conventional catalyst was affected by the dispersed catalyst. Most likely, the dispersed catalyst deposited on the external surface of the catalyst particles and as such had an adverse effect on the diffusion of reactant molecules into the particles interior.

According to Del Bianco et al. (747), the MoS_2 produced in situ from the Mo-naphthenate precursor facilitated active hydrogen required for stabilization of radicals produced by cracking. They used the VR (V + Ni of 300 ppm) derived from Belaym crude. The experiments were conducted between 683 and 723 K at 9 MPa using the Mo concentration ranging from 200 to 5000 ppm. While using the Safania VR (~200 ppm of V + Ni), the Mo-naphthenate was much better MoS_2 precursor than phosphomolybdic acid (748). With respect to HDS, the combination of Mo + Co was better than Mo + Ni in agreement with Lee et al. (749). These authors used the oil soluble compounds of Mo, W, Ni and Co as the precursors for the corresponding dispersed metal sulfide catalysts. For single metals, the best performance was observed for the Mo catalyst. The combination of Co + Mo was the best for HDS, whereas Ni + Mo for HCR.

Various oxidation states of Ni, Co, Mo, Fe and V dispersed in the Heavy Arabian VR using oil soluble compounds were studied by Dabkowski et al. (750). The study was conducted in an autoclave under rather severe conditions, i.e., 713 K and 14 MPa. Under these conditions, with 1000 ppm of the dissolved metal, the HDM of the feed approached 98% at 2 wt% of coke formation during 1 h run. The following order in activity for the coke suppression was established: $Ni^{+2} > Mo^{+6} > V^{+4} \sim Co^{+2} > Fe^{+3}$. Among the total of 15 compounds studied, the Ni^{+2} octoate, Co^{+2} naphthenate and Mo^{+6} naphthenate gave high yields of distillates at low amounts of coke.

Beneficial effects of the addition of P in the form of the oil soluble compounds to conventional catalysts on hydroprocessing of heavy feeds was indicated earlier. Of the particular interest was the study of Kushiyama et al. (160) who studied the effect of P on the activity of the in situ made CoMo catalysts on the conversion of several heavy feeds. There was an optimal amount of P giving the highest conversion. As Figure 5.18 (160) showed, the optimum of the P addition was feed dependent. It is believed that similar optima may also be present for other additives. This suggests that the activity trends may vary from feed to feed.

Rather unique method for the removal of metals from heavy feeds, involving the use of oil soluble phosphorus containing compounds, was developed by Kukes et al. (751). In this case, a phosphorus compound was dissolved in a heavy feed, which was then pumped through the reactor containing the fixed bed of alumina at about 15 MPa and at 673 and 693 K. The Maya heavy crude containing about 360 ppm of V + Ni was investigated in addition to other heavy feeds. Under these conditions, phosphorus (700 ppm) had pronounced effect on the V removal compared with little effect on the removal of Ni and sulfur. In the absence of phosphorus compounds in the feed, the activity of alumina was very low. Significantly enhanced activity in the presence of phosphorus compounds was attributed to their catalytic action after being deposited on the alumina support. The parallel experiments conducted in the batch reactor confirmed that a part of V could also be removed via homogeneous reactions with phosphorus compounds. In this case, the following order of the effectiveness of the additives for V removal was established:

$$(R)_3P > (ArO)_2P(O)H > (ArO)_3P > Ar_3PO > (NH_4)_2HPO_4$$

This order indicates the involvement of steric effects during the V removal. The failure to remove Ni compared with V may be attributed to the presence of the V=O group in vanadyl porphyrins. It is believed that the interaction of the V=O group with phosphorus in a phosphin-like form may assist the cleavage of the V−N bond in porphyrin and subsequently the V=O bond yielding phosphate and releasing V for the deposition on the surface.

12.2.1.2 Water soluble precursors

It has been noted that the water soluble catalyst precursors have attracted less attention compared with the oil soluble precursor. It is believed that for the latter, little problem should be encountered, while adding the aqueous solution of precursor to heavy feed. There is a number of the water soluble compounds used as the precursors for the in situ catalyst formation, i.e., thiomolybdates, phosphomolybdates, nickel and cobalt nitrates, etc. (752). Similarly as the oil soluble precursors, under hydroprocessing conditions, these compounds decompose and are converted to catalytically active sulfides (e.g., MoS_2), which are in a highly dispersed form. As it was indicated above, a high dispersion of the in situ made catalyst in heavy feed ensures perhaps the most efficient contact with the reactants in the feed resulting in a nearly complete catalyst utilization.

The water soluble catalysts containing Mo, Ni and Fe in the amount of ~1000 ppm of metal were added to the VGO derived from a Chinese crude (753). At 708 K and 10 MPa, the following catalyst activity order was established for the removal of asphaltenes + resins: Mo > Ni > Fe. The content of aromatics in the products decreased in the opposite order. Under these conditions, the submicron size of the sulfides of Mo, Ni and Fe formed in situ was catalytically active phases. It was demonstrated that after the dissolution of Ni nitrate in water, sulfidation could be accomplished by the addition of an alkaline metal sulfide to the solution (754). The catalytic activity of this phase was influenced by the method used for its co-slurrying with heavy feed. The water soluble bimetallic catalyst containing MoS_2 and $Fe_{(1-x)}S$ was prepared from $(NH_4)_6Mo_7O_{24} \cdot 4H_2O$ and $Fe(NO_3)_3 \cdot 9H_2O$, respectively, using a high dispersion method in the AR derived from the Chinese crude (755,756). The most active combination comprised 1100–1300 μg/g of MoS_2 and 25 μg/g of $Fe_{(1-x)}S$. With this combination, the highest conversion to distillates and lowest coke formation was observed. About 200 μg/g of Mo added in the form of the soluble Mo compound gave a high conversion of the VR to distillates (757).

12.2.2 Finely dispersed catalysts

Dispersing a mineral form of the finely divided transition metal compounds based catalysts with heavy feeds have been tested as well (758). Under conditions applied during hydroprocessing, as well as in the presence of H_2 and H_2S, these compounds are converted to catalytically active metal sulfides (e.g., MoS_2) which are finely dispersed in the feed. In this regard, the active phase is either in a micron or less than micron size. This depends on the extent of micronization of the solid catalysts. Among the type of inorganic compounds used, salts, oxides, sulfides and alloys could be successfully dispersed in heavy feeds (759). For example, the unsupported Mo sulfide prepared by mechanical milling slowed down the coke formation via quenching radical reactions during hydroprocessing of heavy feeds (760–762). The micronized coal, petroleum coke and AC particles impregnated with the water soluble metal salts have also been tested.

In this regard, AC can provide numerous impregnating sites for active metals because of the large surface area (763,764).

The costs of micronization of solids is significantly greater than that of the pulverization. Therefore, choice of the method for the former has to be carefully assessed. In this regard, a number of other methods have been used for catalyst dispersion. For example, rather unique method was based on the plasma discharge spark between the two electrodes immersed in a hydrocarbon liquid (765,766). The resulting product was a carbonaceous solid containing highly dispersed metals which evaporated from the electrodes. In this case, traditional transition metals have been of the primary interest. For example, Ni in concentration as low as 100–200 ppm decreased the formation of gases and coke and increased the HYD of aromatics in a heavy feed.

Matsumura et al. (767) used the VR derived from several heavy crudes to compare a powder form of the $CoMo/Al_2O_3$ catalyst with the oil soluble MoNiS catalyst. For both catalysts, the yield of liquid products was similar. The $NiMo/Al_2O_3$ catalyst crushed to less than 60 μm size particles was added to the VR derived from the Ural crude for hydroprocessing in the continuous system between 683 and 723 K and between 10 and 12 MPa (768). The VR contained almost 200 ppm of V + Ni, 5.2% asphaltenes and 15.8 CCR. The most beneficial effect was on HDS, whereas that on the conversion of asphaltenes and CCR was much less evident unless the temperature of 703 K or more was approached. It is generally known that at temperatures exceeding 700 K, thermal effects become an important contributor to the overall conversion of asphaltenes. At the same time, the rapid coke formation may affect the overall conversion.

The super oil cracking (SOC) process being developed in Japan, has some unique features, i.e., it uses horizontal furnace tubes as the reaction zone (769). A finely ground catalyst to promote cracking was slurried with a VR. The process operated under severe conditions such as H_2 pressure above 20 MPa and temperature of 753 K. Under such conditions, high conversion of asphaltenes was achieved at a relatively short residence time. The process has been demonstrated on the 3500 b/d scale using a VR.

A series of the transition metals supported on different supports was prepared by Fujimoto et al. (758). The fine particles of these catalysts were slurried with a heavy feed and their activity determined in the autoclave. For the catalysts containing 1 wt% of metal on Al_2O_3, the following order in the hydrogen consumption and HYD activity was established: Ni > Mo > Co > Cr > V > Cu > Fe > W > Al_2O_3 ~ no catalyst. However, Mo was more active than Ni when the comparison was made on the number of moles of metals basis. It is suggested that the relative activity order may change for different heavy feeds and different experimental conditions. In the similar study, Ramirez and Galarraga (770) used the finely divided macroporous SiO_2–Al_2O_3 for the in situ preparation of catalyst with the total amount of active metals approaching 300 ppm. The best performance was observed with Mo, Co and Ni.

As it was reported by Sakanishi et al. (771), activated carbon was also found to be the suitable support. These authors prepared the in situ Mo/AC catalyst by dispersing the finely divided mesoporous AC in a VR together with the oil soluble Mo-dithiocarbamate. The active catalyst was formed under the operating conditions as result of the precursor decomposition followed by the Mo deposition on AC. However, the involvement of the catalyst produced directly from the precursor together with the one made by the deposition of the feed metals on AC could not be ruled out. Most likely, hydroprocessing reactions involving both types of the catalyst were occurring simultaneously. In addition, the AC without active metals could also be involved.

Table 12.2 Properties of catalysts and yield of residual fraction (187)

Parameter	Catalyst		
	NiMoC-A	NiMoC-B	$NiMo/Al_2O_3$
Surface area, m^2/g	1856	335	179
Pore vol., mL/g	0.96	0.17	0.31
Mesopore vol., mL/g	0.71	0.05	0.31
Pore diam., nm	2.07	2.03	6.72
Yield, wt%			
at 623 K	41.3	42.7	57.0
at 723 K	1.3	4.8	5.6
Recycle at 713 K			
1st	9.1	–	10.2
2nd	8.3	–	10.6

The direct comparison of the NiMo catalysts supported on AC with that supported on γ-Al_2O_3 during the slurry bed hydroprocessing of the AR derived from a Middle East crude was conducted by Kouzu et al. (187). The surface properties of these catalysts and yields of residual fraction are shown in Table 12.2. A significant difference in surface properties of the AC supported catalysts should be noted, although their pore diameters were similar. The AC supports were impregnated simultaneously with the Mo and Ni precursors to give 15 wt% of Mo and 3 wt% of Ni. The commercial $NiMo/Al_2O_3$ catalyst used for comparison contained similar amounts of Mo and Ni. Before the experiments, the catalysts were presulfided ex situ. The experiments were performed in an autoclave between 623 and 723 K at 5 MPa of H_2 and 2 h duration. In terms of the yield of the residual fraction (811 K+), the best performance was exhibited by the NiMo/C-A catalyst. At the end of the first experiment, the catalyst was isolated and recycled for the second experiment. The increased yield of the residual fraction indicated a catalyst deactivation. However, little catalyst deactivation was observed after the subsequent recycle. The study of Kouzu et al. (187) may have important implications on the slurry bed hydroprocressing providing that an additional investigation with the focus on catalyst deactivation and reuse is conducted. Thus, the loss of activity after the first test is rather low. Moreover, the activity seemed to stabilize after the first recycle. It is believed that a more efficient activity recovery could be achieved at H_2 pressure higher than 5 MPa used by Kouzu et al. (187).

Table 12.2 (187) indicates a significant difference between the surface properties of the NiMo/C-A and NiMo/C-B catalysts. At the same time, the difference in the activity of these catalysts is less evident, particularly at 623 K. All evidence suggests the absence of diffusion limitations for the finely divided catalysts which were co-slurried with a heavy feed (40,41). A slightly better performance of the NiMo/C-A catalyst may be attributed to the greater amount of active surface hydrogen compared with the NiMo/C-B catalyst. The markedly higher surface area of the former indicates the presence of the much greater concentration of irregularities compared with the NiMo/C-B catalyst. It is believed that such irregularities are the sites for hydrogen activation. Because the rate of hydrogen activation increases with increasing H_2 pressure, the difference in activity in favor of the catalyst NiMo/C-A would increase with increasing H_2 pressure as well (63). This, of course, would have to be experimentally confirmed.

12.2.3 Recovery of dispersed/dissolved catalysts

The drawback of the methods employing dispersed/dissolved catalysts is the uncertainty regarding the catalyst recovery for reuse. For example, the plant processing about 50 000 bbl/d (~8000 t/d) of a heavy feed requiring the continuous addition of ~1000 ppm of the metal catalyst would require the addition of about 8 tons per day of the metal. Therefore, it is crucial that most of the metal can be recovered for reuse. In practical situation, dispersed catalyst ends up in the residue after the fractionation of the products. In an ideal case, part of the VR with metals remaining in, may be recycled and blended with the feed. However, this would require conditions ensuring reactivation of the catalytically active metal phases at the entrance of the reaction zone unless the recycled catalyst still possesses an adequate activity. For example, a sufficiently high H_2 pressure would ensure that the coke deposition on the catalyst particles would be low. In other words, most of the original activity of the catalyst would be still retained. The results published by Kouzu et al. (187) shown in Table 12.2 provide some experimental evidence for this assumption.

There are some indications of the attempts to extract the catalyst from VR by a solvent for subsequent reuse (772,773). Without extraction, the final separation of catalyst depends on the residue utilization option. Thus, the metals of interest will end up in the ash and/or slag in a concentrated form providing that the final residue utilization involves combustion and/or gasification. In this case, conventional methods (e.g., leaching out, extraction, etc.) are available for the separation of metal from the ash in a pure form. However, a high concentration of metals (V, Ni and Fe) in the ash suggests that the latter may be utilized directly. Apparently, the catalyst recovery for reuse requires an additional attention before the processes employing dispersed catalysts can be used on the commercial scale. In any case, such a process would have to operate in a continuous and/or semicontinuous mode near or on the site of petroleum refinery.

With respect to the metal recovery for reuse, the study of Lee et al. (749) deserves an attention. These authors used the oil soluble compounds of Mo, W, Ni and Co as the precursors for dispersed metal sulfide catalysts. In this study, the fixed bed of extrudates made either from a microporous AC or γ-Al_2O_3 was placed downstream of the reaction zone with the aim to remove metals from the product streams. A high efficiency of the metal removal was achieved using the AC extrudates. It is believed that there is a number of methods which are suitable for the recovery of metals which were trapped by the AC, e.g., dissolution/extraction. In this study, the AR containing ~26 ppm of V + Ni was used as the feed.

12.3 BIO-CATALYTIC UPGRADING OF HEAVY FEEDS

Bio-catalysis is based on the ability of microorganisms and their enzymes to use certain substrates and convert them to products of a higher quality. While discussing the down-hole upgrading above it was indicated that various microbes have been found in the reservoirs of the conventional and heavy crude oils, and depending on conditions, they could have either beneficial or an adverse effect on the recovery of heavy crudes. However, the complexity of these effects in the reservoir has been noted.

The preparation of bio-catalysts is usually carried out by the companies producing various bio-products rather than by petroleum refineries. The preparation involves the use of the conventional fermentation techniques, as well as the novel methods developed by various groups involved in genetic research. A bio-catalyst has to be stable under processing conditions. Bio-catalytic upgrading involves contacting petroleum feed with a bio-catalyst (usually dissolved in water), followed by the separation of the product and bio-catalyst for reuse. It is essential that after completion of the operation, the bio-catalyst can be recovered for reuse. This may be rather complex process because of the presence of several types of stable emulsions in the product mixture. Thus, bio-catalytic reactions require the presence of water without which bio-catalysts could not function. In this regard, microorganisms require much more water than enzymes.

The available information suggests that there are five key areas of the heavy feed upgrading where biological treatments can have an impact, i.e., the viscosity reduction, composition improvement, metal and sediments deposition control, de-emulsification and naphthenic acids removal (774). For example, bio-degradation has been finding some practical applications in the viscosity reduction during the enhanced oil recovery, as well as remediation of the soils and waters contaminated with petroleum products. The interests in bio-catalysis for petroleum refining applications has been noted for more than 50 years. This is supported by the extensive information in the scientific literature including the detailed reviews published recently (775,776). This included discussion on the origin and methods of the cultivation of numerous bacteria and enzymes suitable for bio-catalysis. The database enabling comparison of the microbes with enzymes as the bio-catalysts for upgrading petroleum feeds was also established (777). However, in spite of the decades of research, the downstream upgrading of petroleum fractions in petroleum refinery, using this method, is still in the preliminary stage of the development (775–779).

In any case, the available information indicates on the significant complexity of the bio-degradation processes. This can be illustrated using the mechanism of the HDN of carbazole proposed by Bressler and Gray (778). The similar complexity was supported by the metabolic pathway for the desulfurization of DBT (779). Thus, rather than to proceed to a simple primary product, numerous compounds possessing more complex structures than that of the reactant were identified in the product mixture. It should be noted that the complex mechanism is evident even for the relatively simple reactants such as DBT and carbazole. Definitely, the complexity will be much more evident for the structures such as the resins, asphaltenes and metal porphyrins. With respect to the bio-degradation mechanism, little is known about the network involving such complex structures. Therefore, for the bio-degradation of resins, asphaltenes and porphyrins, only a speculative network may be developed with great difficulties.

The anticipation of rather mild operating conditions was perhaps the main reasons that bio-catalysis has been attracting attention as the method for upgrading petroleum feeds. However, even for lighter feeds than those considered in the present review, this method is still in very early stages of development, in spite of the decades of research activities. Therefore, with respect to the upgrading of heavy petroleum feeds, only a brief account of bio-catalysis will be given to indicate the current state of the art and level of the uncertainties involved. However, several extensive reviews on various aspects of bio-catalysis published recently may be recommended to those who have more interest in this field (775,779–781).

The use of bio-catalysis for upgrading of heavy petroleum feeds has been based on the observation that some microbes selectively react with heteroatoms and metals, which are part of the porphyrin structures. The resultant species containing heteroatoms and metals are soluble in the aqueous phase and as such can be separated from the hydrocarbon phase. Also, the fused aromatic rings can be cleaved bio-catalytically and as such converted to smaller molecules (775,781–783). Furthermore, there has been the evidence that for paraffinic crudes, the content of the long chain alkanes can be decreased via bio-catalysis (781,784). As a consequence, the cold flow properties of the petroleum products (e.g., diesel oil, lubricants, etc.) could be improved. In addition, the viscosity of heavy feeds was reduced. This observation was the main reason for using bio-catalysis in the enhanced oil recovery (784). The drawback of bio-catalysis is the presence of the unwanted reactions such as the loss of carbon skeleton (780,785,786). On the other hand, microbes which do not cause the loss of carbon are not very effective. This was demonstrated in the early work published by Patras and Webster (787). These authors tested the model heavy feed consisting of the heteroring compounds and porphyrins which are typical of those present in heavy feeds. Although the rate of microbial bio-desulfurization (BDS) exceeded that of the microbial denitrogenation (BDN) by a factor of 900, these rates were much lower than the overall rates of the HDS and HDN observed during the hydroprocessing using conventional techniques.

The state of the art in bio-catalysis with emphasis on application in the upgrading of petroleum feeds has been recently reviewed by Le Borgne and Quintero (788). It was evident that the emphasis has been on the development of new and more efficient bio-catalysts. In this regard focus has been on model compounds which are present in heavy feeds. For example, DBT and alkylated DBTs have been used to study BDS (778,789,790), carbazoles to study BDN (791–794) and metal porphyrins to study bio-demetallization (794). For BDS, most of the work was carried out on the laboratory scale, although there are some reports of the operation on the pilot plant scale (795,796). In the latter case, only light petroleum fractions such as LCO, LGO, diesel oil, etc. have been used as the feeds (778).

The interest in BDN was prompted by the poisoning effect of the N-compounds present in the feed on hydroprocessing catalysts. Thus, the level of HDS can be significantly increased if the N-compounds could be removed via BDN (793). The aim of the experiments on BDN was to produce the N-containing product which could be readily converted to the nitrogen-free hydrocarbons. However, in spite of the large number of bio-catalysts screened, the formation of such product could not be achieved without the loss of carbon and the loss of the fuel value associated with it (780). Also, the interests in the bio-degradation of aromatic structures in distillates resulted from the limits on the content of aromatics in transportation fuels imposed by the specifications (784,795). In spite of all these efforts, the experimental evidence suggests that bio-degradation is commercially not viable even for the relatively light distillate feeds.

The bio-catalytic conversion of distillation residues has been receiving less attention. The recent studies published by Fedorak et al. (781,782,795,796) have focused on the screening of several bio-catalysts for the BDS, BDN and viscosity reduction. Testing was conducted using the Sandflat and Ratawi crude oils which were spiked with the model compounds such as DBT and phenathrene, in comparison with hexadecane, which was used as the carrier. The bio-transformation was dependent on the origin of the carrier, i.e., the conversions were greater in hexadecane than in crude oils. This was attributed to the greater viscosity of the latter indicating the involvement of diffusion effects. Some bio-catalysts were capable of splitting the $C_{AL}-S$ bonds typical of those present in asphaltenes, indicating on their potential applications for deasphalting of heavy feeds. (797).

The bio-catalytic conversion of heavy crude to light crude has been reported in several studies (798,799). For this purpose, the extremophilic bacteria which were stable at high temperatures, pressures and salt concentration, i.e., *Thiobacillus*, *Achromobacter*, *Pseudomonas* and *Sulfolobus*, have been tested (788). The *Caldariomyces fumago* chloroperoxidase enzyme and several hemoproteins were used for demetallization of asphaltenes (795). Some bacteria were capable of removing heteroatoms and metals, as well as transforming asphaltenes into lighter fractions. Generally, the transformations were dependent on the origin of heavy crude and they were different for different microorganisms. It is believed that the rate of bio-degradation will decrease with increasing viscosity of heavy crude because of the presence of diffusion phenomena. Thus, it is more difficult for bio-catalyst to access the reactive sites of reactants in more viscous medium.

In conclusion of this section it is again emphasized that even for a pure model compound, the biochemical pathways involved rather complex mechanism (778). This suggests that a significant additional research is necessary for the elucidation of this mechanism. The complexity is further enhanced for real feeds. Thus, even for distillate feeds, the metabolic pathways for bio-degradation are not clearly understood. It is, therefore, believed that considering the present "state of art" potential for the commercial application of bio-catalysis in upgrading heavy feeds in the petroleum refineries is rather remote.

Chapter 13

RESIDUES UPGRADING BY CATALYTIC CRACKING

The conversion of large molecules to distillate fractions can also be achieved thermally without gaseous H_2 being present as the reactant. For this purpose, both catalytic and non-catalytic processes have been used commercially. The former processes have been dominated by various modifications of the fluid catalytic cracking (FCC) process. In this process, H_2 is being produced as the unwanted by-product rather than being used as reactant. Initially, the virgin VGO was the predominant feed for the FCC process. The main objective was the conversion of such feeds to a high octane number gasoline. This could be achieved by increasing the content of aromatics during FCC. Gradually, DAO has been introduced as the feed, either without or with pretreatment. In some cases, pretreatment of the feed was necessary to extend catalyst life in the FCC units. It usually involved hydroprocessing to remove asphaltenes and metals, as well as to decrease the content of aromatics and heteroatoms. Detailed accounts of the conventional FCC process, with emphasis on the feeds such as virgin VGO and pretreated DAO, was given by Venuto and Habib (800).

Frequent references to the FCC process over the preceding chapters were made with regard to testing and development of the catalysts for hydroprocessing of VGO, HGO, DAO and other heavy feeds to produce feeds which can be subsequently processed using the conventional FCC process. So, hydroprocessing was usually conducted to avoid extensive modifications of the conventional FCC process without which the direct utilization of the unpretreated feeds would not be possible. Therefore, the feasibility of RFCC process has to take into consideration the costs of all necessary modifications of the FCC process and that of the addition of hydroprocessing units upstream of the latter.

The FCC process can be revamped with the aim to increase production of alkenes to satisfy the demand from petrochemical industry. In this case, the H_2-rich gases as by-product, may be a valuable feed for production of the high concentration H_2 and other products. Continuous modifications and improvements of the FCC process and catalysts have been made with the aim to process the conventional and/or heavier feeds directly, without requiring any pretreatment. No pretreatment was necessary for VGO or even for some DAOs derived from a sweet crude. The direct processing of the asphaltenes and metals containing feeds such as AR has been attempted as well. However, besides significant modifications of the FCC process, this would also require entirely new formulations of the FCC catalysts. So far, there is a limited information indicating a use of the unpretreated AR and VR as the feeds for the commercial RFCC process.

The extensive information on the FCC process is available in the textbook literature (1,655,800–806). For the purpose of this book, basic features of the FCC process are only given with emphasis on the modifications required to accommodate the non-traditional feeds. An objective of the discussion of the properties of the FCC and RFCC catalysts is to distinguish their functionalities from those of hydroprocessing catalysts, as well as to indicate all efforts, to develop novel RFCC catalysts.

13.1 CATALYTIC CRACKING PROCESSES

From commercial point of view, these processes have been dominated by various modifications of the conventional FCC process to suit feeds containing resins, asphaltenes, metals and other coke-forming precursors. The feasibility study indicated a viability of the asphalt residue treatment (ART) process, although so far such process is not being used on a commercial scale. There is a number of processes in various stages of development. However, these processes are beyond the scope of the book because the chances for their commercialization are rather remote. Thus, it was felt that some clarifications were necessary to compare the capabilities of the commercial RFCC units to upgrade "residues" with those upgrading real residues using hydroprocessing. In this regard, some classification of heavy feeds for the RFCC process is desirable. For this purpose, focus is only on commercial and/or nearly commercial processes.

13.1.1 FCC/RFCC process

The detailed flow sheet of the FCC complex, including all downstream and upstream units, is shown in Figure 13.1. This flow sheet was developed by Venuto and Habib (800). The process comprises three main sections, i.e., reactor-riser, regenerator with flue gas handling system and fractionation tower. The conventional FCC process using VGO as the feed is the heat-balanced operation where the heat released in the regenerator is sufficient to vaporize and catalytically crack the feed in the riser-reactor system. After preheating to a desirable temperature, the feed is mixed with hot regenerated/fresh catalyst at the base of the riser. Most of the cracking conversion occurs in the riser. The cracking is conducted at a near atmospheric pressure without H_2 being present. Because of the temperatures employed (e.g., above 500°C), the residence time of catalyst in the riser reactor must be short. Otherwise, an excessive coke deposition on the catalyst surface would complicate stripping of volatile products from catalyst and subsequent catalyst regeneration in the regenerator.

The reactor section on the top of the riser serves as the separator of catalyst dust from the products. Spent catalyst is passed through a stripping zone countercurrent to a flow of steam. The conventional design of the stripper used for FCC may require modifications to accommodate heavy feeds used for RFCC. For the latter, much greater amount of coke and carryovers are deposited on the catalyst. Also the volatility, solubility and reactivity of this coke and carryovers differ from those observed during the conventional FCC operations. After stripping volatile products, the deactivated catalyst is transferred from the reactor to regenerator and then, after regeneration, returned back to the former. A near atmospheric pressure in the reactor assembly is closely coupled with the operation of fractionator.

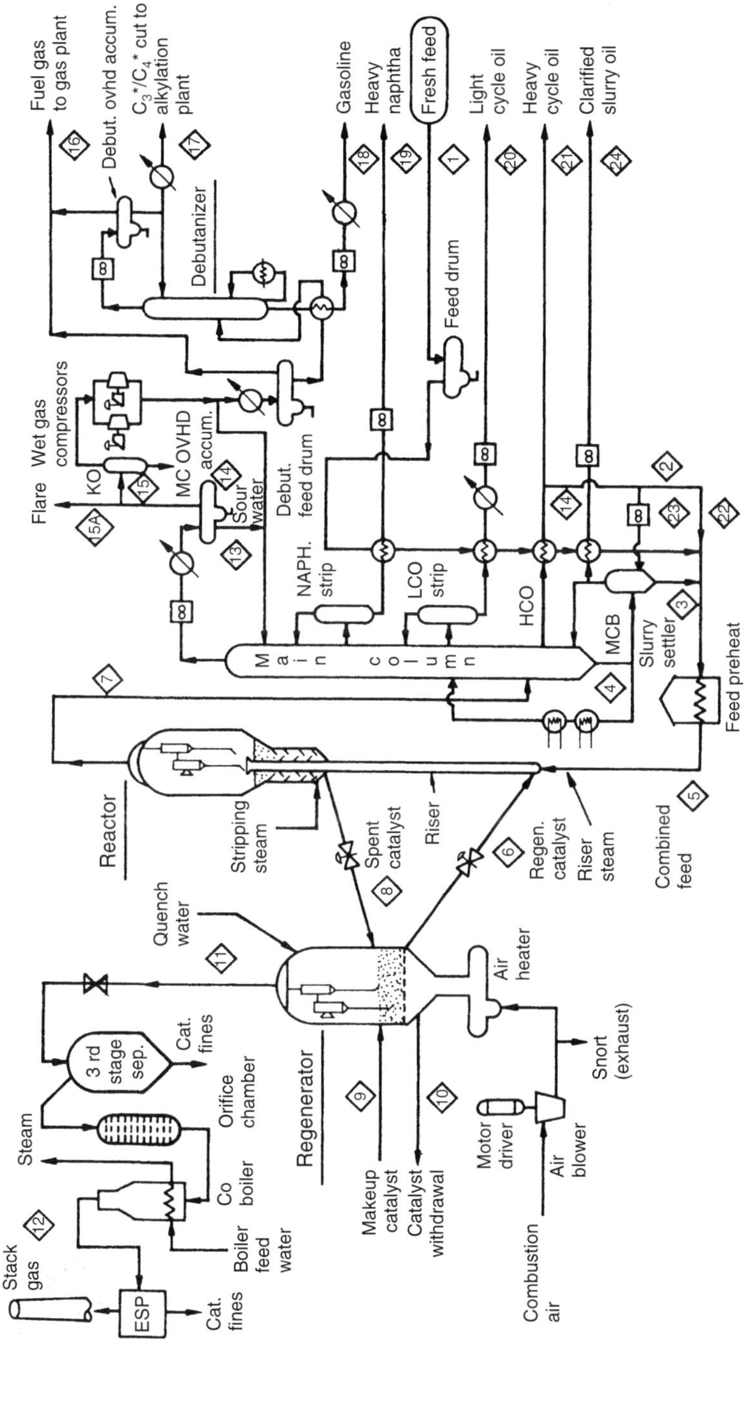

Figure 13.1 Flow sheet of FCC plant (800).

The catalyst regeneration involves the burn-off of coke deposited on the catalyst particles using either air or a mixture of air. Steam may be added to air to control temperature excursions. The organic coke on catalyst consists of the carryovers still left after stripping, delta-coke most of which originated from the CCR precursors in the feed and part of the coke formed during cracking reactions. With introducing more difficult feeds for FCC, the design of regenerator had to be modified because larger amounts of coke had to be removed. This could be achieved using a two-stage combustion (807). Otherwise, a desirable level of the catalyst activity recovery could not be attained. As the result of combustion heat, the temperature of particles is increased. Hot catalyst particles are then returned to the riser-reactor from regenerator. Beside their catalytic function, hot particles also carry the heat necessary for cracking reactions occurring in the riser.

Several modifications of the FCC/RFCC process have been used commercially. In every case, the process is modified with the aim to increase the feedstock flexibility. The extent of modifications increases with increasing content of contaminants, particularly metals, asphaltenes and the CCR forming constituents in the feed. Figure 13.2 compares one of the early FCC unit configuration with that modified for processing heavier feeds (808). For the latter, new design of stripper and separator of the catalyst dust from products can be noted. The rapid disengagement of the catalyst from the products is necessary to minimize secondary reactions causing the yield of dry gas to increase and that of liquid products to decrease. The design of regenerator has been evolving from bubbling beds to fluidized beds operating at much higher temperatures and gas velocities. Moreover, the reactor and regenerator are set to achieve a maximum pressure

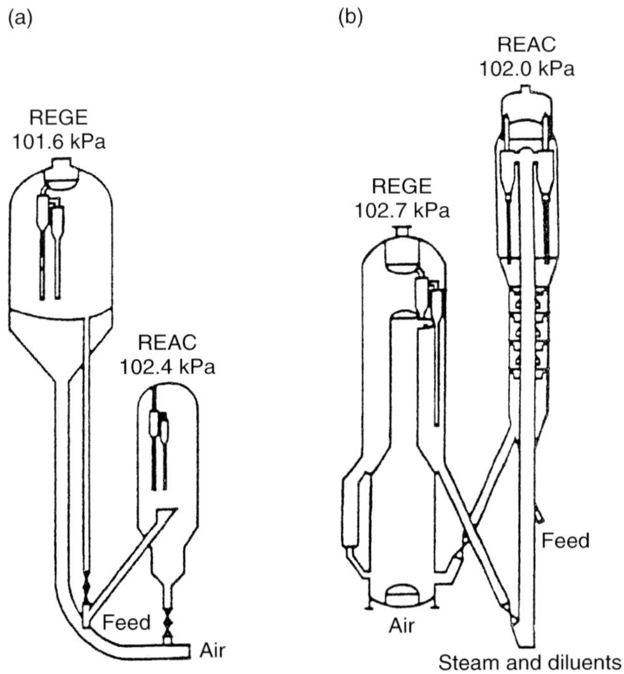

Figure 13.2 Early (a) and modified (b) FCC assembly (808).

differential. This enhances the coke combustion in regenerator and increases the yield of products in reactor. The details of revamping conventional FCC units with aim to improve efficiency and the feed flexibility were given by Wilson and Ross (809). In this case, attention was paid to most of the components which are part of the reactor and regenerator and which are requiring modifications.

13.1.1.1 Effect of delta-coke

As it was indicated above, in the case of the unconventional feeds, both modifications of the FCC/RFCC process and more active catalysts are equally important for achieving improved and steady performance. In this regard, O'Connor et al. (810,811) identified two main constraints during the operation of the RFCC units, i.e., the activity-limited conversion and delta-coke limited conversion. The former reflects the limitation in conversion and/or yield of the unwanted gaseous products in the riser, as well as a higher coke formation due to low catalyst activity. This problem can be alleviated either by using a more active catalyst or changing the catalyst/feed ratio. The delta-coke limitations reflect the situation in regenerator caused by the excessive amount of coke deposited on catalyst in the riser. Thus, the delta-coke is the difference between the amount of coke on spent catalyst entering the regenerator and that on the regenerated catalyst exiting the regenerator and entering the riser. The increased values of delta-coke require larger air supply to regenerator which may cause unwanted temperature excursions. To certain extent, this problem can be controlled in the riser-reactor either by decreasing the dehydrogenation activity of catalyst (due to Ni) or by increasing the catalyst/feed ratio. This suggests that a smooth and prolonged operation of FCC process using the feeds which are heavier than VGOs requires a fine-tuning and careful monitoring of several units and equipment, which are part of the substantially modified FCC process.

The coke on catalyst is a critical factor influencing operation of the FCC units. The content of CCR in the feed may be a measure of the coke forming tendency during the FCC operation providing that for the same amount of CCR in different feeds, the coke-forming tendency is similar. Apparently, most of the CCR in the feed ends up on the catalyst surface as delta-coke. Because the crackability of the different feeds having the same amount of CCR may be different, the total amount of CCR deposited on the catalyst as delta-coke may vary from feed to feed. However, practical experience suggests that these changes may not be so evident. Nevertheless, general trends indicate that the increased CCR in the feed causes the delta-coke to increase. Moreover, the content of CCR in the feed may be an indication of the presence of metals in the feed, i.e., higher content of CCR higher content of metals. This is indicated by the results in Tables 2.1 (25). In this table, a significant difference between the content of CCR and metals of different DAOs should be noted.

The catalyst/feed ratio is an important parameter for controlling coke formation. For example, in FCC units, where the reactor temperature is controlled by a valve on the regenerated catalyst standpipe, the amount of coke formed remains nearly constant because the catalyst/feed ratio can be automatically adjusted to maintain the system in the heat balance (801,802). Thus, as the coke formation was increased because of the increasing amount of CCR in the feed, the catalyst/feed ratio had to be reduced to offset the temperature rise in regenerator caused by the increased amount of coke on catalyst. Consequently, the overall conversion of the feed to desirable products was decreased. Other type of the FCC units employ feed preheat to control the reactor temperature. In

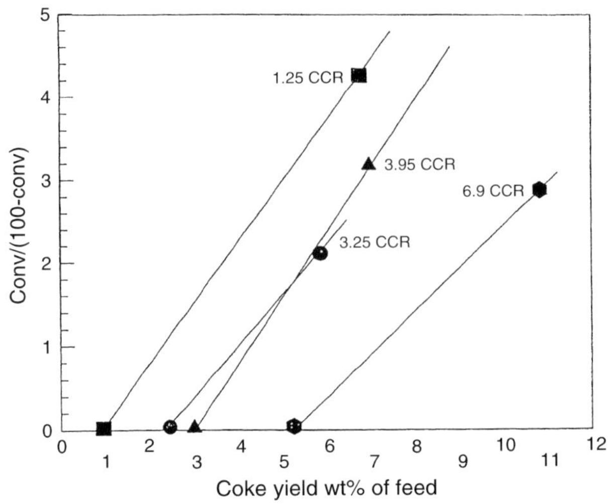

Figure 13.3 Effect of CCR content in feed on coke yield and conversion (801).

these units, an increase in the content of CCR in the feed resulted in the increase in regenerator temperature, a decrease in the feed preheat and an increase in coke yield, although little change in the overall conversion was usually observed.

It was evident that the increased content of CCR in the feed affected the overall conversions. The performance correlations in Figure 13.3 (801) show that for the same conversion, the yield of coke increased with increasing CCR content in the feed. The intercepts on these correlations represent about 75% of the CCR content in the feed suggesting that most of the CCR deposited on the catalyst surface as delta-coke. The increased delta-coke with the increasing CCR content translates into entirely different heat balance, i.e., for high delta-coke values, a catalyst cooler may be necessary. Moreover, a desirable level of combustion of coke on catalyst may not be achievable without modifications of the regenerator. This may include a two stage combustion with modified heat recovery system. A gradual revamping of the FCC units between 1950 and 1990 in response to the increasing content of CCR in the feeds was discussed by Evans and Quin (812). The same trends are also indicated by Figure 13.2 (808).

Attempts to use specific gravity of the feed for predicting the operation of RFCC units have been noted (813). However, rather poor correlations involving parameters such as conversion, yield of products and coke with specific gravity were obtained. The UOP K factor which takes into consideration specific gravity and the cubic average boiling point of the feed is more reliable because it correlates well with the paraffinicity and/or aromaticity of the feeds (814). Thus, the amount of coke on catalyst is expected to increase with increasing aromaticity of the feeds. At the same time, K factor decreased. Therefore, the value of the K factor is a more reliable indication of the delta-coke forming tendency of the feeds than specific gravity.

13.1.1.2 Effect of feed properties

It was indicated that RFCC process is the extension of the FCC process to accommodate heavier feeds. Thus, during early stages, FCC feeds were dominated by virgin VGOs with the CCR content well below 1 wt%. Therefore, such feeds could be used directly without any pretreatment. Later, the HGOs obtained by fractionation of the primary products obtained by coking VRs and heavy crudes/bitumen became important feeds for RFCC. The content of the refractory aromatic and heteroaromatic compounds in HGO was generally greater than in VGO, in spite of the similar boiling range of these feeds. Because of a higher aromaticity, the K factor of the HGO feeds was lower. This resulted in a higher content of CCR in HGO than in VGO. Therefore, for some HGO, a hydroprocessing step may be necessary before it can be used as the feed for conventional FCC units. A direct use of the residues such as ARs and VRs as the RFCC feeds is hampered by the presence of contaminants such as metals (Ni, V and others), asphaltenes and other components possessing a high CCR-forming propensity. It is, however believed that for some unique crude (e.g., North Sea Ekofisk), the content of metals is so low (Table 2.3) that the AR obtained from such crude can be used directly as the feed for RFCC.

13.1.1.2.1 Classification of feeds for RFCC

There seems to be a significant difference between properties of the residues which can be processed by RFCC and those suitable for hydroprocessing, e.g., AR, VR, topped heavy crude/bitumen. In fact, for most part, the former may not be residues in true sense. In spite of this, until now, the "R" in RFCC refers to residues. According to the original classification, a residue used as the feed for RFCC contained at least 5% of the fraction boiling above ~540°C (~1000°F). This classification was later modified by Stokes and Mott (801) who recognized that processing difficulties in commercial FCC units became evident when the content of CCR in the feed exceeded 1 wt%. Below this limit, little difficulty was experienced. However, there might be some uncertainty regarding the upper limit of CCR in the feed, although an information indicates that heavy feeds containing as much as 10 wt% of CCR could be processed (807). However, this required significant modifications of the regenerator-reactor assembly. In fact, Meyers (815) suggested that a major redesign, particularly that of the heat management system, was necessary as soon as the amount of CCR in the feed approached 4 wt%.

According to Table 2.1 (25), the DAOs obtained by deasphalting the AR or VR derived from heavy Zuata crude could be classified as residues (for RFCC) unless they were pretreated to decrease the content of CCR below 1%. At the same time, the VGO obtained from the same heavy crude was not residue, whereas the blend of VGO with DAO was. This indicates the need for modification of the FCC process to suit the blend, as well as the DAO. Table 13.1 compares DAO and VGO obtained from the Heavy Arabian crude (816). It is evident, that based on the content of CCR, the DAO from the propane deasphalting could be classified as a residue, whereas VGO not.

Another classification assigns an upper limit of metals (V + Ni) in the feed which can be still directly processed by RFCC. This limit is approached when the amount of metals in the feed reached 30 ppm (817). This could include either ARs derived from a high-quality sweet crudes or blends of AR with lighter fractions to maintain the metal content below 30 ppm limit. For example, among the crudes shown in Table 2.3 (29), only for the AR derived from the Light Arabian and North Sea Ekofisk crude, the content of metals may be within this limit. Published information indeed confirmed that the AR obtained from another North Sea crude could be used as the feed in RFCC units without any pretreatment (818,819). However, as the properties of this AR showed,

Table 13.1 Properties of VGO and DAO obtained from Heavy Arabian crude (816)

	VGO	DAO	
		C_3	C_4
Density at 15°C (Kg/L)	0.940	0.940	0.965
Viscosity at 100°C, cSt	11.6	na	89
CCR, wt%	0.75	2.6	5.3
Sulfur, wt%	3.1	2.7	2.6
Nitrogen, ppm	870	860	1440
Vanadium, ppm	< 1	0.9	3.9
Nickel, ppm	< 1	0.7	1.7

this was rather untypical AR obtained from a unique crude. Today, the availability of such crude is rather limited. Generally, the amount of metals in the conventional FCC feeds such as VGO is less than 1 ppm, while that in DAO depends on the deasphalting solvent (Table 2.1 and Figure 2.3). It is believed that this introductory comparison of the feeds was necessary in order to distinguish the capabilities of RFCC process to upgrade petroleum residues from those which can be readily upgraded via hydroprocessing. It is also necessary to emphasize that so far it may not be accurate to refer to heavy feeds for RFCC as residues.

Modifications of the process and catalyst development have been progressing with the aim to increase the allowable limits of metals and asphaltenes in the feeds for RFCC. In this regard, the extensive pilot plant testing using the AR containing almost 50 ppm of V + Ni over seven different FCC catalysts conducted by Gilbert (820) showed some promising results. The catalyst giving relatively low yield of coke and gas could be identified. In another study, an AR containing 27 ppm of V + Ni and 7 ppm of Fe was used together with several lighter feeds to study the effect of the feed properties on the RFCC catalyst stripping (821). It was observed that the efficiency of stripper increased with increasing content of the naphtha fraction in products, which was dependent on the properties of the feed. These studies (820,821) indicate that, using more advanced catalysts, the feeds with the content of metals and CCR above 30 ppm and 10 wt%, respectively, may be suitable for the direct upgrading using the RFCC process. However, this may still need to be confirmed on a commercial scale during long trials.

13.1.1.2.2 Pretreatment of feeds for RFCC

During preparation of the FCC feeds by hydroprocessing, the consumption of H_2 is an important parameter. This parameter indicates the level of HDAr. Thus, a high level of HDAr is necessary to decrease the content of the CCR precursors, as well as to achieve desirable yields of the distillate fractions of interest during FCC. It should be also noted that increasing the octane number by increasing content of aromatics in gasoline from FCC may not be a preferred option because of the change in fuel specifications. Among hydrocarbon groups, paraffins, naphthenes and aromatics exhibit a high-, intermediate- and low-cracking reactivity, respectively. Moreover, aromatics in the feed contribute to the formation of delta-coke on the catalyst surface. Therefore, the removal of aromatics from the FCC feeds may be desirable, particularly for the feeds of the coking origin (e.g., HGO). Also, the removal of nitrogen from the FCC feed to less than 1000 ppm may be necessary to diminish catalyst poisoning. In addition, the removal of nitrogen and sulfur is desirable for controlling the NO_X and SO_X emissions from regenerator.

Metals in the feed are effective catalyst poisons as well. In this regard, Ni is more than four times more effective poison than V. Depending on the origin of catalyst, poisoning can be evident at as low as few ppm concentrations of metals in the feed. Therefore, the catalyst used for hydroprocessing to produce suitable FCC feeds should possess adequate activities for HDAr, HDN, HDS and HDM. In some cases, the preparation of the FCC feeds may involve a multistage hydroprocessing.

The residues containing between 50 and 150 ppm of metals and 10–20 wt% of CCR would require entirely new catalytic phases and extensive modification of the process. The information on catalytic cracking of heavy feeds with a metal content exceeding 50 ppm is rather limited (822,823). Therefore, such residues would have to be subjected to either deasphalting or hydroprocessing step to be suitable for RFCC. For the former, the quality of DAO can be controlled by the type of hydrocarbons used for deasphalting as it is shown in Figure 2.3. This issue was discussed in detail by Bousquet and Laboural (816) who demonstrated that the deposition of delta-coke on catalyst in the riser increased with the increasing content of heavy fractions in the feed. Several options involving deasphalting for production of the FCC/RFCC feeds from various crudes were compared by Perries et al. (824). Figure 13.4 (825) shows the effect of the feed quality on the amount of delta-coke and distribution of coke on the catalyst. With the decreasing feed quality, i.e., increasing content of CCR, metals heteroatoms, etc., the amount of CCR coke and that of the contaminant coke, reflecting the amount of the feed metals deposited on the catalyst surface, increased. The amount of carryovers defined as the cat/oil coke exhibited little change, whereas the catalytic coke slightly decreased indicating increased catalyst deactivation with decreasing quality of the feed.

The blends of residues with lighter fractions may, at least, partially alleviate the problems encountered during the RFCC of the former. According to Alkemade and Paloumbis (825), residues containing as much as 50 ppm, 150 ppm and 15 wt% of Ni, V and CCR, respectively, could be successfully blended and used as the RFCC feed. Obviously, the amount of residue in the blend will decrease with the increasing

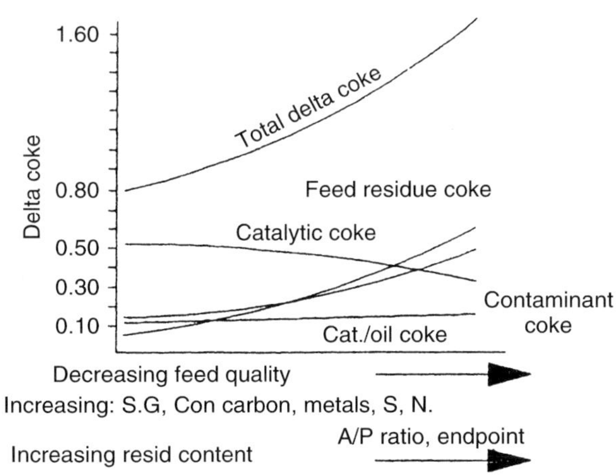

Figure 13.4 Effect of feed quality on distribution of coke (825).

Table 13.2 Properties of Kuwait AR and its products after hydroprocessing (345)

	AR	Product	
		A	B
Density, kg/L	0.997	0.961	0.944
Viscosity, cSt at 50°C	871		
Sulfur, wt%	4.30	2.62	0.56
Nitrogen, ppm	2670	2486	1960
CCR, wt%	14.2	8.5	4.8
Ni, ppm	21	8	6
V, ppm	69	11	12

Table 13.3 Methods for upgrading heavy petroleum feeds to RFCC feeds (817)

Metals (V + Ni)	CCR	Upgrading method
ppm	wt%	
<30	1–10	Can be used directly
30–150	10–20	Deasphalting; hydroprocessing (fixed bed multistage process)
150–500	>20	Deasphalting; hydroprocessing (moving/ebullated bed process)
>500	>20	Coking

content of contaminants in the former. Otherwise, ARs and VRs must be subjected to the multistage hydroprocessing to be suitable as the feeds for FCC. The results in Table 13.2 (345) showed the difference in properties of the AR and its products after a multistage hydroprocessing of the former. These products appear to be suitable as the feed for RFCC. The case A and B involved different catalysts suggesting that the catalyst selection for upgrading AR to the product suitable as the feed to be used directly in RFCC, units requires attention.

It appears that most of the petroleum residues cannot be used directly as the feeds for RFCC process. The extent of pretreatment to upgrade such residues to the feeds suitable for RFCC process increases with increasing content of CCR and metals. Based on the content of metals and CCR, Table 13.3 identifies four types of petroleum residues and/or heavy feeds, as well as some methods suitable for their upgrading. This table takes into consideration the guiding values proposed by Dale (817). Obviously, in the case of feeds, which can be used directly, the extent of revamping of the RFCC process would be significantly greater for the feed containing 10 wt% CCR than that containing 1 wt% CCR. Therefore, the objective of the upgrading is to decrease the content of metals and CCR below 30 ppm and 10 wt%, respectively, to minimize costs of revamping.

13.1.2 Asphalt residue treatment (ART) process

The ART process is included to indicate another extreme of catalytic cracking of heavy feeds. This process was designed for a partial upgrading of heavy crude at wellhead

(822,823). The objective of the process is the decrease in viscosity of heavy crude required for pipelining and at the same time to produce enough heat for EOR. The process has similar features as RFCC process. Thus, it consists of a reactor-regenerator assembly and fractionator. The feasibility study assumed the VR obtained from Cold Lake bitumen (Table 2.3) as the feed containing more than 300 ppm of metals. The nature of catalyst (ARTCAT) was not given, although it was assumed that 25% of the catalyst would have to be disposed. It is believed that the regenerability of this catalyst would be rather low because most of the metals in the feed would deposit on the catalyst surface. In this situation, the main role of the ARTCAT particles may be the transfer of heat from regenerator to reactor. Moreover, the V- and Ni-containing dust formed by attrition of the ARTCAT particles in regenerator would require special attention. Because of the limited operating data, the ART process is still in an early stage of development. Also, it is not clear if the ART process may be classified as the catalytically cracking process because of little information on the structure, composition and durability of the ARTCAT catalyst is available.

13.2 FCC/RFCC CATALYSTS

The development of catalysts for FCC/RFCC has been focusing on the acidic solids such as alumino-silicates and zeolites. At the beginning (early 1940s), various modifications of the amorphous silica–alumina were used in the first commercial operations. The significant enhancement in selectivity and regenerability of the FCC catalysts was achieved by incorporating zeolites in the matrix of silica–alumina, silica and silica clay, as well as by introducing pure zeolite as FCC catalysts. The benefits of these developments included an increase in selectivity for the fractions of interest (e.g., gasoline or diesel fuel), minimization of the production of gaseous by-products, increased resistence to coke deposition, good catalyst regenerability and long catalyst life (826). Because of the countless high severity utilization-regeneration cycles, FCC catalysts must be resistant to deactivation due to the loss of crystallinity caused by high temperatures and steam. In this regard, developments in the structure of FCC catalysts, from very early stages until now, were reviewed by O'Connor et al. (827). Figure 13.5 (828) shows continuous improvements in the performance of FCC catalysts from the beginning until 1990, i.e., before the RE-USY type zeolites were introduced. It is believed that during this period, the VGOs derived from conventional crudes were predominant FCC feeds. Further catalyst developments, with paying attention to RFCC, has taken off in early 1990s.

13.2.1 Structure of catalysts

The activity and stability of FCC catalysts was enhanced by several orders of magnitude by incorporating zeolites in matrix. The former provides cracking and hydrogen transfer activity. The matrix component of the catalysts is less catalytically active than the zeolite component. However, the presence of macropores in the matrix component favors conversion of large molecules via cracking. Matrix also serves as a binder of the catalyst components to increase the resistance to attrition and to attain a desirable density of the catalyst particles. Figure 13.6 depicts all components of the RFCC catalyst (825). In addition, RFCC catalysts contain various additives (804), i.e., Sb, Sn and $CaTiO_2$ as metal

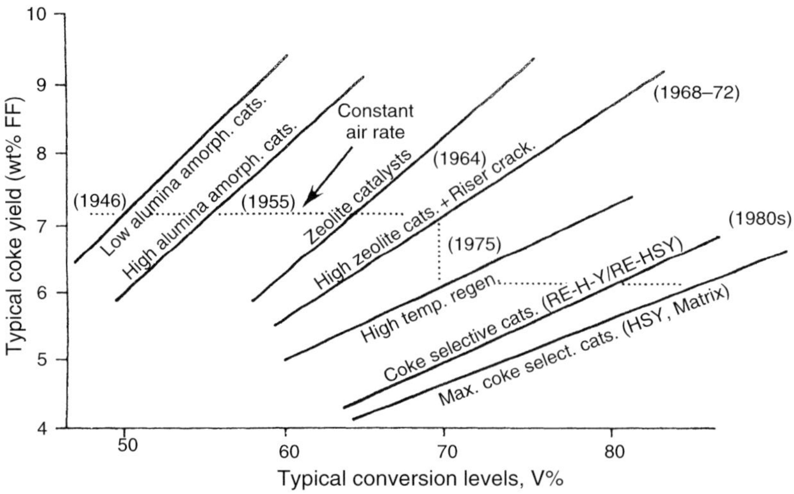

Figure 13.5 Evolution in structure of FCC catalysts before 1990 (828).

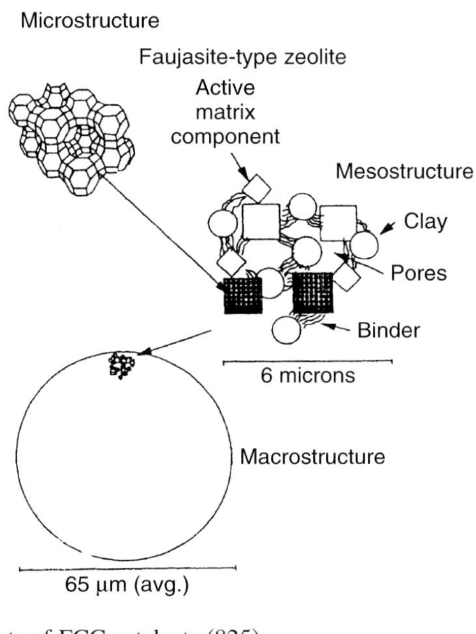

Figure 13.6 Components of FCC catalysts (825).

passivators and/or traps, Pt or Pd as the CO oxidation promoters and MgO, Pt/Al_2O_3, CeO_2/spinel, etc, as the SO_x reduction promoters. Additives such as ten-ring-pentasil (ZSM-5) are being added to increase octane number of products (829). If present in catalysts, this additive facilitates isomerization of *n*-paraffins to iso-paraffins. A typical example of the procedure used for preparation of the RFCC catalysts was described by Woltermann et al. (830).

The activity and selectivity of zeolites can be related to the strength and concentration of acidic sites. The isolated Al tetrahedra are the strongest acidic sites. This strength is decreasing as the number of the next nearest neighbors of the Al tetrahedra increases. Then, the ratio of strong acid sites to weak acid sites depends on the distribution of the Al tetrahedra in unit cell. The zeolite-containing catalysts for applications in RFCC units require a balanced number and strength of acidic sites. For example, a sufficient cracking activity is required to convert high molecular. At the same time, the amount of catalytic coke has to be kept at minimum. This may be achieved with the zeolite possessing an intermediate unit cell size. Thus, the unit cell size of the zeolite component of catalyst is an important parameter for controlling the activity and selectivity.

The unit cell size can be controlled by hydrothermal dealumination to obtain a wide range of the ultra stable (US) zeolites Y and rear earth (RE) exchange of zeolite. The USY zeolite is less active than the same amount of the REY zeolite. Based on the activity, stability and resistence to deactivation, the best performance is exhibited by RE-USY zeolites. For RFCC applications, an optimal unit cell size of this type of zeolites is in the range of 24.27–24.30 Å. The unit cell size may be controlled by the extent of dealumination (831). For example, lower unit cell size zeolites have higher Si/Al ratio. The activity and stability of zeolite also depends on the level of the rare earth exchange (829). For example, for the RFCC catalysts containing RE-USY zeolite, level of the rare earth exchange may range from 25 to 75%. The optimal level of exchange depends on the origin of feed and process conditions. Therefore, for the feed of interest, it has to be determined experimentally.

Apparently, alumina and silica–alumina were matrices of the choice. A clay was often added to provide desirable density of the catalyst. Moreover, when kaoline (as the clay) was added to RFCC catalyst, the passivation of the detrimental effect of V was significantly improved (832). For RFCC applications, pore size distribution of matrix requires a special attention (833). Thus, matrix must possess macropores (>150 preferable between 500 and 1000 Å) to allow deposition of metals and to assist in the cracking of large asphaltenic molecules. Although macropores may have a low activity, but they act as a guide for large molecules to access active sites in micropores. A sufficient mesoporosity (50–150 Å) must also be maintained to ensure cracking of the naphthenic structures and alkyl substituents from aromatic rings, whereas microporosity (20–50 Å) is necessary to crack straight alkyl chains. A novel matrix with the distributed matrix structures (DMS) is a non-porous material (834). Then, surface area and mesoporosity are associated with the external surface and spaces between crystallites. The aggregate morphology is responsible for macroporosity.

A superior performance of the RFCC catalyst was observed when the RE-USY zeolite component was exchanged with the trivalent rare earth cation (826). Such formulation of the catalyst was much more resistant to hydrothermal deactivations than the H-USY zeolite (835). In addition, the zeolite-containing RFCC catalysts exhibited excellent resistence to deactivation by metals such as V, Ni, Fe and others. At least 30% of US zeolite were necessary to achieve desirable catalyst performance. The selection of the optimal amount of zeolite possessing required acid strength and density may be influenced by the origin of feed (836). Generally, the amount of delta-coke on catalyst decreases with the increasing zeolite/matrix ratio, as it is shown in Figure 13.7 (825). Similar decline in coke yield and dry gas yield was also observed when the zeolite/matrix surface area ratio increased (801). Consequently, for matrix alone, both coke and dry gas

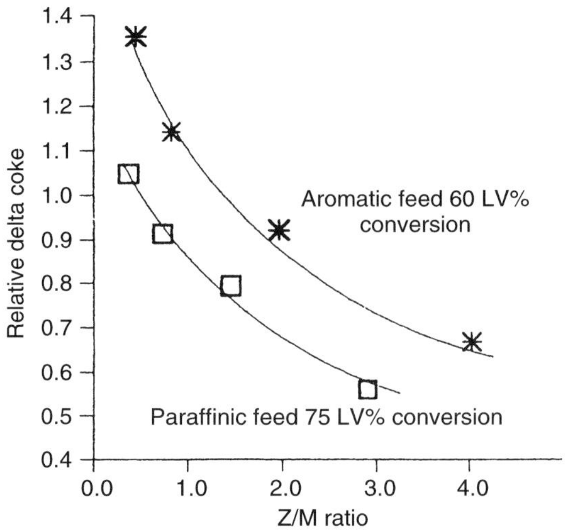

Figure 13.7 Effect of zeolite/matrix ratio on delta-coke (825).

yields increased with increasing surface area. The coke selectivity of an active matrix shall be balanced with the zeolite activity to minimize adverse effect on heat balance, conversion and yield of desirable products. The selectivity of catalysts (e.g., to LPG, gasoline and LCO) can be adjusted by catalyst structure as well (837). For example, a high rare earth (either high matrix or low matrix), high accessibility and high matrix catalysts produce more LCO. High rare earth catalysts gave always lower yields of dry gases. The selectivity for light alkenes and iso-alkanes can be significantly enhanced by some additives (829,833). For example, Figure 13.8 shows effect of the addition of ZSM-5 on octane number of the gasoline fraction obtained from the FCC products.

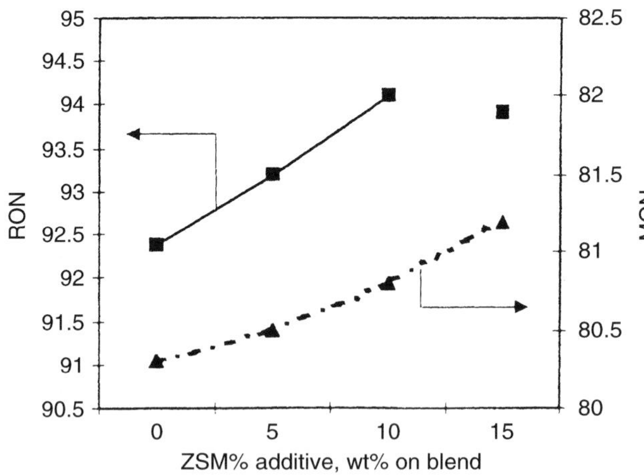

Figure 13.8 Effect of ZSM-5 zeolite on octane number of gasoline (833).

13.2.2 Selection of catalysts for RFCC

There is a wealth of information on the design and selection of catalysts for FCC using conventional VGO (800–806). Based on the decades of practical operation of commercial units, the selection of catalysts for such applications is rather straightforward. Complications began with the introduction of more problematic feeds. An optimal matching the properties of feeds with RFCC catalysts is necessary to minimize yields of coke and maximize that of the products of interest. Otherwise, excessive amount of coke would affect performance of the regenerator. In the following text, the focus is only on the feeds in their properties approaching AR.

Detailed evaluations of RFCC catalysts may be necessary to achieve an optimal match of the catalyst properties with that of the feed to maximize yields and quality of the products. It is essential that procedures used for catalyst evaluation provide reliable data before a final decision for catalyst selection is made. With the progress from FCC toward RFCC, the testing and selection of catalysts has reached new level (838). For example, in the case of virgin VGOs, a reliable database could be established using the conventional microactivity test (MAT). However, for RFCC, more reliable data on catalyst evaluations are obtained using the circulating bed pilot plant. This method provided database necessary for catalyst selection. Morever, the results from accelerating aging obtained under these conditions correlated well with those obtained in commercial units, whereas those obtained by MAT did not. Another method for catalyst testing, i.e., short contact time residue test, consists of the fluidized bed of hot (~1000 K) catalyst particles into which a preheated feed is injected (839). The design of reactor ensured an optimal mixing and cracking at contact time approaching that in commercial units. Compared with MAT, this test could properly rank catalysts with the different zeolite-to-matrix ratios. Modifications of the testing protocol was also necessary to obtain reliable data on the resistence of RFCC catalysts to deactivation (840). Thus, testing under conditions typically used for FCC catalysts provided erroneous results.

It has been indicated that the design of RFCC catalysts has to take into consideration the presence of diffusion limitations due to large molecules in the feed (841). This was confirmed by the decrease in conversion with increasing CCR content of the feed (842). Such limitations are much less evident during the conventional FCC. Among nine RFCC catalysts investigated by Lu et al. (843), the best performance, measured by the yield and quality of products, as well as by the amount of coke on catalyst, was exhibited by the catalyst with the highly accessible large pores matrix. This was in agreement with the results of Yanick et al. (844) who also observed that the amount of coke on catalyst increased with the increasing portion of small pores in the matrix.

A correlation between the pore volume and the matrix surface area, could be established and used for the optimizing catalyst performance for different residues (845,846). At the same time, there was neither correlation between pore volume and zeolite surface area nor between total surface area and the yields of products. For the AR containing 60 ppm of V + Ni and 9.7 wt% CCR, a little decrease in the gasoline yield with surface area decrease from ~300 to ~120 m^2/g was observed (847). At the same time, the amount of coke on catalyst almost tripled. However, a rapid decrease in the gasoline yield was observed below 120 m^2/g. The gasoline yield decrease could not be offset by increasing the catalyst/feed ratio.

Another study on catalyst development for RFCC of AR was published by Tian (848). In this case, several ARs of varying content of VR were investigated. The catalysts were

Table 13.4 Composition of RFCC catalysts used for testing (819)

	A	B	C	D	E	F	G	H
Surface area, m^2/g	138	100	193	164	155	152	109	134
Matrix area, m^2/g	75	41	86	40	36	30	22	38
Zeolite area, m^2/g	63	59	108	124	119	122	87	96
Zeolite/matrix ratio	0.84	1.44	1.28	3.10	3.31	4.07	3.95	2.53

Table 13.5 Properties of AR obtained from North Sea crude (819)

Density, kg/L	0.922
CCR, wt%	2.8
Sulfur, wt%	0.48
Nitrogen, ppm	2000
Vanadium, ppm	3
Nickel, ppm	2

tested in a commercial unit by replacing the in-use catalyst by regular makeup rate until new catalyst reached 70% of the inventory. A high activity and tolerance to metals in the feed (~40 ppm of V + Ni) of the RFCC catalyst was attributed to the presence of the alumina-based matrix possessing a weak acidity in macropores allowing an easy deposition of metals. In addition, a strong acidity in micropores of the zeolite component aided cracking reaction.

Table 13.4 shows composition of the eight catalysts which were tested with the aim to select the most suitable for the RFCC of the AR (375+) properties of which are shown in Table 13.5 (818,819). So, this was rather unique AR obtained from a high paraffinic sweet crude. The test was conducted in a pilot plant unit. The objective of this study was to find an optimum balance between the zeolite surface area and matrix surface area to maximize the yield of gasoline. The results showed that the best catalyst should have a high zeolite surface area and a low matrix surface area. Thus, with respect to the yield of gasoline, catalysts E and F were most active. However, there was a minimum allowable matrix surface area below which the catalyst did not function. This was observed for catalyst G. Therefore, the common advice that the RFCC catalyst should have a large matrix surface area and a moderate zeolite surface area was not confirmed. This may not be so surprising when the origin of the AR is taken into consideration.

13.2.3 Deactivation/regeneration of RFCC catalysts

The increased levels of coke and carryovers in the spent RFCC catalysts compared with the spent FCC catalysts require significant modifications both of the reactor-riser-stripper and regenerator. It is, therefore, crucial that an optimal operation of the stripper part of the reactor is maintained. Also, the greater content of metals in the RFCC feeds than that in FCC feeds adds to the complications of regeneration. In some situations it became more beneficial to transfer a portion of the spent RFCC catalyst to an offsite company for regeneration. For this purpose, the DEMET process was developed and successfully used on a commercial scale (849). The SO_2 and NO_X emissions from regenerator could be minimized by using novel catalysts as well (850). The catalysts must also be resistant to mechanical attrition to minimize losses due to the formation of fines. All these requirements could only be fulfilled with the entirely new materials compared with the traditional catalysts used for FCC of the virgin VGOs. With the advanced catalysts employed, the contact time in the riser of less than 4 s was sufficient to achieve desirable conversions.

13.2.3.1 Effect of coke

The vigorous research on deactivation/regeneration of the FCC catalysts, predominantly focusing on VGO feeds, has been conducted for decades (851,852). Both, deactivation by coke and metals, were investigated. The former started with the work of Voorhies in 1945 (362). Initially, the coke deposition on catalyst is rapid. However, for FCC catalysts and VGO feeds, the amount of coke was not time dependent (853). Thus, the amount of coke on catalyst after 0.05 s contact time was similar as that after 4.6 s (854). Consequently, the catalyst activity exhibited little change. For such feeds, a large portion of the coke originated from cracking reactions. It is believed that in this case, both free radicals and carbocations were involved during the catalytic coke formation. For high CCR content feeds, a part of the delta-coke may be formed by fouling in addition to the catalytic coke. Aromatic compounds and high molecular weight components in the feed, with boiling point exceeding the temperatures usually employed in RFCC reactors, are the cause of the formation of this portion of delta-coke.

The acidic nature of the RFCC catalysts supports their strong interaction with N-bases in the feed. It has been established that N-bases adsorb on active sites and as such cause the loss of activity. Because of the prolonged adsorption and low reactivity under typical RFCC conditions, N-bases may convert to coke. This form of deactivation of the RFCC catalysts may be minimized by maintaining the content of nitrogen in the feed below 1000 ppm. Then, N-bases in the feed may be considered as a source of coke in addition to catalytic coke, CCR coke and carryovers.

As one would expect, the FCC catalyst deactivation by coke deposition in the riser differed from that of a RFCC catalyst (855). For the latter, the portion of feed which could not evaporate was an important contributor to the amount of coke on catalyst. As it was indicated earlier, the CCR content is a semi-quantitative measure of this portion of the feed. During FCC, coke on catalyst increased along the height of riser; however, during RFCC, the coke reached a maximum near the mix zone. A linear correlation between the content of CCR in the feed and the amount of the insoluble (CH_2Cl_2) coke deposited on catalyst was established (856). Moreover, the insoluble coke from heavier feeds was more alkylated.

Gilbert (847) reported that the effect of coke on the products yield from FCC may differ from that of RFCC. For the latter, an AR was used as the feed. In this work, the yield of gasoline increased with increasing amount of coke on catalyst to a maximum followed by the decline in the yield. The increase in the gasoline yield with coke build-up was complemented by the decrease in the formation of gaseous products. However, a decline in conversion with increasing coke formation was observed, although it could be more than offset by increasing the catalyst/feed ratio.

13.2.3.2 Effect of metals

In some studies (825), the reference is made to a contaminant coke, although it reflects the amount of metals which are deposited on catalyst surface from the feed. It is believed that it is more appropriate to refer to this portion of "coke" as to metal deposits.

The optimal amount and type of matrix should be selected to provide adequate resistance to deactivation by metals, particularly by V and Ni. Matrix is usually resistant to deactivation by V, but in combination with Ni and a high surface area matrix, an effective dehydrogenation catalyst can be formed. As the result of metal deposition, the unit cell size and surface area of zeolite decline (857). In some feeds, the presence of Cu and Fe was noted (858). In this case, the deactivating effect of Cu approached that of Ni. In the FCC mode, the catalyst may tolerate as much as 3000 ppm of metals (V + Ni), whereas in the RFCC mode the amount of metals on the novel catalyst, e.g., containing RE-USY zeolite component, may exceed 10 000 ppm (858).

Deactivation of the RFCC catalysts by coke and metals occur simultaneously. For the advanced RFCC catalysts, the amount of deposited coke represents only a fraction of that for amorphous catalysts when the same amount of V + Ni deposited on catalyst was considered. This suggests that with progress in developing more efficient catalysts, the contribution of metals relative to coke, to the overall activity loss became more evident. This was supported by an increased attention being paid to the deposition of metals on the surface of RFCC catalysts and the associated loss of activity and selectivity.

A close examination of the performance lines in Figure 13.3 (801) shows that besides different intercepts, the lines also have slightly different slopes. This is caused by the metals in the feed which deposited on catalyst surface. For example, for the feeds containing 1.25 and 3.25 wt% of CCR, the content of metals (V + Ni) on the equilibrium catalyst was 3100 and 6400 ppm, respectively. However, similar slopes for 1.25 and 3.95 wt% CCR feeds would indicate a similar content of metals in spite of the different CCR content in the feed. Therefore, metals contribute to the loss of conversion expressed per unit of the coke yield.

After deposited on the catalyst, V, Fe and Ni increase coke formation by catalyzing dehydrogenation reactions (859). As the result of this, the H_2/CH_4 ratio in dry gas increased. It was noted above that V in combination with Ni increased the dehydrogenation activity of matrix, although V alone had little detrimental effect. With respect to dehydrogenation, Ni is about four times more efficient than V (800). However, it was established that V caused a permanent deactivation by destroying the structure of zeolite component of the RFCC catalysts. In this regard, the RE-USY zeolite was much more resistant to this form of deactivation than the REY and USY zeolites.

During regeneration, V is converted to V_2O_5 which in contact with steam reacts to give H_3VO_4 (860). The adverse effect of these compounds on catalyst activity can be minimized by the traps containing alkaline earth metals. The fluidization and attrition properties of the traps' particles must be similar as those of the RFCC catalyst particles. The traps can react with H_3VO_4 to give corresponding salts, i.e., MgO and CaO to $Mg_2V_2O_7$ and $Ca_2V_2O_7$, respectively (861). Among several traps (e.g., MgO, CaO, CeO_2, $MgTiO_3$, $CaTiO_3$, $Li_2Ti_2O_7$ and $ZnTiO_3$) the best efficiency was exhibited by MgO and CeO_2 (862). The dual particle approach developed by GRACE Davison employs the V trap particles together with the catalyst particles (825). The effectiveness of this method depends on the ability of V to migrate from the catalyst particles to the trap particles. The performance of these traps may be affected by sulfur competition (804). For some traps, the enhancement in the selectivity for gasoline and light olefins was noted (863).

Attempts have been made to develop methods for passivating the deleterious effect of metals on the catalyst performance by injecting additives into the feed. The one developed by Phillips Petroleum Co. involved deposition of antimony on the FCC catalyst by injecting the antimony-containing compounds into the feed (817). This resulted in a significant reduction in the production of H_2 and coke. At the same time, the yields of distillate fractions was increased. The percent change of the mean values for 10 months before and after passivation are shown in Table 13.6 (817). It is rather evident that, for the latter, all important operating parameters improved. A similar observation was made by Zheng et al. (864) after the addition of 20 wt% of kaolinite to the base RFCC catalyst. Because of the excellent passivating effect, the conversion was increased from 46 to 68%. At the same time, the yield of coke and dry gas decreased. The results obtained on a laboratory scale were confirmed during commercial trials using a residue containing almost 10 wt% CCR and about 50 ppm of metals (V, Ni and Fe). In fact, introducing the novel catalyst resulted in a decreased catalyst consumption in the commercial unit.

The extent of deleterious effect caused by Fe in the feed depends on the structure of RFCC catalysts (865,866). Generally, Fe has been observed to deposit on the exterior of catalyst particles. If present in catalyst, Fe mixed with silica, Ca and other contaminants forming low melting temperature phases, which accelerated sintering and slowed down diffusion of large molecules into particles interior. On the other hand, in the presence of alumina, the deleterious effect was not evident. Zhu et al. (867) showed that the distribution of Fe is more important factor than total amount in determining deactivation by Fe. Thus, large Fe molecules deposition occurring predominantly on the exterior, affected crystallinity, surface area and pore volume of the catalyst. This resulted in a diminished performance of the catalyst. At the same time, small Fe-containing molecules were distributed more evenly and as such their detrimental effect was much less evident.

A simultaneous removal of coke and metals in the regenerator would be an ideal case of regeneration of the RFCC catalysts. One information suggests that this may be accomplished by adding a metal-getter additive with a higher settling velocity than RFCC catalyst (868). Metals are then removed by solid–solid interaction. The uncontaminated portion of the catalyst, forming a discrete zone on top of a more dense zone of the additive, is recycled to the operations. The additive, with the adsorbed metals is withdrawn from the lower part of the fluidized zone.

Table 13.6 Change in operating parameters after passivation (817)

Parameter	Change, %
Fresh feed rate	+11.9
Conversion	+5.4
Yield of gasoline	+20.4
Octane number	+1.3
Yield of coke	−9.1
Yield of H_2	−61.2

13.3 EMISSIONS FROM RFCC PROCESS

Both gaseous and solid emissions from RFCC require attention. The former include CO, SO_x and NO_x from regenerator. Most of the gaseous emissions from the reactor, i.e., H_2S and NH_3, are removed in the process by scrubbing liquids. The spent RFCC catalyst represents a part of the solid emissions besides the fines from reactor and regenerator which passed the system of cyclones. Most of such fines can be removed by water scrubbing.

13.3.1 Gaseous emissions

The amount of gaseous emissions from RFCC depends on the properties of feeds. Higher content of sulfur and nitrogen in the feed translates into higher content of these contaminants in the coke on catalyst. Compared with conventional FCC, the content of sulfur and nitrogen in coke on RFCC catalysts is several times greater. Therefore, for the latter catalysts, entirely different catalyst formulations and the design of regenerator are necessary for achieving a desirable emission control.

13.3.1.1 CO emissions

As it was indicated above, the excessive amount of delta-coke on RFCC catalysts requires an increased air supply and modification of the regenerator to achieve a complete combustion, i.e., yielding CO_2 as the only combustion product. Usually, the CO_2/CO ratio of the gas exiting dense part of the fluidized bed in the regenerator would approach 1.0. Therefore, gas phase after-burning of the gas mixture is necessary to maintain CO emissions within the acceptable limits. This, however, could lead to the unwanted temperature increases which would have adverse effects on the material of regenerator. The problem could be alleviated by combining various additives with RFCC catalysts which would enhance conversion of CO to CO_2 at temperatures usually employed in regenerator This topic was recently reviewed by Chester (869).

During some early attempts, transition metals (e.g., Co, Ni, Fe, Cr and others) were added to enhance combustion, although some of these metals catalyzed dehydrogenation reactions in the reactor. Therefore, their overall beneficial effect was not evident. The addition of zeolite component to FCC catalysts coincided with the attempts to use noble metals such as Pt, Rh, Ru and Pd as the additives to RFCC catalysts enhancing CO oxidation. In this case, the best economic value was obtained with Pt. The addition of Pt included either exchange with zeolite using an exchangeable form of Pt or a direct addition of a reactive Pt compound to the feed. In the latter case, the Pt content in catalyst gradually increased until about 2 ppm level of Pt was attained (870). The gradual increase in the Pt content was complemented by the increase in the CO_2/CO ratio until almost complete elimination of CO was achieved. Table 13.7 (869) compares the effect of various additives on the CO_2/CO ratio. The superior activity of Pt compared with the other metals is rather evident to the catalyst.

13.3.1.2 SO_X emissions

In the reactor-riser, sulfur in the feed ends up in liquid products, gaseous effluent from reactor and in the coke on catalyst. For sulfur, 50–60% appears in liquid products and 35–45% appears as H_2S in gases from the reactor. The H_2S is scrubbed from the gases and is converted to elemental sulfur. This suggests that the SO_X emissions from reactor

Table 13.7 Effect of additives on reactor/regenerator performance (869)

	Base		Pt		Cr	Ni	Mn
Metal, ppm		0.5	1	5	10 000	1000	10 000
Conver., vol.%	71.8	75.2	75.8	71.0	60.8	76.5	56.4
Coke, wt%	3.2	4.6	3.3	3.0	3.0	5.7	2.2
H_2 factor[1]	26	23	27	23	23	142	23
CO_2/CO at 944 K	1.7	3.4	11	80	3.8	1.9	1.8

[1] H_2 factor $= 100 \times H_2/(C_1 + C_2)$

are near zero even in the case that the gaseous products were used as fuel. Therefore, only 2–5% of sulfur in the feed end up in coke on catalyst.

The article published by Vassalos et al. (850) summarizes the situation in SO_X emissions from FCC plants before the stringent environmental regulations were introduced. These authors concluded that conventional methods for SO_X removal from flue gas, such as scrubbing and flue gas desulfurization added to the costs of the operation and as such were not economical. Since that time, significant progress in curtailing SO_X emissions have been made by developing novel methods. Efforts have been focusing on the agents, which after adding to FCC catalysts, can enhance the rate of the following reactions (804,871)

Regenerator	Reactor/stripper
$S + O_2 = SO_2$	$MSO_4 + 4\ H_2 = MS + 4\ H_2O$
$SO_2 + 0.5\ O_2 = SO_3$	$MSO_4 + HC = MO + H_2S$
$SO_3 + MO = MSO_4$	$MSO_4 + 4\ H_2 = MO + H_2S + 3\ H_2O$
	$MS + H_2O = H_2S + MO$

These reactions represent the scheme according to which SO_3 can be fixed in regenerator when the FCC catalyst contains an additive with a high affinity for SOx. The product of these reactions is MSO_4. Subsequently, MO is regenerated in the reactor/stripper assembly with the aid of H_2 and hydrocarbons (HC). The agent can be either incorporated with catalyst or blended with catalyst as separate particles. The summary of information on this topic in the scientific and patent literature on the pre-1990 agents was given be Scherzer (804). It included magnesia either impregnated or added with catalyst particles, rare earth such as yttrium and lanthanum, rare earth mixtures and rare earth minerals, cerium promoted spinel, etc. The results published by Hirschberg and Bertolacini (872) are shown in Table 13.8. They show that combination of catalysts with MgO and Ce components was the most efficient for SO_X removal. However, the recent information indicates that an agent containing $Ce/V/MgO \cdot MgOAl_2O_4$ formulation yielded the most active and stable SO_X removing agent (873). It should be noted that in some cases promising results obtained on a laboratory scale could not be verified on a commercial scale. Therefore, the choice of testing method used for the additive development requires attention, as it was pointed out by Magnabosco (874).

Table 13.8 Effect of addition of cerium, alumina and magnesia on SO_X emissions (872)

Catalyst	SO_X removed at 900 K (μL)
CAT baseline	0
CAT + 25 ppm Ce	44
CAT + 5% Al_2O_3	54
CAT + 25 ppm Ce + 5% Al_2O_3	127
CAT + 25 ppm Ce on Al_2O_3 + 5% Al_2O_3	130
CAT + 500 ppm MgO	113
CAT + 500 ppm MgO + 25 ppm Ce on MgO	235

A comprehensive approach to the SO_X emissions control was developed by Francis et al. (875). In this case, focus was on both reactor and regenerator. A specially developed agent (RESOLVE 950) added to the catalyst increased the sulfur removal in the reactor. Thus, more sulfur removed as H_2S translates into less sulfur in liquid products and in the coke on catalyst. Moreover, the agent was capable of trapping SO_X as a sulfate in the regenerator. After returning to reactor, the sulfates are converted to H_2S and the original agent according to the scheme shown above.

13.3.1.3 NO_X emissions

During FCC, about half of nitrogen in the feed appears in liquid products, whereas less than 10% leaves reactor with gaseous products in the form of NH_3. After water scrubbing of the gas, more than half of this NH_3 remains in the water. The unaccounted nitrogen is part of the coke on catalyst. Therefore, the nitrogen content of coke is comparable to that of sulfur, in spite of the significantly lower content of the former in the feed. Moreover, the removal of nitrogen from the coke in regenerator is slower than that of carbon and sulfur (876). Consequently, the content of nitrogen in coke may gradually increase during repeated cracking-regeneration cycles.

The emissions of NO_X consist of NO, NO_2 and N_2O. They all originate from the fuel nitrogen. Thus, little NO can be formed via oxidation of the nitrogen in air. The NO is the main NO_X component in the flue gas exiting regenerator. Part of the NO is formed by oxidation of the precursors such as NH_3 and HCN, which were released during coke pyrolysis. Another part of the NO is formed during the gas–solid reaction of O_2 with the nitrogen in coke. Most of NO_2 is formed via oxidation of NO after being released into atmosphere. In commercial RFCC regenerators, the fraction of nitrogen released as N_2 varies from about 3 to 25%, depending on the design of regenerators and operating parameters (877). A portion of the N_2 was released from coke directly, whereas the other was formed during the reduction of NO using either CO or unconverted carbon on catalyst. Small amounts of N_2O exit regenerator as well. Presence of the unconverted NH_3 and HCN in flue gas would indicate malfunctioning of regenerator.

Experimental evidence confirmed that chemistry of the NO_X emissions can be controlled by relative concentration of CO and O_2, as well as the design of regenerator (878–880). General observations show that NO_X emissions increased with increasing O_2 concentration in the oxidation gas. Some reports suggest that even under complete

CO combustion conditions at the grid of the regenerator, the NO_x emissions could be decreased (878). Thus, in the O_2-rich zone at the grid, almost complete conversion of CO to CO_2 was achieved. However, the axial variation of the O_2, CO and CO_2 concentrations showed that above this zone, the CO concentration increased and that of O_2 decreased. Consequently, the removal of NO via reduction with CO was enhanced. It is believed that this option for the NO_X control is rather dubious because it requires nearly ideal involvement of several operating parameters at the same time. On the other hand, the O_2 enrichment improved SO_x removal via using a SO_x transfer agent (870). A negative effect of the O_2 enrichment may be an increased mobility of V (881).

Other studies showed that the NO_X emissions increased after the addition of Pt to RFCC catalysts with the aim to enhance CO oxidation (882). This resulted from the diminished availability of CO required for NO_X reduction. This suggests that catalyst development for the NO_X emission control should focus on the material which promote NO_X reduction simultaneously with those promoting CO oxidation. In this regard, Stockwell and Kelkar (882) observed a significant NO_X reduction by the addition of active oxides of a proprietary nature. To identify the most suitable metals, Stockwell (883,884) used a Gibbs energy minimization principle to determine the driving force for the following reaction:

$$NH_3 + MO_x = MO_{x-2.5} + NO + 1.5\ H_2O$$

This reaction takes into consideration NH_3 as the main primary product and/or precursor to NO_X formation. The metals for which $\wedge G < 0$ are promoters of oxidation, whereas those with $\wedge G > 0$ tend to promote NH_3 conversion to N_2. The former metals comprise the Pt group metals. An enhanced rate of oxidation leads to overheating the catalyst particles. This favored NO_X formation. At the same time, base metals such as those present in the water–gas–shift catalysts promote conversion to N_2. For these metals, the overheating effect was much less evident. In a similar study, Rainer et al. (885) compared the commercial Pt-containing RFCC catalyst with several catalysts of a proprietary nature and observed about 30% NO_X reduction, whereas CO emissions remained at the same level. The testing on a laboratory scale followed by commercial trials which confirmed the improved performance.

Another primary product formed in regenerator simultaneously with NH_3 includes HCN. The approach used by Stockwell (883,884) is very useful providing that the conversion of HCN to NO occurs at a similar rate as that of NH_3. Under oxidation conditions of regenerator, this indeed may be the case. Then, the development of the $deNO_x$ catalysts should take into consideration the following scheme:

$$2\ NH_3/(2\ HCN) + 1.5\ O_2/(1.5\ O_2) \rightarrow N_2 + 3\ H_2O/(N_2 + 2\ CO + H_2O) \quad [13.1]$$

$$2\ NH_3/(2\ HCN) + 2.5\ O_2/(2.5\ O_2) \rightarrow 2\ NO + 3\ H_2O/(2\ NO + 2\ CO + H_2O) \quad [13.2]$$

A good deNO_X catalyst would enhance the rate of reactions (13.1) relative to reactions (13.2). Based on this scheme, the involvement of CO during NO reduction could be minimized. Another reaction leading to N_2 formation involves the SCR of NO with the aid of NH_3. Additional benefits could be obtained using a catalyst possessing high activity for the reduction of NO by CO. Such complexity in developing efficient deNO_X catalysts was realized by Barth et al. (886). In accordance with these requirements, these authors developed a novel catalyst comprising layers (2.5 nm thickness) of zeolite MCM-22 spaced apart by clusters of MgO–Al_2O_3. This additive could be used in combination with the Pt-based promoters for simultaneous control of CO and NO_X emissions.

13.3.2 Solid emissions

Solid emissions from RFCC process include fines removed from the regenerator off-gas using an electrostatic precipitator. Such fines cannot be reused in refinery. Another portion of fines, passing the reactor cyclones may appear in the main column bottom as a clarifying slurry oil. Unless recovered by cyclonic separators, these fines end up in the tank bottom. Part of the spent catalyst is removed during the operation and replaced by fresh catalyst to maintain a steady catalyst activity. During withdrawal, the equilibrium catalyst is mixed with a small portion of fresh catalyst. Attempts have been made to apply a procedure for separating fresh catalysts from the equilibrium catalyst and return the former to the riser (887,888). Continuous efforts aimed at containing solid emissions within the plant are the reason that today's units emit only a fraction of the solids compared with the first units (812). Further reduction is anticipated as national standards governing the emissions of particulates have been evolving (41).

13.3.2.1 Properties

Properties of the spent catalysts from FCC/RFCC units differ markedly from those of spent hydroprocessing catalysts. Thus, particle size of both fines and spent catalysts withdrawn from the operation is much lower than that of the spent hydroprocessing catalysts. The flammability of the fines and spent catalysts from FCC/RFCC units is very low because of the much more refractory nature of coke. Moreover, the amount of coke on catalysts is much lower. It is also unlikely that such catalysts can release toxic gaseous compounds in contact with water. However, the fines passing the reactor cyclones and subsequently removed from the clarifying slurry oil require de-oiling before utilization and/or storage.

With respect to the storage and disposal, leachability of the spent catalysts from FCC/RFCC units requires attention, although for some time, such catalysts were classified as non-toxic solid waste (849,889). However, with progress from FCC toward RFCC, the content of contaminant metals in spent catalysts significantly increased, i.e., from about 3000 ppm to more than 10 000 ppm of V + Ni. The TCLP procedure applied to the former catalysts in early 1990s indeed confirmed non-hazardous nature according to the environmental regulations which were in effect at that time, although the elevated amount of Ni, V and Sb in leachates should be of a concern (889). Moreover, the presence of various additives (e.g., Sb, Sn, Ce and others) is another factor requiring attention. Inevitably, for the spent RFCC catalysts, the amount of metals in the leachates is expected to be significantly greater. It is, therefore desirable that spent RFCC catalysts are handled as toxic wastes, although such a designation may not be currently in effect.

Tables 10.3 and 10.5 show the analysis of spent catalysts from hydroprocessing of heavy feeds. These results show that the content of contaminant metals is more than an order of magnitude greater than that in the spent RFCC catalysts. Table 10.7 shows the analysis of leachates from spent hydroprocessing catalysts. A high content of some metals in leachates deserves attention, although these metals may not be regulated. In comparison with RFCC catalysts, particle size of the spent hydroprocessing catalysts is significantly greater. Moreover, the amount of coke on the later catalyst is at least an order of magnitude greater. Smaller particle size and amount of coke favor leachability of metals from the RFCC catalysts compared with spent hydroprocessing catalysts. It is therefore desirable that the former catalysts are handled according to the procedures prescribed by environmental authorities to avoid future liabilities.

13.3.2.2 Disposal and utilization

The lesser contamination of the spent RFCC catalysts compared with the spent hydroprocessing catalysts should not be a reason for a more relaxed approach to the disposal of the former. A detailed analytical evaluation of the spent RFCC catalysts prior to storage/disposal should always be undertaken. In this regard, different procedures may need to be applied for the evaluation of fines and that of the equilibrium catalysts. As general practice, the spent RFCC catalysts are usually disposed of in sanitary landfills. This required a pretreatment to avoid spent catalyst fines flying around. Also, lime had to be added to keep pH of the ground water within acceptable limits. Other treatments prior to landfilling may be necessary. It appears that landfilling has been increasingly more difficult because of the associated liabilities. Thus, the refiner's liability only ends after the solid waste was taken over by the authorized disposal facility. The detailed account of the environmental and safety apsects as well as utilization schemes of the RFCC catalysts was given by Pavel and Evans (889).

A predominant use of the spent FCC/RFCC catalysts is in the area of construction materials such as the filler for asphalt, as well as the production of bricks and cement. This, however, requires that the spent catalyst has no adverse effect on the final properties of the product. The industrial utilization of spent FCC catalysts may be influenced by location of the refinery. In this case, proximity of the source of spent catalysts to the point of utilization is important because it determines transportation costs (890). An information suggests that in one region the consumption of spent FCC catalysts exceeded their supply (891).

Novel utilization schemes may be necessary to deal with the spent RFCC catalysts containing more than 10 000 ppm of metals. Because of the predominant deposition of metals on the external surface of particles, utilization of the spent catalysts in the production of construction materials may require some attention unless most of the metals were removed. Indeed, magnetic separation methods have been evaluated as potential tools for removing metals from the spent RFCC catalysts (886,887).

13.4 PATENT LITERATURE

The patent applications submitted to US Patent Office since 2000 until now complement the information on the development of catalysts for RFCC available in scientific literature. For the purpose of this book, patent selection focused on the new catalyst formulations designed with the aim to increase activity, selectivity and stability of the catalysts. Significant efforts have been devoted to the design of catalysts for some specific applications, i.e., passivation and emissions control.

It is evident that new sources of materials for the catalyst preparation have been identified. For example, several inventions of Stockwell et al. (892–894) focused on zeolite microsphere FCC catalysts with a unique morphology. One formulation comprised a porous matrix and crystallized zeolite freely coating the walls of the pores of the matrix. The catalysts are formed from microspheres containing a *meta*-kaolin. However, the invention is not restricted to macroporous catalysts having a non-zeolite matrix derived solely from kaolin and *meta*-kaolin. Thus, any alumina source which has the proper combinations of porosity and reactivity during zeolite synthesis and can generate the desired catalyst macroporosity and morphology can be used. The desired morphology comprises a matrix which is well dispersed throughout the catalyst, and the macropore walls of matrix are lined with zeolite and are free of binder coatings. Another high-porosity formulation can be formed by the in situ crystallizing an alumino-silicate zeolite from reactive microsphere comprising *meta*-kaolin and hydrous kaolin. Any calcination of the reactive microsphere before reaction with a zeolite-forming solution is done at low temperatures to ensure that hydrous kaolin is not converted to *meta*-kaolin. A similar catalyst formulations was also disclosed by Hurley (895).

A catalyst for RFCC comprises a fluoride salt of a divalent or trivalent metal (896). This salt is dispersed in an inorganic oxide matrix together with a crystalline alumino-silicate zeolite. A divalent metal is at least one selected from the group consisting of Mg^{2+}, Ca^{2+}, Sr^{2+} and Mn^{2+}, and the trivalent metal is at least one selected from the group consisting of La^{3+}, Ce^{3+}, Mn^{3+} and Y^{3+}. Another catalyst formulation consisted of a carbonate of bivalent or trivalent metals, crystalline alumino-silicate zeolite and a mixture of Al and Si compounds (897).

Yaluris and Rudesill (898) disclosed several formulations of an oxidative catalyst/additive having a mean particle size of between 50 and 200 μm. One consisted of an acidic oxide support promoted with at least one metal promoter selected from the group of alkali and/or alkaline earth metals, transition metals, rare earth metals, Group Ib metals, Group IIb metals and metals such Ge, Sn, Bi and Sb. Another contained either a hydrotalcite or spinel promoted with Pt-group metals in addition to the same metals used to promote the acidic oxide support.

13.4.1 Metal passivation

Deleterious effects of metals (V, Ni, Fe, etc.,) in the feed can be offset by adding various passivators either directly to the catalyst or blending the adsorbent particles with catalyst particles for simultaneous circulation. In the latter case, properties of the adsorbent particles have to be selected to ensure that their fluidizing properties approaches those of the catalyst particles. These additives combine with the metals from the feed and therefore act as “traps” or “sinks” so that the active component of the cracking catalyst is protected (899). These metal contaminants are removed along with the catalyst withdrawn from the system during its normal operation and replaced with fresh metal trap and makeup catalyst. Several inventions involved development of absorbents with a high affinity for metals such as V and Ni, have been noted. For example, the metal trapping material developed by Vierheilig (900,901) consisted of a mixed metal oxide compound comprising Mg and Al compounds of a non-hydrotalcite origin. The ratio Mg to Al may vary from about 0.6:1 to about 10:1.

The blend of adsorbent and spent catalyst may be transferred to regenerator either directly or separated before regeneration. The invention of Pankaj et al. (902) was designed for the latter option. In this case, it is possible to handle up to 30 ppm of Ni in the feed and up to 10 000 ppm Ni in the equilibrium catalyst. The spent catalyst and coked adsorbent are separated in a stripper-separator located below the reactor, at a low temperature, as two distinct layers of spent catalyst and coked adsorbent. This depends on the particle size, density and differences in their minimum fluidization velocity, by using steam. Thus, heavier particles of spent catalyst are settled at the bottom of the separator and the lighter particles of coked adsorbent at the intermediate location of the separator. Each stream is then transferred to separate regenerator.

The RFCC catalysts with a high resistence to deactivation by metals have been focus of the attention as well. The one invented by Shibuya et al. (903) had a high resistence to deactivation by V and Ni and inhibited the formation of H_2 and coke. Moreover, it had excellent cracking activity and bottom oil-treating ability. The RFCC catalyst comprised a compound of bivalent and/or trivalent metals in the form of a carbonate and an inorganic oxide matrix and the compound dispersed therein together with a crystalline alumino-silicate zeolite.

A novel procedure for preparation of the RFCC catalyst which can passivate deleterious effect of V and Ni was described by Madon et al. (904,905). The catalyst was made from microspheres containing kaolin, a dispersible boehmite alumina and a sodium silicate and/or silica sol binder. The kaolin portion contains a calcined hydrous kaolin. Calcination to *meta*-kaolin and formation of in situ zeolite by treatment with sodium silicate yielded a catalyst containing Y-faujasite. At the same time, the dispersible boehmite was transformed into a transitional alumina. The transitional alumina may contain a delta alumina phase as well. The catalyst can be used to crack residues or residues-containing feeds. In these applications, the alumina phase formed from the dispersible boehmite passivates Ni and V contaminants.

13.4.2 Sulfur removal during FCC

The SO_X emissions from regenerator may be decreased by maximizing the removal of sulfur in riser to produce a low sulfur product. In this regard, a number of inventions have been reported. For example, the sulfur removal agent disclosed by Chester et al. (906,907) consists of a porous molecular sieve containing a metal in an oxidation state above zero within the interior of the pore structure of the sieve. The molecular sieve may be a zeolite beta or zeolite USY or intermediate pore size zeolites ZSM-5. Non-zeolitic molecular sieves such as MeAPO-5, MeAPSO-5, as well as the mesoporous crystalline materials such as MCM-41, may be used as the sieve component of the catalyst. Metals such as V, Zn, Fe, Co and Ga are effective. The metal-containing sieves or zeolites are used in combination with the active catalytic cracking catalyst (normally a faujasite such as zeolite Y). For the same purpose, Vierheilig and Keener (908,909) described calcined hydrotalcite like compounds and/or mixed metal oxides as additives. The additives can also comprise one or more metallic oxidants and/or supports, as well as a mixed metal oxide compound containing Mg and Al in the ratio of of Mg/Al between 1 and 10. Beswick et al. (910) showed that a zinc-containing FCC cracking catalyst consisting of a zeolite and a matrix, which contains silica and Zn incorporated by the matrix, can be used to produce the products with a low sulfur content. Zinc in the form of Zn oxide supported on a silica-alumina matrix was able to reduce the sulfur level of the cracked product in the presence of a zeolite catalyst. Hu et al. (911) disclosed a cracking catalyst composition comprising a zeolite containing component and about 0.2% Na_2O or less. A 15% reduction of sulfur was achieved because of the Na_2O presence. Another catalyst composition reducing the content of sulfur in FCC products comprised 5–55 wt% metal-doped anionic clay, 10–50 wt% zeolite, 5–40 wt% matrix alumina, 0–10 wt% silica, 0–10 wt% of other ingredients, and balance kaolin (912).

Roberie et al. (913) disclosed that the sulfur content of liquid cracking products could be reduced using a catalyst containing a high content of V. In this case, a V compound was introduced into the feed under steady-state conditions and containing an equilibrium fluid cracking catalyst. The sulfur reduction is achieved even in the presence of other metal contaminants (e.g., Ni and Fe) on the equilibrium catalyst.

The SO_X emissions can also be controlled by various solids which are introduced into the reactor with catalyst. Vierheilig (914) identified a wide range of anionic clay compounds as potential additives to control SO_X emissions. For example, hydrotalcite-like compounds can be made by a heat treating a non-hydrotalcite-like compound and then hydrated to form hydrotalcite-like compounds with an increased hardness and/or density. Another invention includes a trifunctional catalyst consisting of CeO_2 and V_2O_5 which act as oxidative catalyst and cerium oxyfluoride acting as structural promoter (915). The catalyst and promoter are dispersed over the absorbent. The absorbent is spinel-based composite oxides having a general formula of $MgAl_{2-x}Fe_xO_yMgO$, where the x is 0.01–0.5 and y is 0.2–1.2. The raw material for forming the chemical compound containing rare earth cerium is hamartite powder. The trifunctional catalyst of was highly efficient for absorption and desorption of SO_X in flue gas from regenerator. At the same time it promoted oxidation of CO and reduction of NO_X.

13.4.3 Catalysts for CO and NO_X emission control

The development of catalysts and/or agents for simultaneous control of NO_X and CO emissions undertaken by Yaluris et al. (916–918) assumed an incomplete combustion in regenerator to ensure the presence of enough CO to achieve a high conversion of NO_X to N_2. The effluent from regenerator was then transferred to boiler for a complete conversion of CO to CO_2. Under these conditions, little back oxidation of N_2 to NO_X was observed. The one composition contained no zeolite. It consisted of about 5.0 wt% of acidic metal oxide (e.g., silica–alumina, lanthana–alumina and zirconia–alumina). In this case, a metal component included an alkali metal and alkaline earth metal. In addition, an oxygen storage metal oxide containing at least 0.1 ppm of Pd and about 0.5 wt% Rh and/or Yb oxide. The compositions could be used as separate additive particles circulated along with the catalyst inventory. The mean particle size of the additive was between 50 and 200 μm. In similar manner the composition containing a Y-type zeolite and a NO_x reduction component consisting of particles of a zeolite with pore size ranging from about 3 to about 7.2 Å and the SiO_2/Al_2O_3 molar ratio of less than 500 (919) can be used. As the alternative, the NO reduction zeolite particles can be incorporated into the cracking catalyst as an integral component of the catalyst. Another formulation contained a Y-type zeolite, and a particulate NO_X reduction composition containing ferrierite zeolite particles bound with an inorganic binder (920,921). The ferrierite zeolite particles can also be incorporated into the cracking catalyst as an integral component of the catalyst. Other forms of zeolites consisted of beta, MCM-49, mordenite, MCM-56, Zeolite-L, zeolite Rho, errionite, chabazite, clinoptilolite, MCM-22, Offretite, A, ZSM-12, ZSM-23, omega and their mixtures (922).

Different compositions of the catalysts for controlling NO_X emissions during FCC were invented by Kelkar et al. (923–926). They comprised an acidic oxide support, cerium oxide, praseodymium oxide and an oxide of a metal from Groups Ib and IIb such as Cu, Ag and Zn, as well as precious metals such as Pt and Pd. In the case that the acidic support contained alumina, precious metal was distributed in the central interior of particulate additive, whereas metal oxides as a shell around the precious metal. These compositions could be used either as an integral part of the FCC catalyst particles or as separate particles admixed with the FCC catalyst (927).

An increased NOx formation in the presence of Pt, which promotes CO oxidation could be suppressed by the addition of small amounts of Rh and Ir (928). In addition, promoters comprising Pd and Ru promoted the complete combustion of CO in regenerator without causing the additional formation of NO_X. The ratio of Pd/Ru varied from 0.1 to about 10. Similar effects were observed when Rh and Ir or their mixture were deposited on an alumina support (929). Similarly, the composition invented by Peters et al. (930) increased the oxidation of CO, while having little effect on NO_X emissions. It consisted of an acidic support, an alkali or alkali earth metals, Ce oxide and Pd. A good control of NO_X emissions during cracking of the above-average nitrogen content feeds could be maintained using the solid invented by Peters et al. (931,932). It consisted of an acidic support, an alkali or alkali earth metals, Ce oxide, a transition metal (e.g., Cu or Ag) and Pd. The composition disclosed by Kelkar et al. (933) contained a mixed oxide of Ce and Zr in addition to the same components which were part of the solid described by Peters et al. (931).

Some non-noble metal catalysts were found to be effective for simultaneous NO_X and CO control. One consisted of Cu and/or Co and a carrier (934). The carrier can be, for example, hydrotalcite-like compounds, spinels, alumina, zinc titanate, zinc aluminate and zinc titanate/zinc aluminate. The other contained a Cu–Al and Ce–Al complex oxides and Al_2O_3 support (935).

Chapter 14

CARBON-REJECTING PROCESSES

In every chapter of this book, the reference was made to the non-catalytic processes used for the upgrading heavy petroleum feeds. In this case, the VRs, particularly those obtained from the most problematic heavy crudes were the feeds of the primary interest. The carbon rejection from such feeds can be achieved either thermally or via deasphalting. The detailed review of these processes is out of the scope of this book. It is, however, believed that a cursory account of these processes may aid readers in identifying the essential factors and/or reasons for selecting non-catalytic route over catalytic upgrading such as hydroprocessing. It has been indicated that there is a limit on the metals and asphaltenes contents above which the catalytic upgrading of heavy feeds becomes economically unattractive. Thus, above such limit, the costs of the catalyst inventory cannot be offset by the increased yields of liquid products obtained using the hydrogen addition processes compared with that obtained from the carbon-rejection processes.

The scale of the operation is an important factor which has to be considered while comparing the hydroprocessing processes with the carbon rejection-processes. For the most problematic heavy feeds, practical experience suggests that the carbon-rejecting processes have been the processes of choice for large scale operations, e.g., more than 100 000 lbb/d upgrading. However, in the case of the Syncrude plant in Canada, the EXXON fluid-coking process comprising two coker-burner assemblies has been operating in parallel with two ebullated bed reactor systems with the daily production of more that 200 000 b/d and 50 000 b/d, respectively. Such integration of hydroprocessing with coking may result in the increased production of liquids. For this purpose, the VR obtained after vacuum distlillation of the products from the ebullated bed reactor may be additionally processed in the coker.

14.1 THERMAL PROCESSES

Thermal processes used for the carbon rejection from heavy feeds are based on the thermal treatment of heavy feeds yielding distillates, as well as the H_2 and hydrocarbons containing gases. During thermal treatment, part of the asphaltenes is converted to distillates, whereas the other part is rejected together with most of the metals in the solid residue referred to as coke. The lower stability of the primary products obtained during thermal processing requires a catalytic hydroprocessing step, particularly when such products have to be pipelined to refineries. This has to be considered in the overall comparison of the schemes for carbon-rejecting processes with those used for hydroprocessing. The former operate at temperatures approaching 800 K either in the absence or presence of H_2. At the same time, the temperatures employed during hydroprocessing rarely exceed 700 K. In the case of the carbon-rejecting processes, the operation is conducted at a near atmospheric pressure. The heat required for thermal cracking is supplied by the combustion of by-products such as the H_2 and hydrocarbons containing gas and/or of the solid residue such as coke.

The direct rejection of the carbon from heavy feeds in the absence of catalyst can be achieved thermally using visbreaking, hydrovisbreaking/hydrocracking and coking processes. It has been, generally, observed that the large-scale upgrading plants usually employ coking technology. In this case, gaseous by-products may be valuable feedstocks for the production of hydrogen, synthesis gas, electricity, etc. Moreover, the petroleum coke may no longer be a refinery waste. In fact, its value can be significantly enhanced via gasification to synthesis gas which can be converted to similar commercial products as those obtained from natural gas and/or gaseous by-products (e.g., H_2, NH_3, chemicals, etc.). In addition, the synthesis gas may be used as the feed for the FT synthesis to produce the blending component having a high cetane number. At the same time, the wax by-product of the FT synthesis may be hydroisomerized to a high-quality lube base oil and middle distillates. This suggests that coke may displace natural gas as the source of H_2, chemicals and high-quality blending components for transportation fuels. Then, the additional natural gas can become available for more noble applications.

Table 14.1 (936) compares the yields of distillates from thermal upgrading processes. The extension of visbreaking through hydrovisbreaking to thermal hydrocracking resulted in the significant increase in the yield of distillates. The yield was further increased by deasphalting of the solid pitch from the high conversion hydrocracking. With respect to the yield of distillates, flexi-coking process is more efficient than the delayed coking process. Moreover, the former has a provision for the conversion of the most of coke to fuel gas which can be used on-site for generation of electricity and steam. It is believed that only traces of asphaltenes and metals were present in the distillates obtained from thermal processes except for the HCR-deasphalting case as shown in Table 14.1. For the latter, the distillates also include DAO suggesting that their hydroprocessing may require more severe conditions because of the presence of asphaltenes and metals (~50ppm).

Table 14.1 Yield of products (wt%) from thermal processes (936)

Yield	Visbr.	Hydrovisbr.	HCR	Coking		HCR/DEA[1]	Ebullated
				Delayed	Flexi		
Gas	1.7	2.5	3.0	10.7	10.0	3.0	4
Naphtha	6.1	3.0	3.7	13.9	13.7	3.7	13
Gas oil	10.4	34.5	53.3	42.6	54.4	67.3	66
Total dist.	16.5	37.5	57.0	56.5	68.1	71.0	79
Residue	81.8	60.0	40.0	32.8	22.0	26.0	17
Form of res.	tar	Liquid oil pitch	Solid pitch	Coke lumps	Coke powder	Solid pitch	Vacuum residue

[1] DEA – deasphalting

For comparison, the product distribution from ebullated bed reactor is also shown in Table 14.1 (936). With respect to the yield and distribution of the products, the advantages of hydroprocessing compared with the thermal processes are rather evident. Moreover, the VR from the ebullated bed may be subjected to additional treatments, i.e., by coking or deasphalting, suggesting that the yield of distillates may exceed 80%. However, for the assessment of technical and economic feasibility, several other factors have to be considered when comparing the hydroprocessing option with thermal options.

A comparison of the catalytic processes with non-catalytic processes for heavy feeds upgrading was made by Qabazard et al. (937). It was evident that among the non-catalytic processes, visbreaking and coking processes have been gaining on importance. With respect to the yields of distillates, the latter may be considered as high conversion processes, whereas visbreaking is a low conversion process as it is shown in Table 14.1. Among coking processes, the delayed coking and fluid-flexi-coking processes have been used on the commercial scale for decades (1,2,19,939). With respect to the overall conversion of heavy feeds, hydrovisbreaking and hydrocracking are between visbreaking and coking processes. The simplified schematics of the visbreaking and coking processes are shown in Figure 14.1. It is noted that the features of the hydrovisbreaking and hydrocracking reactors may be similar as those of the slurry bed reactors shown in Figures 8.1 (437) and 8.10 (464).

14.1.1 Visbreaking and hydrovisbreaking

In the case of visbreaking, a VR may be converted to distillates (about 20 wt%) and tar. According to Figure 14.1, visbreaking involves passing the VR at about 750 K through a flasher to produce two fractions. The conditions in the flasher, i.e., temperature and residence time, ensure relatively low conversion to distillates. The lighter fraction from the flasher enters the atmospheric distillation to produce the gasoline and gas oil fractions, whereas heavy fraction enters vacuum distillation to produce VGO and visbreaking tar. The latter is sometimes termed as the heavy fuel oil. Viscosity of the visbreaking tar is much lower than that of the corresponding VR. This simplifies utilization of the former either during combustion or gasification because of the improved pumpability.

During visbreaking, there may be a limit on the yield of liquid products beyond which their stability may be affected. This limit can be extended by conducting visbreaking in the presence of H_2, so called hydrovisbreaking. In this case, both the yield and stability of liquid products are enhanced. Thermal hydrocracking, an extension of hydrovisbreaking, is conducted in the absence of catalyst and at higher temperatures and H_2 pressures that during the latter. This results in the increased yields of distillates. This is confirmed by the results in Table 14.1, i.e., the gradual increase of the yield of distillates in the following sequence: visbreaking < hydrovisbreaking < HCR. The VR termed as pitch is the by-product of thermal hydrocracking. The pitch may be further processed using either coking or deasphalting to produce additional liquids. However, because of the different colloidal properties compared with the virgin VR, its utilization may require a special attention.

14.1.2 Coking

There has been decades of practical experience in the operation of the delayed coking and fluid-flexi-coking processes. These coking processes dominate petroleum industry

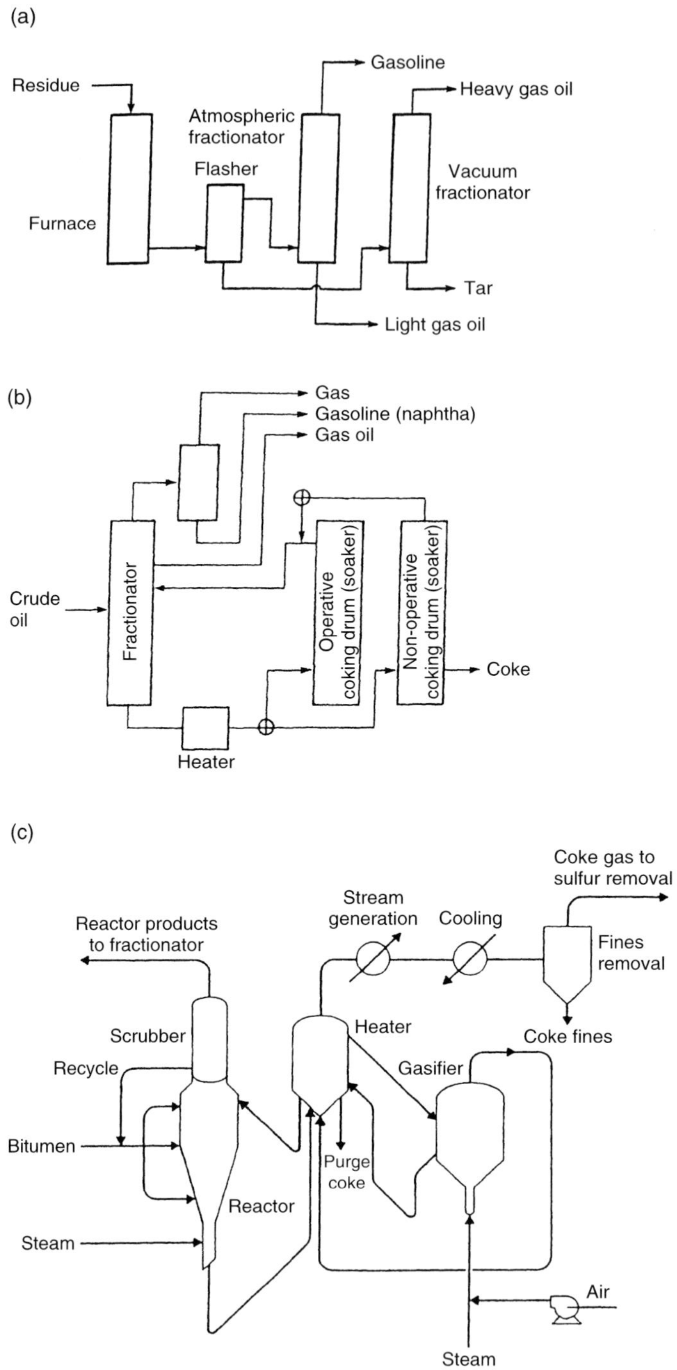

Figure 14.1 Simplified schematics of thermal processes used for upgrading heavy petroleum feeds; (a) visbreaking, (b) delayed coking and (c) flexicoking.

until now, although there have been other configurations of coking processes in various stages of development. The choice between the delayed and fluid-flexi-mode of coking process depends on the scale of the operation. Generally, the latter process is chosen for large-scale production compared with a small- or medium-scale production for the delayed coking process.

Limited outlets for the utilization of coke may be a drawback of coking processes. However, recent trends around the world indicate on the growing interests in the integration of the coke utilization with petroleum refining. For example, in a site-specific case, the demand for electricity may be the driving force for such integration. Electricity and steam are the products of combustion of petroleum coke. At the same time, gasification of petroleum coke can lead to a slate of final products (937,938). In this case, the choice of the final product depends on the market demands. For the production of electricity and/or steam, gasification of petroleum coke has advantages over combustion because of the higher overall thermal efficiency and much better environmental performance (939). This was confirmed by Ikbal et al. (940) who compared five different configurations of petroleum refinery to obtain the most efficient utilization of the residues from carbon-rejecting processes.

A coking process installed at the wellhead could address challenges facing heavy oils producers such as the energy requirements to produce steam for the EOR. Moreover, the cost of diluent, which otherwise would have to be used to provide the crude mixture suitable for transportation by pipeline, can be eliminated (941).

14.1.2.1 Delayed coking process

This process comprises two reactors/drums, one operating in the coking mode while another is simultaneously decoked (Figure 14.1). This ensures the semicontinuous operation. For the operation, the coking drum is filled with VR exiting from the bottom of fractionator. The heat required for coking is supplied by the internal heat source. The gaseous and liquid products exit at the top and enter fractionator. The final boiling point of the gas oil is determined by the temperature of coking (~800 K). More than 50% of the gas oil fraction boils above 650 K. The coke removed from the drum at the end of coking cycle is in the form of lumps. Therefore, it has to be crushed before its utilization via combustion or gasification. The delayed coke is suitable for preparation of various carbon products.

14.1.2.2 Fluid-flexi-coking process

The fluid-flexi-coking process operates in a continuous mode. The diagram of the fluid-flexi-coking process is shown in Figure 14.1. In the flexi-coking mode, the process consists of the coker where the conversion is achieved by injecting heavy feed in a liquid form into a fluidized bed of the hot coke particles. Subsequently, the coke particles with a thin layer (~5 μ) of the fresh deposit on the external surface are withdrawn from the coker and transferred to the heater where their temperature is increased by contacting with hot gasification products. Part of the coke from heater enters gasifier where it is converted to fuel gas by reacting with steam. In the fluid-coking mode, the heater is replaced by burner. In this case, the coke particles withdrawn from the coker are combusted with air under the oxygen-starving conditions. The aim of the partial combustion is to increase the temperature of the coke particles (to about 900 K) which are then returned to the coker. With this arrangement, all heat required for coking reactions is generated in the process. Most of the produced coke has particle size less than 1 mm (942). Because of this fact, little preparation is necessary prior to coke utilization via either combustion or gasification. During coking, most if not all metals are concentrated in coke. Also, part of the asphaltenes is converted to coke and the other to distillates. This ensures that primary liquids from coking contain little metals and asphaltenes. The coking temperature ensures that most of the liquid products are in the distillates range consisting of the naphtha and HGO fractions. However, more than 50% of HGO boil above 350°C. If the objective of coking operation is the synthetic crude, the primary products have to be stabilized via hydroprocessing before pipelining.

14.1.2.3 EUREKA process

An alternative to conventional coking processes for heavy feed upgrading is the commercially proven EUREKA process consisting of the semibatch reactor system (943). The source of heat used for the EUREKA process differs markedly from that used in coking processes. In the former case, a heavy feed (usually VR) is preheated and mixed with the recycle oil before being fed into another preheater. Subsequently, the mixture is injected into the reactor together with the superheated steam. The latter provides heat necessary for thermal cracking. It is unlikely that at temperatures employed (~850 K) the reaction of steam with the feed yielding H_2 can occur to any great extent. The yield of distillates from EUREKA process is higher than that from delayed coking but lower than from fluid-flexi-coking. Moreover, the residue, i.e., EUREKA pitch, has much better combustion and/or gasification properties than delayed coke. It can be also used as a binder for the preparation of metallurgical coke from a lower quality coals.

14.2 CARBON REJECTION BY DEASPHALTING

The physical separation of asphaltenes and metals from heavy petroleum feeds, termed as deasphalting, has already been mentioned over the previous chapters. However, it is believed that a brief account may be necessary for a better understanding of this process in relation to the separation using distillation methods. The former, conducted under rather mild conditions, is based on the difference in solubility of petroleum constituents (e.g., oil, resins and asphaltenes) in various solvents, while distillation methods are based on the difference in boiling range of hydrocarbon fractions. Deasphalting involves the formation of two phases as the result of mixing heavy feed with a solvent. One phase comprises dissolved portion of heavy feed and the other consists of the solid or semisolid residue, i.e., predominantly asphaltenes and metals. To allow countercurrent flow of the two phases in the industrial reactors, the specific gravity of the solvent must be different than that of the feed. This would also ensure an easy separation of the solvent from the dissolved and precipitated portions of heavy feed. In any case, the solvent recovery may account for relatively large part of the energy requirements of deasphalting process. It was reported that from the energy demand point of view, the solvent recovery under supercritical conditions reduces utilities compared with conventional method (944).

In refinery practice, normal alkanes such as propane, butane, pentane, hexane and heptane have been the solvents of choice. However, the use of a light paraffinic naphtha fraction has also been noted. The early deasphalting units used propane as the solvent and VGO and AR as the feeds to produce the base oil for lubricants preparation.

Gradually, VR and heavy crudes have been included as the feeds and higher alkanes as the solvents. Ability of the alkane solvents to dissolve asphaltenes increases with the increase in their molecular weight. Consequently, the quality of DAO is decreasing. However, this is compensated by a higher yield of DAO. These events are shown in Figure 2.3 and Table 2.1 (25). The decreased quality (because of increased yield) requires more extensive hydroprocessing of DAO. Therefore, the optimization of the deasphalting operation involves trade-offs between the higher yield of the more contaminated DAO and the additional cost of hydroprocessing associated with it.

A desirable efficiency of deasphalting may be achieved by optimizing the operating parameters, such as the degree of dilution (solvent/feed ratio), temperature and contact time (1,2). Generally, the yield of asphaltenes increased and that of DAO decreased with increase in the ratio of mass of solvent/mass of feed to a maximum. For example, for pentane as solvent and Athabasca bitumen as heavy feed, the maximum was reached when the ratio of about 20 was approached. However, as it is shown in Figure 14.2 (945), the optimal ratio may depend on the origin of the feed, although the trends appear to be similar for the different feeds. As one would expect, the largest yield of the residue (asphalt) was obtained from the heavy feed derived from thermal cracking (case A). At the same time, the yield of asphalt from the corresponding VR before thermal cracking (case B) was lower. The increased yield of the asphaltenes precipitate was complemented by decreased yield of DAO. At the same time, the quality of DAO was improved, as it was indicated by the decreasing content of V shown in Figure 14.2. In fact, the linear correlation between the yield of asphaltenes and removal of metals was established (946,947). The effect of deasphalting on the parameters such as the H/C ratio, overall molecular weight, viscosity and softening point of asphalt were quantified as well (948). In this case, Cold Lake bitumen was investigated. However, it is believed that the similar trends may be established for other heavy feeds.

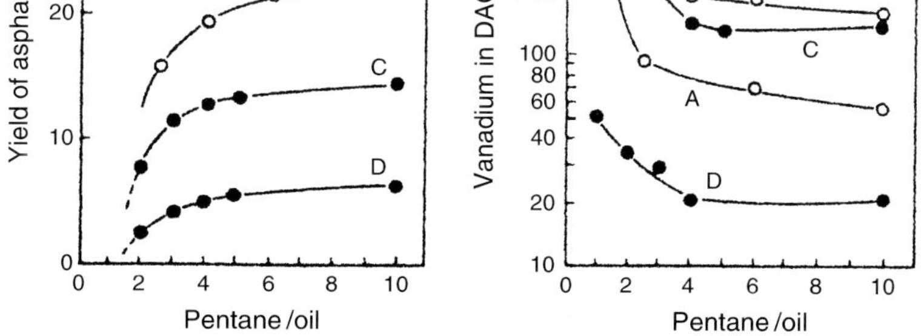

Figure 14.2 Effect of pentane/oil ratio on yield of asphalt and content of vanadium in DAO (9451). A – hydrocracked pitch, B – VR to produce A, C – AR from Venezuelan crude and D – AR from Middle East crude.

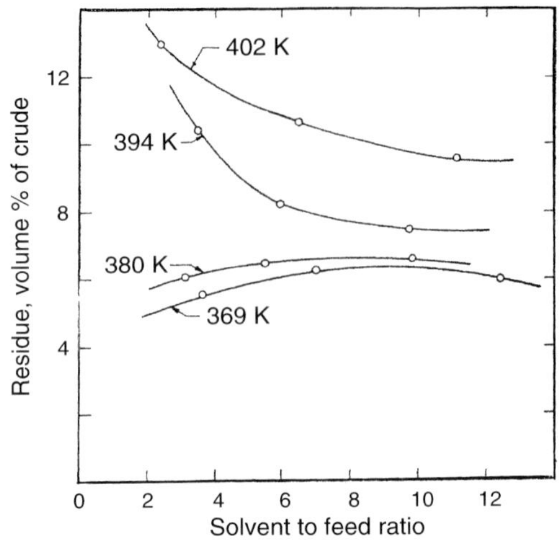

Figure 14.3 Effect of *i*-butane/feed ratio and temperature on yield of asphalt (15% reduced Kansas crude) (949).

For the most widely used solvent such as propane, the optimal temperature for the oil dissolution and asphaltenes separation was between 40 to 60°C (949). This suggests that the operation must be conducted at elevated pressures. At such temperatures, efficient mixing of the solvent with the feed may only be achieved for AR. However, it is unlikely that VR and heavy crudes can be efficiently mixed with any solvent and as such be deasphalted in commercial units at an ambient temperature. Resins began to enter the precipitate together with asphaltenes, when temperature is increased above the ambient temperature. Therefore, the yield of precipitate increases and that of DAO decreases with increasing temperature. At the same time, the quality of the latter is improving. Apparently, there may be an optimal combination of temperature and the solvent/feed ratio, giving the best yield and quality of the DAO. This is shown in Figure 14.3 using the VR derived from a US feed and *i*-butane as solvent (950).

The optimal combination of temperature and the solvent/feed ratio depends on the type of solvent and the feed origin. For example, for the Boscan heavy crude, the yield of DAO using the solvents such as LPG, pentane, petroleum ether and hexane was 52, 73, 75 and 78 wt%, respectively (33). At the same time, the content of metals in DAO was 55, 234, 450 and 537 ppm, respectively. Viscosity and specific gravity of the DAO exhibited similar increasing trends. Then, rather different conditions and type of catalyst will be required for hydroprocessing of the DAO obtained using the LPG solvent and that using hexane.

The contact time is an important parameter, particularly for heavy feeds such as VRs. For such feeds, it takes more time for solvent molecules to access the asphaltenes entities and initialize their flocculation. For alkane solvents, it may take several hours before a filtrable form of the asphaltenes agglomerates can be formed. The situation may be partly alleviated by using a molten and/or liquid forms of VRs. Therefore, deasphalting of such feeds cannot be conducted at ambient temperatures. Several methods of mechanical mixing have been tested with the aim to improve contact of the solvent with the feed (951). Among these methods, rotating disc contactor (RDC) has been shown a superior

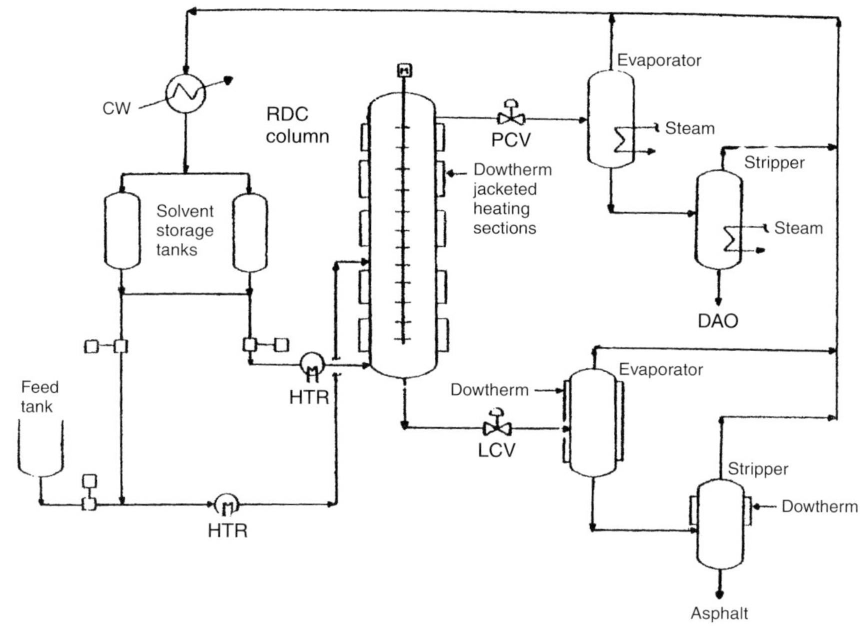

Figure 14.4 Simplified flowsheet of deasphalting process.

efficiency (952). Schematics of the deasphalting plant employing the RDC is shown in Figure 14.4. The RDC consists of more than 10 rotors and stators enclosed in a column. Rotation of the rotors and design of the stators ensure an efficient contact of the solvent with the feed in the countercurrent flow, i.e., the solvent with dissolved oil exits at the top and the agglomerates of asphaltenes at the bottom. With this arrangement, a multistage deasphalting can be achieved. Thus, a single-stage process is not applicable for heavy feeds such as VRs.

A number of solvents other than alkanes have been tested for deasphalting of heavy feeds as well (1). They include alcohols, ketones, aldehydes, ethers, acids, amides, nitrates, etc. Among gaseous solvents, CO_2 has been receiving an attention (953,954). In this case, CO_2 has to be present in a liquid form. This can only be achieved by conducting deasphalting at higher pressure. This would require a redesign of the conventional deasphalting reactor, as well as the upstream and downstream units. Nevertheless, refinery practice indicates that hydrocarbons have been used predominantly as the solvents for deasphalting.

The asphalt as the deasphalting by-product may be used as the fuel for production of steam and electricity. In this case, either combustion or gasification technologies have been used commercially. However, a dilution with a gas oil fraction may be required to achieve desirable pumpability to enable feeding to the combustion and/or gasification reactor. Depending on the solvent used for deasphalting, the asphalt by-product can also be used as the feed for a coking process to generate additional liquid fuels and the coke by-product. In this regard, several potential options were compared by Elliot et al. (955). Other options for the utilization of asphalt may be identified, i.e., carbon products, construction material, etc.

Chapter 15

UNCOMMON METHODS FOR UPGRADING HEAVY FEEDS

Rather unusual approach used for upgrading heavy petroleum feeds involves the use of ultrasonic energy, electric field and magnetic field. It should be noted that these methods are considered to be of an exploratory nature. Thus, it seems unlikely that any of these methods will be used on a commercial scale in a foreseeable future. However, their brief account may be useful because they differ markedly from the catalytic and non-catalytic methods which have been either under investigation or used for upgrading heavy petroleum feeds on a commercial scale.

The ultrasonic treatment was used to treat HGO at a near atmospheric pressure in the absence of additives (956). The lighter gaseous hydrocarbons produced were identified as methane, ethylene, ethane and propylene. The basic nitrogen-containing compounds in HGO were more easily cleaved by the cavitational energy than non-basic compounds. Maximum of the nitrogen and sulfur conversion of 11 and 7%, respectively, as well as a 5% reduction in viscosity, were obtained at the optimized sonochemical conditions. A radical chain mechanism was proposed to explain mechanism of the reactions of hydrocarbons which were initiated by ultrasound.

The information indicating the use of the ultrasonic treatment for upgrading the metals- and asphaltenes-containing feeds is limited, although the weak chemical bonds present in asphaltenes and porphyrin molecules may serve as potential sites for the cleavage followed by depolymerization (21,39). For the asphaltenes, ^{1}H-NMR analysis before and after sonification revealed a significant decrease in the naphthenic structure (957). This involved both cracking and dehydrogenation. In the case of Athabasca bitumen alone, about 12% reduction in viscosity on sonification was achieved (958). The additives, such as toluene, NaOH and water, had no beneficial effects. However, in the presence of reducing agent such as sodium borohydride, the reduction of asphaltenes was more than three times greater than that in the absence of the reducing agent (959). In the study conducted by Kang et al. (960), Athabasca asphaltenes were ultrasonically treated using five discrete ultrasonic frequencies. Also, the suitability of four saturating gases were explored. Kinetics of the reduction in the AMW were determined by vapor pressure osmometry measurements and UV/visible spectrophotometric measurements. The significant decrease in the AMWs was observed after rather short treatment times, i.e., from ~1200 to ~470 g/mol in 15 min. The observed reduction in the AMWs of asphaltenes was greatest at the 358 kHz frequency. At the frequency of 205 kHz, the observed reduction in AMW was fastest with the mixture of Ar + H_2 (50%/50% v/v), whereas the saturation with hydrogen, air and oxygen gases exhibited minimal effect on kinetics. The heptane-soluble fractions of asphaltenes increased by more than 50% when the sonochemical treatment was used as an upgrading process.

A limited information suggests that the conversion of porphyrins via sonification was also confirmed (961,962). In this case, the model compound such as VO-TPP was used. Relatively large conversion was observed at room temperature. However, this was achieved in the presence of H_2O_2. Nevertheless, for heavy feeds containing metals and asphaltenes, the effect of ultrasonic treatment has not yet been investigated in sufficient details.

The viscosity reduction of crude oil was achieved by applying either an electric or magnetic field (963). Paraffinic and intermediate crudes, as well as a heavy crude oil, were investigated. The method had little effect on the temperature of crude oil. However, the viscosity reduction was not permanent, although the reduced viscosity was maintained for several hours. Among the crude oil tested, the most pronounced effect was observed for the paraffinic crude oil.

Chapter 16

CONCLUSIONS AND FUTURE PERSPECTIVES

It is inevitable that in decades to come, heavy feeds will dominate petroleum refining. For this purpose, both carbon rejection and hydrogen addition processes have been used commercially and will continue to be used. The content of metals (Ni + V) in the most problematic feeds such as VRs and ARs derived from heavy crudes (e.g., Boscan, Orinoco, Maya, Zuata) may exceed 1000 ppm. The present state-of-art of the upgrading processes suggests that the enormous costs of catalyst inventory required for hydroprocessing of such feeds could not be avoided. In this regard, the carbon rejection routes, i.e., thermal cracking and hydrocracking as well as the direct coking and deasphalting, may alleviate the problem. Practical experience suggests that direct coking (delayed and fluid/flexi coking) has been the preferred route for large plants processing more than 100 000 barrel per day of VR. By-products of coking such as hydrocarbon gases and solid coke may be valuable feeds for the production of electricity, steam, H_2 and chemicals on the site of refinery. A growing interest in the integration of the processes producing such products with petroleum refinery has been noted. The viability and flexibility of such integration improve with increasing size of refinery. Coincidently, trends around the world, supported by economic parameters, indicate a departure from small refineries toward large refineries. Nevertheless, hydroprocessing will remain an integral part of petroleum refinery in spite of the fact that the primary upgrading of VR was conducted using a carbon rejection process. In any case, the primary products derived from VRs and topped heavy crudes (e.g., VGO and HGO) by carbon rejection technologies will always require additional upgrading using hydroprocessing methods.

Significant advances have been made in the understanding of the role of conventional catalysts during hydroprocessing of heavy petroleum feeds. The new knowledge has been built on the many years' experience gained during the hydroprocessing of conventional distillates. The effects of additives such as fluoride, phosphate, borate and others were clarified to the point that some of them have been already added to the commercial hydroprocessing catalysts. Their beneficial effect was confirmed by improved catalyst activity and selectivity, as well as by slowing down deactivation, thus prolonging catalyst life. In some cases, a marked improvement in the performance of the modified catalysts was observed. However, in many studies, the results were obtained during the runs lasting less than 1 day on stream. It is desirable that these catalysts are tested during long runs, e.g., lasting several months, in order to confirm their commercial viability. More reliable results are obtained from the testing carried out in pilot plants than in bench scale and laboratory scale units. It is inevitable that the promising results obtained in microscale systems are verified using reactors which can accommodate the catalysts in their operating size and shape.

An extensive information suggests that there is a significant potential for further improvement in catalyst activity by using supports other than the traditionally used γ-Al_2O_3. For example, the supports which are more acidic than γ-Al_2O_3 (e.g., amorphous silica-alumina, zeolites) improve the activity for HCR, but they also may increase the rate of coke formation. Therefore, with respect to heavy feeds, additional research is needed to achieve the optimal balance between the HCR activity and catalyst deactivation. In this regard, the combination of TiO_2 with γ-Al_2O_3 resulted in the supports which have been used for the preparation of the commercial catalysts. In this case, the method of the TiO_2 addition to γ-Al_2O_3 had pronounced effect on catalyst activity. More acidic supports used either alone or in the combination with γ-Al_2O_3 include various zeolites, SiO_2–Al_2O_3. The carbon may be the support of choice when the acidity of catalyst is not desirable. The published results show that the development of novel supports is the area where still some significant improvements in catalyst performance may be realized. Again, these results still have to be demonstrated on a larger scale during long runs, rather than in bench scale units. Moreover, the data obtained for one heavy feed may not be representative of the other, suggesting that the feeds' selection for testing deserves attention.

Modifications of the methods used for the addition of active metals to supports resulted in the enhancement of catalyst activity. For example, impregnation of the supports with aqueous solutions of the salts containing active metals in the presence of chelating agents diminished the unwanted interaction of the latter with the support. As a result of this, a more active and stable type II active phase was formed during subsequent sulfidation of the catalyst. Apparently, further enhancement in catalyst activity can be achieved by studying a wider range of chelating agents to identify the most suitable agent. Also, the additional efforts may be necessary, with the aim to optimize experimental conditions used during the catalyst preparation.

Among the non-conventional catalysts, metal (Mo, W, Co, Ni, Fe, etc.) carbides and nitrides have been tested for hydroprocessing. Most of the information on the use of metal carbide and nitrides involved model feeds, although a gradual incorporation of real feeds in hydroprocessing studies has been noted. However, there is little evidence supporting commercial utilization of these catalysts. This may result from rather unusual trends in their intrinsic activity, i.e., increase in the activity with increasing size of the catalyst particles. It is suggested that the high HYD activity of metal carbides and nitrides at rather low temperatures makes such catalysts suitable for multistage hydroprocessing. In this case, heavy feeds such as VGO, HGO and DAO would be prehydrogenated under mild conditions in the first stage over metal carbides and nitrides. The prehydrogenated feed would be further processed in the next stages. It is believed that the subsequent stages, particularly those focusing on HCR and HDN of the prehydrogenated feed, could be conducted under milder conditions, provided a suitable catalyst for every stage was selected. The experimental database on the performance of metal phosphides is incomplete for making an assessment of their suitability as catalysts for hydroprocessing heavy feeds. Thus, most of the published results were obtained in microreactors or bench scale reactors using model compounds rather than real feeds. It is therefore, desirable that the additional attention is paid to metal phosphide, with focus on their suitability as the catalysts for hydroprocessing applications.

The noble metal sulfide-based catalysts are known for exhibiting very high HYD activity, whereas their activity for hydrocracking and hydrogenolysis may be rather low. This may be overcome by combining noble metal sulfides with acidic supports such as zeolites. After being used in hydroprocessing of heavy feeds, the recovery of noble metals from spent catalysts is believed to be difficult and costly. Therefore, there is little potential for using noble metal-based catalysts for hydroprocessing of heavy feeds except for that of VGO, HGO and DAO. These feeds are used as starting materials for production of diesel oil and lube base stock. The isomerization and HCR of *n*-paraffins, as well as the HYD of aromatics, are necessary to achieve desirable cold flow properties and VI. Noble metals supported on zeolites and amorphous silica–alumina have been found suitable for these applications. Further developments in this field with focus on better catalysts and processes are anticipated with the aim either to increase yields and improve quality of the products or to decrease the number of catalytic stages which are part of the processes involved.

The importance of physical and mechanical properties of catalysts increases with the increasing content of asphaltenes and metals in heavy feed. Requirements of a suitable porosity combined with the optimally balanced catalyst functionalities impose challenges on the design of the catalysts for hydroprocessing of heavy feeds compared with that for light feeds. For the most problematic feeds, this can be overcome by conducting the operation in several stages. The catalyst for the up-stream stage must possess a high metal storage capacity. To achieve this, the catalyst has to be active for HDM and HDAs and at the same time has to possess a desirable metal storing capacity. It has been indicated that a high level of HDM cannot be achieved without a high level of the asphaltenes conversion. Practical experience suggests that for this purpose, i.e., the first stage of a multistage system, the low active metal-loaded support (e.g., less than 4 wt% of Mo) possessing a suitable combination of meso- and macroporosity may exhibit a desirable performance. The catalysts for downstream units must be tailor-made to take into consideration the change in properties of the original heavy feed after every stage. Moreover, the results obtained for one heavy feed may not be applicable for other feeds. Therefore, the testing and selection of catalysts for every stage must involve the feed which in properties approaches that of the product from the previous stage, which is used as the feed in the subsequent stage. This indicates a significant complexity in hydroprocessing of high asphaltene and metal feeds in multistage systems. Therefore, for such feeds, the experimental testing, preferably on a pilot plant scale, may be necessary before optimal operating parameters can be identified with the aim to ensure a synchronized operation of the process comprising several reactors in a series.

It has been noted that high conversions of the most problematic feeds were achieved when fine catalyst particles were slurried/dispersed in the feed. Low-cost solids occurring either naturally or disposed from various industrial processes exhibited a good activity. In this case, a "once-through" option has been in a near-commercial stage. The feasibility of this method depends on the availability of suitable solids in the proximity of petroleum refinery. It is believed that with further improvement and/or optimization, competitiveness of the "once-through" method used for hydroprocessing the most problematic feeds can be increased.

A micronized form of the transition metal oxides and sulfides slurried with heavy feeds was, in their activity and selectivity, comparable or better than conventional catalysts. Similar performance was exhibited by the oil-soluble organometallic compounds and the water-soluble inorganic salts containing transition metals. In spite of this, the processes employing soluble/dispersed catalysts have not reached a commercial stage. This results from difficulties in recovery of catalysts for reuse. After fractionation, the catalyst remains concentrated in the heaviest part of the products, i.e., VR. The VR can be further processed by coking to obtain additional distillates. In the case that the heavy feed processing option to "extinction" was practiced on the site, e.g., via either combustion or gasification of VR and petroleum coke, most of the metals will be concentrated in the solid residue such as ash and/or slag, together with the V and Ni contained in the original feed. The conventional methods of hydrometallurgy can be used to isolate these metals and catalytically active metals in a pure form for reuse. The trends around the world indicate the growing interest in the integration of these processes with petroleum refineries. Then, the viability of the soluble/dispersed catalysts for hydroprocessing of the most problematic feeds may improve when all these parameters which result from integration are taken into consideration.

The catalytic dewaxing of VGO and DAO for the production of diesel oil and lubricants is conducted under conditions approaching those used during hydroprocessing, although the catalyst formulation may differ from those found in conventional catalysts. Because of the necessity to remove *n*-paraffins (via isomerization or hydrocracking), these catalysts must consist of acidic supports such as zeolites. In addition, the content of aromatics has to be decreased to meet specifications of commercial products. An ideal catalyst for production of diesel oil and lube base stock should exhibit a high activity for HYD and/or HDAr, as well as for hydroisomerization. The recent developments in this field indicate that combinations of a suitable zeolite with noble metals (e.g., Pt and Pd) may exhibit desirable performance. Moreover, significant improvements in the dewaxing efficiency can be achieved by different process configurations. It has been noted that information on the development of dewaxing catalysts is predominantly in the patent literature.

From the commercialization point of view, bio-upgrading of heavy feeds is still in very early stages of development in spite of the decades of research in this field. The available information obtained predominantly using either model compounds or light feeds indicates significant complexity of the bio-upgrading reactions. Moreover, there are some uncertainties regarding recovery of bio-catalysts for reuse. On the cost basis, bio-catalysts cannot compete with the conventional hydroprocessing catalysts. At the same time, the emerging processes using dissolved/dispersed catalysts are much more viable than bio-upgrading when heavy petroleum feeds are taken into consideration. It is, therefore, unlikely that among the competitive technologies, bio-upgrading will make any impact on the upgrading of heavy feeds in petroleum refining in a foreseeable future.

To a certain extent, some events occurring during hydroprocessing can be quantified by kinetic measurements. However, for heavy feeds, the complexity involved prevents a full utilization of the published results which are influenced by numerous experimental parameters, i.e., range of temperature and H_2 pressure, LHSV, H_2/H_2S ratio, type of catalyst, origin of feed, time on stream when the measurement was performed, type of experimental system, form of equations used to treat data. This complexity prevents rationalization of the results obtained in different studies. In some cases, the results have a limited value because only a brief description of the experimental systems was given. The results are also affected for model feeds and light feeds, unless all necessary information is available. Therefore, for heavy feeds, it is not easy to establish even general trends on the basis of kinetic data, which for most part only reflect very specific situation, e.g., one feed and one catalyst under given conditions. This becomes evident during the development of the models for predicting catalyst performance. For most part, such models cannot be applied generally to various feeds and catalysts. However, it has been shown that for best known models, the situation may be alleviated by a slight modification of the model developed for one feed using the parameters determined experimentally for a heavy feed of interest. In this regard, the results obtained in pilot plant are more suitable than those in bench scale units. Apparently, a caution is required while using the results obtained during the accelerating aging experiments. The results from commercial operations, as the most reliable results, have been used to verify the validity of models as well.

It may be concluded that, in decades to come, hydroprocessing of heavy petroleum feeds containing asphaltenes and metals will be dominated by the conventional modified Co(Ni)Mo(W) catalysts. As has been indicated, there is still the possibility for further improvement in catalyst performance, i.e., by introducing novel supports or by modifying γ-Al_2O_3 support. For the latter, beneficial effects of additives and the conditions applied during preparation should be noted. The different process configurations is another alternative for improving hydroprocessing conversions. For the asphaltenes and metals containing feeds, multistage processes modified to suit a particular feed will dominate catalytic upgrading. In this regard, a desirable conversion can be obtained in the moving and ebbulated bed reactors. Using these reactors, the upper limits of metals and asphaltenes in heavy feeds, which can still be catalytically processed, were increased to about 700 ppm.

An alternative option for residue upgrading appears to be the RFCC process. However, the reference to this process as one being suitable for residue upgrading compared with conventional and advanced hydroprocessing may be misleading. Therefore, a brief account given in the book was necessary to clarify the relation between RFCC and hydroprocessing from the feed properties and process requirements' point of view. For the former, limitations on the content of metals and asphaltenes in the feed are rather evident. Thus, significant modifications of the FCC for RFCC process and novel catalyst formulations are required for the feed with the metal content approaching 30 ppm, whereas such feed can by upgraded via hydroprocessing without any problems. The costs of these modifications have to be compared with the costs of a hydroprocessing unit added upstream of RFCC unit to justify such modifications. The pretreated feed could then be further upgraded using conventional FCC process.

The non-catalytic upgrading processes (coking and deasphalting) deserve serious considerations for heavy feeds, with the content of metals exceeding 500 ppm of V + Ni. Potential for integration of technologies, which can convert the final refinery residues (e.g., petroleum coke, asphalt, sludge, oily emulsions) to marketable products (electricity, hydrogen, alcohols, fertilizers, etc.), may be the driving force for non-catalytic upgrading of the most problematic heavy feeds. However, this option may be site specific. Thus, there might be a merit for the integration in one location, whereas little merit in the other. For example, integration may be justified by the demand for electricity and other products in certain regions.

The potential integration of the Fischer–Tropsch (FT) process with petroleum refining may be an attractive option deserving reevaluation. The process consists of gasification of the final refinery residues (petroleum coke, asphalt from deasphalting, refinery sludge, etc.) to synthesis gas which after purification and water–gas shift step may be converted to hydrocarbons, predominantly in diesel oil and lube base oil boiling range. This represents a conversion of low-value feeds to high-value commercial products such as lubricants and transportation fuels. Because all sulfur and nitrogen are removed during purification of gasifier products, no hydroprocessing of the products from the FT process is necessary. Apparently, the drawback of this process is the production of the by-product wax. However, after additional treatment such as hydroisomerization and/or HCR, the wax can be converted to lube base oil, as well as valuable blending components with transportation fuels. This suggests that with such integration, little residue would be left behind. In recent years, renewed interests in the development of improved catalysts for water–gas shift and FT process have been noted.

REFERENCES

1. J.G. Speight, The Chemistry and Technology of Petroleum, Marcel Dekker, New York, 1999.
2. J.J. McKetta, Petroleum Processing Handbook, Marcel Dekker, New York, 1991.
3. B. Schuetze and H. Hoffman, Hydroc. Proc. 2 (1984) 75.
4. R.L. Dickenson, F.E. Biasca, B.L. Schulman and H.E. Johnson, Hydroc. Proc. 2 (1997) 57.
5. H. Beuther and B.K. Schmid, Proc. World Petr. Congr., 1963, Section III, Paper 20.
6. J.M. Oelderik, S.T. Sie and D. Bode, Appl. Catal. 47 (1981) 1.
7. W.I. Beaton and R.J. Bertolacini, Catal. Rev.-Sci. Eng. 33 (1991) 281.
8. P.W. Tamm, H.F. Harnsberger and A.G. Bridge, Ind. Eng. Chem. Proc. Des. Dev. 20 (1981) 262.
9. L.O. Oyekunle and R. Hughes, Ind. Eng. Chem. Res. 26 (1987) 1945.
10. C.J. Pereira and J.W. Beeckman, Ind. Eng. Chem. Res. 28 (1989) 422.
11. C.J. Pereira, W.-C. Cheng, J.W. Beeckman and W. Suarez, Appl. Catal. 42 (1988) 47.
12. J.C. Plumail, R. Marseau, G. Martino and H. Toulhoat, Am. Chem. Soc. Div Petr. Chem. Prep. 33 (1988) 103.
13. C.A. Maier, J.P. Peries and A. Quignard, Upgrading Residues by Conversion and Refining with a New Generation Catalysts, Rapport IFP 34 297, 1986.
14. A. Billon, J. Bousquet and J. Rossarie, Proc. 1988 NPRA Ann. Meet., March 1988, San Antonio, TX.
15. K.W. Robschlager, W.J. Deelen and J.F. Naber, Fourth UNITAR/UNDP Int. Conf. on Heavy Oils and Tar Sands, August 1988, Edmonton, Canada, Vol. 5, p. 249.
16. C.T. Adams, A.A. Del Paggio, H. Schaper, W.H.J. Stork and W.K. Shiflett, Hydroc. Proc. 9 (1989) 57.
17. J.D. Carruthers and D.J. DiCamillo, Appl. Catal. 43 (1988) 253.
18. J.G. Speight, Asphaltenes and the structure of petroleum, in Petroleum Chemistry and Refining, J.G. Speight (Ed.), Taylor and Francis Publ., Washington, DC, 1998, p. 103.
19. M.R. Gray, Upgrading Petroleum Residues and Heavy Oils, Marcel Dekker, New York, 1994.
20. I. Wiehe and K.S. Liang, Fluid Phase Equilib. 117 (1996) 201.
21. T. Suzuki, M. Itoh, W. Takegami and T. Watanabe, Fuel 61 (1982) 402.
22. K.A. Pilcher and J.M. Winterbottom, Chem. Eng. Technol. 11 (1988) 89.
23. H.D. Sherwood, Oil Gas J. 76 (1978) 148.
24. A. Billon, G. Heinrich, Y. Jacquin and J.P. Peries, Proc. 11th World Petr. Congr., 1983 Vol. 4, p. 35.
25. K.Z. Lynch, R.L. Hood, O. Gomez and L.G. Aquino, UNITAR Conf., February 1995, Houston, TX, p. 245.
26. M.M. Boduszynski, Am. Chem. Soc. Div. Petr. Chem. Prep. 20 (1983) 1376.
27. M.M. Boduszynski, Am. Chem. Soc. Div. Petr. Chem. Prep. 30 (1985) 626.
28. C. Takeuchi, Y. Fukui, Y. Shiroto and M. Nakamura, Proc. 11th World Petr. Congr., 1983 Vol. 4, p. 187.

29. F.E. Biasca, B.L. Schulman and R.L. Dickenson, Proc. 3rd UNITAR/UNDP Int. Conf. on Heavy Crudes and Tar Sands, 1985, Long Beach, CA.
30. F.W. Wenzel, Proc. 4th UNITAR/UNDP Int. Conf. on Heavy Crudes and Tar Sands, 1988, Edmonton, Canada.
31. B. Riediger, Erdoel Kohle, 26 (1975) 9.
32. G.E. Dolbear, A. Tang and E.L. Moorhead, in Metal Complexes in Fossil Fuels, R.H. Filby and J.F. Branthaver (Eds), Am. Chem. Soc. Symp. Ser. 344, 1987, p. 220.
33. D.A. Viloria, J.H. Krasuk, O. Rodriguez, H. Buenafama and J. Lubkowitz, Hydroc. Proc. 3 (1977) 109.
34. J.J. Heithaus, J. Inst. Petr. 48 (1962) 45.
35. J.J. Altamirano, M.A. Flores, O. Pie, N. Panivino, C. Arzete, G. Kappelmann, M.T. Lopez, S. Espinoza and S. Rosales, Rev. Inst. Mex. Petr. 18 (1986) 32.
36. L. Carbognani and S.A. Intevep, Caracas, Venezuela Tech. Rep#INT0255992, 1992.
37. C. Dekkers, Oil Gas J. 1 (1999) 44.
38. J.D. Carruthers, J.S. Brinen, D.A. Komar and S. Greenhouse, in Catalytic Hydroprocessing of Petroleum, M. Oballa and S.S. Shih (Eds), Marcel Dekker, 1994, p. 175.
39. J.F. Branthaver, in Metal Complexes in Fossil Fuels, R.H. Filby and J.F. Branthaver (Eds), Am. Chem. Soc. Symp. Ser. 344, 1987, p. 188.
40. E. Furimsky and F.E. Massoth, Catal. Today 52 (1999) 381.
41. E. Furimsky, Catal. Today 30 (1996) 223.
42. A.C. Jacobsen, B.H. Cooper and P.N. Hannerup, Proc. 12th Petrol. World Congr., 1987, Vol. 4, p. 97.
43. J. Goulon, A. Retournard, P. Friant, C. Ginet-Goulon, C. Berte, J.F. Muller, J.L. Poncet, R. Guilard, J.C. Escalier and B. Neff, J. Chem. Soc. Dalton Trans. 6 (1984) 1095.
44. T.F. Yen, Chemical Aspects of Metals in Petroleum, Ann Arbor, MI, 1975, p. 221.
45. R.H. Filby and D. Strong, in Tar Sands and Oil Upgrading Technology, S.S. Shih and M. Oballa (Eds), AIChE Symp. Ser. 282, 1991 Vol. 87, p. 1.
46. J.G. Reynolds, W.R. Biggs and S.A. Bezman, in Metal Complexes in Fossil Fuels, R.H. Filby and J.F. Branthaver (Eds), Am. Chem. Soc. Symp. Ser. 344, 1987, p. 205.
47. P.C.H. Mitchell, Catal. Today 7 (1990) 439.
48. R.H. Filby and G.J. van Berkel, in Metal Complexes in Fossil Fuels, R.H. Filby and J.H. Branthaver (Eds), Am. Chem. Soc. Symp. Ser. 344, 1987, p. 2.
49. E.L. Sughrue, D.W. Hausler, P.C. Liao and D.J. Strope, Ind. Eng. Chem. Res. 27 (1988) 397.
50. K. Sakanishi, N. Yamashita, D.D. Witehurst and I. Mochida, Am. Chem. Soc. Div. Petr. Chem Prep. 42 (1997) 373.
51. R.D. Grigsby and J.B. Green, Energy Fuels 11 (1997) 602.
52. E. Furimsky and P.J. Champagne, Fuel Proc. Technol. 6 (1982) 269.
53. F. Vandeneeckoutte, R. Hubaut, S. Pietrzyk, T. Des Courrieres and J. Grimblot, React. Kinet. Catal. Lett. 45 (1991) 191.
54. L. Rankel, in Metal Complexes in Fossil Fuels, R.H. Filby and J.F. Branthaver (Eds), Am. Chem. Soc. Symp. Ser. 344, 1987, p. 257.
55. A. Cohen, G. Di Bernardo and D. Decroocq, Rev. Inst. Fr. Petr. 43 (1988) 281.
56. J.P. Peries, P. Renard, T. Des Courieres and J. Rossarie, Proc. of 4th UNITAR/UNIDP Conf. on Heavy Crudes and Tar Sands, 1988, Edmonton, Canada, Paper 95.
57. Y. Jacquin, H. Toulhoat, A. Quignard and J.-F. Le Page, Rev. Inst. Fr. Petr. 38 (1983) 371.
58. F.E. Massoth, Adv. Catal. 27 (1978) 265.
59. B.C. Gates, J.R. Katzer and G.C.A. Schuit, Chemistry of Catalytic Processes, McGraw-Hill, New York, Chapter 5, 1979.
60. P. Grange, Catal. Rev. Sci. Eng. 21 (1980) 135.
61. H. Topsoe, B.S. Clausen and F.E. Massoth, in Catalysis, Science and Technology, J. Anderson and M. Boudart (Eds), Springer, 1996, Vol. 1, p. 1.
62. P. Ratnasamy and S. Sivashanker, Catal. Rev. Sci. Eng. 22 (1980) 401.

63. M. Breysse, E. Furimsky, S. Kasztelan, M. Lacroix and G. Perot, Catal. Rev. Sci. Eng. 44 (2002) 651.
64. D.L. Trimm, Design of Industrial Catalysts, Elsevier, Amsterdam, 1980.
65. G. Poncelet, P. Grange and P. Jacobs, Preparation of Catalysts III, Elsevier, Amsterdam, 1983.
66. V.J. Lostaglio and J.D. Carruthers, Chem. Eng. Prog. 82 (1986) 46.
67. D.L. Trimm and A. Stanislaus, Appl. Catal. 21 (1986) 215.
68. E. Furimsky and F.E. Massoth, Catal. Today 17 (1993) 537.
69. S. Kasztelan and B. MacGarvey, J. Catal. 147 (1994) 476.
70. B. MacGarvey and S. Kasztelan, J. Catal. 148 (1994) 149.
71. P. Zeuthen, P. Blom, B. Muegge and F.E. Massoth, Appl. Catal. 68 (1991) 117.
72. P. Dufresne, N. Brahma, F. Labruyere, M. Lacroix and M. Breysse, Catal. Today 29 (1996) 251.
73. E. Furimsky, Appl. Catal. 171 (1998) 177.
74. M. Absi-Halabi, A. Stanislaus, A. Qamra and S. Chopra, in Catalysis in Petroleum Refining and Petrochemical Industries 1995, M. Absi-Halabi et al. (Eds), 1996, p. 243.
75. S.K. Maity, J. Ancheyta, L. Soberanis and F. Alonso, Appl. Catal. 253 (2003) 125.
76. E. Furimsky and F.E. Massoth, Catal. Rev. Sci. Eng. 47 (2005) 297.
77. A.Corma, A. Martinez and V. Martinez-Sori, J. Catal. 200 (2001) 259.
78. S.K. Maity, J. Ancheyta, L. Soberanis, F. Alonso and M.E. Llanos, Appl. Catal. 244 (2003) 141.
79. M.S. Rana, J. Ancheyta, P. Rayo and S.K. Maity, Catal. Today 98 (2004) 151.
80. J. Ancheyta, M.S. Rana and E. Furimsky, Catal. Today 109 (2005) 3.
81. J. Ancheyta, G. Betancourt, G. Centeno, G. Marroquin, F. Alonso and E. Garciafigueroa, Energy Fuels 16 (2002) 1438.
82. B.M. Moyse, B.H. Cooper and A. Albjerg, NPRA Ann. Meet. Paper AM-84-59, March 1984, San Antonio, TX.
83. A. de Bruijn, I. Naka and J.W.M. Sonnemans, Ind. Eng. Chem. Proc. Des. Dev. 20 (1981) 40.
84. S. Kobayashi, S. Kushiyama, R. Aizawa, Y. Koinuma, K. Inoue, Y. Shimizu and K. Egi, Ind. Eng. Chem. Res. 26 (1987) 2245.
85. J. Bartholdy and B.H. Cooper, Am. Chem. Soc. Div. Petr. Chem. Prep. 38 (1993) 386.
86. C. Hung, R.L. Howell, D.R. Johnson, Chem. Eng. Prog. 3 (1986) 57.
87. J.C. Plumail, H. Toulhoat, Y. Jacquin and G. Martino, Proc. 8th World Congr. Catal., 1984, Berlin, Vol. 11, p. 317.
88. M.J. Macias, J. Ancheyta, Catal. Today 98 (2004) 243.
89. K. Onuna, Stud. Surf. Sci. Catal. 31 (1987) 543.
90. J. Walendziewski and J. Trawczynski, Appl. Catal. 119 (1994) 45.
91. J.C. Dart, Chem. Eng. Prog. 71 (1975) 46.
92. T. Takatsuka, Y. Wada, S. Hirohama and S. Komatsu, AIChE Symp. Ser. 85 (1989) 49.
93. B.H. Cooper, B.B.L. Donnis and B. Moyse, Oil Gas J. 84 (1986) 39.
94. S.K. Bej, Energy Fuels 16 (2002) 774.
95. A. Chander, A. Kundu, S.K. Bej, A.K. Dalai and D.K. Vohra, Fuel 80 (2001) 1043.
96. S.K. Bej, A.K. Dalai and S.K. Maity, Catal. Today 64 (2001) 333.
97. S.K. Bej, R.P. Dabral, P.C. Gupta, K.K. Mittal, G.S. Sen, V.K. Kapoor and A.K. Dalai, Energy Fuels 14 (2000) 701.
98. I. Pitault, P. Fongarland, M. Mitrovic, D. Ronze and M. Forissier, Catal. Today 98 (2005) 31.
99. M.-H. Yang, P. Grange and B. Delmon, Appl. Catal. 154 (1997) L7.
100. G. Gualda and S. Kasztelan, J. Catal. 161 (1996) 319.
101. M.H. Al-Dahhan, F. Larachi, M.P. Dudukovic and A. Laurent, Ind. Eng. Chem. Res. 36 (1997) 3292.
102. H. Korsten and U. Hoffmann, AIChE J. 42 (1996) 1350.
103. W. Garcia and J.M. Pazos, Chem. Eng. Sci. 37 (1982) 1589.

104. S.T. Sie, AIChE J. 42 (1996) 3498.
105. S.M. Richardson, H. Nagaishi and M.R. Gray, Ind. Eng. Chem. Res. 35 (1996) 3940.
106. J.A. Mahoney, K.A. Robinson and E.C. Meyers, Chemtech 8 (1978) 758.
107. R.H. van Dongen, D. Bode, H. Van der Eijk and J. Van Klinken, Ind. Eng. Chem. Proc. Des. Dev. 19 (1980) 630.
108. M.S. Rana, M.L. Huidobro, J. Ancheyta and M.T. Gomez, Catal. Today 109 (2005) 346.
109. M.R. Gray, Y. Zhao, C.M. McKnight, D.A. Komar and J.D. Carruthers, Energy Fuels 13 (1999) 1037.
110. L. Reyes, C. Zarpa and J.H. Krasuk, Catalyst Deactivation B. Delmon and G.F. Froment (Eds), Elsevier Science, Amsterdam, 1994, Vol. 88.
111. A. Stanislaus, M. Absi-Halabi and Z. Khan, in Catalysis in Petroleum Refining and Petrochemical Industry, M. Absi-Halabi (Eds), Elsevier Science, Amsterdam, 1996, p. 189.
112. M. Nunez and M. Villamizar, Appl. Catal. 252 (2003) 51.
113. R. Galiasso, R. Blanco, C. Gonzales and N. Quinteros, Fuel 62 (1983) 817.
114. J. Bartholdy and P.N. Hannerup, Am. Chem. Soc. Div. Petr. Chem. Prep. 35 (1990) 619.
115. J. Bartholdy and P.N. Hannerup, Stud. Surf. Sci. Catal. 68 (1991) 273.
116. M. Absi-Halabi, A. Stanislaus, T. Al-Mughni, S. Khan and A. Qamra, Fuel 74 (1995) 1211.
117. P. Zeuthen, B.H. Cooper, F.T. Clark and D. Arters, Ind. Eng. Chem. Res. 34 (1995) 755.
118. J.M. Ammus, G.P. Androutsopoulos and A.T. Tsetsekou, Ind. Eng. Chem. Res. 26 (1987) 1313.
119. S. Kobayashi, S. Kushiyama, R. Aizawa, Y. Koinuma, K. Inoue, Y. Shimizu and K. Egi, Ind. Eng. Chem. Res. 26 (1987) 2241.
120. H. Koyama, E. Nagai, H. Torii and M. Kumagai, Stud. Surf. Sci. Catal. 100 (1996) 147.
121. M. Cheng, F. Kumata, T. Saito, T. Komatsu and T. Yashima, Stud. Surf. Sci. Catal. 117 (1998) 485.
122. H. Homan Free and H.J. Lovink, Erdoel Kohle 40 (1987) 346.
123. V. Sundaramuthy, A.K. Dalai and J. Adjai, Catal. Lett. 102 (2005) 299.
124. J. Hebert, V. Santes, M.T. Cortez, R. Zarate and L. Diaz, Catal. Today 109 (2005) 559.
125. C. Botchwey, A.K. Dalai and J. Adjaye, Ind. Eng Chem. Res. 43 (2004) 5854.
126. M. Nuńez Isaza, A. Pachon, V. Kafarov and D.E. Resasco, Appl. Catal. 199 (2000) 263.
127. T. Takahashi, H. Higashi and T. Kai, Catal. Today 104 (2005) 76.
128. Y.W. Chen, M.C. Tsai, C. Li and B.C. Kang, Can. J. Chem. Eng. 72 (1994) 854.
129. A. Marafi, H. A. Stanislaus, M. Absi-Halabi, A. Hauser and K. Matuishita, Petr. Sci. Technol. 23 (2005) 385.
130. A. Marafi, A. Hauser, A. Stanislaus and K. Matsushita, Petr. Sci. Technol. 23 (2005) 395.
131. A. Marafi, A. Hauser and A. Stanislaus, Energy Fuels 20 (2006) 445.
132. A. Marafi, A. Stanislaus, A. Hauser, S. Fukase, K. Matsushita, A. Al-Barood and M. Absi-Halabi, Am. Chem. Soc. Div. Fuel Chem. Prep. 46 (2001) 238.
133. C.J.J. den Ouden, O.K. Bhan, S. Boardman, R.D. Street and S.E. George, Am. Chem. Soc. Div. Petr. Chem. Prep. 42 (1997) 378.
134. M.S. Rana, J. Ancheyta, S.K. Maity and P. Rayo, Catal. Today 104 (2005) 86.
135. I. Mochida, X.Z. Zhao and K. Sakanishi, Ind. Eng. Chem. Res. 29 (1990) 2324.
136. M. Marafi, A. Al-Barood and A. Stanislaus, Petr. Sci. Technol. 27 (2005) 899.
137. M. Marafi, A. Al-Barood and A. Stanislaus, 18th World Petr. Congr. 2005, Johannesburg.
138. M. Marafi, A. Stanislaus and A. Al-Barood, Proc. 18th North Am. Catal. Soc. Meet., June 2003, Cancun, Mexico 87a.
139. J. Ancheyta-Juarez, S.K. Maity, G. Betancourt-Rivera, G. Centeno-Nolasco, P. Rayo-Mayoral, M.T. Gomez-Perez, Appl. Catal. 216 (2001) 195.
140. J. Ancheyta, G. Centeno and F. Trejo, Petr. Sci. Technol. 22 (2004) 219.
141. J. Ancheyta-Juarez, G. Betancourt-Rivera, G. Marroquin-Sanchez, A.M. Perez-Arellano, S.K. Maity, Ma.T. Cortez and R. Del Rio-Soto, Energy Fuels 15 (2001) 120.

142. J. Ancheyta, G. Betancourt, G. Marroquin, G. Centeno, L.C. Castaneda, F. Alonso, J.F. Munoz, Ma.T. Gomez and P. Rayo, Appl. Catal. 233 (2002) 159.
143. J. Ancheyta, G. Centeno, F. Trejo and G. Marroquin, Energy Fuels 17 (2003) 1233.
144. P. Rayo, J.Ancheyta, J. Ramirez and A. Gutierrez-Alejandre, Catal. Today 98 (2004) 171.
145. S.K. Maity, J. Ancheyta, F. Alonso and M.S. Rana, Catal. Today 98 (2004) 193.
146. C. Takeuchi, Y. Fukui, M. Nakamura and Y. Shiroto, Ind. Eng. Chem. Proc. Des. Dev. 22 (1983) 236.
147. J. Trawczynski and J. Walendziewski, Appl. Catal. 119 (1994) 59.
148. S.M. Richardson and M.R. Gray, Energy Fuels 11 (1997) 1119.
149. J.M. Jones, R.A. Kydd, P.M. Boorman and P.H. van Rhyn, Fuel 74 (1995) 1875.
150. P. Zeuthen and K. Andersen, Am. Chem. Soc. Div. Petr. Chem. Prep. 43 (1998) 80.
151. D. Ferdous, A.K. Dalai and J. Adjaye, Can. J. Chem. Eng. 83 (2005) 855.
152. D. Ferdous, A.K. Dalai and J. Adjaye, Appl. Catal. 260 (2004) 153.
153. D. Ferdous, A.K. Dalai, J. Adjaye and L. Kotlyar, Appl. Catal. 294 (2005) 80.
154. Y.W. Chen, W.C. Hsu, C.S. Lin, B.C. Kang, S.T. Wu, L.J. Leu and J.C. Wu, Ind. Eng. Chem. Res. 29 (1990) 1830.
155. Y.W. Chen and W.C. Hsu, Ind. Eng. Chem. Res. 36 (1997) 2526.
156. M.C. Tsai, Y.W. Chen and C. Li, Ind. Eng. Chem. Res. 32 (1993) 1603.
157. S.K. Maity, J. Ancheyta, L. Soberanis and F. Alonso, Appl. Catal. 250 (2003) 231.
158. S.K. Maity, J. Ancheyta, M.S. Rana, Energy Fuels 19 (2005) 343.
159. S.K. Maity, J. Ancheyta, M.S. Rana and P. Rayo, Catal. Today 109 (2005) 42.
160. Kushiyama, R. Aizawa, S. Kobayashi, Y. Koinuma, I. Uemasu and H. Ohuchi, Ind. Eng. Chem. Res. 30 (1991) 107.
161. M.C. Tsai, Y.W. Chen, B.C. Kang, J.C. Wu and L.J. Leu, Ind. Eng. Chem. Res. 30 (1991) 1801.
162. Y.W. Chen, M.C. Tsai and C. Li, Ind. Eng. Chem. Res. 33 (1994) 2040.
163. Y.W. Chen and M.C. Tsai, Ind. Eng. Chem. Res. 36 (1997) 2521.
164. D.S. Yang, R. Dureau, J.P. Charland and M. Ternan, Fuel 75 (1996) 1199.
165. M. Breysse, P. Afanasiev, C. Geantet and M. Vrinat, Catal. Today, 86 (2003) 5.
166. G.M.K. Abotsi and A.W. Scaroni, Fuel Proc. Technol. 22 (1989) 107.
167. S.K. Maity, M.S. Rana, S.K. Bej, J. Ancheyta-Juarez, G. Murali Dhar and T.S.R. Prasada Rao, Appl. Catal. 205 (2001) 215.
168. M.S. Rana, S.K. Maity, J. Ancheyta, G. Murali Dharand and T.S.R. Prasada Rao, Appl. Catal. 253 (2003) 165.
169. B. Caloch, M.S. Rana and J. Ancheyta, Catal. Today 98 (2004) 91.
170. J.W. Gosselink and W.H.J. Stork, Ind. Eng. Chem. Res. 36 (1997) 3354.
171. M.A. Ali, T. Tatsumi and T. Masuda, Appl. Catal. 233 (2002) 77.
172. S. Ahmed, S.A. Ali, H. Hamid and K. Honna. Stud. Surf. Sci. Catal. 145 (2003) 295.
173. W.C. Choi, K.H. Lee, K. Choi and B.H. Ha, Stud. Surf. Sci. Catal. 127 (1999) 243.
174. R.S. Mann, I.S. Sambi and K.C. Khulbe, Ind. Eng. Chem. Res. 27 (1988) 1788.
175. M.G. Yang, I. Nakamura and K. Fujimoto, Catal. Today 43 (1998) 273.
176. A.V. Abramova, E.V. Slivinskii, Y.Y. Goldfarb, A.A. Panin, E.A. Kulikova and G.A. Kliger, Kinet. Catal. 46 (2005) 758.
177. A.A. Esener and I.E. Maxwell, Stud. Surf. Sci. Catal. 50 (1989) 263.
178. H. Makishima, Y. Tanaka, Y. Kato, S. Kure, H. Shimada, N. Matsubayashi, A. Nishijima and M. Nomura, Catal. Today 29 (1996) 267.
179. T.J. Gardner, L.I. McLaughlin, J.E. Miller, R.S. Sandoval, G.J. Antos and L. Wang, Am. Chem. Soc. Div Petr. Chem. Prep. 47 (2002) 73.
180. V. Santes, J. Hebert, M.T. Cortez, R. Zarate, L. Diaz, P. Narayana, S. Swaimi, M. Aquine and M. Vrinat, Appl. Catal. 281 (2005) 121.
181. V. Santes, J. Hebert, M.T. Cortez, R. Zarate, L. Diaz and B. Medellin, Petr. Sci. Technol. 22 (2004) 103.

182. R. Silva-Rodrigo, C. Calderon-Salas, J.A. Melo-Banda, J.M. Dominguez and A. Vasquez-Rodriguez, Catal. Today 98 (2004) 123.
183. M. Ushio, J. Futigami, R. Kuroda, T. Ida, K. Hayashida and H. Ishihara, Am. Chem. Soc. Div. Petr. Chem. Prep. 50 (2005) 347.
184. L.A. Rankel, Energy Fuels 7 (1993) 937.
185. K. Honna, K. Sato, Y. Araki, Y. Miki, N. Matsubayashi and H. Shimada, Stud. Surf. Sci. Catal. 127 (1999) 427.
186. I. Nakamura, K. Aimoto and K. Fujimoto, AIChE Symp. Ser. 273, 85 (1989) 15.
187. M. Kouzu, Y. Kuriki, K. Uchida, K. Sakanishi, Y. Sugimoto, I. Saito, D. Fujii and K. Hirano, Energy Fuels 19 (2005) 725.
188. Y. Hashimoto, T. Enomoto, K. Honna, H. Aizono, H. Ueki, N. Ohshilo, M. Yoshimoto and H. Shimada, Am. Chem. Soc. Dive Petr. Chem. Prep. 47 (2002) 351.
189. N. Oshiho, T. Enomoto, K. Honda, H. Ueki, Y. Hashimoto, H. Aizono, M. Yoshimoto and H. Shimada, Fuel 83 (2004) 1895.
190. K. Madhusudan, B. Wei and C. Song, Catal. Today 43 (1998) 261.
191. B.C. Kang, S.T. Wu, H.H. Tsai and J.C. Wu, Appl. Catal. 45 (1988) 221.
192. A. Stanislaus, K. Al-Dolama and M. Absi-Halabi, J. Mol. Catal. 181 (2002) 33.
193. Z.S. Ying, B. Gevert, J.-E. Otterstedt and J. Sterte, Ind. Eng. Chem. Res. 34 (1995) 1566.
194. K.J. Smith, C. Ng, K. Badree and H. Meng, AIChE Symp. Ser. 282, 87 (1991) 106.
195. M.A. Altajan, J.F. Kriz and M. Ternan, Catalyst Deactivation, C.H. Bartholomew and J.B. Butt (Eds), Elsevier, Amsterdam, 1991, p. 315.
196. A. Segawa, K. Watanabe and M. Yoshimoto, Am. Chem. Soc. Div Petr. Chem Prep. 45 (2000) 605.
197. H. Fukuyama, S. Terai, M. Uchida, J.L. Lano and J. Ancheyta, Catal. Today 98 (2004) 207.
198. E. Lopez-Salinas, J.G. Espinosa, J.G. Hernandez-Cortez, J. Sanchez-Valente and J. Nagina, Catal. Today 109 (2005) 69.
199. Z.-S. Ying, B.S. Gevert, J.-E. Ottersted and J. Sterte, Appl. Catal. 153 (1997) 69.
200. F.X. Long and B.S. Gevert, Am. Chem. Soc. Div. Petr. Chem. Prep. 50 (2005) 379.
201. S.K.Maity, J. Ancheyta, M.S. Rana and P. Rayo, Energy Fuels 20 (2006) 427.
202. M.S. Rana, J. Ancheyta, S.K. Maity and P. Rayo, Catal. Today 109 (2005) 61.
203. J. Ramirez, P. Rayo, A. Gutierrez-Alejandre, J. Ancheyta and R.S. Mohan, Catal. Today 109 (2005) 54.
204. C.E. Ramos-Galvan, G. Sandoval-Robles, A. Castillo-Mares and J.M. Dominguez, Appl. Catal. 150 (1997) 37.
205. S. Inoue, T. Takatsuka, Y. Wada, S. Nakata and T. Ono, Catal. Today 43 (1998) 225.
206. E. Furimsky, Appl. Catal. 204 (2003) 1.
207. J. Trawczynski, Appl. Catal. 197 (2000) 289.
208. J.A. Melo-Banda, J.M. Dominguez and G. Sandoval-Robles, Catal. Today 65 (2001) 279.
209. S.T. Oyama, J. Catal. 216 (2003) 343.
210. C.W. Ingram and K. Ghebreyessus, Am. Chem. Soc. Div. Petr. Chem. Prep. 49 (2004) 92.
211. S.T. Oyama, X. Wang, F.G. Requejo, T. Sato and Y. Yoshimura, J. Catal. 209 (2002) 1.
212. N. Escalona, M. Yates, P. Avila, A. Lopez Agudo, J.L. Garcia Fierro, J. Ojeda and F.J. Gil-Llambias, Appl. Catal. 240 (2003) 151.
213. N. Escalona, J. Ojeda, R. Cid, G. Alves, A. Lopez Agudo, J.L.G. Fierro and F.J. Gil Llambias, Appl. Catal. 234 (2003) 45.
214. M.M. Hossain, M.A. Al-Saleh, M.A. Shalabi, T. Kimura and T. Inui, Appl. Catal. 278 (2004) 65.
215. M.A. Al-Saleh, M.M. Hossain, M.A. Shalabi, T. Kimura and T. Inui, Appl. Catal. 253 (2003) 453.
216. T. Isoda, K. Kusakabe, S. Morooka and I. Mochida, Energy Fuels 12 (1998) 493.
217. D.K. Lee, I.C. Lee and S.I. Woo, Appl. Catal. 109 (1994) 195.

218. Y. Miyatani, S. Yasuda, Y. Su, K. Kaneda, S. Urata and M. Nomura, J. Jpn. Petr. Inst. 42 (1999) 246.
219. K. Usui, K. Kidena, S. Murata, M. Nomura and W. Trisunaryanti, Fuel 83 (2004) 1899.
220. S. Terai, H. Fukuyama, K. Uehara and K. Fujimoto, J. Jpn. Petr. Inst. 43 (2000) 17.
221. D.K. Lee, W.L. Yoon and S.I. Woo, Fuel 75 (1996) 1185.
222. S.B. Jaffe, Ind. Eng. Chem. Res. 31 (1992) 2483.
223. S.B. Jaffe, H. Freund and W.N. Olmstead, Ind. Eng. Chem. Res. 44 (2006) 9840.
224. J. Ancheyta, S. Sanchez and M.A. Rodriguez, Catal. Today 109 (2005) 76.
225. F.M. Dautzemberg and J.C. de Deken, in Metal Complexes in Fossil Fuels, Am. Chem. Soc. Symp. Ser. 344, 1987, p. 233.
226. W.E. van Zijll Langhout, C. Ouwerkerk and K.M.A. Pronk, Proc. Natl Meet., June 8–12, 1980, Am. Inst. Chem. Eng. Philadephia.
227. M.A. Rodriguez and J. Ancheyta, Energy Fuels 18 (2004) 789.
228. H.R. Reinhoudt, C.H.M. Boons, A.D. van Langeveld, J.A.R. van Veen, S.T. Sie and J.A. Moulijn, Appl. Catal. 207 (2001) 25.
229. G. Marroquin, J. Ancheyta and C. Esteban, Catal. Today 104 (2005) 70.
230. X. Ma, K. Sakanishi and I. Mochida, Ind. Eng. Chem. Res. 35 (1996) 2487.
231. X. Ma, K. Sakanishi, T. Isoda and I. Mochida, Ind. Eng. Chem. Res. 34 (1995) 748.
232. L.Trytten, M.R. Gray and E.C. Sanford, Ind. Eng. Chem. Res. 29 (1990) 725.
233. D. Ferdous, A.K. Dalai and J. Adjaye, Ind. Eng. Chem. Res. 45 (2006) 544.
234. S.M. Yui and E.C. Sanford, 1985 Proc. Ref. Dept: 50th API Midyear Meet. Kansas City, 1985, API, Washington, DC, p. 290.
235. S.M. Yui and E.C. Sanford, Ind. Eng. Chem. Res. 28 (1989) 1278.
236. C. Botchwey, A.K. Dalai, J. Adjaye, Can. J. Chem. Eng. 82 (2004) 478.
237. J. Ancheyta, M.J. Angeles, M.J. Macias, G. Marroquin and R. Morales, Energy Fuels 16 (2002) 189.
238. G. Valavarasu, M. Bhaskar, B. Sairam, K.S. Balaraman and K. Balu, Petr. Sci. Technol. 23 (2005) 1323.
239. K. Aboul-Gheit, Erdol Erdgas Kohle 105 (1989) 319.
240. M.R. Gray, Ind. Eng. Chem. Res. 29 (1990) 505.
241. M. Sau, K. Basak, U. Manna, M. Santra and R.P. Verma, Catal. Today 109 (2005) 112.
242. M.A. Callejas and M. T. Martinez, Ind. Eng. Chem. Res. 38 (1999) 3285.
243. A. Marafi, S. Fukase, M. Al-Marri and A. Stanislaus, Energy Fuels 17 (2003) 661.
244. A. Marafi, H. Al-Bazzaz, M. Al-Marri, F. Maruyama, M. Absi-Halabi and A. Stanislaus, Energy Fuels 17 (2003) 1191.
245. O.K. Bhand and S.E. George, Stud. Surf. Sci. Catal. 100 (1996) 135.
246. F.X. Long and B.S. Gevert, J. Catal. 200 (2001) 91.
247. S. Trasobares, M.A. Callejas, A.M. Benito, M.T. Martinez, D. Severin and L. Brouwer, Ind. Eng. Chem. Res. 37 (1998) 11.
248. M.A. Callejas and M.T. Martinez, Energy Fuels 14 (2000) 1304.
249. M.A. Callejas and M.T. Martinez, Energy Fuels 14 (2000) 1309.
250. M.A. Callejas and M.T. Martinez, Energy Fuels 13 (1999) 629.
251. R. Galiasso, J. Garcia, J. Caprioli and A. Souto, Am. Chem. Soc. Div. Petr. Chem. Prep. 30 (1985) 50.
252. S. Trasobares, M.A. Callejas, A.M. Benito, M.T. Martinez, D. Severin and L. Brouwer, Ind. Eng. Chem. Res. 38 (1999) 938.
253. M.T. Martinez, M.A. Callejas, E. Carbo and A. Hernandez, Stud. Surf. Sci. Catal. 109 (1997) 565.
254. H. Nagaishi, E.W. Chan, E.C. Sanford and M.R. Gray, Energy Fuels 11 (1997) 402.
255. G. Gualda and S. Kasztelan, in Catalyst Deactivation, B. Delmon and G.F. Froment (Eds), 1994, Vol. 88, p. 145.
256. S. Sanchez, M.A. Rodriguez and J. Ancheyta, Ind. Eng. Chem. Res. 44 (2005) 9409.

257. F. Mosby, R.D. Buttke, A. Cox and C. Nikolaides, Chem. Eng. Sci. 41 (1986) 989.
258. R. Ayasse, H. Nagaishi, E.W. Chan and M.R. Murray, Fuel 76 (1997) 1025.
259. R. Prins, Adv. Catal. 46 (2001) 399.
260. E. Furimsky, Appl. Catal. 199 (2000) 147.
261. A. Stanislaus and B. H. Cooper, Catal. Rev. Sci. Eng. 36 (1994) 75.
262. R.F. Sullivan, M.B. Boduszynski and J.C. Fetzer, Energy Fuels 3 (1989) 603.
263. P.L. Jokuty and M.R. Gray, Energy Fuels 5 (1991) 791.
264. P.L. Jokuty and M.R. Gray, Ind. Eng. Chem. Res. 31 (1992) 1445.
265. L.B. Sun, Z.M. Zong, J.H. Kou, L.F. Zhang, Z.H. Ni, G.Y. Yu, H. Chen, X.Y. Wei and C.W. Lee, Energy Fuels 20 (2006) 19 (2005) 1.
266. F. Khorasheh, H.A. Rangwala, M.R. Gray and I.G. Dalla Lana, Energy Fuels 3 (1989) 716.
267. S. Asaoka, S. Nakata, Y. Shiroto and C. Takeuchi, Ind. Eng. Chem. Proc. Des. Dev. 22 (1983) 242.
268. R. Zou and L. Liu, in Asphaltenes and Asphalt, T.F. Yen and G.V. Chilingarian (Eds), Elsevier Science BV, New York, 1994, p. 339.
269. Izquierdo, L. Carbognani, V. Leon and A. Parisi, Fuel Sci. Technol. Int. 1 (1988) 561.
270. R.B. Solari, in Asphaltenes and Asphalt, F.T. Yen and G.V. Chilingarian (Eds), Elsevier Science, New York, 2000, Vol. 2, p. 149.
271. J. Ancheyta, G. Centeno, F. Trejo and J.G. Speight, Catal. Today 109 (2005) 162.
272. E.C. Sanford, Energy Fuels 9 (1995) 549.
273. J.G. Steer, K. Muehlenbachs and M.R. Gray, Energy Fuels 6 (1992) 540.
274. F. Ali, N. Ghaloum and A. Hauser, Energy Fuels 20 (2006) 231.
275. Y. Shiroto, S. Nakata, Y. Fukui and C. Ta keuchi, Ind. Eng. Chem. Proc. Des. Dev. 22 (1983) 248.
276. I. Gawel, D. Bociarska and P. Biskupski, Appl. Catal. 295 (2005) 89.
277. F. Trejo, J. Ancheyta, G. Ceteno and G. Maroquin, Catal. Today 109 (2005) 178.
278. A. Hauser, A. Stanislaus, A. Marafi and A. Al-Adwani, Fuel 84 (2005) 259.
279. H. Seki, S. Sato and M. Yoshimoto, J. Jpn. Petr. Inst. 44 (2001) 154.
280. S. Asaoka, S. Nakata, Y. Shiroto and C. Takeuchi, in Metal Complexes in Fossil Fuels, R.H. Filby and J.F. Branthaver (Eds), Am. Chem. Soc. Symp. Ser. 344, 1987, p. 275.
281. L.A. Rankel and L.D. Rollmann, Fuel 62 (1983) 46.
282. R. Galiasso Tailleur and L. Caprioli, Catal. Today 109 (2005) 185.
283. H. Knozinger, D. Cordischi and B. Vielhaber, Catal. Today 7 (1990) 447.
284. M. Loos, I. Ascone, C. Goulon-Ginet, J. Goulon, C. Guillard, M. Lacroix, M. Breysse, D. Faure and T. des Courieres, Catal. Today 7 (1990) 515.
285. J.P. Janssen, A.D. Van Langeveld and J.A. Moulijn, Appl. Catal. 179 (1999) 229.
286. P.C.H. Mitchell and C.E. Scott, Catal. Today 7 (1990) 467.
287. C.-W. Huang and J. Wei, Ind. Eng. Chem. Proc. Des. Dev. 19 (1980) 250.
288. C.-W. Huang and J. Wei, Ind. Eng. Chem. Proc. Des. Dev. 19 (1980) 257.
289. R. Agrawal and J. Wei, Ind. Eng. Chem. Proc. Des. Dev. 23 (1984) 505.
290. R.A. Ware and J. Wei, J. Catal. 93 (1985) 100.
291. J.P. Janssen, G. Elst, E.G. Schrikkema, A.D. van Langeveld, S.T. Sie and J.A. Moulijn, Recl. Trav. Chim. Pays Bas 115 (1996) 465.
292. A.J. Garcia-Lopez, R. Cuevas, J. Ramirez, J. Ancheyta, A.A. Vargas-Tah, R. Nares and A. Gutierres-Alejandre, Catal. Today 109 (2005) 545.
293. H. Knozinger, D. Cordischi and B. Vielhaber, Catal. Today 7 (1990) 447.
294. P.C.H. Mitchell and C.E. Scott, Catal. Today 7 (1990) 467.
295. H. Hirao, S. Shaik and P.M. Kozlowski, J. Phys. Chem. A 110 (2006) 6091.
296. C.H. Tsai, F.E. Massoth, S.Y. Lee and J.D. Seader, Fuel Proc. Technol. 29 (1991) 153.
297. F.X. Long and B.S. Gevert, J. Catal. 222 (2004) 1.
298. F.X. Long, B.S. Gevert and P. Abrahamsson, J. Catal. 222 (2004) 6.

299. B.P. Embaid, F. Gonzales-Jimenez and C.E. Scott, Appl. Catal. 287 (2005) 75.
300. A. Morales and R. Galiasso, Fuel 61 (1982) 13.
301. G. Gualda and H. Toulhoat, Rev. Inst. Fr. Petr. 43 (1988) 567.
302. E. Furimsky, Ind. Eng. Res. Prod. Res. Dev. 17 (1977) 329.
303. E. Furimsky, Erdol Kohle 35 (1982) 455.
304. M.A. Callejas, M.T. Martinez, T. Blasco and E. Sastre, Appl. Catal. 218 (2001) 181.
305. H. Mizutani, K. Kiriyama, K. Kondo, K. Idei and I. Mochida, J. Jpn Petr. Inst. 47 (2004) 107.
306. J. Bartholdy and B.H. Cooper, Am. Chem. Soc. Div. Petr. Chem Prep. 38 (1993) 386.
307. J. Bartholdy and B.H. Cooper, Stud. Surf. Sci. Catal. 100 (1996) 117.
308. M. De Wind, Y. Miyauchi and K. Fujita, Stud. Surf. Sci. Catal. 199 (1996) 157.
309. A. Al-Nasser, S.R. Chaudhari and S. Bhattacharya, Stud. Surf. Sci. Catal. 100 (1996) 171.
310. M.S. Rana and J. Ancheyta, Catal. Today 109 (2005) 24.
311. Y. Yokoyama, N. Ishikawa, K. Nakanishi, K. Satoh, A. Nishijima, H. Shimada, N. Matsubayashi and M. Nomura, Catal. Today 29 (1996) 261.
312. Y. Tanaka, H. Shimada, N. Matsubayashi, A. Nishijima and M. Nomura, Catal. Today 45 (1998) 319.
313. S. Gamez, R. Bacaus and M. Vrinat, Am. Chem. Soc. Div. Petr. Chem. Prep. 50 (2005) 375.
314. H. Toulhoat, J.C. Plumail, G. Martino and Y. Jacquin, Am. Chem. Soc. Div. Petr. Chem. Prep. 30 (1985) 85.
315. J. Devanneaux, J.P. Gallez, L. Mariette, P.A. Engelhard, Am. Chem. Soc. Div. Petr. Chem. Prep. 32 (1987) 84.
316. S. Dejonghe, R. Hubaut, J. Grimblot, J.P. Bonnelle, T. Des Courieres and D. Faure, Catal. Today 7 (1990) 569.
317. C. Guillard, M. Lacroix, M. Vrinat, M. Breysse, B. Mocaer, J. Grimblot, T. Des Courieres and D. Faure, Catal. Today 7 (1990) 587.
318. C.E. Scott, B.P. Embaid, F. Gonzalez-Himenez, R. Hubaut and J. Grimblot, J. Catal. 166 (1997) 333.
319. W. Kearnbach, W. Kisielow, L. Warzecha, K. Miga and R. Klecan, Fuel 69 (1990) 221.
320. D. Dong, S. Jeong and F.E. Massoth, Catal. Today 37 (1997) 267.
321. B.D. Muegge and F.E. Massoth, Stud. Surf. Sci. Catal. 68 (1991) 297.
322. M. Anemia, Y. Korai and I. Mochida, J. Jpn Petr. Inst. 46 (2003) 99.
323. N. Koizumi, Y. Urabe, K. Inamura, T. Itoh and M. Yamada, Catal. Today 106 (2005) 211.
324. P. Zeuthen, J. Bartholdy, P. Wiwel and B.H. Cooper, Stud. Surf. Sci. Catal. 88 (1994) 1998.
325. W. Khanda, I. Siu, J. Adjaye, A.E. Nelson and M.R. Gray, Energy Fuels 18 (2004) 539.
326. K.P. de Jong, Ind. Eng. Chem. Res. 33 (1994) 821.
327. K.P. de Jong, D. Reinalda and C.A. Emeis, in Catalyst Decativation, B. Delmon and F.G. Froment (Eds), Elsevier, Amsterdam, 1994, Vol. 88, p. 155.
328. K.P. de Jong, H.P.C.E. Kuipers and J.A.R. van Veen, in Catalyst Deactivation 1991, C.H. Bartholomew and J.B. Dutt (Eds), Elsevier, Amsterdam, 1991, p. 28.
329. J. Van Dorn, J.A. Moulijn and G. Djega-Mariadassou, Appl. Catal. 60 (1990) 77.
330. J. Van Dorn and J.A. Moulijn, Fuel Proc. Technol. 35 (1993) 275.
331. A.D. van Langeveld, E.L.J. Ariens, J.A. Moulijn and R. De Jong, Am. Chem. Soc. Div. Petr. Chem. Prep. 43 (1998) 55.
332. P. Wiwel, P. Zeuthen and A.C. Jacobsen, Stud. Surf. Sci. Catal. 68 (1991) 257.
333. K. Matsushita, R. Koide, S. Fukase, A. Stanislaus, A. Al-Barood, F. Al-Jasem and M. Absi-Halabi, Am. Chem. Soc. Div. Petr. Chem. Prep. 45 (2000) 655.
334. C.H. Yang, S.F. Chen, H. Chen and F. Du, Am. Chem. Soc. Div. Petr. Chem. Prep. 47 (2002) 318.
335. J.P. Janssen, Y.M. Xu, G. Witte, A.D. Van Langeveld, S.T. Sie and J.A. Moulijn, Fuel 77 (1998) 1367.
336. E. Furimsky, A. Siukola and A. Turenne, Ind. Eng. Chem. Res. 35 (1996) 440.

337. T.E. Myers, B.L. Myers, T.H. Fleisch and G.W. Zajac, AIChE Symp. Ser. 273, 85 (1989) 21.
338. A. Morales and R.B. Solari, Stud. Surf. Sci. Catal. 100 (1996) 125.
339. Y. Yamamoto, Y. Mizutani, Y. Shibata, Y. Kitou and H. Yamazaki, Stud. Surf. Sci. Catal. 100 (1996) 181.
340. A.W. Aldag, AIChE Symp. Ser. 273, 85 (1989) 32.
341. C. Takeuchi, S. Asaoka, S. Nakata and Y. Shiroto, Am. Chem. Soc. Div. Petr. Chem. Prep. 30 (1985) 96.
342. D. Alvarez, R. Galiasso and P. Andreu, J. Jpn Petr. Inst. 22 (1979) 234.
343. T.H. Lindstrom, F. Levesque and M. Cathcart, Hydroc. Proc. Febr. 82 (2003) 49.
344. T.C. Xiao, J.X. Su, H.T. Wang, Y.L. Lu, Z.F. Yan, A.P.E. York, J.B. Claridge, J. Sloan and M.L.H. Green, Ind. Eng. Chem. Res. 39 (2000) 3679.
345. Hauser, H. Marafi, A. Stanislaus, A. Al-Abduani, A. Energy Fuels 19 (2005) 544.
346. K. Al-Dalama and A. Stanislaus, Chem. Eng. J. 120 (2006) 33.
347. C. Yang, F. Du, H. Zheng and K.H. Chung, Fuel 83 (2004) 1907.
348. M. Yumoto, S.G. Kukes, M.T. Klein and B.C. Gates, Ind. Eng. Chem. Res. 40 (2001) 131.
349. M. Yumoto, S.G. Kukes, M.T. Klein and B.C. Gates, Ind. Eng. Chem. Res. 35 (1996) 3203.
350. H. Koyama, E. Nagai and H. Kumagai, Am. Chem. Soc. Div. Petr. Chem. Prep. 42 (1995) 446.
351. A. Marafi and A. Stanislaus, Appl. Catal. 159 (1997) 259.
352. A. Marafi, A. Hauser and A. Stanislaus, Energy & Fuel 20 (2006) 1145.
353. P.R. Fozard, J.W. McMillan and P. Zeuthen, J. Catal. 129 (1995) 43.
354. P. Zeuthen, J. Bartholdy and F.E. Massoth, Appl. Catal. 152 (1995) 103.
355. J. Bartholdy, P. Zeuthen and F.E. Massoth, Appl. Catal. 129 (1995) 33.
356. M. Ternan, E. Furimsky and B.I. Parsons, Fuel Proc. Technol. 2 (1979) 45.
357. H. Higashi, T. Takahashi and T. Kai, Stud. Surf. Sci. Catal. 145 (2003) 339.
358. J. Ancheyta, G. Betancourt, G. Centeno and G. Marroquin, Energy Fuels 17 (2003) 462.
359. F. Kumata, H. Seki, T. Saito and M. Yoshimoto, J. Jpn. Petr. Inst. 44 (2001) 252.
360. M. Seki and M. Yoshimoto. J. Jpn. Petr. Inst. 44 (2001) 259.
361. A. Stanislaus, A. Hauser and M. Marafi, Catal. Today 109 (2005) 167.
362. A. Voorhies, Ind. Eng. Chem. 37 (1945) 318.
363. Y.W. Chen, W.C. Hsu, C.S. Lin, B.C. Kang, S.T. Wu, L.J. Leu, J.C. Wu., Ind. Eng. Chem. Res. 29 (1990) 1830.
364. S.Y. Lee, J.D. Seader, C.H. Tsai and F.E. Massoth, Ind. Eng. Chem. Res. 30 (1991) 29.
365. T. Takatsuka, Y. Wada and S. Inoue, in Deactivation and Testing of Hydrocarbon-Processing Catalysts, Chapter 30, p. 414, Am. Chem. Soc., Publ. 1996.
366. H.A.H. Al-Adwani, H.M.S. Lababidi, I.M. Alatigi and F.S. Al-Dafferi, Can. J. Chem. Eng. 83 (2005) 281.
367. I.C. Lewis and L.S. Singer, Am. Chem. Soc. Div. Petr. Chem. Prep. 30 (1985) 834.
368. I.C. Lewis, Carbon 20 (1981) 519.
369. J.W. Bunger, Am. Chem. Soc. Div. Petr. Chem. Prep. 30 (1985) 549.
370. D. Nohara and T. Sakai. Ind. Eng. Chem. Res. 31 (1992) 14.
371. M. Absi-Halabi, A. Stanislaus and D.L. Trimm, Appl. Catal. 72 (1991) 193.
372. J. Kubo, H. Higashi, Y. Ohmoto and H. Arao, Energy Fuels 10 (1996) 478.
373. K.P. de Jong, Ind. Eng. Chem. Res. 33 (1994) 3141.
374. E. Furimsky, Fuel Proc. Technol. 6 (1981) 1.
375. P. Kovacic and J. Oziomek, J. Org. Chem. 29 (1964) 100.
376. M. Meot-Ner, J. Phys. Chem. 88 (1980) 2724.
377. B.D. Flockart, I.M. Sesay and R.C. Pink, J. Chem. Soc. Chem. Commun. 10 (1980) 439.
378. B.C. Gates, J.R. Katzer and G.C.A. Schuit, Chemistry of Catalytic Processes, McGraw Hill, New York, 1979.
379. A.W. Scaroni, R.G. Jenkins, Am. Chem. Soc. Div. Petr. Chem Prep. 30 (1985) 544.

380. P. Magnoux, A. Rabeharitsara and H.S. Cerqueira, Appl. Catal. 304 (2006) 142.
381. H.A. Abdullah, A. Hauser, F.A. Ali and A. Al-Adwani, Energy Fuels 20 (2006) 320.
382. S.K. Sahoo, S.S. Ray, and I.D. Singh, Appl. Catal. 278 (2004) 83.
383. H. Seki and F. Kumatu, Stud. Surf. Sci. Catal. 126 (1999) 357.
384. H. Seki and F. Kumata, Energy Fuels 14 (2000) 980.
385. A. Fonseca, P. Zeuthen and J.B. Nagy, Fuel 74 (1995) 1267.
386. A. Fonseca, P. Zeuthen and J.B. Nagy, Fuel 75 (1996) 1363.
387. A. Fonseca, P. Zeuthen and J.B. Nagy, Fuel 75 (1996) 1413.
388. K. Matsushita, A. Hauser, A. Marafi, R. Koide and A. Stanislaus, Fuel 83 (2004) 1031.
389. J. Van Dorn, J.L. Bosch, R.J. Bakkum and J.A. Moulijn, in Catalyst Deactivation 1987, B. Delmon and G.F. Froment (Eds), Elsevier Science, Amsterdam, 1987, p. 391.
390. J. Van Dorn, H.A.A. Barbolina and J.A. Moulijn, Ind. Eng. Chem. Res. 31 (1992) 101.
391. Y. Yoshimura and E. Furimsky, Fuel 65 (1986) 1388.
392. F.A. Ali, A. Hauser, H.A. Abdullah and A. Al-Adwani, Energy Fuels 20 (2006) 45.
393. K. Matsushita, A. Marafi, A. Hauser and A. Stanislaus, Fuel 83 (2004) 1669.
394. J. Bartholdy and S.I. Andersen, Energy Fuels 14 (2000) 52.
395. F. Ali, G. Narjes and A. Hauser, Energy Fuels 20 (2006) 231.
396. S. Li, C. Liu and W. Liang, Am. Chem. Soc. Div. Petr. Chem. Prep. 46 (2001) 259.
397. W. Yanfei, C. Jiang, J. Shengsheng and S. Benxia, Energy Fuels 17 (2003) 344.
398. X. Zhang, M. Chodakowski and J.M. Shaw, Energy Fuels 19 (2005) 1405.
399. X. Zhang and J.M. Shaw, Energy Fuels 20 (2006) 473.
400. I. Mochida, X.Z. Zhao and K. Sakanishi, Ind. Eng. Chem. Res. 29 (1990) 334.
401. H. Beuther, O.A. Larson and A.J. Perrotta, in Catalyst Deactivation, B. Delmon and G.F. Froment (Eds), Elsevier, Amsterdam, 1980, p. 271.
402. J.C. Fetzer, Stud. Surf. Sci. Catal. 100 (1996) 263.
403. R.F. Sullivan, M.M. Boduszinski and J. Fetzer, Energy Fuels 3 (1989) 603.
404. M.A. Bannayan, H.L. Lemke and W.K. Stephenson, Stud. Surf. Sci. Catal. 100 (1996) 273.
405. V.A. Munoz, S.V. Ghorpadkas and M.R. Gray, Energy Fuels 8 (1994) 426.
406. M.R. Gray, Y. Zhao and C.M. McKnight, Fuel 79 (2000) 285.
407. P. Dufresne, Appl. Catal. (2006).
408. J. Guitian, A. Souto, L. Ramirez, R. Marzin and B. Solari, Proc. of Fourth UNITAR/UNDP Int. Conf. on Heavy Crudes and Tar Sands, August 1988, Edmonton, Canada, Vol. 4, p. 237.
409. C.S. Kim and F.E. Massoth, Fuel Proc. Technol. 35 (1993) 289.
410. R.L.C. Bonne, P. van Steenderen, A.D. van Langeveld and J.A. Moulijn, Ind. Eng. Chem. Res. 34 (1995) 3801.
411. D. Ronze, P. Fongarland, I. Pitault and M. Forissier, Chem. Eng. Sci. 57 (2002) 547.
412. M. Sato, N. Takayama, S. Kurita and T. Kwan, Nippon Kagaku Zasshi, 92 (1971) 834.
413. E. Newson, Ind. Eng. Chem. Proc. Des. Dev. 14 (1975) 27.
414. J. Wei, Catalyst Deactivation 1991, C.H. Bartholomew and J.B. Butt (Eds), Elsevier Science, 1991, p. 333.
415. R.R. Melkote and K.F. Jensen, Chem. Eng. Sci. 44 (1989) 649.
416. E.L. Sughrue, R. Adarme, M.M. Johnson, C.J. Lord and M.D. Phillips, Catalyst Deactivation 1991 C.H. Bartholomew and J.B. Butt (Eds), Elsevier Science, 1991, p. 281.
417. L.O. Oyekunle and O.B. Ikpekri, Ind. Eng. Chem. Res. 43 (2004) 66.
418. L.O. Oyekunle, O.B. Ikpekri and A. Jaiyola, Catal. Today 109 (2005) 128.
419. C.J. Pereira, Ind. Eng. Chem. Res. 29 (1990) 512.
420. C.J. Pereira, J.W. Beeckman, W.C. Cheng and W. Suarez, Ind. Eng. Chem. Res. 29 (1990) 520.
421. P. Bourseau, G. Muratet, C. Sailloour, H. Toulhoat, Rev. Inst. Fr. Petr. 40 (1985) 595.
422. H. Toulhoat, R. Szymanski and J.C. Plumail, Catal. Today 7 (1990) 531.
423. F.M. Dautzemberg, J. van Klinken, K.M.A. Pronk, S.T. Sie and J.B. Wijffels, ACS Symp. Ser. 65 (1978) 254.

424. P.N. Hannerup and A.C. Jacobsen, ACS Symp. Ser. 28 (1983) 576.
425. S.J. Khang and J.F. Mosby, Ind. Eng. Chem. Proc. Des. Dev. 25 (1986) 437.
426. S. Togawa, E.K.T. Kam and M. Al Shamali, Am. Chem. Soc. Fuel Div. Prep. 46 (2001) 3.
427. E.K.T. Kam, M. Al-Shamali and M. Al-Jeraidan, Studies on Deactivation Mechanism of Hydroprocessing Catalysts Using Kuwaiti Feedstocks, Report KISR 7037, December 2003.
428. E.K.T. Kam, M. Al-Shamali, M. Juraidan and H. Qabazard, Energy Fuels 19 (2005) 753.
429. M. Juraidan, M. Al-Shamali, H. Qabazard and E.K.T. Kam, Energy Fuels 20 (2006) 1354.
430. J.P. Janssens, B.J. Bezemer, A.D. van Langeveld, S.T. Siie and J.A. Moulijn, Catalyst Deactivation 1994, B. Delmon and F.G. Froment (Eds), Elsevier Science, Amsterdam, 1994, p. 335.
431. J.P. Janssens, A.D. van Langeveld, S.T. Sie and J.A. Moulijn, ACS Series 1997, Chap. 18, p. 238.
432. B.M. Vogelaar, R.J. Berger, B. Bezemer, J.-P. Janssen, A.D. van Langeveld, S. Eisbouts and J.A. Moulijn, Chem. Eng. Sci. 61 (2006) 7463.
433. H. Toulhoat, D. Hudebine, P. Raybaud, D. Guillame and S. Kressmann, Catal. Today 109 (2005) 135.
434. S.H. Kodama, H. Nitta, T. Takatsuka and T. Yokoyama, J. Jpn. Petr. Inst. 23 (1980) 310.
435. E.K.T. Kam, M.M. Al-Mashan and H. Dashti, Stud. Surf. Sci. Catal. 100 (1996) 283.
436. V.T. Sinha, M.S. Butensky and D. Hyman, Ind. Eng. Chem. Proc. Des. Dev. 25 (1986) 321.
437. F. Morel, S. Kressman, V. Harle and S. Kasztelan, Stud. Surf. Sci. Catal. 106 (1997) 1.
438. H. Qabazard and R. Adarme, in Catalysts in Petroleum Refining 1989, D.L. Trimm (Ed.) Elsevier Science, Amsterdam, 1990, p. 61.
439. M. Absi-Halabi, A. Stanislaus and H. Qabazard, Hydroc. Proc. 2 (1997) 45.
440. A. Kundu, K.D.P. Nigam, A.M. Duquenne and H. Delmas, Rev. Chem. Eng. 19 (2003) 531.
441. F.M. Nooy, Oil Gas J. 82 (1984) 152.
442. G.L. Hamilton, M. Fitzgerald and V. D'Amico, Proc. 1985 AIChE Natl Meet., November 2–7, 1986, Miami, FL, p. 1.
443. J.W. Gosselink, A. Van den Paverd and W.H.J. Stork, in Catalysis in Petroleum Refining 1989, D.L. Trimm (Ed.), Elsevier Science, Amsterdam, 1990.
444. R.T. Pennind, A.G. Vickers and B.R. Shah, Hydroc. Proc. 5 (1982) 145.
445. J.G. Sikonia, F. Stolfa, L.E. Hutchings, W.L. Jacobs and V.P. Burton, Oil Gas J. 79 (1981) 141.
446. H.R. Jackson and K.E. Whithead, Oil Gas J. 75 (1977) 66.
447. A.M. Henke, Oil Gas J. 97 (1970) 68.
448. S.G. Paradis, G.D. Gould, D.A. Bea and E.M. Reed, Chem. Eng. Prog. 67 (1971) 57.
449. B.J. Young and R.L. Richardson, Hydroc. Proc. Sept. 56 (1977) 103.
450. A. Billon, F. Morel, M.E. Morrison and J.P. Peries, Rev. Inst. Fr. Petr. 49 (1994) 495.
451. S. Kressmann, F. Morel, V. Harle and S. Kasztelan, Catal. Today 43 (1998) 203.
452. A. Al-Nasser, S.R. Chaudhuri and S. Battacharya, Stud. Surf. Sci. Catal. 100 (1996) 171.
453. M. Skripek, E. De la Fuente and W.H. Smith, Fifth UNITAR/UNDP Int. Conf. on Heavy Oils and Tar Sands, August 1991, Caracas, Venezuela, Vol. 4, p. 37.
454. W.C. Van Zijl Langhout, C. Ouwerkek and K.M. Pronk, Oil Gas J. 1 (1980) 120.
455. R.L. Howell, C. Hung, K.R. Gibson and H.C. Chen, Oil Gas J. 83 (1985) 121.
456. G.L. Scheuermann, D.R. Johnson, B.E. Reynolds, R.W. Bachtel and R.S. Threlkel, Fuel Proc. Technol. 35 (1993) 39.
457. W.J. Beaton, N.K. McDaniel, W.E. McWirter, R.D. Petersen and R.P. van Driesen, Oil Gas J. 84 (1986) 47.
458. B. Scheffer, M.A. van Koten, K.W. Robschlager and F.C. de Bocks, Catal. Today 43 (1998).
459. R.P. van Driesen, J. Caspers, A.R. Campbell and G. Lunin, Hydroc. Proc. 5 (1979) 107.
460. R.M. Eccles, Fuel Proc. Technol. 35 (1993) 21.
461. K. Tanabe and M.R. Gray, Energy Fuels 11 (1997) 1040.

462. K. Niemann, Proc. Fourth UNITAR/UNDP Int. Conf. Heavy Crudes and Tar Sands, 1988, Edmonton, Canada, p. 3.
463. K. Niemann and F. Wenzel, Fuel Proc. Technol. 35 (1993) 1.
464. F.W. Wenzel and A. Herrera, Proc. UNITAR/UNDP Int. Conf. on Heavy Crudes and Tar Sands, 1988, Edmonton, Canada.
465. R. Marzin, J. Guitian, J. Krasuk, F. Silva, A. Suoto and R.B. Solari, Proc. Fourth UNITAR/UNDP Int. Conf. on Heavy Crude and Tar Sands, August 7, 1988, Edmonton, Canada.
466. F. Silva, G. Guitian, R. Galiasso and J. Krasuk, 11th World Petr. Congr., 1983, London, UK, Vol. 4, p. 199.
467. R.B. Solari, F. Silva, A. Souto, G. De Drago, J. Guitian, R. Marzin and J.C. Gonzales, Proc. Fifth UNITAR/UNDP Int. Conf. on Heavy Crudes and Tar Sands, 1991 Caracas, Venezuela, Vol. 4.
468. T. Kondo, W.F. Souza and H.T. Sakurai, Am. Chem. Soc. Div. Petr. Chem. Prep. 48 (2003) 143.
469. J.G. Gatsis, US Patent 4,194,967, March 1980.
470. J.A. Dunn, J.B. McLeod, R.D. Myers and R. Bearden Jr, Energy Fuels 17 (2003) 45.
471. T. Itoh and H. Sue, Fifth UNITAR/UNDP Int. Conf. on Heavy Oils and Tar Sands, August 1991, Caracas, Venezuela, Vol. 4, p. 47.
472. F.L. Platanega, S. Eijsbouts and M.B. Cerfontain, US Patent 6,860,987, March 2005.
473. A.G. Gabrielov, O. K. Bhan, US Patent 6,218,333, April 2001.
474. D.E. Sherwood Jr, B.H. Bartley, L.D. Neff and P.S.E. Dai, US Patent 5,047,142, September 1991.
475. D.E. Sherwood Jr, B.H. Bartley, L.D. Neff and P.S.E. Dai, US Patent 5,094,994, March 1992.
476. H. Toulhoat, Y. Jacquin and T. Dupin, US Patent 4,499,203, 1985.
477. P.S.E. Dai and D.E. Sherwood Jr., US Patent 5,397,456, March 1995.
478. P.S.E. Dai, G.V. Nelson, G. Nongbri, R.E. Pratt and D.E. Sherwood Jr, US Patent 5,399,259, March 1995.
479. R.A. Kemp, US Patent 4,820,680, April 1989.
480. M. de Boer, US Patent 6,030,915, February 2000.
481. D.C. Artes and S.G. Kukes, US Patent 5,322,829, June 1994.
482. W.I. Beaton, A.L. Hensley and A.J. Evans, US Patent 4,707,466, November 1987.
483. W.I. Beaton, A.L. Hensley and A.J. Evens, US Patent 4,657,665, April 1987.
484. A.J. Evans and A.L. Hensley, US Patent 4,657,664, April 1987.
485. L.B. Peck, A.L. Hensley and G.L. Ott, US Patent 4,746,419, May 1988.
486. K. Fujita, T. Kamo, M. Shimowake and Y. Inoue, US Patent 5,888,380, March 1999.
487. D.R. Johnson, US patent 4,976,848, December 1990.
488. D.R. Johnson, US Patent 5,089,463, February 1992.
489. A.Q.M. Boon, C.J.J. Den Ouden and O.P. Bhan, US Patent 5,744,025, April 1998.
490. C.J. Pereira and W.C. Cheng, US Patent 4,908,344, March 1990.
491. J. Pereira and W.C. Cheng, US Patent 4,791,090, December 1988.
492. K.R. Gibson, US Patent 4,606,815, August 1986.
493. J.E. Duddy, L.M. Abrams and S.J. Hildebrandt, US Patent 5,925,238, July 1999.
494. S.F. Abbo, H.D. Simpson and P.S. Iyer, US Patent 5,906,731, May 1999.
495. J.P. Angevine, P. Chu, T.F. Degnan and G.W. Kirker, US Patent 4,834,869, May 1989.
496. H.D. Simpson, US Patent 5,403,806, April 1995.
497. K. Fujita, S. Masafumi and T. Kamo, US Patent 6,655,349, December 2003.
498. D.E. Sherwood Jr, US Patent 5,968,348, October 1999.
499. D.E. Sherwood Jr and P.S.E. Dai, US Patent 5,928,499, July 1999.
500. H.D. Simpson, M.J. Schwedock and J.W. Ward, US Patent 5,334,307, August 1994.
501. S. Abe, Y. Takahashi, A. Hino and M. de Boer, US Patent 6,893,553, May 2005.

502. P.S.E. Dai, C.N. Campbell II, B.R. Martin and R.H. Petty, US Patent 5,453,411, September 1995.
503. P.S.E. Dai, C.N. Campbell II, J.A. Durkin and D.E. Sherwood Jr, US Patent 5,322,616, June 1994.
504. P.S.E. Dai, C.N. Campbell II, J.A. Durkin and D.E. Sherwood Jr, US Patent 5,436,215, July 1995.
505. P.S.E. Dai, C.N. Campbell II, B.R. Martin and D.E. Sherwoof Jr, US Patent 5,308,472, May 1994.
506. P.S.E. Dai, C.N. Campbell II, B.R. Martin and D.E. Sherwood, Jr, US Patent 5,342,507, August 1994.
507. D.F. Best, G.N. Long R.J. Pellet, J.A. Rabo and E.T. Wolynic, US Patent 5,019,240, May 1991.
508. M.M. Habib and A.J. Dahlberg, US Patent 5,393,410, February 1995.
509. M.L. Occelli, US Patent 5,328,590, July 1994.
510. T.F. Degnan Jr, K.M. Keville, M.E. Landis and D.O. Marler, US Patent 5,290,744, March 1994.
511. Nakaoka, US Patent 5,686,374, November 1997.
512. R.J. Quann, US Patent 4,587,012, May 1986.
513. H. Fukuyama, K. Ohtsuka, S. Terai and S. Sawamoto, US Patent 6,797,153, September 2004.
514. S.J. Miller, US Patent 4,859,312, August 1989.
515. P.J. Angevine, T.E. Degnan Jr and M.E. Landis, US Patent 4,812,228, March 1989.
516. C.A. Audeh and T.Y. Tan, US Patent 4,582,595, April 1986.
517. R.H. Wolk and W.C. Rovesti, US Patent 3,819,509, June 1974.
518. R.H. Wolk, G. Nogbri and W.C. Rovesti, US Patent 3,901,792, August 1975.
519. T.T. Hilbert, N.D. Mazzone and M.S. Sarli, US Patent 5,510,016, 23 April 1996.
520. J.A. Howell, US Patent 4,659,452, April 1987.
521. Y. Fukui, Y. Shirota, M. Ando and Y. Homma, US Patent 4,166,026, August 1979.
522. L.E. Hutchings, US Patent 4,239,614, December 1980.
523. E.A. Jones, US Patent 4,234,314, November 1980.
524. L.E. Gardner and S.G. Kukes, US Patent 4,828,683, May 1989.
525. S.G. Kukes and K.K. Brandes, US Patent 4,767,523, August 1988.
526. S.G. Kukes, T. Davis and K.K. Brandes, US Patent 4,778,587, October 1988.
527. K.K. Brandes, S.G. Kukes and T. Davis, US Patent 4,778,588, October 1988.
528. H.J. Trimble and D.R. Cash, US Patent 5,879,642, March 1999.
529. C.W. Hung and B.E. Reynolds, US Patent 4,830,736, May 1989.
530. M.M. Habib, P.L. Winslow and R.O. Moore Jr, US Patent 5,593,570, January 1997.
531. M.M. Habib, P.L. Winslow and R.O. Moore Jr, US Patent 5,439, 860, August 1995.
532. T.C. Ho, US Patent 4,973,397, November 1990.
533. T.S. Chou, C.R. Kennedy and S.S. Shih, US Patent 4,997,544, March 1991.
534. P.L. Winslow and R.F. Sullivan, US Patent 5,071,805, December 1991.
535. R. Gupta, S.J. Rossetti, D.C. Dankworth, J.L. Kaufman, D.L. Vannauker and J.P. Bailor, US Patent 6,846,469, January 2005.
536. M.D. Ackerson and S.B. Byars, US Patent 6,881,326, April 2005.
537. J. Bartholdy, US Patent 5,368,722, November 1994.
538. T.S. Chou, C.R. Kennedy and S.S. Shih, US Patent 4,997,544, March 1991.
539. T.N. Kalnes, US Patent 6,451,197, September 2002.
540. T.N. Kalnes and D.B. Gates, US Patent 6,387,245, May 2002.
541. T.N. Kalnes, US Patent 6,361,683, March 2002.
542. T.N. Kalnes, US Patent 6,328,879, December 2001.
543. J.W. Myers, S.L. Parrott and E.L. Sughrue, US Patent 4,619,759, 28 October 1986.
544. F. Banta, US Patent 4,447,314, May 1984.
545. S.S. Shih and D. Milstein, US Patent 4,421,633, December 1983.

546. W.D. Gillespie, US Patent 5,232,578, August 1993.
547. J.S. Buchanan, US Patent 6,017,443, January 2000.
548. R.D. Cash, US Patent 4,430,203, February 1984.
549. A.G. Bridge and B.E. Stangeland, US Patent 4,615,789, October 1986.
550. A.J. MacLean, R.L. Holloway, V.A. Lawson and J.W. Cronen, US Patent 4,902,405, February 1990.
551. L.E. Gardner, R.J. Hogan, E.L. Sughrue and J.W. Myers, US Patent 4,657,663, April 1987.
552. A.L. Hensley Jr, L.M. Quick, US Patent 4,431,525, February 1984.
553. L.E. Gardner, R.J. Hogan, E.L. Sughrue II and J.W. Myers, US Patent 4,657,663, April 1987.
554. J.W. Van der Meer and M.C. Zonnevylle, US Patent 7,070,745, July 2006.
555. T.N. Kalnes, S.R. Dunne and V.P. Thakkar, US Patent 6,312,586, November 2001.
556. L.L. Laccino, E.S. Ellis, R. Gupta and B.A. Raich, US Patent 6,579,442, June 2003.
557. L.L. Laccino and J.J. Schorfheide, US Patent 6,4097,810, December 2002.
558. B.E. Stangeland, D.C. Kramer, D.S. Smith and J.T. McCall, US Patent 5,599,440, February 1997.
559. R. Gupta, D.C. Dankworth, D.D. Fleming, E.S. Ellis and J.J. Scharfheide, US Patent 5,985,131, November 1999.
560. R. Gupta, H. Jung, E.S. Ellis, J.J. Schorfheide and L.L. Iaccino, US Patent 6,153,086, November 2000.
561. R. Gupta, H. Jung, E.S. Ellis and E.G. Markley, US Patent 6,036,844, March 2000.
562. R.H. Wolk, M.C. Chervenak and S.B. Alpert, US Patent 3,788,973, January 1973.
563. A.J. Dahlberg and J.F. Mayer, US Patent 6,783,660, August 2004.
564. G.V. Nelson and G.C. Wray, US Patent 3,891,538, June 1974.
565. G.P. Hamner, US Patent 3,977,961, August 1976.
566. G.P. Hamner, US Patent 4,014,821, March 1977.
567. W.F. Arey Jr, G.P. Hamner and W.H. Sawyer, US Patent 3,977,962, August 1976.
568. G.P. Hamner, US Patent 4,073,718, February 1978.
569. P.J. Angevine, US Patent 4,415,436, November 1983.
570. A.M. Henke and J.D. McKinney, US Patent 3,936,370, February 1976.
571. D.B. Gillis, US Patent 5,980,732, November 1999.
572. D.C. Kramer and B.E. Strangeland, US Patent 6,086,749, July 2000.
573. G.L. Scheuerman, US Patent 5,916,529, June 1999.
574. W. Mounce, US Patent 3,830,728, August 1974.
575. G.P. Hamner and K.R. Clem, US Patent 3,887,455, June 1975.
576. J.G. Kunesh, US Patent 4,344,840, August 1982.
577. K.R. Gibson, R. Threlkerl and P. Leung, US Patent 7,074,740, July 2006.
578. R.M. Steinberg, J.G. Niccum and J.C. Strickland, US Patent 5,298,151, March 1994.
579. J.F. Mosby, L.B. Peck, J.L. Taylor and W.I. Beaton, US Patent 5,013,427, May 1991.
580. D.C. Artes and S.G. Kukes, US Patent 5,322,829, June 1994.
581. F. Morel, S. Kressman and J. Colyar, US Patent 6,620,311, September 2003.
582. F. Morel, P.H. Bigeard, S. Kressman, D. Duee and J.L. Duplan, US Patent 6,207,041, March 2001.
583. P.H. Bigeard, F. Morel, C. Gueret, P. Briot and P. Marion, US Patent 6.153,087, November 2000.
584. F. Morel, G. Heinrich, S. Kressman, A. Billon, J.L. Duplan and T. Chapus, US Patent 6,017,441, January 2000.
585. F. Morel, J.L. Duplan, A. Billon and S. Kressman, US Patent 6,171,477, January 2002.
586. F. Morel, S. Kressman and J.L. Duplan, US Patent 6,447,671, September 2002.
587. J.L. Rockwell, US Patent 7,033,488, April 2006.
588. D.L. Trimm, in Catalysis in Petroleum Refining 1980, D.L. Trimm (Ed.), Elsevier Science, Amsterdam, 1980, p. 41.
589. P. Dufresne, F. Valeri and S. Abotteen, Stud. Surf. Sci. Catal. 100 (1996) 253.

590. T. Suzuki and P. Dufresne, Stud. Surf. Sci. Catal. 92 (1995) 215.
591. Y. Noguchi, T. Itoh, Y. Obayashi and K. Komine, Am. Chem. Soc. Div. Petr. Chem. Prep. 38 (1993) 50.
592. P. Dufresne, Am. Chem. Soc. Div. Petr. Chem. Prep. 38 (1993) 54.
593. C.S. Kim, F.E. Massoth and E. Furimsky, Fuel Proc. Technol. 32 (1992) 39.
594. P.T. Clark, A.L. Hensley, J.Z. Shyu, J.A. Kaduk and G.J. Ray, Stud. Surf. Sci. Catal. 62 (1991) 417.
595. G. Berrebi, P. Dufresne and Y. Jacquier, Environ. Prog. 12 (1993) 97.
596. I. Guibard, S. Kressmann, F. Morel, V. Harle and P. Dufresne, Hydr. Eng. 30 (2000) 54.
597. E. Furimsky, J. Houle and Y. Yoshimura, Appl. Catal. 33 (1987) 97.
598. H. Beuther and R.A. Flinn, Ind. Eng. Chem. Prod. Res. Dev. 2 (1963) 53.
599. M. Marafi, A. Stanislaus, C.J. Mumford and M. Fahim, Appl. Catal. 47 (1989) 85.
600. M. Marafi, A. Stanislaus, C.J. Mumfors and M. Fahim, in Catalysis in Petroleum Refining, D.L. Trimm et al. (Eds), Elsevier, Amsterdam, 1989, p. 213.
601. A. Stanislaus, M. Marafi and M. Absi-Halabi, Appl. Catal. 100 (1993) 195.
602. A. Stanislaus, M. Marafi and M. Absi-Halabi, Am. Chem. Soc. Div. Petr. Chem. Prep. 39 (1994) 627.
603. A. Stanislaus, M. Marafi, and M. Absi-Halabi, Am. Chem. Soc. Div. Petr. Chem. Prep. 38 (1993) 62.
604. M. Marafi, A. Stanislaus and C.J. Mumford, Catal. Lett. 18 (1993) 141.
605. M. Marafi, A. Stanislaus and F. Jassem, Am. Chem. Soc. Div. Petr. Chem. Prep. 43 (1998) 486.
606. M. Marafi, A. Stanislaus and M. Absi-Halabi, Appl. Catal. B, Environ. 4 (1994) 19.
607. M. Marafi and A. Stanislaus, J. Mol. Catal. 202 (2003) 117.
608. A. Stanislaus, M. Marafi and M. Absi_Halabi in Hydrotreating Technology for Pollution Control, M.L. Ocelli and R. Chianeli (Eds), Marcel Dekker, 1996, p. 327.
609. M.R. Reda, Ind. Eng. Chem. Res. 30 (1991) 2148.
610. W.R. Ernst, L.H. Hiltzik, A.R. Garcia, M.D. Franke, A.S. Myerson and J.D. Carruthers, Miner. Metall. Process. 4 (1987) 78.
611. K.W. Babcock, L.H. Hiltzik, W.R. Ernst and J.D. Carruthers, Appl. Catal. 47 (1989) 295.
612. S.J. Hildebrandt, R.O. Koseoglu, J.F. Duddy and D.E. Sherwood, Am. Chem. Soc. Div.Petr. Chem. Prep. 38 (1993) 40.
613. M. Seapan and Z. Guohui, Am. Chem. Soc. Symp. Ser. 411 (1989) 9.
614. J.S. McPartland and R.G. Bautista, Sep. Sci. Technol. 25 (1990) 919.
615. M. Marafi and E. Furimsky, Erdol Kohle 121 (2005) 93.
616. J.S.M. Jocker, Am. Chem. Soc. Div. Petr. Chem Prep. 38 (1993) 74.
617. R.E. Siemens, B.W. Jong and J.H. Russel, Conserv. Recycl. 9 (1986) 189.
618. B.W. Jong, S.C. Rhoads, A.M. Stubbs and T.R. Stoelting, Recovery of Principal Metals from Spent Hydroprocessing Catalysts, US Bur. Mines Report 9252 (1988).
619. M.A. Olazabal, L.A. Fernandez and J.A. Madariaga, J. Met. Solvent Extr. Ion Exch. 9 (1991) 735.
620. M.A. Olazabal, M.M. Orive, L.A. Fernandez and J.A. Madariaga, J. Met. Solvent Extr. Ion Exch. 10 (1992) 623.
621. T.N. Angelidis, E. Tourasanidis, E. Marinou and G.A. Stalidis, Resour. Conserv. Recycl. 13 (1995) 269.
622. S. Kelebek and P.A. Distin, J. Chem. Technol. Biotechnol. 44 (1989) 309.
623. N.K. Nadirov and S.A. Petrashow, Chem. Technol. Fuels Oils 25 (1990) 424.
624. S. Gryglewicz and M. Rutkowski, Erdol Kohle 47 (1994) 34.
625. I. Tsuboi, T. Hashimoto, E. Kunugita and I. Komasawa, Kagaku Kogaku Ronbushu, 19 (1993) 722.
626. N. Mutsukawa, Shigen 1 (1989) 36.

627. N. Rokukawa, J. Mining Metall. Inst. Jpn 103 (1987) 593.
628. S. Toda, J. Mining Metall. Inst. Jpn 105 (1989) 261.
629. R.A. Ference and R.F. Sibenik, Am. Chem. Soc. Div. Fuel Chem. Prep. 27 (1982) 71.
630. B.B. Kar, B.V.R. Murthy and V.N. Misra, Int. J. Miner. Process. 76 (2005) 143.
631. G. Berrebi, Proc. Ketjen Catal. Symp., May 1984, Amsterdam, Netherlands.
632. R.K. Biswas, M. Wakihara and M. Taniguchi, Hydrometallurgy, 14 (1985) 219.
633. I. Gaballah and M. Djona, Metall. Mater. Trans. B 25 (1994) 481.
634. I. Gaballah and M. Djona, Metall. Mater. Trans. B 2 (1995) 49.
635. B. Blaustein, J.T. Hauck, G.J. Olson and J.P. Baltrus, Fuel 72 (1993) 1613.
636. P. Joffe and G.T. Sperl, DOE Report PC-92119-T4 (1993).
637. L.J. Silva, L.A. Bray and D.W. Matson, Ind. Eng. Chem. Res. 32 (1993) 2485.
638. V.V. Lunin, I.I. Miroschnichenko, D.M. Panteleev and Y.I. Solovetskii, Kinet. Catal. 34 (1993) 646.
639. K. Inoui, P. Zhang and H. Tsuyama, Am. Chem. Soc. Div. Petr. Chem. Prep. 38 (1993) 77.
640. P. Zhang, K. Inoui and H. Tsuyama, Energy Fuels 9 (1995) 231.
641. G.M. Ritcey, Handbook of Solvent Extraction, Wiley Interscience, New York, 1981.
642. Y. Sato, K. Kondo and F. Nakashio, J. Chem. Eng. Jpn 22 (1989) 200.
643. M. Nishizawa, T. Yokoyama, T. Kimura and T.M. Suzuki, Bull. Chem. Soc. Jpn 57 (1994) 2859.
644. F.M. Lee, R.D. Knudsen and D.R. Kidd, Ind. Eng. Chem. Res. 31 (1992) 487.
645. T. Itoh and H. Sue, Fifth UNITAR/UNDP Int. Conf. on Heavy Crudes and Tar Sands, August 1991, Caracas, Venezuela, Vol. 4, p. 43.
646. R.J. Schreiber and C.P.E. Yonley, Am. Chem. Soc. Div. Petr. Chem. Prep. 38 (1993) 97.
647. A. Stanislaus, G.R. Gouda and S. Al-Fulaij, Am. Chem. Soc. Div. Petr. Chem. Prep. 43 (1998) 491.
648. A. Matsumura, T. Kondo, Y. Sasaki and K. Unegawa, Erdol Kohle 47 (1994) 280.
649. T. Sakabe and T. Yagi, Hydroc. Proc. 12 (1979) 103.
650. E. Furimsky and M. Biagini, Fuel Proc. Technol. 46 (1996) 17.
651. E. Furimsky, Appl. Catal. 156 (1997) 207.
652. R. Schmitt, Oil Gas J. 89 (1991) 101.
653. E. Furimsky, L.C.G. Janke and R. Dureau, Fuel Proc. Technol. 33 (1993) 77.
654. E. Furimsky, M. Nielsen and P. Jurasek, Energy Fuels 9 (1995) 439.
655. S. Raseev, Thermal and Catalytic Processes in Petroleum Refining, Marcel Dekker, New York, 2003, p. 741.
656. J. Scherzer and A.J. Gruia, Hydrocracking Science and Technology, Marcel Dekker, New York, 1996, p. 223.
657. T.R. Farrel and J.A. Zakarian, Oil Gas J. 86 (1986) 49.
658. R. Bertram and F. Danzinger, NPRA Annual Mtg., San Antonio TX, March 1994, AM-94-50.
659. E. Honde, in Hydrocarbon Technology International, P. Harrison (Ed.), Sterling Publ., 1993.
660. J.G. Bendoraitis, A.W. Chester, F.G. Dwyer and W.E. Garwood, Proc. 7th Int. Zeol. Conf., 1986, Tokyo, p. 669.
661. R.J. Taylor and A.J. McCormack, Ind. Eng. Chem. Res. 31 (1992) 1731.
662. A. Hennico, A. Billon, P.H. Bigeard and J.P. Perries, Rev. Inst. Fr. Petr. 48 (1993) 261.
663. N.Y. Chen and W.E. Garwood, Ind. Eng. Chem. Res. 25 (1986) 641.
664. N.Y. Chen, W.E. Garwood and F.G. Dwyer, Shape Selective Catalysis in Industrial Applications, Marcel Dekker, New York, 1989.
665. D.J. O'Rear, Ind. Eng. Chem. Res. 26 (1987) 2337.
666. D.J. O'Rear and J.A. Biscardi, US Patent 6,773,578, August 2004.
667. R.B. LaPierre and R.D. Partridge, US Patent 5,284,573, February 1994.
668. R.N. Reynolds Jr, US Patent 6,793,803, September 2004.
669. A.W. Chester, W.E. Garwood and J.C. Vartuli, US Patent 4,574,043, March 1986.

670. P.J. Angevine, K.M. Mitchell, S.M. Oleck and S.S. Shih, US Patent 4,612,108, September 1986.
671. G. Barre, J.-P. Darnanville and L.G. Huve, US Patent 7,077,948, July 2006.
672. S.J. Stephen and J.M. Rosenbaum, US Patent 6,962,651, November 2005.
673. S.J. Miller, US Patent 7,053,254, May 2006.
674. T.S. Chou, N.Y. Chen, G.G. Karsner, C.R. Kennedy, R.B. Lapierre, M.G. Malconian, R.J. Quan and S.S. Wong, US Patent 6,416,654, July 2002.
675. E. Benazzi, N. Mrachal-George, J. Martens and M. Claude, US Patent 6,984,309, January 2006.
676. N. Marchal-George, E. Benazzi, T. Cseri and S. Kasztelan, US Patent 6,660,897, December 2003.
677. M.B. Carroll, F.A. Smith and J.C. Bixel, US Patent 6,517,704, February 2003.
678. R.C. Dougherty, D.N. Mazzone, R.F. Socha and H.K.C. Timken, US Patent 6,294,077, September 2001.
679. T.R. Forbus Jr and C.P. Kyan, US Patent 6,287,454, September 2001.
680. D.P. Leta, S.L. Soled, G.B. McVicker and S. Hantzer, US Patent 6,475,374, November 2002.
681. E. Benazzi, P. Marion, A. Billon, C. Gueret, J.-C. Hipeaux and J.P. Gouzard, US Patent 6,884,339, April 2005.
682. E. Benazzi, N. Marchal-George, T. Cseri, P. Marion, C. Gueret and S. Kasztelan, US Patent 6,603, 402, August 2003.
683. E. Benazzi, N. George-Marchal, C. Guret, P. Briot, A. Billon and P. Marion, US Patent 6,350,370, February 2002.
684. B.R. Cook, J.W. Johnson, G. Cao, R.A. McEvoy and R.H. Ernst, US Patent 6,284,127, September 2001.
685. I.A. Cody, W.J. Murphy, S. Hantzer, D.W. Larkin, J.E. Galagher Jr and J.T. Kim, US Patent 7,077,947, July 2006.
686. R.D. Gillespie US Patent 6,818,589, November 2004.
687. R.D. Gillespie, US Patent 7,041,866, May 2006.
688. R.D. Gillespie, US Patent 6,977,322, December 2005.
689. J.G. Vassilakis, R.R. Rosin, S.A. Bradley, R.D. Gillespie, M.J. Cohn and F. Xu, US Patent 7,015,175, May 2006.
690. J.A. Rodriguez, J.P. Pariente, A.C. Lara, A.C. Canos, I.A. Cody, W.J. Murphy and S.J. Linek, US Patent 6,294,081, September 2001.
691. S.J. Miller, US Patent 4,921,594, May 1990.
692. E. Benazzi and N. Marchal-George, US Patent 6,736,959, May 2004.
693. K. Fujihara and Y. Morishima, US Patent 6,365,037, April 2002.
694. I.A. Cody, D.R. Boate, S.J. Linek, W.J. Murphy, J.E. Gallagher and G.J. Harting, US Patent 6,325,918, December 2001.
695. I.A. Cody, D.R. Boate, S.J. Linek, W.J. Murphy, J.E. Gallagher, A. Ravella and R.A. Demmin, US Patent 6,322,692, November 2001.
696. P. Briot, C. Gueret, J.-C. Hipeaux, E. Benzazzi, P. Marion and A. Billon, US Patent 6,783,661, August 2004.
697. E.E. Unmuth, J.A. Mahoney and R.J. Bertolacini, US Patent 2,636,299, January 1987.
698. I.A. Cody, W.J. Murphy, J.E. Gallagher, J.P. Boyle, A.M. Zinicola, C.J. May, J.T. Kim, J.A. Groestch and S.S. Hantzer, US Patent 6,974,535, December 2005.
699. W.E. Garwood, Q.N. Le and S.S. Wong, US Patent 5,037,528, August 1991.
700. D. Farshid, US Patent 6,787,026, September 2004.
701. K.L. Riley, W.J. Murphy, I.A. Cody, S.L. Soled, G.B. McVicker and S. Miseo, US Patent 6,383,366, May 2002.
702. K.R. Albinson, S. Mizrabi and D.J. Neuman, US Patent 4,913,797, April 1990.
703. C.P. Kyan and P.J. Oswald, US Patent 5,603,824, February 1999.
704. I.A. Cody and W.J. Murphy, US Patent 6,951,605, October 2005.

705. J.F. Grootjans and P.J. Breadel, US Patent 4,810,356, March1989.
706. S.M. Oleck and R.C. Wilson, US Patent 4,437,976, March 1984.
707. S.M. Oleck and R.C. Wilson, US Patent 4,490,242, December 1984.
708. N.Y. Chen and T.F. Degnan, US Patent 4,604,261, August 1986.
709. M.B. Carroll, G.P. Schleicher and J.P. Boyle, US Patent 6,569,313, May 2003.
710. M.B. Carroll, F.A. Smith and J.C. Bixel, US Patent 6,659,312, May 2003.
711. W.J. Murphy, I.A. Cody and B.G. Silbernagel, US Patent 6,620,312, September 2003.
712. D.J. O'Rear, US Patent 6,833,065, December 2004.
713. N.Y. Chen, W.E. Garwood, T.J. Huang, Q.N. Le and S.S. Wong, US Patent 4,919,788, 1990.
714. R.B. LaPierre, R.D. Partridge, N.Y. Chen and S.S. Wong, US Patent 4,419,220, 1983.
715. J. Miller, US Patent 4,689,138, 1987.
716. W.E. Garwood, J.L. Schienker and J.C. Vartuli, US Patent 4,773,987, 1987.
717. P.J. Berlowitz, J.J. Habeeb and R.J. Wittenbrink, US Patent 6,475,960, November 2002.
718. J.W. Johnson, A.R. Bishop, W.B. Genetti, L.L. Ansell and R.A. Fiato, US Patent 6,846,778, January 2005.
719. J.A. Biscardi and D.P. Fong, US Patent 6,579,441, June 2003.
720. S.C. Fung, Am. Chem. Soc. Div. Petr. Chem. Prep. 38 (1993) 105.
721. J.N. Beltramini, in Catalytic Naphtha Reforming, G.J. Antos, A.M. Aitani and J.M. Perera (Eds), Marcel Dekker, New York, 1995.
722. N. Pehrson, Oil Gas J. 88 (1988) 55.
723. J.P. Rosso, Chem. Eng. Prog. 12 (1992) 66.
724. J.P. Rosso and M.I. El Guindy, in Catalytic Naphtha Reforming, G.J. Antos, A.M. Aitani and J.M. Parera (Eds), Marcel Dekker New York, 1995, p. 395.
725. P. Scholtz, Chem. Technol. 40 (1988) 451.
726. R.T. Jacobsen, Chem. Eng. Prog. 4 (2005) 41.
727. R.T. Jacobsen, Chem. Eng. Prog. 3 (2005) 20.
728. R.T. Jacobsen, Chem. Eng. Prog. 2 (2005) 22.
729. R.K. Mishra, Proc. 17th Precious Metals Conf., 1993, New Port, Rhode Island, p. 449.
730. D.R. Tyson and R.G. Bautista, Sep. Sci. Technol. 22 (1987) 1149.
731. F.T. Awadalla, R.E. Molnar and G.M. Ritcey, US Patent 5,304,233, 1994.
732. G.B.J. de Boer and M. Hausselaar, US Patent 5,302,183, 1994.
733. N. Ezawa, H. Inoue, S. Takada and H. Masuda, US Patent 5,252,305, 1995.
734. G.-Q. Li, S. Konishi, K. Saito and T. Sugo, J. Membr. Sci. 95 (1994) 63.
735. L. Vallejos, T. Vasquesz and C. Ovalles, Am. Chem. Soc. Div. Petr. Chem. Prep. 45 (2000) 591.
736. K. Fujiwara, Y. Sugai, N. Yazawa, K. Ohno, C.X. Hong and H. Enomoto, Stud. Surf. Sci. Catal. 151 (2004) 405.
737. J.G. Weisseman, R.V. Kessler, R.A. Sawicki, J.D.M. Belgrave, C.J. Lauresham, S.A. Mehta, R.G. Moore and M.G. Ursenback, Energy Fuels 10 (1996) 883.
738. J.G. Weissman and R.V. Kessler, Appl. Catal. 140 (1996) 1.
739. M. Greaves, A. El-Saghr and T.X. Xia, Am. Chem. Soc. Div. Petr. Chem. Prep. 45 (2000) 595.
740. G. Ondrey, Chem. Eng. 12 (2005) 13.
741. A. Del Bianco, N. Panariti, S. Di Carlo, J. Elmouchnino, B. Fixari, P. Le Perchec, Appl. Catal. 94 (1993) 1.
742. R. Bearden and C.L. Aldridge, Energy Prog. 1 (1981) 44.
743. G.T. Hager, X.X. Bi, F.J. Derbyshire, P.C. Eklung and J.M. Stencel, Am. Chem. Soc. Div. Fuel Chem. Prep. 36 (1991) 1901.
744. K. Sato, Y. Iwata, K. Honda, Y. Araki, Y. Miki and H. Shimada, Sci. Catal. 1998, 69 (1999) 411.
745. B. Shi, D.L. Lin, L.Q. Wang and G.H. Que, Am. Chem. Soc. Div. Petr. Chem. Prep. 46 (2001) 410.

746. D. Kennepohl and E. Sanford, Energy Fuels 10 (1996) 229.
747. A. Del Bianco, N. Panariti, S.D. Carlo, P.L. Beltram and P. Carniti, Energy Fuels 8 (1994) 593.
748. N. Panariti, F. Bazzano, A. Delbianco, M. Marchionna, R. Montanari and S. Rosi, Am. Chem. Soc. Div. Petr. Chem. Prep. 46 (2001) 78.
749. D.K. Lee, S.K. Park, W.L. Yoon, I.C. Lee and S.I. Woo, Energy Fuels 9 (1995) 2.
750. M.J. Dabkowski, S.S. Shih and K.R. Albinson, AIChE Symp. Ser. 282, Vol. 87, 1991, p. 53.
751. S.G. Kukes, A.W. Aldag and S.L. Parrott, in Metal complexes in Fossil Fuels, R.H. Filby and J.F. Branthaver (Eds), Am. Chem. Soc., Washington, DC, 1987, Chapter 17, p. 265.
752. C.E. Snape, C. Bolton, R.G. Dosch, and H.P. Stephens, Energy Fuels 3 (1989) 421.
753. D. Liu, W. Guo, W. Deng, J. Zhou and G. Que, Am. Chem. Soc. Div. Petr. Chem. Prep. 50 (2005) 391.
754. D. Liu and K. Ma, Am. Chem. Soc. Div. Petr. Chem. Prep. 50 (2005) 295.
755. C. Guan, Z. Wang, S. Yu and A. Guo, Am. Chem. Soc. Div. Petr. Chem. Prep. 48 (2003) 150.
756. C. Guan, Z. Wang, S. Yu, A. Guo and A. Jiang, Am. Chem. Soc. Div. Petr. Chem. Prep. 48 (2003) 110.
757. J. Wang, W. Deng and G. Que, Am. Chem. Soc. Div. Petr. Chem. Prep. 48 (2003) 344.
758. K. Fujimoto, J. Chang and N. Tsubaki, J. Jpn. Petr. Inst. 43 (2000) 25.
759. D. Gar, and E.N. Givens, Am. Chem. Soc. Div. Fuel Chem. Prep. 28 (1983) 200.
760. C. Song, A.K. Saini and Y. Yoneyama, Fuel 79 (2000) 249.
761. K. Hirano, M. Kouzu, T. Okada, M. Kobayashi, N. Ikenaga and T. Suzuki, Fuel 78 (1999) 1867.
762. T. Kaneko, T. Tazawa, N. Okuyama, M. Tamura and K. Shimasaki, Fuel 79 (2000) 263.
763. R. Bearden and C.L. Aldridge, Energy Prog. 1 (1981) 44.
764. F. Rodriguez-Reinoso, Carbon 36 (1998) 159.
765. L. Rouleau, R. Bacaud, M. Breysse, and J. Dufour, Appl. Catal. 104 (1993) 137.
766. L. Rouleau, R. Bacaud, M. Breysse, and J. Dufour, Appl. Catal. 104 (1993) 149.
767. A. Matsumura, T. Kondo, Y. Sasali, and K. Unegawa, Erdol Kohle 47 (1994) 280.
768. J. Mosio-Mosiewski, and I. Morawski, Appl. Catal. 283 (2005) 147.
769. M. Seko, 1988 NPRA Ann. Meet., March 20–22 March, San Antonio, TX.
770. M.M. Ramirez de Agudelo and C. Galarraga, Am. Chem. Soc. Div. Petr. Chem. Prep. 38 (1993) 45.
771. K. Sakanishi, I. Saito, I. Watanabe and I. Mochida, Fuel (2004).
772. L. Del Bianco, N. Panariti, and M. Marchionna, Am. Chem. Soc. Div. Petr. Chem. Prep. 40 (1995) 743.
773. B. Fixari, S. Penreux, J. Elmouchniuo, P. Le Perchec, M. Vrinat and F. Morel, Energy Fuels 8 (1994) 588.
774. K.M. Kirkwood, J.M. Foght and M.R. Gray, Stud. Surf. Sci. Catal. 151 (2004) 113.
775. H.K. Kotlar, O.G. Brakstad, S. Markussen and A. Winnberg, Stud. Surf. Sci. Catal. 151 (2004) 1.
776. J.J. Kilbane and S. Le Borgne, Stud. Surf. Sci. Catal. 151 (2004) 29.
777. M. Ayala and R. Vazquez-Duhalt, Stud. Surf. Sci. Catal 151 (2004) 67.
778. D.C. Bressler and M.R. Gray, Energy Fuels 16 (2002) 1076.
779. K.M. Kirkwood, J.M. Foght and M.R. Gray, Stud. Surf. Sci. Catal 151 (2004) 113.
780. G. Hamer and N. Al-Awadhi, Acta Biotechnol. 20 (2000) 335.
781. K.M. Kirkwood, S. Ebert, D. Kharbanda, J.M. Foght, P.M. Fedorak and M.R. Gray, Am. Chem. Soc. Div. Petr. Chem. Prep. 48 (2003) 50.
782. Q. Wu, M.R. Gray, M.A. Pickard, P.M. Fedorak and D. Foght, Am. Chem. Soc. Div. Petr. Chem. Prep. 48 (2003) 47.
783. R.M. Miller and R. Bartha, Appl. Environ. Microbiol. 55 (1989) 269.
784. L.J. Nash, Am. Chem. Soc. Div. Petr. Chem. Prep. 49 (2004) 251.

785. D.C. Bressler, L.A. Kirkwood, J.M. Foght, P.M. Fedorak and M.R. Gray, Am. Chem. Soc. Div. Petr. Chem. Prep. 48 (2003) 44.
786. L.M. Gieg, A. Otter and P.M. Fedorak, Environ. Sci. Technol. 30 (1996) 575.
787. L.E. Patras and I.A. Webster, Stud. Surf. Sci. Catal. 50 (1989) 273.
788. L. Le Borgne and R. Quintero, Fuel. Proc. Technol. 81 (2003) 155.
789. M.A. Pacheco, E.A. Lange, P.T. Pienkos, L.Q. Yu, M.P. Rouse, Q. Lin and L.K. Linguist, 1999 Ann. NPRA Meet., Paper AM-99-27, San Antonio, TX.
790. A.P. Borole, E.N. Kaufman, M.J. Grossman, V. Minak-Bernero, R. Bare and M.K. Lee, Biotechnol. Progr. 18 (2002) 88.
791. J.J. Kilbane, R. Ranganathan, L. Cleveland, K.J. Kayser, C. Ribeiro and M.M. Linhares, Appl. Environ. Microbiol. 66 (2000) 688.
792. M.J. Benedik, P.R. Gibbs, P.R. Riddle and R.C. Willson, Trends Biotechnol. 16 (1998) 390.
793. S.-I. Sato, N. Ouchiyama, T. Kimura, H. Nojiri, H. Yamane and T. Omori, 179 (1997) 4841.
794. P.M. Fedorak, K.M. Semple, R. Vazquez-Duhalt and D.W.S. Westlake, Enzyme Microb. Technol. 15 (1993) 429.
795. J.M. Foght, Stud. Surf. Sci. Catal. 151 (2004) 145.
796. G. Ondrey, Chem. Eng. 105 (1998) 17.
797. L. Linguist and M. Pacheco, Oil Gas J. 97 (1999) 45.
798. E.T. Premuzic, M.S. Lin, H. Lian, W.M. Zhou and J. Yablon, Fuel Proc. Technol. 15 (1993) 207.
799. E.T. Premuzic, M.S. Lin, M. Bohenek and W.M. Zhou, Energy Fuels 13 (1999) 297.
800. P.B. Venuto and E.T. Habib Jr, Fluid Catalytic Cracking with Zeolite Catalysts, Marcel Dekker, New York, 1979.
801. G.M. Stokes and R.W. Mott, AIChE Symp. Ser. 273, 86 (1988) 58.
802. R.W. Stoke, Oil Gas J. (1988) 67.
803. P.K. Ladwig, in Handbook of Petroleum Refining Processes, R.E. Meyers (Ed.), 1986, McGraw-Hill, New York, 1st Ed.
804. J. Scherzer, Octane Enhancing Zeolitic Catalysts, Marcel Dekker, New York, 1990.
805. G. Ertl, H. Knozinger and J. Weitkamp, Handbook of Heterogeneous Catalysis, Vol. 4, VCH Verlagsgesellschaft, Weinheim, Germany, 1997.
806. J.W. Wilson, Fluid Catalytic Cracking Technology and Operation, Penn Well, Tulsa, Ok, 1997.
807. Anon., Refining Processes 98, Hydroc. Proc. 77 (1998) 53.
808. L.L. Upson, C.L. Hemler and D.A. Lomas, Stud. Surf. Sci. Catal. 76 (1993) 396.
809. J.W. Wilson and C. Ross, Am. Chem. Soc. Div. Petr. Chem. Prep. 44 (1999) 527.
810. P. O'Connor, A.W. Gevers, A.P. Humphries, L.A. Gerritsen and P.H. Desai, ACS Symp. Ser. 452 (1991) 318.
811. P. O'Connor and S.J. Yanik, Stud. Surf. Sci Catal. 100 (1995) 323.
812. R.E. Evans and G.P. Quinn, Stud. Surf. Sci. Catal. 76 (1993) 563.
813. W.S. Letzsch and A.G. Ashton, Stud. Surf. Sci. Catal. 76 (1993) 449.
814. R. Sadeghbeigi, Fluid Catalytic Cracking Handbook, Gulf Publ., Houston, TX, 1995, p. 65.
815. R.A. Meyers, Handbook of Petroleum Refining Processes, McGraw-Hill, New York, 1983, pp. 2–30.
816. J. Bousquet and T. Laboural, Oil Gas J. 85 (1987) 62.
817. G.H. Dale, 11th World Petr. Congr. 1983, Vol. 4, p. 83.
818. S.I. Anderson and T. Myrstad, Am. Chem. Soc. Div. Petr. Chem. Prep. 51 (2006) 401.
819. S.I. Anderson and T. Myrstad, Oil Gas Eur. Mag. 23 (1997) 19.
820. W.R. Gilbert, Stud. Surf. Sci. Catal. 134 (2001) 219.
821. C.M. Alvarenga Baptista and H.S. Cerqueira, Stud. Surf. Sci. Catal. 149 (2004) 287.
822. G.E. Denis, Fourth UNITAR/UNDP Conf. on Heavy Crudes and Tar Sands, August 7–12, 1988, Edmonton, Alts, Paper 13.
823. A.J. Suchanek and A.S. Moore, Oil Gas J. Jan. 23, 87 (1989) 61.

824. J.P. Peries, A. Billon, A. Hennico, E. Morrison and F. Morel, Proc. 6th UNITAR/UNDP Intern. Conf. on Heavy Crude and Tar Sands, 1995, Houston TX, Vol. 10, p. 229.
825. U. Alkemade and S. Paloumbis, Stud. Surf. Sci. Catal. 100 (1995) 339.
826. P. van de Gender, R.M. Benslay, K.C. Chuang, W.E. Cormier, G.M. Wolterman and E.P. Boldingh, AIChE Symp. Ser. 291, 88 (1992) 21.
827. P. O'Connor, P. Imhof and S.J. Yanik, Am. Chem. Soc. Div. Petr. Chem. Prep. 44 (1999) 466.
828. I.E. Maxwell and W.H.J. Stork, Stud. Surf. Sci. Catal. 137 (2001) 775.
829. F.G. Dwyer and T.F. Degnan, Stud. Surf. Sci Catal 76 (1993) 499.
830. G.M. Woltermann, J.S. Magee and S.D. Griffith, Stud. Surf. Sci. Catal. 76 (1993) 106.
831. L. Pine, P.J. Maher and W.A. Wachter, J. Catal. 85 (1984) 466.
832. C. Liu, Y. Deng, X. Gao and Z. Tang, Am. Chem. Soc. Div Petr. Chem. Prep. 48 (2003) 100.
833. C.T. O'Connor, E. Van Steen and M.E. Dry, Stud. Surf. Sci. Catal. 102 (1996) 323.
834. D.M. Stockwell, X. Liu, P.J. Nelson, T.A. Gegan and C.F. Keweshan, Am. Chem. Soc. Div. Petr. Chem. Prep. 48 (2003) 216.
835. E.L. Kugler, T.H. Gardner and J. Panpranot, Am. Chem. Soc. Div. Petr. Chem. Prep. 44 (1999) 447.
836. S.J. Yanik, P. O'Connor, M.F. Brady, D.K. Abner and M.C. Friedrich, AIChE Symp. Ser. 291, 88 (1991) 9.
837. F. Fraga and M. Cesar, Am. Chem. Soc. Div. Petr. Chem. Prep. 44 (1999) 473.
838. M.M. Mitchell and H.F. Moore, Am. Chem. Soc. Div. Petr. Chem. Prep. 33 (1988) 547.
839. K.Y. Yung, P. Imhof and M. Baas, Am. Chem. Soc. Div. Petr. Chem. Prep. 47 (2002) 281.
840. L.O. Almanza, C.A. Vergel and L.J. Hoyos, Am. Chem. Soc. Div. Petr. Chem. Prep. 46 (2001) 381.
841. P. O'Connor and A.P. Humphries, Am. Chem. Soc. Div. Petr. Chem. Prep. 38 (1993) 598.
842. F.H.H. Khouw, M.J.R.C. Nieskens, M.J.H. Borley and K.H.W. Roebschlaeger, NPRA 1990 Ann. Meet., Paper No. AM-90-42.
843. Y. Lu, M. He, J. Song and X. Shu, Am. Chem. Soc. Div. Petr. Chem. Prep. 44 (1999) 453.
844. S.J. Yanik, P. O'Connor, NPRA 1995, Ann. Meet., Paper No. AM-95-35.
845. S.-I. Andersen and T. Myrstad, AIChE, Spring Meet., 9–11 March, 1998, New Orleans, Paper 31.
846. S.-I. Andersen and T. Myrstad, Am. Chem. Soc. Div. Petr. Chem. Prep. 47 (2002) 495.
847. W.R. Gilbert, Am. Chem. Soc. Div. Petr. Chem. Prep. 44 (1999) 534.
848. H. Tian, Am. Chem. Soc. Div. Petr. Chem. Prep. 47 (2002) 306.
849. S.K. Pavel and F.J. Elvin, Am. Chem. Soc. Div. Petr. Chem. Prep. 40 (1995) 198.
850. I.A. Vasalos, E.R. Strong, C.K.R. Hsieh and G.J. D'Souza, Oil Gas J. 75 (1977) 141.
851. R. Hughes, Deactivation of Catalysts, Academic Press, London, 1984.
852. M. Guisnet and P. Magnoux, Stud. Surf. Sci. Catal. 88 (1994) 53.
853. G.F. Froment, in Catalyst Deactivation 1997, C.H. Bartholomew and G. Fuentes (Eds), Elsevier, Amsterdam, 1997.
854. M.A. Den Hollander, M. Makkee and J.A. Moulijn, Appl. Catal. 187 (1999) 3.
855. H.H. Shan, C.H. Yang, G.L. Niu and J.F. Zhang, Am. Chem. Soc. Div. Petr. Chem. Prep. 47 (2002) 321.
856. C.M.L. Alvarenga Baptista and H.S. Cerqueira, Am. Chem. Soc. Div. Petr. Chem. Prep. 47 (2002) 239.
857. F. Hernandez-Beltram, R. Garcia-de-Leon, E. Mogica, J.C. Moreno and R. Gonzales, Am. Chem. Soc. Div. Petr. Chem. Prep. 44 (1999) 516.
858. A.A. Lappas, Z.A. Tsagrasouli, J.A. Vasalos and A. Humphries, Am. Chem. Soc. Div. Petr. Chem. Prep. 44 (1999) 507.
859. R.E. Ritter and G.W. Young, 1984 NPRA Ann. Meet., San Antonio, TX.
860. T.J. Dougan, U. Alkemade, B. Lankhampal and L.T. Boock, Oil Gas J. 92 (1994) 81.

861. R.F. Warmsbecher, A.W. Peters and H.M. Masseli, J. Catal. 100 (1986) 130.
862. H.J. Jeon, S.K. Park and S.I. Woo, Appl. Catal. 306 (2006) 1.
863. C. Li, S. Yan, Z. Qiu, L. Tang and M. Gu, Am. Chem. Soc. Div. Petr. Chem. Prep. 44 (1999) 537.
864. S. Zheng, S. Sun, Z. Tan, X. Gao and X. Xu, Am. Chem. Soc. Div. Petr. Chem. Prep. 51 (2006) 421.
865. G. Yaluris, W.C. Cheng, M. Peters, L.T. McDowell and L. Hunt, Am. Chem. Soc. Div. Petr. Chem. Prep. 48 (2003) 183.
866. S.J. Foskett, Hydroc. Proc. 11 (2001) 71.
867. Y. Zhu, Q. Du, W. Lin, L. Tang and J. Long, Am. Chem. Soc. Div. Petr. Chem. Prep. 51 (2006) 457.
868. M.N. Harandi, J.A. Herbst, H. Owen, P.H. Schipper and R.T. Thomson, US Patent 528 669, 1994.
869. A.W. Chester, Am. Chem. Soc. Div. Petr. Chem. Prep. 51 (2006) 389.
870. F.D. Hartzell and A.W. Chester, Hydroc. Proc. July (1979) 137.
871. W.R. Gilbert, R.E. Roncolatto and Y.L. Lam, Am. Chem. Soc. Div. Petr. Chem. Prep. 47 (2002) 155.
872. E.H. Hirschberg and R.J. Bertolacini, ASS Symp. Ser. 375 (1988) 114.
873. L.M. Magnabosco, Am. Chem. Soc. Div. Petr. Chem. Prep. 51 (2006) 388.
874. L.M. Magnabosco, Am. Chem. Soc. Div. Petr. Chem. Prep. 48 (2003) 209.
875. J. Francis, L. Luo and J.A. Gonzales, Am. Chem. Soc. Div. Petr. Chem. Prep. 51 (2006) 428.
876. X. Zhao, A.W. Peters and G.W. Weatherbee, Ind. Eng. Chem. Res. 36 (1997) 4535.
877. R.B. Miller, T.E. Johnson, C.R. Santner, A.A. Avidan and J.H. Beech, 1996 Ann. NPRA Meet., 1996, San Antonio, TX.
878. K. Limbach, S. Tamhankar, S. Ganguly, R. Menon, R. Ramachandran,
879. L. Henton and H. Moore, Am. Chem. Soc. Div. Petr. Chem. Prep. 44 (1999) 449.
880. A. Battacharaya and J.S. Yoo, Stud. Surf. Sci. Catal. 76 (1993) 531.
881. W.C. Cheng, G. Kim, A.W. Peters, X. Zhao and K. Rajagopalan, Catal. Rev. Sci. Eng. 40 (1998) 39.
882. G. Krishnaiah, L.V. Langan, J.A. Rudesill and W.C. Cheng, Am. Chem. Soc. Div. Petr. Chem. Prep. 48 (2003) 211.
883. D.M. Stockwell and C.P. Kelkar, Am. Chem. Soc. Div. Petr. Chem. Prep. 48 (2003) 191.
884. D.M. Stockwell, Am. Chem. Soc. Div. Petr. Chem. Prep. 51 (2006) 402.
885. D.M. Stockwell, Am. Chem. Soc. Div. Petr. Chem. Prep. 51 (2006) 404.
886. D.R. Rainer, L. Luo and J.A. Gonzales, Am. Chem. Soc. Div. Petr. Chem. Prep. 51 (2006) 431.
887. J. Barth, A. Jentys and J.A. Lercher, Stud. Surf. Sci. Catal. 154 (2004) 2441.
888. W. Schertz, NPRA Meet., February 1994, Houston, TX.
889. W.P. Hettinger Jr, Catal. Today 13 (1992) 157.
890. S.K. Pavel and F.J. Elvin, in Extraction and Processing for the Treatment and Minimization of Wastes, J. Hager et al. (Eds), Miner. Met. and Mater. Soc., 1993, p. 1015.
891. M.K. Spearman, AIChE J. Natl Meet., Seatle 1993, Report 4d, p. 11.
892. R. Schmitt, Oil Gas J. 89 (1991) 101.
893. D.M. Stockwell, R.P. Brown and S.H. Brown, US Patent. Appl., 0199386, October 2003.
894. D.M. Stockwell, S.H. Brown and J.Y. Ryn, US Patent Appl. 0220046, November 2004.
895. L. Xu and D.M. Stockwell, US Patent Appl. 0235642, November 2004.
896. M.T. Hurley, US Patent Appl. 0221991, December 2003.
897. T. Shibuya, J. Naito and N. Sekine, US Patent Appl. 0090379, April 2005.
898. T. Yamada, T. Shibuya and N. Sekine, US Patent Appl. 0256290, December 2005.
899. G. Yaluris and J.A. Rudessil, US Patent, 226024, September 2005.
900. G. Yaluris and J.A. Rudesill, US Patent Appl. 0006100, January 2006.

901. A.A. Vierheilig, US Patent Appl. 0060504, March 2006.
902. A.A. Vierheilig, US Patent Appl. 0027485, February. 2006.
903. K. Pankaj, R.R. Mari, D.J. Kumar, S.L. Lal, S. Mandal S. Makhija and S. Ghosh, US Patent Appl. 0211017, November 2003.
904. T. Shibuya, J. Naito, H. Yamada and N. Sekine, US Patent Appl. 0035750, February 2004.
905. R. Madon, D.H. Harris, M. Xu, D.M. Stockwell, B. Lerner and G.W. Dodwell, US Patent Appl. 0130107, July 2003.
906. R. Madon, D.H. Harris, M. Xu, D.M. Sockwell, B. Lerner and G.W. Dodwell, US Patent Appl. 0089640, May 2003.
907. A.W. Chester, T.G. Roberie, H.K.C. Timken and M.S. Ziebart, US Patent Appl. 0189260, September 2005.
908. W. Chester, H.K.C. Timken, T.G. Roberie and M.S. Ziebart, US Patent Appl. 0124485, June 2005.
909. A.A. Vierheilig and B. Keener, US Patent Appl. 0121363, June 2005.
910. A.A. Vierheilig and B. Keener US Patent Appl. 0121362, June 2005.
911. C.L. Beswick B.K. Speronello, J.B. McLean and M.R. Schmalfeld, US Patent Appl. 0205466, September 2005.
912. R. Hu, X. Zhao, R.F. Wormsbecher and M.S. Ziebert, US Patent Appl. 0205464, September 2005.
913. P. O'Connor, G.A. Pearson, J.J. Springs and D. Stamires, US Patent Appl. 0029717, February 2004.
914. T.G. Roberie, R. Kumar, M.S. Ziebart, W.C. Cheng, X. Zhao and N. Bhore, US Patent Appl. 0099573, May 2004.
915. A.A. Vierheilg, US Patent Appl. 0203806, October 2003.
916. Z. Liu, X. Li, H. Chen and Y. Liu, US Patent Appl. A0100096, May 2006.
917. G. Yaluris and J.A. Rudesill, US Patent Appl. 0042158, February 2005.
918. G. Yaluris and J.A. Rudesill, US Patent Appl. 0074811, April 2004.
919. G. Yaluris, J.A. Rudesill and W. Suarez, US Patent 0077492, April 2004.
920. G. Yaluris, M.S. Ziebart and X. Zhao, US Patent Appl. 0232839, October 2005.
921. G. Yaluris, M.S. Ziebart and X. Zhao, US Patent Appl. 0100494, May 2005.
922. G. Yaluris, M.S. Ziebart and X. Zhao, US Patent Appl. 0100493, May 2005.
923. G. Yaluris, M.S. Ziebart and X. Zhao, US Patent Appl. 0230285, October 2005.
924. C.P. Kelkar, D.M. Stockwell and S. Tauster, US Patent Appl. 0098259, May 2003.
925. C.P. Kelkar, D.M. Stockwell and S. Tauster, US Patent Appl. 0141897, July 2004.
926. C.P. Kelkar, D.M. Stockwell and S. Tauster, US Patent Appl. 0141898, July 2004.
927. C.P. Kelkar, Y. Li, R.J. Madon and M. Vaarkamp, US Patent Appl. 0072675, April 2004.
928. M. Vaarkamp and D.M. Stockwell. US Patent Appl. A0204420, September 2006.
929. D.M. Stockwell, US Patent Appl. 0040823, February 2006.
930. M. Xu, US Patent Appl. 0067322, March 2005.
931. A.W. Peters, E.F. Rakiewicz, G.D. Weatherbee and X. Zhao, US Patent Appl. 0092796, July 2002.
932. A.W. Peters, J.A. Rudesill, G.D. Weatherbee, E.F. Rakiewicz and M.J.A. Barbato-Grauso, US Patent Appl. 0092795, July 2002.
933. A.W. Peters, J.A. Rudesill, G.D. Wheatherbee, E.F. Rakiewicz, US Patent Appl. 0205465, September 2005.
934. C.P. Kelkar, D.M. Stockwell and S.J. Tauster, US Patent Appl. 0163691, April 2005.
935. A. Vierheilig, US Patent Appl. 0086442, May 2004.
936. B. Lin, W. Zhang, Y. Liu, S. Li and N. Li, US Patent Appl. 0004024, January 2004.
937. B.L. Schulman and R.L. Dickenson, Proc. 5th UNITAR/UNDP Int. Conf. on Heavy Crudes and Tar Sands, 1991, Caracas, Venezuela, Vol. 4, p. 105.
938. H. Qabazard, R. Adarme and B.L. Crynes, in Catalysis in Petroleum Refining 1989, D.L. Trimm (Ed.) Elsevier Science, Amsterdam, 1989, p. 61.

939. E. Furimsky, Rev. Inst. Fr. Petr. 54 (1999) 597.
940. E. Furimsky, Fuel Proc. Technol. 56 (1998) 263.
941. R. Ikbal, V. Patel and H.D. Sloan, Proc. Sixth UNITAR/UNDP Int. Conf. on Heavy Crude and Tar Sands, 1995, Houston, TX, p. 10.
942. J.P. Schaack and F. Chan, Oil Gas J. 87 (1989) 55.
943. E. Furimsky, Fuel Proc. Technol. 67 (2000) 205.
944. T. Takatsuka, R. Watari and H. Hayakawa, in Catalysts in Petroleum Refining and Petrochemical Industries 1995, M. Absi-Halabi et al. (Eds) Elsevier Science, Amsterdam, 1996, p. 293.
945. J.Y. Low, R.L. Hood and K.Z. Lynch, Am. Chem. Soc. Div. Petr. Chem. Prep. 40 (1995) 780.
946. B. Eckermann and A. Vogelpohl, Erdol Kohle 43 (1990) 224.
947. A.S. Farag, O.I. Sif El-Din, M.F. Youseff, S.I. Hassan and S. Farmawy, Hung. J. Ind. Chem. 17 (1989) 289.
948. A. Vogelpohl and B. Eckermann, Proc. 5th UNITAR/UNDP Int. Conf. on Heavy Crude and Tar Sands, 1991, Caracas, Venezuela, Vol. 4, p. 145.
949. G. Brons and J.M. Yu, Energy Fuels 9 (1995) 641.
950. V.A. Adewusi, T. Oshinowo and B. Ademodi, Energy Sour. 14 (1992) 135.
951. P.H. Johnson, K.L. Mills Jr and B.C. Benedict, Ind. Eng. Chem. 47 (1955) 1578.
952. J. Hocker and A. Vogelpohl, Chem. Eng. Technol. 10 (1987) 125.
953. H.M. Feintuch, M.J. McGrath and C. Alzamora, Proc. AIChE Natl Meet., 1987, Houston, TX, Session 16, p. 1.
954. B. Obah and H.-J. Neumann, Erdol Kohle 40 (1987) 486.
955. H.-J. Neumann, I. Rahimian and G. Zenke, Erdol Kohle 40 (1987) 119.
956. J.D. Elliott, R.L. Godino and M.J. McGrath, Proc. Third UNITAR/UNDP Int. Conf. on Heavy Crude and Tar Sands, 1983 Long Beach, CA 195, p. 1699.
957. R. Gopinath, A.K. Dalai and J. Adjaye, Energy Fuels 20 (2006) 271.
958. K. Dunn and T.F. Yen, Am. Chem. Soc. Div. Petr. Chem. Prep. 44 (1999) 213.
959. A. Chakma and F. Berruti, Fifth UNITAR/UNDP Int. Conf. of Heavy Oils and Tar Sands, August 1991, Caracas, Venezuela, Vol. 4, p. 101.
960. T.F. Yen and J.R. Lin, Fifth UNITAR/UNDP Int. Conf. on Heavy Oils and Tar Sands, August 1991, Caracas, Venezuela, Vol. 4, p. 75.
961. N. Kang, I. Hua and C. Xiao, Ind. Eng. Chem. Res. 45 (2006) 5239.
962. S.P. Tu and T.F. Yen, Am. Chem. Soc. Div. Petr. Chem. Prep. 44 (1999) 209.
963. J.R. Lin and T.F. Yen, Energy Fuels 7 (1993) 112.
964. R. Tao and X. Xu, Energy Fuels 20 (2006) 2046.

INDEX